Y0-BQJ-830

TREATISE ON COATINGS

Volume 4

Formulations

PART I

TREATISE ON COATINGS

(In Five Volumes)

EDITED BY

RAYMOND R. MYERS AND J. S. LONG

VOLUME 1: **Film-Forming Compositions**
VOLUME 2: **Characterization of Coatings:** Physical Techniques
VOLUME 3: **Pigments**
VOLUME 4: **Formulations**
VOLUME 5: **Qualitative and Quantitative Analysis**

TREATISE ON COATINGS

VOLUME 4 (In Two Parts)

Formulations

PART I

Edited by

RAYMOND R. MYERS

Department of Chemistry
Kent State University
Kent, Ohio

J. S. LONG

Department of Chemistry
University of Southern Mississippi
Hattiesburg, Mississippi

MARCEL DEKKER, New York

COPYRIGHT © 1975 by MARCEL DEKKER, INC. ALL RIGHTS RESERVED

Neither this book nor any part may be reproduced or transmitted in any form or by any means, electronic or mechanical, including photocopying, microfilming, and recording, or by any information storage and retrieval system, without permission in writing from the publisher.

MARCEL DEKKER, INC.

270 Madison Avenue, New York, New York 10016

Library of Congress Catalog Card Number: 67-21701

ISBN: 0-8247-1477-6

Current Printing (last digit):
10 9 8 7 6 5 4 3 2 1

PRINTED IN THE UNITED STATES OF AMERICA

PREFACE

The reader who has studied the previous volumes of this Treatise is familiar with the types of vehicles and pigments available, and also with high-level techniques used in testing which serve as a guide to fundamental ideas of properties of films and how they form. With this fundamental knowledge one is ready to consider what is called Formulation, i.e., design of a formula for a coating to be used for a specific purpose. Coatings for the exterior of a frame house certainly differ from a baked automotive enamel or an artist's "oil color"; and to a lesser extent they differ from interior flat paints, floor enamel for concrete floors, and so on.

Before starting in the business of formulating, one should do plenty of fundamental thinking. What are the basic properties needed, in order of importance; and what are the agents that will tear down the film as time goes on? Interior finishes are not subjected to the ultraviolet wave lengths of sunlight; window glass stops these rays. Flat wall paints and gloss or semi gloss enamels on doors and other woodwork are only occasionally subjected to water, and do not require the ability to withstand these elements as well as do exterior finishes. The coatings industry is a high technology effort requiring a fine balance of competing requirements or interrelated parameters.

Linseed oil, an ester of glycerol combined with a mixture of unsaturated 18 carbon chain length fatty acids having 1, 2, and 3 double bonds, was discovered about the 9th century. On exposure to air in a thin film it changed spontaneously from a liquid to a solid film, so it began to be used in coatings. No one knew its composition, and at that time the composition of air was unknown; as a consequence its use in films was not directed by intelligent factors. Yet unsaturated esters have persisted in house paints to the present day even though ultraviolet radiation and rain cause esters to hydrolyze readily, especially in the presence of metallic soaps of the type used as catalysts. If they did not hydrolyze, one could not digest oils and fats which, like linseed oil, are esters of glycerol with fatty acids. Ethers, on the other hand, are not so easily hydrolyzed. Had this factor been considered at the time paint chemists were developing oleoresinous systems, the discovery of epoxies may have occurred decades earlier than it did.

Planning of another sort is receiving major emphasis at the start of this new volume of Treatise on Coatings. Henceforth it is necessary to consider the impact on the environment of new formulations; and the strings are being drawn tightly around certain existing formulations, particularly those which contain heavy metals. Here the question of toxicity is particularly important for interior paints, and stringent limitations have been imposed on all coatings which will come in contact with children. As a consequence, this volume was compiled during an exceptionally sensitive period of time. Problems which once were solved on the scientific and technical level have given way to problems of safety, burdened with political overtones. While paint scientists labor over this new array of formulation variables and consider their environmental impact, it is incumbent on the industry to persuade the general public that the problems of deteriorating quality may exceed those which are in the process of being solved.

It is fortunate that the volume on formulation is being launched at the time when these drastic changes are occurring. Despite our intentions to the contrary, we are aware that some formulations will be on their way to obsolescence by the time the volume is completed, and submit that they represent the technical high water mark, to be used as a target for future formulations made with a restricted list of allowed ingredients.

R. R. M.
J. S. L.

CONTRIBUTORS TO VOLUME 4, PART I

MURRAY ABRISS, E.I. du Pont de Nemours Co., Marshall Laboratory

CHARLES A. BURGER, Interstab Chemicals, Inc., New Brunswick, New Jersey

G. W. GERHARDT, Mobil Chemical Company, Pittsburgh, Pennsylvania

LYMAN P. HUNTER, Bennett's, Salt Lake City, Utah

DAVID M. JAMES,* International Red Hand Marine Coatings, Gateshead Co., Durham, England

J. B. G. LEWIN,[†] Pinchin Johnson Paints, Kingston-on-Thames, England

DONALD K. LUTES, SR., Technical Manager, Coil Coatings, The Sherwin-Williams Company, Chicago, Illinois

BRUCE N. McBANE,[‡] PPG Industries, Inc., Coatings and Resins Division, Springdale, Pennsylvania

M. J. MASCIALE, Mobil Chemical Company, Edison, New Jersey

G. G. SCHURR, The Sherwin-Williams Company, Chicago, Illinois

G. W. SEAGREN, Mobil Chemical Company, Pittsburgh, Pennsylvania

E. GUSTAVE SHUR,[†] Inmont Corporation, Clifton, New Jersey

FRED B. STIEG, Titanium Pigments Division of NL Industries, Inc., South Amboy, New Jersey

OLIVER VOLK, E.I. du Pont de Nemours Co., Marshall Laboratory

---

*Present address: 14 Beaconsfield Crescent, Lowfell, Gateshead 9, Tyne & Wear, England.

[†]New company title: International Pinchin Johnson.

[‡]Present address: PPG Industries, Inc., Coatings and Resins Division, Allison Park, Pennsylvania.

[†]Deceased.

CONTENTS OF VOLUME 4, PART I

CONTENTS OF OTHER VOLUMES

VOLUME 1, PART I

VOLUME 1, PART II

VOLUME 1, PART III

VOLUME 2, PART I

CONTENTS OF OTHER VOLUMES

TREATISE ON COATINGS

Volume 4

Formulations

PART I

Chapter 1

AIRCRAFT FINISHES

J. B. G. Lewin

Pinchin Johnson Paints
Kingston-on-Thames, England

1-1. INTRODUCTION

The technology of aircraft finishes has required greater specialization than is customary in the surface coatings industry. Even from the earliest days of flying, coatings have been considered an integral part of an aircraft structure, and thus their performance contributes significantly to the safety of the aircraft and its occupants. Formulations, proving, and inspection are based to a large extent on this consideration.

Aircraft construction has progressed from the wood and fabric of the early pioneers to the aluminum, titanium, and steel alloys of present day usage. In this transition, it is possible to trace three distinct phases which have dictated the qualities required of aircraft finishes. The first phase was from the early twentieth century to the mid nineteen-thirties when the coating of fabric and wood was of greater consequence than the protection of the lesser used metals. The introduction of all metal aircraft, almost coincidental with the advent of alkyd resins, marked the second phase, which a decade later merged into the third, that of the jet-engined aircraft. The surface coating problems then encountered were formidable enough; it is fortunate, however, that the advancing technology of the coatings industry has kept pace with these demands on it and it is of necessity a partner in the rapid progress being made by the aviation industry. The following chapter describes finishes for aircraft in the sequence outlined above, though some of the products of early usage are still in minor demand today.

1-2. PHASE I: FABRIC AIRCRAFT

A. DOPES

For many years use has been made of the fact that many colloids behave like stretched membranes when applied to fabrics. It was natural therefore that the early pioneers of aircraft should first turn to natural products such as starch and sago to induce tension in the fabric-covered surfaces in order to provide taut and airproof surfaces. Less water sensitivity was an obvious requirement and the use of cellulose acetate solutions in the years prior to the First World War was the first significant advance in achieving this. The original solvent mixtures were rich in chlorinated compounds causing great discomfort to the operators. It is possible that the term "dopes" for such coatings derived from this feature. A heritage from these early years is still to be seen in British Standard Specifications X26 and X27 which ban the use of chlorinated solvents in doping and finishing schemes for aircraft.

The merits of the cellulose acetate dopes were their high initial tautening properties and their fire retardance qualities particularly if they were used in combination with the phosphate plasticizers. Their defects were: (1) their susceptibility to continual exposure to humidity which resulted in a progressive loss of tension in the fabric, (2) careful application and formulation of solvent mixtures were necessary to avoid "blushing," and (3) a limited range of viscosity grades was available in the United Kingdom. Nevertheless their use continued on a diminishing scale to specification DTD. 26 until about 1930 by which time cellulose nitrate dopes to specification DTD. 83 were in general use in the U.K. Their use continues today to BSX. 26 and 27 despite the introduction of cellulose acetate butyrate which combines some of the virtues of both cellulose nitrate and cellulose acetate for use in dopes. Many alternative materials have been evaluated, including chlorinated compounds such as chlorinated rubber and polyvinyl chloride copolymers, but the cellulose derivatives appear most suitable for giving the strained orientations during film formation necessary for tautening the fabric.

Gourlay (1) has studied the strain energy relations of colloid films deposited on linen fabrics. His method was essentially that defined in Appendix D of BSS.X26. Wooden frames isolating a circular orifice 10 in. in diameter were covered with linen to BSS. 6 F1. In order to ensure a plane surface, the linen was fixed under a uniform tension of 1 lb/linear in. Various colloids in suitable solvent mixtures were applied to the fabric by brushing, thus giving a circular area 10 in. in diameter of coated, unsupported fabric firmly fixed around the circumference. In order to measure the tension in the coated fabric, a uniform pressure was applied to one face of the fabric and the height of the sagitta of the resulting spherical cap measured carefully to 0.0005 in. The tension was then calculated by means of the relation $T = PR/2$ where T is the tension, P is the pressure difference, and R is the radius of curvature. R is related to the height x of the sagitta by $R = 25/2x$. A first step, using a solution of cellulose nitrate in butyl acetate, was to show that the relationship between the tension and weight of cellulose nitrate per unit area is linear (Fig. 1-1). This relationship is expressed by the equation

$$T = Kw + T_0$$

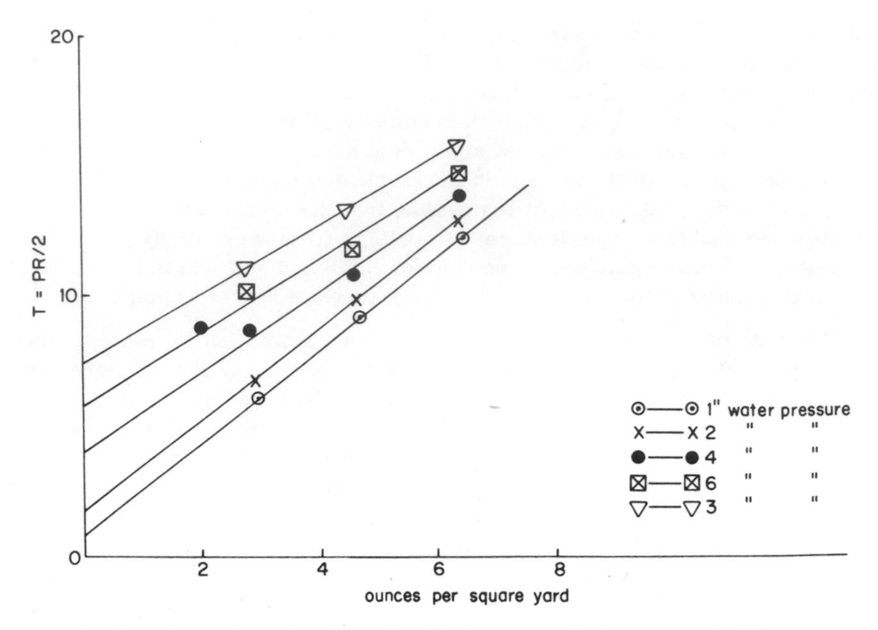

FIG. 1-1. Fabric tension of cellulose nitrate impregnated linen.

where K is a constant, W the weight of colloid per unit area, and T_o the extrapolated tension when W is zero. Figure 1-2 shows that both K and T_o are linear functions of the pressure. The behavior of the colloid is therefore described by the relation $T = (-\alpha P + k_o)w + \beta P$, where α and β are constants vanishing at zero pressure and k_o is a constant characteristic of the colloid. As the value of k_o at 1 in. water pressure is very close to the value at zero pressure, this pressure was chosen for convenience in determining k_o, defined as the strain energy constant of the colloid and expressed in pounds per inch per ounce of deposited film per square yard of surface.

Gourlay (1) determined this strain energy constant for cellulose triacetate, cellulose diacetate, cellulose nitrate, cellulose acetate proprionate, cellulose acetate butyrate, ethyl cellulose, benzyl cellulose, and chlorinated polyvinyl chloride at temperatures of 24° C, 35° C, and 50° C with relative humidity at these temperatures of 30%, 80%, and 100%. Benzyl cellulose and chlorinated polyvinyl chloride showed a marked relaxation after heating to 50°C; ethyl cellulose showed a slight relaxation. The remaining colloids did not show this relaxation. The cellulose acetates though having higher initial tensions than cellulose nitrate and cellulose acetate butyrate, relaxed more sharply with increasing humidity and cellulose nitrate was slightly less susceptible to both temperature and humidity than cellulose acetate butyrate.

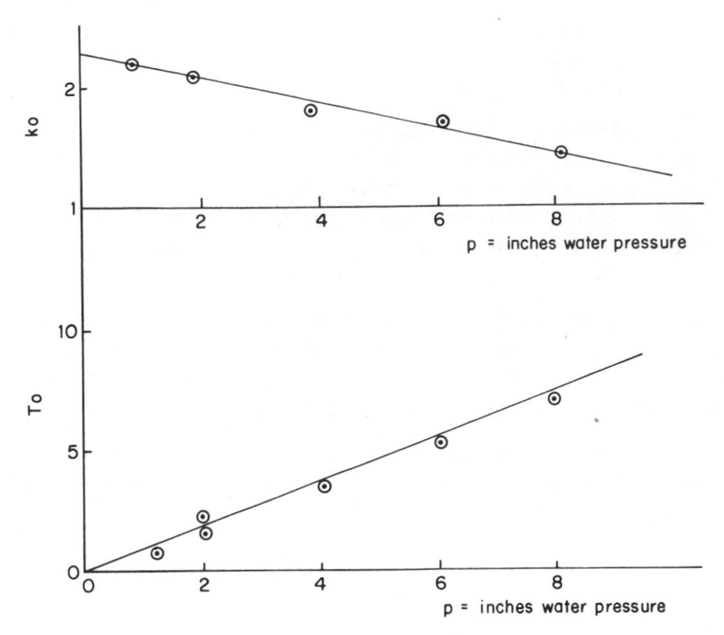

FIG. 1-2. Reprinted from the Journal of the Oil Colour Chemists Association, 1946, Vol. XXIX, p. 96, by courtesy of the Oil Colour Chemists' Association.

Other factors studied were (1) the effect of the viscosity of the colloid, (2) the effect of the degree of esterification or etherification of the cellulose molecule, (3) the effect of the volatile and nonvolatile solvents from which the colloid is deposited, (4) variation of k_o over a wide range of temperature, and (5) the value of k_o against moisture content of films.

High, medium, and low viscosity grades of cellulose nitrate gave similar values of k_o at 20°C and 30% relative humidity (RH) and two differing viscosity grades of chlorinated polyvinyl chloride of the same chlorine content gave an identical value at 20°C and 60% RH from which the author concluded that viscosity is not a material factor affecting k_o.

Little variation in their k_o values at 20°C and 30% RH was obtained from determinations on grades of cellulose nitrate varying in nitrogen content between 11.0 and 13.4%, though the highest level gave the lowest k_o. Cellulose acetate containing 62% of acetic acid gave a slightly higher k_o than that containing 55% and in the case of ethyl cellulose the value of k_o did not vary greatly over an ethoxy content from 42-49%.

Ethyl acetate, butyl acetate, and amyl acetate were examined as solvents for nitrocellulose and were without effect on the value of k_o so long as the test frames were allowed to age until maximum tension was observed (400 hrs. at 65% RH and 21°C). Earlier determinations gave higher tensions in favor of the more volatile solvents.

The use of plasticizers in dopes and the colloid/plasticizer ratio as a determinant to control tensioning of fabric is well known, and the author showed the rapid reduction in the strain energy constant of nitrocellulose as increasing amounts of a true solvent plasticizer, dibutyl phthalate, or a nonsolvent plasticizer, castor oil, were made. The effects were practically identical.

Values of k_o for ethyl cellulose and cellulose nitrate were plotted over a range of temperatures from -80°C to 50°C. Similar percentage increases occurred at -80°C; the ethyl cellulose, however, was more susceptible to heat relaxation at temperatures above 20°C than nitrocellulose.

Limited data indicated that the moisture content of the films was more significant in reducing the strain energy constant in the case of cellulose nitrate than the other colloids studied.

1. Doping Schemes

Current British doping schemes are based on cellulose nitrate to Defence Specification Def-1432 and conform to the requirements of British Standard Specification X26. Both cellulose nitrate and cellulose acetate butyrate systems are in use in the United States and typical specifications for the former are MIL-D-5552, MIL-D-5553, MIL-D-5554, and MIL-D-5555; examples of the latter are MIL-D-5549, MIL-D-5550, and MIL-D-5551. Some French usage is made to the American specifications, although the French specification AIR.8102 covers a series of nitrocellulose based dopes and finishes for use on fabric. * British practice for Service use is for the pigmented colored finishes for camouflage or other purposes to be matte; a final coat of transparent nontautening finish may then be applied if a glossy finish is required. American specifications permit of the dopes themselves to be pigmented to give flat or glossy finishes according to design requirements.

British Standard Specification X26 gives details of doping schemes of low, medium, and high tautness, selection being governed by the performance or construction of the fabric covered aircraft or component to be coated. Appendix A defines the test frame to be used. This shall be a strong rect-

---

*These perform well on nylon as well as linen, although wetting is more difficult. However some French jet aircraft have fabric rudders; cellulose acetate butyrate systems are preferred on these as having better ester lubricant resistance.

angular wooden frame reinforced with metal to prevent warping, measuring 10 in. × 10 in. internally with two holes $\frac{3}{16}$ in. in diameter bored through one of the sides and having a piece of $\frac{1}{2}$-in. 5-ply wood with a central hole 10 in. in diameter screwed to one face and covered on the plywood face with linen fabric to British Standard F1 under a tension of approximately 2 lb/in. width in the warp and 1 lb/in. width in the woof. Appendix D gives the test procedures for measuring tautness and film weights of dopes. The dope is applied uniformly by brush or spray to give an added film weight of 3.5 ± 0.5 oz/yd^2, this being the added weight per square yard after drying to constant tautness with both surfaces of the fabric exposed to the air. After then conditioning for not less than 2 hr at 65% RH and $18.5 \pm 0.5^\circ$C, the frame, still in the conditioned atmosphere, is fitted with an airtight back and connected through the holes in its side with a sensitive and accurate manometer. Any device such as a water pump is adjusted to reduce the air pressure inside the frame by an amount equivalent to 2 in. of water, a large air reservoir being included in the system to smooth out variations in pressure. The resulting depression, d, of the center of the circle of unsupported fabric is measured using an instrument accurate to within 10^{-3} in. that does not impose a load exceeding 5 g on the area of unsupported fabric. The tautness, T, is given by the equation

$$T = \frac{451}{d} \text{ lb/in.}$$

where d is in 10^{-3} in. The weight of dope used is determined by cutting a minimum area of 36 in.^{2} from the circle of unsupported doped fabric, weighing it, removing the dope by suitable solvents, and weighing it again. A piece of undoped fabric of the same area is cut from the same length as was used to cover the test frame and is weighed. After the same solvent treatment as before under the same atmospheric conditions, the fabric is weighed again. From the data obtained, the weight in ounces of dope per square yard of fabric is calculated.

Use is made of the ratio of the tautness in pounds force per inch to the weight of dope in ounces per square yard to define the tautness properties of each of the tautening dopes used in the low tautness scheme (No. 751), medium tautness scheme (No. 752), and high tautness scheme (No. 753). The following requirements apply:

	Ratio	
Scheme	Min.	Max.
No. 751	1.0	1.4
No. 752	1.4	1.8
No. 753	1.4	1.8

The systems are built up by a combination of the appropriate dopes and nontautening finishes to give the desired properties of tautness and protection. The low tautness schemes were developed for use on gliders and light aircraft where for example a heavy application of high tautness dope could cause serious distortion of the structure. The medium tautness schemes are principally used on control surfaces such as ailerons and elevators, and high tautness schemes were formulated for specialized constructions, e.g., geodetic, that is, an alloy lattice covered by large areas of fabric. All three schemes have an intermediate coating pigmented with aluminum to reduce the deteriorative effects of light and, in the case of the medium and high tautness schemes, the first coat is red-oxide tinted both to assist in fabric protection and to enable inspection to ensure that good penetration has been achieved and adequate coats have been applied. Table 1-1 indicates the number and types of coats used in each system. It should be noted that the nontautening finishes, which conform to BSS.X27 where their use is in a scheme for application to metallic materials, are such that when applied in the number of coats and to the film weights given in Table 1-1 over fabric already tautened by the use of tautening dopes, applied as indicated in Table 1-1, do not reduce the tautness of the system by more than 20% compared with the tautness value determined before their application.

Other clauses give methods of test for resistance to high temperatures and to natural weathering and for freedom from film defects. The individual tautening dopes and the relevant full doping schemes are applied to fabric in the specified manner and strips cut from the frames for heat ageing tests (4 days at $>95^{\circ}C$) before bend tests are applied using specified mandrels. Natural weathering tests require a fully coated frame without the plywood but with a waterproof backing to be exposed at $45^{\circ}C$ facing south for six months including two months of maximum sunshine in the United Kingdom. The tensile strength of cut-outs is then determined and should not be less than the undoped fabric under the identical specified conditions. The "Freedom from Film Defects" clause is designed primarily to ensure that blushing during application of the doping schemes should not occur. The test conditions (RH 65-70% at 18-21°C and a circulation air speed of 3 ft/sec) are specified; test frames, dopes, finishes, and brushes are stored in this ambience for not less than 2 hr. The tautening dopes are then applied as one full coat, and the nontautening dopes as one full coat to a test piece previously tautened with an appropriate doping scheme. After drying under the test conditions, the frames are examined visually.

Although minimum nonvolatile contents and minimum aluminum and red oxide contents are specified, specification BSS.X26 is primarily a performance specification. By contrast, comparable U.S. military specifications give guide formulations, carefully defining the composition and percentage of the ingredients to be used, as well as including physical tests, as governed by Federal Standard 141 Paint Varnish Lacquer and Related Materials Methods of Inspection, Sampling and Testing. Table 1-2 gives typical formulations for cellulose acetate butyrate dopes. Appropriate thinners are MIL-T-6096 for normal use and MIL-T-6097 for use in adverse conditions. American cellulose nitrate dopes to MIL-D-5552, 3, 4, and 5 also give guide formulations. An example is given in Table 1-3. The appropriate thinner is T1-T-266 for normal use and MIL-T-6095 for use in adverse humidity conditions. A typical British dope formulation is given in Table 1-4. The appropriate thinners conform to Specification DEF-1216.

B. WOOD AND METAL PROTECTION

In the U.K. prior to the late 1930's, the painting of wooden and metal components was much in line with the practices observed in good quality coach painting. Tung oil and linseed-oil-based varnishes with copal and, later, phenolic resins served to provide varnishes and enamels for the aircraft industry. Much use was made of seaplane varnish (BX.17)* often lightly pigmented with aluminum for the treatment of wood. Pigmented oleoresinous enamels were supplied as "matte or glossy pigmented oil varnishes," a nomenclature which remained long after such formulations had been converted to an alkyd resin basis. Use was also made of cellulose nitrate lacquers containing, e.g., damar resin, these also being applied over nonpickling oleoresinous primers. British Specifications of this era were DTD. 62 and DTD. 260 (oleoresinous) and DTD. 63, 83, and 308 (cellulose). A nitrocellulose lacquer (DTD. 103) was supplied for wooden airscrews.

However, the growing use of light alloys, for weight reduction and the need to protect, in particular, seaplanes, from corrosion, emphasized the value of chromates as corrosion inhibitors for use in the aircraft industry. Much valuable work in this field was carried out at Royal Aircraft Establishment Farnborough by such workers as Mardles, Le Brocq, and Sutton (2). Cole (3) related the choice of chromate pigment to chromate solubility. If the chromate chosen is too soluble, it is soon expended; if insufficiently soluble the concentration of chromate ions is too low to prevent corrosion. Strontium chromate found favor at Farnborough as striking a happy balance between having a reserve of chromate undissolved and a useful amount in solution although little use was made industrially of this until later when technical difficulties prevented the use of the much cheaper but otherwise similar potassium zinc chromate. Good protection of magnesium alloys was obtained with oil-based primers pigmented with zinc and strontium chromates (4). Barium chromate was the preferred chromate for use in oleoresinous jointing compounds applied between dissimilar metals, its low leaching rate serving to maintain a long-term reservoir of chromate ions. Specification DTD. 369 covered such products.

---

*Indicates specification to which material conforms.

TABLE 1-1. Number and Types of Coats in Low, Medium, and High Tautness Schemes

Designation	No. 751		
Tautness	Low		
Description of components	Dry weight		Normal number of coats
For all finishes other than aluminum	oz/yd^2	g/m^2	
Red oxide tautening dope	—	—	—
Transparent tautening dope	2.0 ± 0.4	68.0 ± 13.5	3 or 4
Aluminum nontautening finish	1.0 ± 0.2	34.0 ± 7.0	. 2
Aluminum tautening dope	—	—	—
Pigmented nontautening finish[a]	1.0 ± 0.2	34.0 ± 7.0	1 or 2
Transparent nontautening finish[b]	1.0 ± 0.2	34.0 ± 7.0	1 or 2
For an aluminum finish			
Red oxide tautening dope	—	—	—
Transparent tautening dope	2.0 ± 0.4	68.0 ± 13.5	3 or 4
Aluminum nontautening finish	1.0 ± 0.2	34.0 ± 7.0	2
Transparent nontautening finish[b]	1.0 ± 0.2	34.0 ± 7.0	1 or 2

[a] For color and finish matching purposes only, a weight addition of not more than 2 oz/yd^2 (68 g/m^2) will be permitted for yellow, white and sky finishes.

[b] Only where a glossy finish is required.

The protective properties of lanoline also received much attention at Farnborough, and compositions on resin-hardened lanoline, and containing zinc chromate, were developed for use on Alclad sheet, duralumin, plain and cadmium-plated steel as well as magnesium-rich alloys. Specification DTD. 279 resulted from this work as did DTD. 420, a matte pigmented finish based on resin-hardened lanoline which could be removed by solvent washing after long exposure and when pigmented with zinc oxide, etc., had value as an antifouling for seaplanes. Another lanoline resin solution (dye tinted) used as a temporary protective appeared as specification DTD. 663.

High contents of zinc chromate continued in favor for use in primers, particularly those for magnesium, and again varnish media were used as binders. Specifications were becoming more diverse, and the introduction of alkyds was a noticeable step forward. The philosophy of the "universality" of

TABLE 1-1. — Continued

No. 752 Medium			No. 753 High		
Dry weight		Normal number of coats	Dry weight		Normal number of coats
oz/yd^2	g/m^2		oz/yd^2	g/m^2	
2.0 ± 0.4	68.0 ± 13.5	3	0.75 ± 0.15	25.5 ± 5.0	1
—	—	—	4.75 ± 0.95	161.0 ± 32.0	6 or 7
—	—	—	—	—	—
1.0 ± 0.2	34.0 ± 7.0	2	1.0 ± 0.2	34.0 ± 7.0	2
1.0 ± 0.2	34.0 ± 7.0	1 or 2	1.0 ± 0.2	34.0 ± 7.0	1 or 2
1.0 ± 0.2	34.0 ± 7.0	1 or 2	1.0 ± 0.2	34.0 ± 7.0	1 or 2
3.0 ± 0.6	102.0 ± 20.5	4	0.75 ± 0.15	25.5 ± 5.0	1
—	—	—	5.75 ± 1.15	195.0 ± 39.0	8
1.0 ± 0.2	34.0 ± 7.0	2	1.0 ± 0.2	34.0 ± 7.0	2
1.0 ± 0.2	34.0 ± 7.0	1 or 2	1.0 ± 0.2	34.0 ± 7.0	1 or 2

primers was adopted in the U.K., i.e., a primer for aircraft use is required to perform adequately in a system be the finishing coats applied over it oleoresinous, cellulosic, or synthetic resin based.

It was at this stage that the introduction of metal skinned aircraft began to cause the decline of the cellulose derivatives from their position of preeminence in the aircraft industry to that of an ancillary role.

1-3. PHASE II: METAL SKINNED AIRCRAFT

A. PRETREATMENTS

The aircraft industry more than most industries is aware of the importance of substrate cleanliness and surface preparation to secure adequate paint adhesion and performance. These circumstances have been dictated by the presence of rivet runs as areas of contaminant collection prior to painting and the later exposure of painted parts to possible attack by engine oils, fuel,

TABLE 1-2. Cellulose Acetate Butyrate Dopes

Ingredient	Clear dope to Mil-D-5549, wt.%	Flat white dope to Mil-D-5550, wt.%	Glossy white dope to Mil-D-5551, wt.%
EAB-171-15[a]	9	7.92	8.6
Triphenyl phosphate	1	2.75	6.1
Tricresyl phosphate	—	2.75	—
Chalk resisting titanium dioxide		4.29	6.3
Siliceous flatting agent		4.29	—
Diacetone alcohol	9	7.8	7.9
Ethyl acetate	36	31.2	31.6
Methyl ethyl ketone	31.5	27.3	27.6
Butyl acetate	13.5	11.7	11.9
	100.0	100.0	100.0

[a]Eastman Chemical. This grade of cellulose acetate butyrate should be 16-22 sec in order that these formulas conform to the specification.

TABLE 1-3. MIL-D-5553 Dope, Cellulose Nitrate, Clear

5-6.5 sec Cellulose Nitrate (35% alcohol)	9.8-10.2%
15-20 sec Cellulose Nitrate (35% alcohol)	4.9-5.1%
Castor oil	1%
Dibutyl phthalate	1%
n-Butanol	6.9-7.1%
Butyl acetate	14.7-15.3%
Ethyl acetate	19.6-20.4%
Toluene	32.3-33.7%
Petroleum aliphatic naphtha	7.8-8.2%

TABLE 1-4. Clear Dope No. 751

5-6 sec cellulose nitrate (30% isopropyl alcohol)	1.0%
15-20 sec cellulose nitrate (30% isopropyl alcohol)	10.5%
Dibutyl phthalate	3.5%
Acetone	21.0%
Butanol	9.0%
66 O/P I.M.S.	19.0%
Butyl acetate	3.0%
Toluene	14.0%
Xylene	8.0%
Petroleum aliphatic naphtha (B.R. 70°C-100°C)	11.0%

water, and hydraulic fluids. Even before the advent of the jet engine and the introduction of aggressive synthetic lubricating oils and modern hydraulic fluids, strict codes of practice for the pretreatment of metals prior to painting were necessary. Specifications DTD.901, 902, 911A, and 915 offered guide lines for the rendering of metallic surfaces to be suitable for paint schemes and Ledwith (5) described the metals and pretreatments then in use in the U.K. aircraft industry in the 1940's. Examples of British Specifications then obtaining for the high strength, copper-bearing Duralumin type are DTD.364 or 5L3, for the Alclad type (Duralumin clad with thin layers of pure aluminum) DTD.390, for aluminum zinc alloys DTD.363 and those for magnesium-rich alloys were DTD.118, 259, 285, and 289. These light alloys could be present as rolled sheets, extrusions, forgings, and castings, and their painting treatment depended on their relative corrosion resistance. Some of the pretreatments were:

1. Degreasing by trichlorethylene vapor.
2. Anodizing. The production of a porous but adherent oxide film by anodic oxidation.
3. Acid chromate pickle. Components after degreasing were immersed for 20 min in a solution containing 15% v/v sulfuric acid S.G. 1.82 and 5% w/v chromic acid at 145-150°F followed by cold and hot rinses as specifield in DTD.915A.
4. Alkaline chromate bath. Processes such as the modified Bauer-Vogel (MBV) or Pylumin utilized a 5-min. immersion in a boiling solution of sodium chromate and sodium carbonate, followed by cold and hot rinses in water.
5. Deoxidine process. After degreasing, a suspension of a grease-absorptive pigment in an acid solution was brushed over the surface and allowed to

dry. The pigment deposit was washed off with water leaving the surface clean and slightly etched.

6. Magnesium alloys. Components after cleaning could be immersed for 30 min in a boiling solution containing 3% w/v ammonium sulfate, 1.5% w/v ammonium dichromate, 1.5% w/v potassium dichromate, and about 0.35% w/v ammonia S.G. 0.880 followed by a rinse. Another process in DTD. 911 specified 10 sec to 2 min at room temperature in a bath containing 15% w/v sodium dichromate and 20-25% by volume of nitric acid S.G. 1.42 with the usual subsequent rinse.

Not only did these pretreatments improve the adhesion of priming coats but in most instances corrosion resistance was improved. However, the importance of priming immediately after pretreatment is stressed.

B. "UNIVERSAL" PRIMERS

Mention has been made of the fact that in the U.K. specifications for aircraft finishes are based on performance requirements; where necessary, types of media or pigmentations are described in broad terms that allow the coatings technologist some latitude in his use of raw materials. For a primer to be "universal," be it used on steel or light alloys, in baking or air-drying systems, its performance should conform to the requirements of the separate current DTD specifications covering systems involving different types of fillers and finishes used over the primer. At the period under review, a universal primer would be required to conform to: Specifications DTD. 63, 235, 260, 314, 517, 754, 768, 796, and 911 — a series of cellulose and synthetic finishing schemes. DTD. 911 — a specification for the pretreating and painting of magnesium alloys — required a minimum content of 15% by weight of zinc chromate in the dried primer film. As "universal" primers were required to accept cellulose as well as synthetic finishes, alkyds generally replaced oleoresinous media as binders. However, an exception and one of the most successful of the media used in such primers was a tung oil/phenolic resin/zinc oxide dispersion. A typical formulation of this type — UP4 primer — is given in Table 1-5. Alkyd types were similar to AN-TT-P-656B and current formulations resemble those given in British MIL-P-8585. Such "universal" primers had a wide use in the U.K. until the introduction of ester lubricants and phosphate ester hydraulic fluids demanded a greater level of fluid resistance, but they are still useful in less aggressive environments.

C. ETCH PRIMERS

Bath pretreatments can be costly in time, control, and space and the use of brush pretreatments involves much tedious manual work in applying and

TABLE 1-5. UP4 Primer

Zinc chromate to DTD.377	9.1%
Titanium dioxide	4.5%
Asbestine	27.6%
Bakelite dispersion resin Z3692 (approx 50% solids)	31.4%
Secondary butanol	6.1%
Xylene	19.6%
Pigment/binder: 2.5/1	
15.8% zinc chromate on nonvolatiles	

washing off. It was natural therefore that the etch primers were welcomed in the U.K. aircraft industry in the late 1940's as a means of both pretreating and priming light alloys. An added bonus was the savings in weight effected by their use.

The first etching primers were developed by Whiting and Wangner (6) and were of the well-known WP1 composition given in Table 1-6. A considerable literature exists on investigations into the mechanism of the reactions occurring in the use of etch primers and of variations of the components and their proportions in order to improve the primer properties. Although the mechanism is not clearly understood it is necessary that easily oxidizable primary alcohols should be present, that the water content of the mixed base/acid mixture should be not less than 3.9% if early gelation is to be avoided, that the properties of the polyvinyl butyral resin should be closely defined and that zinc tetroxychromate is favored as the chromate pigment. It is probable that all the components enter a series of complex reactions with each other and the substrate to form the required tenacious coating. Current two-pack U.S. and U.K. specifications (MIL-C-8514, MIL-P-15328, and Defence Specification 1408) are basically similar to the WP1 formulation (as indeed was DTD.808 issued at the period under review). Typical of U.K. work on etch primers are papers by Coleman (7) and Prosser (8).

Numerous attempts to improve the WP1 formulation have resulted in the so-called water-resistant primers and the one-pack types. The two-pack water-resistant types generally include a proportion of, e.g., a phenolic resin with the polyvinyl butyral resin, often as a 30% replacement. Care must be exercized to ensure that such modifications do not lead to primer embrittlement or inadequate adhesion and retarded drying of topcoats. One-pack types can be prepared by substituting zinc chromate or barium chromate for zinc tetroxychromate in WP1 type formulations, though it is probable thereby that some of the complexing reactions of the WP1 type do not occur. Bayliss and Wall (9,10) describe the very good performance obtained on steel with a zinc chromate one-pack type.

TABLE 1-6. WP1 Etch Primer

Base		Parts by weight
Polyvinyl butyral resin (Vinylite XYHL)		7.2
Zinc tetroxy chromate		6.9
Talc		1.1
Isopropanol (99%) or ethanol (95%)		48.7
n-Butanol		16.1
		80
Acid diluent	Phosphoric acid 85%	3.6
	water	3.2
Isopropanol (99%) or ethanol (95%)		13.2
		20

The acid diluent is added to the base component in the ratio of 1 part to 4 parts by weight and thoroughly stirred.

Soon after the introduction of etch primers to the U.K. aircraft industry, Cole (11) at R.A.E. Farnborough evaluated the performance of five two-pack etch primers, a single solution type, and two "universal primers" (one highly chromated) with topcoats of DTD. 260 (alkyd) or DTD. 754 (nitrocellulose) finishes on a variety of substrates under conditions simulating those experienced by an aircraft on board a carrier, on a beach landing strip, or on board ship as deck cargo. A priming system of 1 coat of two-pack etch primer and 1 coat of a "universal" primer was also included in the program. Specimens were suspended in a covered shed open on one side and sprayed three times per working day with natural seawater. Etch and "universal" primers were applied (1) to degreased mild steel, the "universal" primers however being additionally applied to phosphated mild steel; (2) to mild steel, zinc plated ± passivation; (3) to mild steel, cadmium plated ± passivation; (4) to degreased DTD. 646 aluminum alloy, the "universal primers" being additionally applied to phosphoric etched or chromate etched DTD. 646; (5) to degreased DTD. 687 aluminum alloy, the universal primers also being applied to phosphoric or chromate etched DTD. 687; (6) to anodized DTD. 646; (7) to anodized DTD. 687; (8) to degreased DTD. 118 magnesium alloy; (9) to chromated DTD. 118; (10) to degreased DTD. 259 magnesium alloy; (11) to chromated DTD. 259; (12) to degreased brass, the "universal" primers additionally being applied to etched brass.

Corrosion was recorded visually or by weight loss up to a maximum of $4\frac{1}{2}$ years exposure. Conclusions were that the relative performance of the painting schemes varied markedly from metal to metal. On zinc- and cadmium-plated steel the etch primer schemes gave better performance than the conventional schemes applied over the classical methods of surface treatment. On steel and aluminum alloys the use of etch primers is an acceptable substitute for conventional methods of protection provided that the etch primer is followed by a full protective scheme. Their use on magnesium alloys is risky because of the danger of attack on the metal.

The two-pack etch primers have given sterling service to the aircraft industry, and though the use of one-pack types is not now generally accepted for use on nonferrous surfaces, there are still useful one-pack primers available for steel, galvanized and zinc-sprayed surfaces. Peak use of etch primers in the U.K. was prior to the introduction of phosphate ester hydraulic fluids, the resistance of the primers to this type of hydraulic fluid being suspect unless a high-bake schedule is employed. Also they can induce brittleness on some surface conversion pretreatments. Further the degree of degreasing and cleanliness of surface prior to their application are all important for satisfactory adhesion, and they should not be applied under very dry conditions. It is claimed however that polyvinyl formal-based primers have improved fluid resistance, stoving being used for optimum results.

D. FINISHES

The important British finishing schemes from the World War II period to first, the introduction of ester lubricants, and then phosphate ester hydraulic fluids are tabulated in Table 1-7, and U.S. specifications of a similar nature are quoted. In the main their requirements could be met by high performance alkyd or nitrocellulose based finishes used over "universal" or etch primers, and several are in current use today where ester lubricant and hydraulic fluid resistance are not demanded.

Lightweight finishes were required for the protection of heavy Transport Command piston-engined aircraft. A cellulose scheme of DTD.766 utilizing an etch primer was preferred to DTD.802, a synthetic finish applied direct to pretreated metal which gave problems of application and opacity at the low film weight required. During World War II a similar demand for light weight finishes and a shortage of raw materials had resulted in the use of one-coat alkyd-based camouflage colors containing some zinc chromate for application to pretreated metal. Specification DTD.658 covered such finishes at a film weight of 1-1.25 oz/yd^2 but their use was discontinued when a long service life was again expected of military aircraft.

Test methods for the DTD specifications of this period (now embraced in Defence Specification 1053) included bend and scratch tests, drying tests, tests for resistance to sea water, distilled water, and to the currently used

TABLE 1-7. Development of Aircraft Finishes

Specification	No. of coats	Total film weight, oz/yd^2	Type of primer	Type of finish	Uses	American equivalent
DTD.63B (now obsolete — superseded by DTD.899)	1 ct primer 1 ct cellulose enamel	1.25-2.75	Universal or etch	Nitro-cellulose + alkyd	Mainly internal	TT-L-20
DTD.235 (now BSX.31)	1 ct primer 1 ct stoving enamel	0.9-1.4	Universal or etch	Oxidizing alkyd	Prefabricated components	MIL-E-5557 or MIL-E-7729 type III
DTD.260 (now obsolete — superseded by DTD.827)	1 ct primer 1 ct pigmented oil varnish	1.25-2.75	Universal or etch	Alkyd	Mainly internal	MIL-E-7729 type I
DTD.314 (now BSX.28)	1 ct primer 1 ct matte pigmented oil varnish	1.5-2.5	Universal or etch	Alkyd	External skins	MIL-E-5556
DTD.420	1 ct matte pigmented	1.5-2.5	—	Lanoline/ hard resin	Removable finish and for sea-planes	—
DTD.517 (now obsolete — superseded by DTD.827)	1 ct primer 1 ct matte pigmented synthetic finish + 1 ct varnish when gloss required	1.5-2.5 0.75	Universal or etch	Phenol modified alkyd	Naval aircraft	—

DTD.754 (now BSX.26 and 27)	1 ct primer 1-2 ct matte cellulose finish + 1 ct transparent cellulose varnish when gloss required	2-3.25	Universal or etch	Nitro-cellulose	Internal or external	MIL-D-5555
DTD.766 (now BSX.30)	1 ct etch primer 1 ct nitrocellulose aluminum finish	0.8-1.2	Etch	Nitro-cellulose + alkyd	Transport command	—
DTD.772 (now BSX.29 embracing also DTD.768)	1 ct primer 1 ct cellulose or synthetic filler	2.9-4.75	Universal or etch	Nitro-cellulose + alkyd	High gloss finish	TT-L-32
	1 ct gloss cellulose finish (1 ct primer 1 ct high gloss black)	2.0-2.25	Universal	Nitro-cellulose + alkyd	Night bombers	—
DTD.785	Dope system + 1 ct gloss black	0.8-1.2	Universal	Nitro-cellulose + alkyd	Night bombers	—
DTD.785	Dope system + 1 ct gloss black	0.8-1.2	Universal	Nitro-cellulose + alkyd	Night bombers	—
DTD.796 (now obsolete superseded by DTD.827)	1 ct primer 1 ct high gloss black	1.5-2.25	Universal	Alkyd	Night bombers	—
DTD.802 (now obsolete)	1 ct synthetic aluminum finish	0.35-0.6	—	Alkyd	Transport command	—

aviation fuels and lubricants, gloss determinations, cycling tests, and natural weathering tests of long duration during which panels facing south in the open at an angle of 45° to the horizontal were sprayed three times per working day at intervals of three to four hours with artificial sea water.

French practice during this period was also to use etch and chromated alkyd primers, with cellulose and alkyd based undercoats and finishes, the requirements of systems for both wood and metal being defined. Fewer national specifications were issued, the French Air Ministry recognizing outstanding protective materials by official approvals, a method similar to the British DTD. 900 specification procedure. Some use also was made of materials to American specifications.

1-4. PHASE III: MODERN AIRCRAFT FINISHES

A. INTRODUCTION

As described earlier, alkyd and cellulose nitrate/alkyd finishes used over etch or alkyd based primers with a suitable pretreatment provided good protection to U.K. metal-skinned aircraft using the engine oils, hydraulic fluids, and fuels of the 1940s. From about 1950, the use of the ester lubricants of the di(2-ethylhexyl) sebacate type in jet engines introduced an aggressive potential stripper of paint films. Successful attempts to modify existing finishes while preserving their good drying properties were made. More resistant alkyds were produced and adequate resistance to ester lubricants was introduced into cellulose nitrate lacquers by the masking effect of other more resistant nonconvertible film formers of, e.g., the vinyl or acrylic resin families. A further competitor then entered the field with the introduction of the polyamide or amine-cured epoxies, which, although slower drying, offered improved resistance to ester lubricants. Their performance in primers is of particular importance as they possess good adhesion on pretreated surfaces, good water resistance, and good corrosion protection when suitably pigmented. Moreover, the use also from the 1950s of ester hydraulic fluids based on, e.g., tricresyl phosphate, introduced another fluid with a profound deteriorating effect on paint films and many other nonmetallics. Epoxy systems showed up well against this new agent, their chief disadvantages being chalking and yellowing tendencies during exposure. They in turn face competition from the polyurethanes, which although originating in Germany in 1937 as surface coatings, did not, until after further German development work, achieve their due prominence in the U.K. aircraft industry until the 1960s. Their special importance is due to their high durability and resistance to organic fluids. A third dominant finish in the present day U.K. aviation industry is the acrylic family, the advantages of this type being rapid drying, stability to heat and light, fair resistance to contaminants, and ease of removal and touch-up. Thus today in the U.K., the primers of importance are the epoxy/polyamide type used with adequate pretreatments and etch primers considered as pretreatment primers rather than as fully protective primers. Finishes consist chiefly of polyurethane, epoxy, acrylic, vinyl, alkyd, or nitrocellulose types, their selection being determined by the conditions of operation of the aircraft and by the fluids carried by the aircraft. Many aircraft today operate long range and encounter a wide range of climatic conditions.

In intense sunlight, the temperature of a dark coating on a static aircraft can be raised to 100°C; skin temperatures can change rapidly during flight and, on high speed aircraft, can reach temperatures of 130°C by kinetic heating. Considerable erosive influences can be met while flying through rain, dust, and hail. Cyclic condensation occurs during the various temperature and pressure fluctuations and keel areas may contain large quantities of water containing dissolved salts as well as aggressive fluids such as phosphate ester hydraulic fluids, Lockheed 22, ester lubricants, fuels, etc. Further, as the whole structure is vibrating under changing stress, cracking tendencies of coatings are accentuated. Hoey (12) has reviewed modern aircraft finishes against this background.

Although light aircraft are not subject to the total severity of the fluids and environments described and the use of alkyds, nitrocellulose, vinyl, and acrylic finishes continues to some degree, the tendency is toward the use of the higher performance polyurethane finishes used over epoxy primers as their range of operations increases. The constructors and operators of the bigger aircraft however, are faced with most of the operating conditions described and have little option but to use the highest performance schemes available, though their emphasis may vary. The manufacturer's main requirements are to achieve corrosion resistance, particularly of inaccessible areas in the interior and this necessitates schemes of good adhesion and fluid resistance. The carrier however is also concerned with painting his aircraft with high durability exterior finishes, and these should have good adhesion and flexibility. Fluid and corrosion resistance have lesser priorities, but ease of application is important in maintaining a civil airliner. The importance of good adhesion is common to both philosophies and the study of pretreatments with respect to their effect on various alloys and the bonding of coatings to them is fundamental to the aircraft industry. Hence a brief description of the alloys and pretreatments in use today follows before the technology of current finishing schemes is discussed.

B. ALLOYS AND PRETREATMENTS

So vast is this field and so many are the factors influencing design selection that discussion is limited to a brief statement of the aluminum alloys and pretreatments most often encountered in aircraft paint evaluation. A great variety of aluminum alloys is used in the aircraft industry, and their properties are governed by their composition, form, and temper or condition, heat treatment, etc. Properties of obvious importance are mechanical properties such as strength and ductility and corrosion resistance associated with environment. Corrosion failures can be categorized as surface, intergranular, or stress corrosion; anticorrosive paints and pretreatments are of utmost importance to the aircraft engineer. An area of recent study is the influence of paint coatings on stress corrosion, the latter form being generally worse with the aluminum zinc alloys. Evans (13) has reviewed this subject. A description of the main grades of pure aluminum and types of aluminum alloys, their corrosion resistance, the effect of alloying and heat treatments, effects of chemical conversion coatings and recommendations for painting are given by Champion (14). British DTD specifications use pure aluminum as a test substrate for paints to obtain standardized results in paint evaluation, and pretreatment by a chromic sulfuric acid pickle to Def. 1053, Method 2.5b(ii) is specified. The aircraft manufacturers generally prefer assessments to be carried out on their selected alloys and pretreatments. The alloys may be zinc-, copper-, or magnesium-rich aluminum alloys, can be clad or unclad, and may or may not require a pretreatment or anodizing. The more frequently used British high copper-containing types are B.S. Aircraft Specifications L70 and L71, with L72 and L73 as their aluminum clad equivalents. Similar French types are A-U4SG; American practice favors 2024-T3 to QQ-A/250/4 when unclad and to QQ-A/250/5 when clad. Other important alloys of this type are the British B.S. 1470 HS14, which the French A-U4G approximates. The higher zinc-containing alloys of equivalence are A-Z5GU (French), QQ-A 250-12d (7075) (American), L95 and DTD. 687 (British), and of the magnesium-rich aviation aluminum alloys A-G3, BS. 1470 NS5, and QQ-A 250-10 (5454) are similar as is the case with A-G5, BS. 1470, NS8, and QQ-A-250-9 (5456). Examples of composition types are given in Table 1-8. It is not necessarily the case that a paint system giving best performance on one type of alloy or pretreatment will perform equally well if the alloy or pretreatment is varied.

Pretreatments can significantly improve the corrosion resistance of alloys; another prime factor for their use is to present a satisfactory surface for good paint performance. During heat and machine processing and during storage, aluminum alloys are subject to oxide skin formation and contamination with oils and greases that are deterrents to good paint adhesion. Many bath pretreatments are available, their time of etching or surface conversion varying with the type of chemical or alloy used. A further factor in the selection of a pretreatment process depends on the practicability of treating the alloy surface by immersion or power spray. Larger components and assembled structures often require brush or paste application of pretreatment materials.

TABLE 1-8. Normal Composition % (residue aluminum and normal impurities)

Alloy	Cu	Mg	Zn	Si	Mn	Gr	Fe	Ti
L.70 and L.71	4.6	0.7	—	0.8	0.75	—	—	—
L.97	3.8-4.9	1.2-1.8	0.5	—	0.3-0.9	0.1	0.5	—
A-U4G1	3.8-4.9	1.2-1.8	0.25	0.5	0.3-0.9	0.1	0.5	0.2
A-U4G	3.5-4.7	0.4-1.0	0.25	0.3-0.8	0.3-0.8	0.1	0.7	0.2
A-Z5GU	1.2-2.0	2.0-3.5	5-6.5	0.4	0.1-0.9	0.35	0.5	0.2
A-G3	0-0.1	2.6-3.8	0-0.2	0-0.4	0.1-0.6	0-0.4	0-0.5	0-0.2
A-G5	0-0.1	4.5-5.5	0-0.2	0-0.4	0.2-1.0	0-0.4	0-0.5	0-0.2

Though there are many pretreatment processes available to the aircraft industry, those most frequently used throughout the world are known by the Amchem trade mark "Alodine" or as "Alocrom" in Imperial Chemical Industries Ltd. marketing areas. The principal processes in the I.C.I. range for light alloy use are Alocrom 100 (phosphate) or Alocrom 1000 or 1200 (oxychromate) conversion coatings, Deoxidines 202 and 624 cleaners for brush application, and various Ridoline alkaline immersion cleaners. Deoxidizers are also available for oxide removal by immersion prior to Alocrom treatment, spot welding, etc. (15).

Alocrom 1200 is best known as it gives a high level of corrosion resistance, and the principal aircraft constructors are equipped with full process equipment including large Alocrom 1200 tanks. It is characterized by giving a golden yellow coating on aluminum and its alloys and this when freshly formed is quite soft so that care must be taken not to damage it during rinsing and drying. When baths are operating correctly, thin, hard coatings integral with the metal are formed, and these range in color from iridescent gold to golden yellow. Bath temperatures, immersion times, Alocrom concentration, and acidity need to be balanced for optimum results, and a preliminary treatment of the metal with degreasing and deoxidizing operations is necessary for satisfactory surface conversion. Its residual acidity can often assist the adhesion of coatings, but if excessive occasionally may impair the curing and critical adhesion of some two-component materials. The use of etch primers is not recommended over Alocrom as embrittlement may result.

Alocrom 1000 also gives fairly good corrosion resistance and does not significantly change the appearance of the aluminum alloy surface, the formation of a perceptible golden color indicating overtreatment. It requires only one chemical for make-up and replenishment and is finding favor with aircraft constructors where the outside skin will not be painted overall. Adhesion of paint to inside surfaces is good and the coating is free from acidic residues. It does however require a higher operating temperature.

Alocrom 100 gives a protective bluish-green coating on aluminum and its alloys, though the corrosion resistance imparted is less than with the Alocrom 1200 types. It has a low operating temperature, is rapid in coating formation, and can be used on foodstuffs containers as the coating is nontoxic. It has very good adhesion for paints and a wide range of plastics.

Basic equipment consists of simple dip tanks, though the "Alocrom" and "Deoxidizer" tanks should be constructed of stainless steel or mild steel lined with rigid PVC, 'Neoprene," or sheet polythene, rinsing tanks being constructed of mild steel. Steam coil or immersion heaters are fitted as necessary to maintain recommended temperatures. These are at most 70°C as with Alocrom 1000, though Alocrom 100 and 1200 operate satisfactorily at normal room temperatures. Low temperature ovens are desirable for drying the work. Times of immersion are generally of the order of 2-5 min. Mild etching dip solutions of Deoxidizers 1 and 2 are also available for uniformly activating surfaces before using Alocrom 1000 and 1200 particularly. They are prepared from powdered chemicals which are dissolved in acid at the time of use. Deoxidizer 17 is particularly useful for high copper bearing alloys, which are normally prone to corrosion since it greatly improves the corrosion resistance of subsequent Alocrom coatings by eliminating copper salt deposits from the surface.

Removal of grease as a precleaning operation can be carried out by trichlorethylene degreasing, or by spray, or dip application of a suitable alkali Ridoline.

All the bath processes require careful chemical control and maintenance for satisfactory operation and the suppliers of the pretreatment chemicals give full instructions and service to ensure the best results.

Many articles or surfaces are too large to be conveniently dipped and brush pretreatment processes are available for such circumstances. Deoxidine 202 is a brush or spray-gun applied paste material designed for use on riveted, seamed, and similar surfaces prior to painting. Its paste form limits its penetration into these areas where removal of residues would be difficult. It provides a lightly etched surface and removes traces of grease, oil, and other contaminants. It can be used on other metals except zinc and magnesium. After application it is allowed to dry on the surface, this taking 10-20 min, and the powder is then removed by washing down with clean water or by rubbing lightly with a Scotchbrite pad. Painting should follow immediately after the surface is prepared; where it has been water rinsed, it should be allowed to dry thoroughly, seams or crevices being blown out with compressed air. Deoxidine 624 is a brush-on/wash-off acid cleaner for all metals except zinc, cadmium, and magnesium and is particularly suitable for removing grease, oil, corrosion products, and oxide films from aluminum and aluminum alloy surfaces prior to brush Alocrom treatment or etch priming. It is applied liberally, scrubbing well to remove oil or corrosion products. Up to 20 min may be necessary to obtain a clean surface after which rinsing with clean water takes place. If Alocrom treatment is to follow it is essential that a completely water wet surface with no break marks is obtained during the rinsing process. Surplus water is removed with a damp cloth and Alocrom then applied. If however paint is to be applied with no further pretreatment, then the surface is dried by blowing with compressed air or by wiping with clean dry cloths, painting following as soon as possible.

A specially prepared Alocrom 1200 solution may be applied by brush, swab, or low pressure spray to aluminum and its alloys; it has an additional use as a filming treatment over aged anodized or Alocrom treated surfaces and for touching-up damaged areas on such coatings. In such a use, the areas should first be cleaned by solvent wiping only. In other circumstances, solvent wiping, followed by Deoxidine 624 or rubbing with Scotchbrite pads may be necessary for precleaning. The Alocrom 1200 action on a properly prepared surface is to give an iridescent golden to golden yellow coating in 1-5 min, after which excess Alocrom is removed from the surface by flushing with clean water or gently swabbing with a soft sponge, rinsing finally with demineralized water. Drying is again carried out with compressed air or gentle wiping with soft cloths.

Brushing Alocrom 1000 is also used for aluminum and aluminum alloy surfaces and application procedures and process times are similar to those described for Alocrom 1200. The coating however should properly be invisible and thus forms an ideal base for a clear lacquer.

Anodizing

The above processes are nonelectrolytic in character; hence a brief description of the anodizing process is necessary to complete this limited survey of pretreatment processes used for aluminum and its alloys in the aircraft industry. After suitable pretreatment and cleaning, the aluminum components, racked to provide electrical contact, are immersed in a tank containing a dilute solution of acid (chromic or sulfuric) kept at a constant operating temperature and agitated by compressed air. The aluminum is made the anode of the circuit and lead sheet is usually the cathode. Oxygen reacts with the aluminum surface to give a hard oxide film chemically bound to the metal. The thickness of the oxide film produced is proportional to the amount of current passed through the solution at a particular temperature and acid concentration and a minimum average film thickness of 0.001 in. is normal for most applications. As with paint performance there can be a marked difference in response from alloy to alloy and the final appearance of the anodic film can vary according to alloy composition and thermal history. After anodizing, the material is rinsed well in running water to remove excess acid and is then sealed (e.g., by hot water) unless bonding is required when a stronger joint is given by the unsealed surfaces. It is also preferable not to seal completely before the application of paint. For other applications, sealing is considered important to ensure a stable and impervious coating; poorly sealed films can deteriorate on exposure to give whitish powdery coatings. Anodized surfaces should be painted as soon as possible; aged anodized films are often difficult surfaces on which to secure good paint adhesion. Etch primers or Alocrom 1200 treatment may be used to assist in such circumstances. Paint adhesion to various aluminum alloys after different anodizing treatments can also be variable (16).

C. PRIMERS

Epoxy primers have outstripped in importance any other types of primers in the aviation industry in the last two decades, and it is probable this preeminence will remain for many years to come. There are numerous national and "house" specifications (those issued by the world's principal aircraft constructors) which define the qualities required of epoxy primers, and an examination of the rigorous requirements contained in these explains their large scale usage. They have good resistance to aircraft fluids, good resistance to corrosive environments, good acceptance of all types of finishes, good mechanical properties, and good adhesion to pickled or filmed surfaces. They do not have good adhesion to untreated aluminum and are of course, two-pack materials but with a long pot-life. Also, their curing is inhibited at a temperature of below $13^{\circ}C$. A table of properties (Table 1-9) in comparison with that of other primers largely obtained from measurements on anodized panels is given by Hoey (17).

The epoxy resins used in the primers are derived from diphenylol propane and epichlorhydrin to give chains of diphenylol/propane residues with epoxy terminals which can be reacted with polyamines or compounds with primary or secondary amine groups to give a highly cross-linked structure. The formulas for diphenylol propane and epichlorhydrin are

$$\underset{\text{epichlorhydrin}}{\underbrace{CH_2-CH-CH_2Cl}_{\backslash O /}} \qquad \underset{\text{diphenylol propane}}{HO-C_6H_4-C(CH_3)_2-C_6H_4-OH}$$

If the diphenylolpropane is represented as OH-R-OH and the $-CH_2-CH-CH_2$ radical as -G-, then the molecules formed in making the commonly used resins can be represented by the general formula

$$(n+1)\ HO-R-OH + (n+2)\ \underbrace{CH_2-CH}_{\backslash O /}-CH_2-Cl + (n+2)\ NaOH \rightarrow$$

$$\underbrace{CH_2-CH}_{\backslash O /}-CH_2-\left[O-R-O-\underset{|}{G}\atop{\quad\ \ OH}\right]_N-O-R-CH_2-\underbrace{CH-CH_2}_{\backslash O /}$$

These resins therefore contain in their molecules

(a) Nonreactive ether linkages.
(b) Esterifiable hydroxyl groups.
(c) Terminal reactive groups.

Thus the following characteristics are important in formulating with epoxide resins.

TABLE 1-9. Relative Performance of Primers

Primer	Adhesion, lb/in.^{2}	Over coating, hr	Fluid resistance	Scratch	Inhibition	Bend	Two-pack	Pre-treatment
Epoxy	4000	4	Very good	2500	Good	$\frac{1}{4}$	Yes	Req.
PVB etch	2500	1	Good	1500	Slight	$\frac{1}{4}$	Yes	Not req.
PVB/phenol etch	3000 (erratic)	1	Good	1500	Slight	$\frac{1}{4}$	Yes	Not req.
Synthetic	1500	4	Fairly good	1200	Good	$\frac{1}{4}$	No	Req.

1. Epoxide equivalent, i.e., grams of resin containing 1 mole, i.e., 16 g oxygen in the form of epoxy groups.
2. Epoxide equivalent per gram of resin.
3. Hydroxyl equivalent, i.e., gram of resin containing one hydroxyl equivalent, i.e., 16 g oxygen in form of hydroxyl groups.
4. Hydroxyl equivalent per gram resin.
5. Esterification equivalent, i.e., gram resin required to esterify 1 mole monobasic acid.
6. Esterification equivalent per gram.
7. Molecular weight.
8. The index "n."

The following figures are typical of commercially used resins for cold curing applications:

n	m.p.	Epoxide equivalent	Hydroxyl equivalent	Density
2	70° C	450-525	130	1.21
4	100° C	870-1025	175	1.15
8	130° C	1650-2050	190	1.15
12	150° C	2400-4000	200	1.19

In these cold curing applications, use is made of the fact that the epoxy terminals can react with polyamines or compounds containing primary and secondary amine groups, e.g., polyamides to give highly cross-linked structures. The amine reaction can be represented as follows:

$$R'{-}NH_2 + \overset{\;\;\;O}{CH_2{-}CH}{-}CH_2{-}O{-}R \rightarrow R'{-}NH{-}CH_2{-}\underset{OH}{CH}{-}CH_2{-}O{-}R$$

and the secondary amine group reacts with another epoxy group thus:

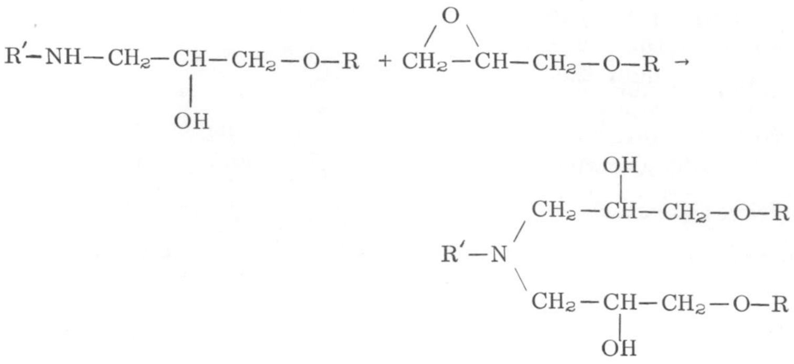

$$R'-NH-CH_2-\underset{\displaystyle OH}{\underset{|}{CH}}-CH_2-O-R + CH_2\overset{O}{-}CH-CH_2-O-R \rightarrow$$

$$R'-N\begin{cases} CH_2-\overset{\displaystyle OH}{\overset{|}{CH}}-CH_2-O-R \\ CH_2-\underset{\displaystyle OH}{\underset{|}{CH}}-CH_2-O-R \end{cases}$$

To obtain cross-linked structures, polyamines having at least four active hydrogen groups per molecule are used, di-ethylene triamine and tri-ethylene tetramine being typical. The amount of polyamine required for any epoxy resin is calculated on the basis of each epoxy group theoretically consuming one amino-hydrogen atom according to the following equation:

Amount of polyamine in grams per 1000 of resin = epoxy content + H-active equivalent.

For diethylene triamine [$H_2N-(CH_2)_2-NH-(CH_2)_2-NH_2$ mol wt = 103] the H-active equivalent is 21 and for tri-ethylene tetramine [$H_2N-(CH_2)_2-NH-(CH_2)_2-NH-(CH_2)_2-NH_2$ mol wt = 150] the H-active equivalent is 25. Epoxy contents for resins are expressed as mole epoxide per 1000 g resin. In practice, a 10-25% excess of polyamine is used to take into account losses due to evaporation and other practical factors. The use of free polyamines especially under cold and damp atmospheric conditions, may result in bloom formation in the film. Further disadvantages of the polyamines are odor, volatility, and irritant action. Hence it is usual to use an adduct where possible. The adduct is prepared on the principle of reacting amine with the terminal epoxy groups of the resin, so theoretically 1 g equivalent of epoxy resin (containing 2 epoxide equivalents) is reacted with 2 g eq of polyamine. In practice, a considerable excess of amine is used, and this is taken into account when calculating the usage proportions of the adduct as a curing agent. The simplest method is to calculate the amount of epoxide resin required for the total amine used and to deduct from this the amount of epoxy resin contained in the adduct solution.

Polyamide resins are condensation products resulting from the reaction of polyamines with dimeric and trimeric unsaturated fatty acids, e.g., linoleic acids, according to the following formula:

$$nHOOC-R-COOH + nH_2N-R'-NH_2 \rightarrow HO(OC-R-CONH-R'-NH)_nH$$

Highly functional polyamines produce polyamides with mainly primary and secondary amine groups and these react readily with epoxy groups. It is possible to calculate theoretical ratios of usage of epoxide resins with polyamides, amine values, and H-active equivalents being used to characterize the polyamides. The H-active equivalent relates only to the primary and secondary amine groups which can form C-N groups. These equivalents for the well-known General Mills "Versamids" are given by Gotze (18) and quoted in the following table:

	Milliequivalent per gram resin				Acid value	SG	H-active equivalent
	Primary amines	Secondary amines	Tertiary amines	Amine total			
Versamid 100	0.61	0.77	0.31	1.69	5.4	0.98	approx 500
Versamid 115	2.89	0.70	0.66	4.25	3.1	0.99	approx 150
Versamid 125	4.77	0.25	0.90	5.92	3.2	0.97	approx 100
Versamid 140	4.49	1.01	0.98	6.48	1.7	0.97	approx 100

The theoretical amount of polyamide resin in parts by weight required for 1000 g of epoxy resin content is the epoxy content xH-active equivalent of the polyamide resin. In practice, however, the ratio of epoxide resin/polyamide resin is less critical than in the case of epoxide resin/polyamine formulations and can be varied within fairly wide limits depending on the film characteristics required. Thus resins of epoxy equivalents of 450-525 may, for example, be used with polyamides of amine values of 210-220 (milligrams equivalent of KOH per gram of resin) between limits of one part to three parts of epoxide resin to one part of polyamide resin to give films of good performance. In general, longer pot lives and cure times are given by the lower polyamide contents as well as a greater tendency to brittleness in the cured films. However, better resistance to water, chemicals, and solvents and greater ultimate hardness are compensating factors accruing at the higher ratios.

The epoxide resin/polyamide systems, in general, give better flexibility, adhesion, gloss retention, and water resistance than the epoxide resin/polyamine systems and though slightly inferior in acid and solvent resistance are preferred for U.K. aircraft usage for these reasons, and provide the media of aircraft epoxy primers. Selection of solvent mixtures is similar to that of the finishes and is described later. Pigmentation is an important feature, corrosion inhibitive chromates being incorporated into the primer base.

Typical guide formulations are quoted in Mil-P-23377; it is necessary however to describe some of the factors determining U.K. usage of epoxy primers.

Both natural and accelerated weathering tests have shown the advantage of primers that leach chromates over similar primers that do not, this being particularly evident where there are breaks in the paint film. In consequence, much work has been carried out in the U.K. to establish the ideal shape of a chromate leaching curve of a primer necessary to give best general protection, particularly with respect to filiform corrosion and with emphasis on epoxy primers. Such leaching rate determinations have been carried out in distilled water, sea water, and dilute sodium chloride solutions. Tuitt (19) summarizes some of these investigations. His apparatus consists of a cell in which a Perspex tube with a fill/drain hole closed with a rubber bung is mounted between primer coated aluminum panels using wax seals at the primer/Perspex interfaces to prevent leakage of the test fluid. This can be removed for analytical examination and replaced by fresh fluid by use of fill/drain hole. Normally the test fluid is changed after 1, 2, 3, 4, 7, 10, 14, 18, 23, 29, and 36 days. The concentration of the hexavalent chromium in the solution is measured by a colorimetric method, based on diphenyl carbazide and calculated as milligrams of chromate (CrO_4) leached per square decimeter of paint film. Proportional adjustments can be made if necessary to the concentrations to correct for deviation of the true paint film weight from the standard 1 oz/yd^2, this relationship having been established from dry film weights over the range 0.8-1.2 oz/yd^2 on panels coated both by spinning and spraying. Spraying with air at 6% or 87% RH, the addition of water (0.1% or 1%) to the primer either immediately before spraying or 3 months previously, and humidities of 20%, 50%, and 85% RH, during the seven days air cure and during panel preparation are factors not affecting the leach rates. Tests carried out at 35 and 45°C show very little variation from those at 40°C, a temperature range of interest in the operation of subsonic aircraft. An ideal leaching rate curve is postulated with distilled water at 40°C as test fluid so that a minimum amount of chromate present in the leachate is equivalent to 4 mg/dm^2 at 1 and 2 days, 0.10 mg/dm^2 at 7-10 days, and 0.05 mg/dm^2 at 29-36 days. The initial fairly high leach rate is considered desirable in order to saturate the metal surface with chromate ions when the first water permeates the primer. Subsequent leaching should be controlled at a sufficient level for effective inhibition; the presence of ample chromate in the film does not always ensure this, other factors influencing its release. Some of these are also examined by Tuitt (19). The shape of the leach rate curve can be varied by altering the pigment volume concentration of the primer, by varying the milling period, and by changing the chromates. By using strontium and barium chromates and applying the other factors, it is possible to make primers with high or low initial leach rates and comparatively high or low leach rates toward the end of the curve. Duplex primer systems are examined, each primer coat being applied at 0.5 oz/yd^2. A strontium chromate primer of high initial leach rate is applied directly to the metal and followed by a strontium chromate primer of a relatively high level leach rate. The leaching curve of such a system, considered ideal for subsonic aircraft, is given. Similar effects may be obtained by overcoating a strontium chromate primer with a barium chromate primer by which the latter effectively suppresses the high initial leach rate of the strontium chromate primer, so that there is a further reserve of chromate (strontium) to boost the level portion of the composite curve (normally barium). This is particularly important in the case of supersonic aircraft where the temperature of leaching could be 100°C for short periods, under which conditions strontium chromate primers can sustain the high initial leach rate for several days, thus rapidly depleting the primer of chromate. Thus for supersonic aircraft either the duplex system of a strontium chromate primer followed by a barium chromate primer, or barium chromate primers only are favored. The effect of changing the test fluid from distilled water to synthetic sea water is also described. The leaching rate of a barium chromate containing primer is quadrupled at the initial part of the curve; that of the strontium chromate primer is increased fourfold in the steady part of the curve. A primer containing equal quantities of strontium and barium chromates gives a very high initial leach rate followed by a fairly high steady rate. Variations in pH were also studied by the preparation of buffered solutions containing acetates, phosphates, and borates to give the desired pH. When the buffer contains phosphate ions, there is an unexpected flattening effect on the leach curve at pH values 5, 6, and 7. The effect of aircraft fluids was examined also by coating test strips with the contaminant before leaching and renewing these fluids every week. Overall the effect of fluid contamination is not considered to be significant as regards the general protection afforded by leachable chromates. Hoey (30) has also discussed chromate leaching rates. Work continues in the U.K. More recently, the effect of medium, pigment volume concentration, and type of chromate have been studied, milling conditions being constant, using a dilute sodium chloride solution as test fluid. The pH of this could be adjusted by additions of dilute hydrochloric acid or caustic soda. As anticipated, the primer medium affected the leaching characteristics and increasing the pigmentation increased the initial leach rates and the total amount leached after prolonged immersion. The chromate pigments used, with the exception of barium chromate, did not have as marked an effect on the leaching characteristics as might have been expected because of the difference of pigment solubilities.

Because strontium and barium chromates proved their worth as corrosion inhibitors, the British Specification DTD. 5567, which covers the requirements of interior and exterior protective finishing schemes of the cold curing epoxide type, makes provision for their inclusion in the primer. It is mandatory to have not less than 20% of strontium chromate in the dried film. Certain extenders and white pigments also are permissible as well as green oxide of chromium and barium chromate. The latter may be used for balancing leaching rates, although currently such a clause is not included in the specification. The specification provides for two schemes, that of primer only where minor condensation is likely to be encountered and that for primer and finish where resistance to aircraft operating fluids or external protection are required. In accordance with British practice, this primer is that required for use in DTD acrylic and polyurethane specifications, and its acceptance is based on rigorous performance tests, no guide formulations being published. Thus the use of either polyamides or amine adducts is permitted as the curing agent in solution form to allow of mixing in equal proportions by volume with the primer base.

By contrast the principal American military aircraft epoxy primer specifications define the required compositions within close limits as well as giving performance requirements.

Much has been asked of such primers, the chief requirements being corrosion protection, resistance to aircraft fluids, and good adhesion to a variety of pretreatments without cracking or flaking during the repeated flexing and straining of the metal substrate. However, emphasis may vary according to experience gathered both from practice and carefully designed laboratory tests. Scott (21), for example, describes dynamic tests in which painted metal specimens were strained either in tension or fatigue before, after, or even during environmental tests involving high or low temperatures and contamination with fluids. A further example already discussed is the influence of leaching rates. If, however, the main requirements are met, practical problems can arise. Thus casual damage may necessitate repair with the consequent difficulty of stripping the existing solvent-resistant coatings without damaging bonding. Incompatibility of the primers may occur with other forms of protection, e.g., sealants, jointing compounds, etc. Further, the philosophy of painting easily examined areas, e.g., exteriors may differ from that of painting inaccessible or sealed structures.

Thus views may differ as to the relative merits of primer properties according to individual design and maintenance experience. Hence primers of very high water and fluid resistance allied to good adhesion might be expected to function well under special circumstances in which damage is unlikely. However, many engineers would expect the filiform corrosion resistance of these types to be less than that of those primers of lesser fluid resistance in which controlled chromate leaching is available to offset possible damage and undercutting. It is also the case that more easily strippable systems may be produced by the use of highly fluid resistant schemes over less fluid-resistant etch primers, an adequate compromise in fluid and corrosion resistance being achieved. A further factor may be the requirements of storage stability. British practice is that materials are expected to meet specification requirements after a years storage in temperate climates; many potentially valuable materials are thus debarred from usage.

Thus in formulating epoxy primers for aircraft usage, there are many possibilities open to the coatings technologist depending on the required performance. The epoxy equivalent of the epoxide resin and the amine values of the polyamide or amine adducts curing agents are of obvious importance as is the ratio in which resin and curing agent are blended. Maintenance of the curing agent's efficacy during storage is also significant; in this respect combinations of amines and polyamide resins may prove useful. Pigment volume concentrations in the cured film are of obvious influence on its tensile strength and flexibility, and chromate leaching rates are dependent on this factor as well as the nature of the chromate and its volume concentration in the pigment mixture.

Specification Mil-P-23377B requires the use of an epoxide resin of epoxy equivalent 425-550, its melting point, viscosity in solution, color, and specific gravity also being specified. It is used in combination with a polyamide resin of amine value 210-220, color, viscosity, and specific gravity values also being defined. The epoxide resin to polyamide resin ratio in the film as applied is approx 1.8/1. Strontium chromate is present at the level of 27% by weight of the dried film and is used with other pigments to give a pigment volume concentration of approx 25% in the cured film.

Anodized alclad panels adhering to Spec. QQ-A-250/5, are used for test purposes, and performances clauses for drying time adhesion, overcoating, and resistance to water and lubricating oil are included in the specification. Water resistance is tested after air-drying the primer for 72 hr and then immersing in distilled water for 18 hr at room temperature. Any slight softening shall be recovered in 4 hr, and loss of adhesion, discoloration, wrinkling, blistering, or any other defect are not permitted. Lubricating oil resistance is tested on similarly prepared and dried panels (a dry film of $0.6 \times 10^{-3} - 0.9 \times 10^{-3}$ in. dried for 72 hr) by immersion in diester lubricating oil composed of 2-ethyl-hexyl sebacate (95%) and tricresyl phosphate (5%) by weight at 25°F for 4 hr. No blistering, softening, or any other film defects are permitted, discoloration being disregarded. The adhesion should be good and the films when cooled to room temperature after the test should exhibit no cracks when bent over $\frac{1}{4}$ in mandrel.

Specification DTD. 5567 is more searching in its test requirements. These include drying time, flexibility, impact resistance, hardness and adhesion, water resistance, resistance to organic solvents (toluene/2.2.4.trimethyl pentane), resistance to synthetic lubricating oils (both cold and hot pyrolized), resistance to phosphate ester hydraulic oil to AFS. 295 (Skydrol), resistance to mineral-based hydraulic oil to DTD. 585, resistance to continuous salt spray on BSL. 70 alloy, resistance to humidity under condensation conditions, resistance to heat, resistance to temperature cycling ("Cold Crack" test), resistance to battery acid and natural weathering. A two-pack epoxy finish conforming also the requirements of the specification when used with the primer is used for overcoating the primer for the battery acid and natural weathering tests. Resistance to cold lubricating and hydraulic oils is tested by partially immersing primer films (0.85 ± 0.15 oz/yd^2) which have dried for seven days at 18-21°C on pretreated aluminum for 1000 hr continuously in the test fluid at 15-21°C. The panels are then subjected to a scratch test under a load of 1200 g after excess oil has been removed by wiping with a soft rag. Similar test panels are examined after similar drying periods for hot pyrolized oil and hot Skydrol 500 resistance. The hot pyrolized oil is applied (0.1 ± .029 g to 5 in.^{2}) by spreading, and the treated panel is placed horizontally in an oven at 70 ± 2°C for 1000 hr, the film of pyrolized oil being renewed in five consecutive days every week. After cooling to room temperature and removing oil, a scratch test of 1200 g is applied. A similar test for hot Skydrol resistance is required, half the area of the test panel (5 in. × 2 in.) being uniformly spread with a film of phosphate ester hydraulic oil by means of a brush and this film being renewed on five consecutive days per week. An important clause is that the materials shall retain the properties detailed after storage in their original containers for 12 months in temperate climates or six months in tropical climates. In the U.K., epoxide resins of epoxide equivalent 450-500 are normally used with polyamide resins of amine value 210-220. Tawn (22) has described the influence of imidazoline content in polyamides used as coreactants with epoxy resins and indicated that moderately high imidazoline contents appear to promote solvent resistance in cured epoxy/polyamide systems. Ratios of epoxide resin to polyamide resin influence flexibility, hardness, and chemical resistance, the lower ratios giving more flexible films of lower chemical resistance and hardness. The requirements of specification DTD.5567 generally demand a higher ratio of epoxide resin to polyamide than does MIL-P-23377 B.

Specification MIL-P-27316A has similar rigorous performance clauses, including hydraulic fluid resistance, corrosion resistance as a single coat, water, humidity, and hydrocarbon resistance, as well as a requirement for storage stability of one year. It nominates the manufacturer of the resin base component (Shell Chemical Company, Epon 1009) and polyamide catalyst (General Mills Co., Versamid 115) and requires certification of the composition from them. The epoxy resin/polyamide ratio is approximately 1.9/1 in the nonvolatiles. These contain approx 30% by weight of calcium chromate, which gives ready release of chromate ions. Titanium dioxide is used with the calcium chromate to increase thermal reflectance and gives a total pigment volume content of approximately 28% in the dried film. Similarity in performance to the British primer (DTD.5567) is shown by the suitability of MIL-P-27316 primer for use with two-component polyurethane coatings MIL-C-38412). DTD.5567 primer is required to meet the performance requirements of the exterior glossy finishing schemes of the cold curing polyurethane type to DTD.5580.

This satisfactory performance of epoxy primers with two pack polyurethanes of good color retention and durability favors their continued use in such systems while the merits of polyurethane primers are assessed with the same thoroughness that has been applied to the evaluation of epoxy primers over many years. Well proved products such as the zinc chromate synthetic primer (MIL-P-8585 Type 1) also give satisfactory results with the two-pack polyurethanes (MIL-C-27227 and MIL-C-38412).

Many types of polyurethane primers are possible, their hydraulic fluid and water resistance commending them for aircraft applications. The higher molecular weight epoxide resins containing 8-12 hydroxyl groups per molecule and having hydroxyl contents of 5-6.8% can be used as media. They may be cured with aromatic or aliphatic isocyanates, e.g., Bayer's Desmodur L — a trifunctional reaction product of tolylene diisocyanate with trimethylolpropane* or Desmodur N† — an aliphatic biuret tri-isocyanate derived from hexamethylene di-isocyanate, using -NCO/-OH ratios of up to 1.3/1 depending on the film properties of chemical resistance and hardness required. Best solvent resistance is given with Desmodur L. These primers will cure at 0°C by reaction of the secondary hydroxyl groups. Alternative possible media are epoxide/polyester resin combinations reacted with Desmodur L to a high crosslink; Desmodur S1 — a higher molecular weight variant of Desmodur L but difunctional in character — is also available for use with epoxide or epoxide/polyester resins. It has properties of elastifying without loss of chemical resistance and may be used for this purpose instead of polyesters of low functionality.

Zinc chromate may be used as the inhibitive pigment to give a dry film content of not less than 12.5% by weight. Pigment/binder ratios between 1.5 and 2.0/1 are used, pigments such as titanium dioxide, talc, and asbestine being preferred. It is generally necessary to check the adhesion to metal of polyurethane primers, the use of an etch primer often assisting in this respect. The corrosion resistance of polyurethane primers related to chromate leaching rates, adhesion to various pretreatments, critical times of overcoating for satisfactory intercoat adhesion not only with polyurethanes but with other types of finish are factors of importance which require long-term study. Some use is currently made in the US of polyurethane primers as intermediate coats between polyurethane finishes and etch primers for the exterior finishing of commercial aircraft. Such systems can be more readily removed for repainting, etc., than the system of an oxychromate conversion coating followed by a chromated epoxide primer and aliphatic urethane topcoat.

---

*Supplied as a 67% solution in oxitol acetate and containing 11.5% NCO or as a 75% solution in ethyl acetate containing 13.0% NCO.

†See p. 50 . Supplied as 100% solids containing 37% NCO.

D. FINISH COATS

It is convenient to separate the more recent DTD Specification materials into those which are resistant to the diester lubricants but not to the phosphate hydraulic fluids, and those that are highly resistant to both agents. In the first category are certain types of nitrocellulose lacquers, alkyd, and acrylic finishes, and in the latter the epoxies and polyurethanes.

1. Nitrocellulose Finishes

Because of the introduction of the diester lubricants — cellulose ester solubilizers — and a desire to retain the quick-drying properties of cellulose lacquers for aircraft usage, much work in the U.K. was carried out to incorporate ester lubricant resistant resins into nitrocellulose containing lacquers to give quick-drying films of adquate resistance. Parallel work in America led to the introduction of nitrocellulose/acrylic resin combinations. Engine operating conditions cause pyrolysis of the ester lubricants giving contaminants acting as powerful paint strippers. Most DTD Specifications calling for resistance to synthetic lubricating oils therefore also include tests for resistance both to hot and to pyrolized oil. This is prepared by heating Aircraft Turbine Engine Synthetic Type Lubricating Oil to Specification D. Eng. R. D. 2487 (RDE/0/463) in a standard distillation apparatus until approximately one-third of the lubricant has broken down and distilled over. The rate of heating is such that the thermometer at the top of the distillation column reaches 175°C at the end of the process although the temperature of the lubricant iself may be approximately 500°C. The distillate and residue are allowed to cool to room temperature and then mixed together to give the necessary test fluid. British work favored the use of, e.g., polyvinyl acetals rather than acrylic resins as maskants to give nitrocellulose based lacquers that were suitably resistant to the hot pyrolized oils, and DTD. 899 comprises a glossy cellulose finishing scheme of good ester lubricant resistance which uses materials of this type. Etch or "universal" primers are used as priming coats depending on whether a pretreatment has been used and are air dried 1 or 4 hr, respectively, before the application of successive coats. These may be a cellulose or synthetic (alkyd based) filler followed by a cellulose-based sealer before application of the finishing coats, in which case the fillers are given 4 hr drying, the sealer 2 hr drying, and the finish 2 hr drying also. Alternatives are to omit the fillers and the sealer if a universal primer is used, or the fillers if an etch primer is used, the purpose of the sealer being to secure satisfactory intercoat adhesion in the system. It is necessary to use an abrasive cutting compound followed by a liquid polish, both free from wax, to obtain a highly glossy surface; film weights can vary from approx 2 oz/yd^2 to 6 oz/yd^2 depending on the level of finish required. Black, white, and gray-green colors are additionally available as matte finishes. Test methods require the systems to be resistant to continuous immersion for 1000 hr in cold ester lubricant (D. Eng R. D. 2487(without a change in scratch test value (after removal of oil by solvent-washing) from the initial value (1000 g BS. 3900 Part E2) determined after 24 hr drying at 18-21°C of the finishing coat of the system. Hot pyrolized ester lubricant resistance is tested by uniformly spreading 0.1 ± 0.02 g of pyrolized ester lubricant over 5 in. of a panel coated with the system. This is carried out 24 hr after application of the finishing coat which has been allowed to dry at 18-21°C. The treated panel is then placed horizontally on the shelf of an oven maintained at 60 ± 2°C for 48 hr; when cooled to room temperature, softening, discoloration, and blistering should be absent. Other test methods specify the toughness, hardness, and adhesion required, the resistance to organic solvents and hot kerosine, resistance to distilled and sea water, and resistance to natural weathering. All items except universal primers and synthetic fillers should be capable of being thinned with Def. 1216 cellulose thinners, DTD. 840 thinners being used for the remainder.

The American nitrocellulose/acrylic resin finishes, approximately equivalent to DTD. 899, are specified by MIL-D-19537 for gloss finishes and MIL-L-19538 for camouflage finishes. Guide formulations require an approximate 11/5/4 ratio of methyl methacrylate copolymers/nitrocellulose/ diiso octyl phthalate or dioctyl phthalate for the medium. Lubricating oil resistance is tested by immersion for 2 hr at 119-123°C in lubricating oil to MIL-L-7808. Similar systems have been successful in France when used with etch primers or primers of the MIL-P-8585 type.

2. Alkyd Systems

Carefully selected alkyds of the appropriate oil length, etc., give adequate resistance to ester lubricants in air-dried systems and the British Specifications for a high gloss ester lubricant resistant synthetic system is DTD. 827. "Universal" or etch primers are used followed by a coat of alkyd or oleoresinous based filler. The etch primer is given 1 hr drying and the universal primer 4 hr drying before the filler is applied. The specification requires the filler to be tested for wet flatting properties after 4 hr drying; it is required that no clogging of A. 320 grade silica carbide paper shall occur. For the general tests, e.g., toughness, hardness, adhesion, organic solvents, sea water and distilled water resistance, however, the filler is given 16 hr drying before overcoating with the finishing coat. Tests on the scheme then follow after seven days drying of the finishing at 18-21°C, although the finish itself is required to be hard dry in 5 hr under these conditions. The ester oil lubricant resistance test, however, foreshortens the drying of the finish in the system build-up to three days, and then requires an immersion test in oil, as specified in D. Eng R. D. 2487 (RDE/0/463) for four days. After solvent wiping to remove oil, a scratch test of 1000 g by method BS. 3900 Part E. 2 should not penetrate the film.

Other tests are required to determine, at various intervals, the resistance to recoating of the finish itself and the standard natural weathering clause requires two years exposure in the U.K. at an angle of 45° facing south with artificial sea water spray three times per day at 3 hr intervals. This system gives a high gloss finish of average film weight of 4 oz/yd and though there is no precise American equivalent, Specification MIL-E-7729 Type I. for gloss air-drying enamels is similar in intent, and is well-known in France.

3. Acrylic Finishes

An attractive property of the acrylic family of resins is their stability to light and heat because of the essentially polyethylenic backbone contained in their structure. This factor is of importance in relation to the high skin temperatures which may arise from kinetic heating; under these circumstances, acrylic finishes which possess good initial color show no significant change. Their fluid resistance is also fair, and in the absence of the more aggressive fluids, their suitability for finishing schemes for supersonic aircraft is indicated. A report on a six-year evaluation for such applications has been published and is relevant to the external finishing of Concorde supersonic airliner. Other advantages of the acrylics are (1) their relatively rapid drying which is not particularly temperature dependent, as is the case with most cold curing systems; (2) they are mostly one-pack materials, (3) they can be readily repaired, (4) they have reasonable durability, and (5) they can be polished to a high gloss.

Hoey (50), in discussing paints for high-speed aircraft, describes the heat problems involved and states that the likely maximum temperature at which acrylics can be useful is 150°C.

Military use in the U.K. of acrylic finishes is governed by specifications DTD. 5599 and DTD. 5602, which are designed to provide rapid drying finishing schemes mainly for use on exterior surfaces of aircraft which have been given an approved pretreatment. The chromated epoxy two-component cold-curing primer (DTD. 5567) is used with both finishes. If a filler is required, this also is a two-component cold-curing epoxide type (DTD. 5555), the systems of primer and finish being known as Schemes I and those of primer, filler, and finish being known as Scheme II.

Film weight additions when dry are 0.85 ± 0.15 oz/yd for primer, 2 ± 0.25 oz/yd for the filler, and 1.25 ± 0.25 oz/yd for the finishes (normally achieved in two coats). Drying times at 20 ± 2°C and 60-70% RH to allow for overcoating the primer or filler with finish are required to be 4 hr and the finishes should be hard dry in 2 hr. However, rubbing properties of the filler are checked after it has been given 16 hr drying in standard conditions by wet flatting with Grade A. 320 silicon carbide paper to BS. 872. A further clause determines the overcoating resistance both of the normally dried and aged schemes, the filler coat being flatted in the Scheme II test process. Tests are carried out to recoat with finish to a weight of 3.75 ± 0.75 g/ft^2 allowing no more than 2 min between successive coats. The periods after application of the finishing coat at which the schemes are tested for recoating with finish are (1) 4 hr, (2) 16 hr, (3) 7 days, and (4) 7 days followed by heating at $100^\circ \pm 2^\circ$C for 72 hr. No defects should show in the overcoated panels. Other tests are designed to reproduce aircraft conditions, and a method for determination of resistance to temperature cycling is included. Panels of the schemes after seven days drying of the finishing coat are subjected to a cycle of humidity (24 hr under conditions in BS. 3900 Part F2), 20 hr storage at -20°C and 4 hr storage at $-2 \pm 2^\circ$C. This cycle is repeated twice weekly for six wecks after which the film should not show signs of cracking, chipping, flaking, blistering, or loss of gloss. Heat resistance is determined by heating panels of both schemes to both specifications for 1000 hr at $130 \pm 2^\circ$C, cooling to $20^\circ \pm 2^\circ$C and carrying out bend tests as specified. Slight yellowing is permitted. Kerosine resistance is tested, e.g., immersing panels in kerosine at 120°C for 2 hr, allowing them to dry for 1 hr for examination, following which the test panels are given a bend test 24 hr later. Test pieces are also immersed in a mixture of 75 parts by volume of 2.2–4 trimethyl pentane and 25 parts by volume of pure toluene for 2 hr at $20 \pm 2^\circ$C, allowed to dry for 30 min for examination, and then given a bend test 24 hr later. Hydraulic fluid resistance is examined by immersion in mineral-based hydraulic oil to DTD. 585 for 1000 hr at 20°C $\pm 2^\circ$C. After wiping the test piece free of oil by the use of a soft rag dipped in the mixture of organic solvents described above, a scratch test under a load of 1500 g is carried out. Artificial sea water resistance and toughness, hardness, and adhesion clauses are also included, both dry and wet film scratch tests being carried out. In the case of DDT. 5599 a wet scratch test of 1000 g is the minimum requirement as against 1200 g in DTD. 5602. Another difference in the materials' specifications, related to their end uses, is in resistances to hot pyrolized ester lubricants, though their performance in resisting cold and hot ester lubricants is expected to be of a similar order. Both specifications require resistance to D. Eng. R.D. 2487 (R. D. E. 0/463) oil to be such that after immersion in the oil at 20°C for 1000 hr and after wiping off oil with a soft rag dipped in organic solvents, scratch test requirements under a load of 1500 g are met. A similar test for resistance to hot oil specifies a period of immersion in the oil of 8 hr at 70°C $\pm 2^\circ$C followed by a scratch test of 1200 g. Hot pyrolized lubricating oil DTD. 5602 tests require a period of test of 1000 hr at 70°C $\pm 2^\circ$C followed by a 600 g scratch test, and in DTD. 5599 a period of 24 hr at $70 \pm 2^\circ$C followed by visual examination. In neither case should the film show blistering; only slight discoloration is permitted. The test procedure is to spread, uniformly, pyrolized lubricant at a given film weight over half the area of test panels and to place them horizontally on the shelf of an oven maintained at $70 \pm 2^\circ$C. In the case of DTD. 5602, the test solution is wiped off and renewed daily on five consecutive days every week for the complete test period. Both specifications contain a two-year natural weathering clause, artificial sea water spraying of the exposed panels being carried out at three to four intervals three times a day. Both specifications provide for the finish to be polished to a highly glossy surface, but additionally allow for a limited range of colors to be available as mattes.

The specification titles indicate their scope and intended applications. DTD. 5599 is described as a selectively strippable acrylic finishing scheme for use on aircraft and DTD. 5602 as an acrylic finishing scheme for general purpose use on aircraft. The finishing coat in DTD. 5599 can if desired be removed by the use of a suitably approved paint remover, e.g., that specified in Def. 1443 without detriment to the underlying primer or filler. Where this selective strippability is required the primer or filler, as appropriate, should be allowed to air dry for 16 hr before overcoating with the finish. This property of strippability without disturbing the underlying epoxide materials is of advantage for a change of camouflage without increasing weight and commends itself to the Fleet Air Arm for the use of such acrylics on carrier-based aircraft.

Resins based mainly on methyl methacrylate polymers provide the media of these formulations, those polymerized to the higher molecular weight or containing small quantities of polar co-monomers being used where the greater resistance to hot pyrolized ester lubricant is required. Considerable dilution is thus necessary for satisfactory spray application; a further consequence of the high molecular weight is that the resins require additions of plasticizers which may be lost by leaching of volatilization during service. A predisposition to cracking is therefore possible because internal plasticization with a more flexible monomer, e.g., ethyl acrylate impairs fluid resistance. This possibility of varying the hardness and flexibility by attention to the monomers used in the copolymers and of introducing reactive groupings for subsequent crosslinking after the application of the finish to the aircraft to give an insoluble polymer retaining the stability advantages of the acrylics is an attractive one to the aircraft finishes technologist. These crosslinking copolymers are often referred to as thermosetting acrylics because the polar co-monomer grouping and crosslinking agent combination are normally not sufficiently reactive for air-drying applications. Resins containing glycidyl methacrylate as the polar co-monomer can however be formulated to give self-cure with catalysts or polyamines at normal temperatures and Hoey (17) describes resins of this type, which are however limited in potlife. Other types of the thermosetting acrylics can be based on acrylic acid, acrylamide, alkoxymethyl acrylamide, or hydroxypropyl methacrylate as polar co-monomer but require stoving with the usual curing agents, e.g., melamine or epoxide resins, to achieve the correct order of crosslinking density. Since the relative rate of reaction of polyisocyanates is greatest with a hydroxyl substituent grouping the largest potential for a crosslinking acrylic/polyisocyanate resin for air-drying purposes would be the uses of hydroxy acrylic copolymers. Barker and Lowe (23) describe the film properties of such resin combinations with polyisocyanates and suggest methods of increasing the degree of reaction. An overlay two pack acrylic/polyisocyanate varnish possessing good ester lubricant resistance is in use in the U.K. It also provides adequate protection to fluorescent finishes based on thermoplastic acrylic resins without detracting from the visibility of the system. However, such two-pack acrylics have not yet achieved the phosphate ester hydraulic fluid resistance of the epoxy/polyamide finishes or the trimethylolpropane/phthalic acid polyesters used with aliphatic polyisocyanates.

4. Epoxy Finishes

Epoxy finishes are one of the most important classes used in the aerospace industry. They have good general chemical and solvent resistance, good protection from corrosion, and good adhesion over suitable pretreatments or primers. They are much used on weapons, missiles and space vehicles and for the interior and exterior protection of many military and civil aircraft.

In formulating epoxy finishes, the selection of epoxide resin/polyamine or polyamide curing agent ratios follows the practice used in primers; thus the amine adduct addition in relation to epoxide resin is controlled within fairly close limits, though the resin/polyamide ratio may be chosen from a fairly wide range. In practice, the best all-round properties are generally obtained by working with an average polyamide content so that an epoxide resin/polyamide ratio of approximately 2/1 is typical. Weathering resistance is not unduly affected by such variations and the finishes have good protective qualities marred by their tendencies to chalk and yellow on ageing. Simple mixing ratios, uniform within a range of colors, are usual for blending the pigmented and clear components. An equal volume mixture is generally preferred. As both glossy and matte finishes may be required, and pigmentation demands for binder are variable, some ingenuity in balancing formulations is necessary. It may be desirable in some instances to use the polyamide component, which, because of its fatty acid chains is a good wetting agent, as the dispersion vehicle for difficult pigments, e.g., carbon black, or for matteing pigments to give a matteing base mixture for use with glossy colors. Suspension aids such as aluminum stearate and flow control agents, e.g., small quantities of a silicone oil or a butylated urea formaldehyde resin to eliminate surface defects such as floating or cratering, may also be required and in most instances careful attention to solvent mixtures is necessary. Mixtures of aromatic hydrocarbons, ketones, and alcohols, particularly isopropyl alcohol and butyl alcohol, are generally used, as well as glycol ethers in small quantities. As with other high boilers, the glycol ethers can be retained in the coating for long periods and if their use is excessive, then water resistance may be impaired. Diacetone alcohol is very prone to this effect. Esters should not be used in amine adduct formulations or in polyamide solutions, and aliphatic hydrocarbons must be avoided. Aromatic hydrocarbons, however, such as toluene, xylene, and the higher boilers with a high trimethyl benzene content are suitable. Greater care is necessary in selecting solvent mixtures for curing with amine adducts, the solvent mixture preferred consisting of about equal volumes of active medium boiling solvents and aromatic hydrocarbons. The pot life of the mixed epoxide resin/curing agent components is also influenced by the solvent mixture, for high alcohol contents shorten pot lives and ketones lengthen them. Pigmentation does not normally present difficulties though organic pigments sensitive to amine should be avoided. Pigment/binder ratios of up to 1/1 are normal, and in the case of heavily pigmented rubbing fillers can be exceeded. These can utilize extenders such as barytes, blanc fixe, talc, and asbestine; fine silicas are generally used for matte finishes.

Specification Mil-C-22750 gives typical formulations for chemical- and solvent-resistant epoxy-polyamide coatings both glossy and matte. An epoxide resin of 425-550 epoxide equivalent is specified for use with a polyamide resin of amine value 210-220 at a ratio of $1.85 \pm 0.5/1$. The two-component systems are formulated for equal volume mixing, the pigmentations for the matte finishes being dispersed in the polyamide content. An induction period of 1 hr before application is allowed after mixing and the mixed components have a pot-life of more than 24 hr (unthinned). Tack-free drying times are 6 and $1\frac{1}{2}$ hr, respectively, for the glossy and matte finishes with dry through times of 24 and 7 hr, these drying times being determined on anodized aluminum. Thinners to Mil-T-19544 are required for spray application and may be added after mixing of the components prior to the induction period. Testing is carried out using alclad panels (QQ-A-362) anodized in according with Specification Mil-A-8625, Type 1. Two sets are sprayed with wash primer (Mil-C-8514) at a dry film thickness of 0.2×10^{-3} or 0.3×10^{-3} in. and are given 45 min air drying. One set is overcoated with two coats of finish with 30 min between coats to a dry film thickness of $1.6 \times 10^{-3} \pm 0.2 \times 10^{-3}$ in. and given eight days drying before testing. After a further 1 hr of drying with the top coat in the manner described, the other set is given a coat of 0.5×10^{-3} - 0.8×10^{-3} in. dry thickness of epoxy primer Mil-P-23377 before overcoating. A third set of panels is also prepared using a coat of epoxy primer (Mil-P-23377), which after 1 hr of drying is overcoated with the top coat, and the system is also given eight days drying before testing. Principal tests require the wash-primed schemes to resist distilled water for 24 hr as indicated by a tape test after immersion of scribed panels, and for the wash primer + topcoat system to resist immersion in synthetic diester lubricating oil for 24 hr at 200° F. The topcoat, however, when applied to Mil-P-23377 or to base anodized metal should resist immersion in the ester lubricant for 24 hr at 300° F.

Specification DTD. 5567 is for an interior and exterior cold-curing epoxide protective finishing scheme; Scheme I is for primer only and is designed for use where minor condensation only is likely to be encountered. For external protection or where resistance to aircraft fluids and heavy condensation is expected, the primer is overcoated with a cold-curing epoxy top coat in accordance with Scheme II. The specification allows the use of either amine adduct or polyamide as curing agent, the resin vehicle and curing agent being mixed either in the ratio of two to one or, alternatively, in equal parts by volume. For spray application Def. 1216A thinners are used; for roller coat application of the primer, the 2/1 mixing ratio type is used with DTD. 5597 thinners. A pot-life of 8 hr at 18-21°C or 4 hr at 32-35°C is required of both primer and finish, all mixtures being discarded after 18 hr. An induction period after mixing of 30 min at temperatures above 13°C is required before application. The drying time of the finish is such as to be surface dry in $1\frac{1}{2}$ hr and hard dry in 8 hr; the primer is given 4 hr drying before application of the finish and the system is given seven days drying at 18-21°C before being subjected to rigorous performance tests for physical characteristics and resistance to heat, humidity, salt spray, water, temperature cycling, and a variety of aircraft fluids, including phosphate ester hydraulic oils and synthetic lubricating oils. These tests have been described in the section dealing with primers.

DTD. 5555 covers the requirements of an exterior glossy finishing scheme based on cold curing epoxide resins. Three schemes are permitted: Scheme I allows the use of an etch or epoxy primer with a finish coat; Scheme II requires an intermediate coat of an epoxide two component filler, and Scheme III has the epoxy primer as intermediate coat. The general properties of the primer and finish are similar to those in DTD. 5567 and the formulation of these and the epoxide filler followed the same principles. However, where an etch primer is used in the schemes, resistance to aircraft fluids is of a lower order, thus resistance to phosphate esters is not specified. Tests include toughness, hardness, and adhesion, the severity of bend tests being less for Scheme II. Resistance to artificial seawater, resistance to synthetic lubricating oils, retention of adhesion, resistance to organic solvents, and resistance to natural weathering are also requirements. The adhesion of the finish to primer which has been allowed seven days drying at 65-70°F and a relative humidity of 60-70% should be satisfactory as shown by a bend test at 0°C seven days after the finish has been applied.

Organic solvent resistance is demonstrated by an immersion test of seven days duration for all schemes followed by a bend test. Scratch tests used for determination of the resistance to cold ester oil on a scribed panel, the period of immersion being 1000 hr and the needle loading being 1200 g. The hot pyrolized ester lubricant resistance test is a smear test requiring resistance for 48 hr at 70 ± 2°C.

5. Epoxy/Isocyanate Finishes

As indicated earlier, the epoxide resins have reactive hydroxyl groups capable of reaction with polyisocyanates. It is possible to increase the functionality of the resins by reacting them with dialkanolamines. If the formulation on p. 00 is represented as

$$\underset{\text{Alkanolamine}}{(HO{-}R')_2NH} + \underset{\text{Epoxide resin}}{\overset{\quad\ O\quad}{CH_2{-}CH}{-}CH_2{-}\left[O{-}R{-}O{-}\underset{OH}{G}\right]_n{-}O{-}R{-}CH_2{-}\overset{\quad O\quad}{CH{-}CH_2}} + \underset{\text{Alkanolamine}}{NH(R'{-}OH)_2}$$

$$\rightarrow (HO{-}R')_2N\,CH_2{-}\underset{OH}{CH}{-}CH_2{-}\left[O{-}R{-}O{-}\underset{OH}{G}\right]_n{-}O{-}R{-}CH_2{-}\underset{OH}{CH}{-}CH_2\,N(R'{-}OH)_2$$

Alkanolamine adduct

it is possible to prepare a range of pale straw colored solid resins based on epoxide resins where n = 2, 4, 8, or 12 by reacting 1 g mole of the resin with 2 g moles of the dialkanolamine. These resins may be used in solution in polyurethane grade esters and ketones with aromatic hydrocarbons for reactions with aromatic or aliphatic polyisocyanates at desired -OH/-NCO ratios. A wide range of two-component products of differing properties may be produced. These may be varied in pot life, hardness, flexibility, and chemical resistance according to requirements. One significant advantage over the amine-cured type of system is the satisfactory hardening of epoxide resins at low temperatures, e.g., 5°C. Finishes can be produced with excellent resistance to phosphate ester hydraulic fluids and synthetic lubricating oils; they compare favorably in many respects with the two-component polyester/polyisocyanate finishes of maximum aircraft fluid resistance. Improved flexibility and good drying can be achieved at some expense of pot life. Since their chalk resistance is comparable to that of the cheaper epoxide/polyamide finishes, their use is usually limited to applications where a high degree of water and acid resistance is required.

6. Polyurethane Finishes

In World War II, German aircraft were finished with polyurethane coatings; their extensive use for this purpose both in the U.K. and elsewhere is, however, relatively recent and is expected to increase. In general terms, the polyester/aliphatic isocyanate types can be formulated to possess excellent gloss, durability, color, and gloss retention as well as excellent water and aircraft fluid resistance. Some of their defects are related to their virtues; thus the harder types of maximum fluid resistance are difficult to strip or recoat and may show "chipping" on impact with sand, hail, etc. Also yellowing can occur with prolonged exposure to temperatures above 120°C, and they are two-pack materials of limited pot life when mixed. Polyurethane technology, however, is vigorous and dynamic, and no other coatings offer so many possibilities to meet present and future aircraft requirements. There is already a vast literature on polyurethanes [see, for example, Saunders and Frisch (24) and, more recently, Taft and Mohar (51)] no more than a survey of currently used U.K. aircraft formulations is attempted and the chemistry is dealt with only briefly.

Polyhydroxylic and polyisocyanate bodies are the reactants commonly used for polyurethane coatings. A number of different reactions take place depending on prevailing conditions. The principal air-curing reactions in the absence of catalysts are

$$\text{R-NCO} + \text{R}'\text{-OH} \rightarrow \text{R-NH-COOR}' \quad \text{urethane} \tag{1}$$

$$\text{R-NCO} + \text{H-OH} \rightarrow \text{R-NH}_2 + \text{CO}_2 \quad \text{amine} \tag{2}$$

$$\text{R-NH}_2 + \text{R-NCO} \rightarrow \text{R-NH-CONHR} \quad \text{urea} \tag{3}$$

$$\text{R-NCO} + \text{R}'\text{COOH} \rightarrow \text{RNH-CO-R}' + \text{CO}_2 \quad \text{amide} \tag{4}$$

$$\underset{\text{urethane}}{\text{R-NH-COOR}'} + \text{R-NCO} \rightarrow \text{R-}\underset{\text{CONHR}}{\underset{|}{\text{N}}}\text{-COOR}' \quad \text{allophanate} \tag{5}$$

$$\underset{\text{urea}}{\text{R-NH-CONH-R}} + \text{R-NCO} \rightarrow \text{R-}\underset{\text{CO-NHR}}{\underset{|}{\text{N}}}\text{CO-NHR} \quad \text{biuret} \tag{6}$$

The speeds of reaction depend on many factors but as a general guide the following relative order applies:

Aliphatic NH_2 > aromatic NH_2 > primary OH > water > secondary OH > tertiary OH > phenolic OH > COOH > $CONH_2$.

Thus Reactions (1), (2), and (3) are the more important reactions at normal temperatures.

The recognized categories of polyurethane coatings are:

1. Urethane oils (isocyanate modified drying oils).
2. Isocyanate terminated adducts or prepolymers (moisture cured).
3. "Blocked" isocyanate coatings (heat cured).
4. Isocyanate-terminated adducts or prepolymers with a hydroxyl-containing compound as a second component.
5. Prepolymers similar to type 2 with a catalyst as a second component.

The most significant types for aircraft usage are 1 and 4, particularly the latter. Type 1 coatings are the reaction products of diisocyanates with hydroxyl-containing drying oil derivatives, and as they are generally formulated to have a hydroxyl/isocyanate ratio greater than 1.0, the final product contains no reactive isocyanate groups. The film is formed by air oxidation and crosslinking of the ethylenically unsaturated fatty acid side chains.

The two-pack systems (type 4), however, are influenced by the presence of moisture both in the atmosphere or in the composition, this moisture being a reactant competing with the hydroxyl groups for the available isocyanate. Hebermehl (25) has shown that low relative humidities influence to a considerable degree the rate of cure of two-component polyester-isocyanate coating compositions. He followed the rate of cure of films at different relative humidities (RH) using infrared to show the disappearance of -NCO groups. A striking acceleration was noted when the RH was increased from 0 up to 33%, further increases having relatively little effect. Infrared data were in agreement with an estimated conversion of 30% of the -NCO groups to urea, and 70% to urethane at a relative conversion of 30% or greater. Berger (26) has also used infrared to follow the decrease of -NCO groups in two-component lacquers over long periods at 0 to 97% RH. The rate of -NCO reduction was greatly accelerated by increases in atmospheric humidity but it was considered that the proportion of -NCO groups reacting with water or hydroxyl groups was unaffected. Lowe (27) has presented data showing the relationship between the amounts of water present in a typical series of urethane lacquers at the beginning of its application as a coating, and the proportion of -NCO groups required for reaction with it, the isocyanate added to the lacquers being in the stoichiometric relationship of one -NCO for one -OH group. Although the moisture contents of the coatings were from only 0.2-0.36%, significant percentages of the isocyanate contents of the films could be expected to react with water, formulations containing few hydroxyl groups being effected to a greater extent in the competition for -OH groups. No account was taken of adventitious moisture which could also be significant. It has been shown that a typical polyester resin used for such purposes will take up 2% of its own weight of water when exposed in a film of 0.5 cm thickness in air at 60% RH for 4 days. The presence of water was not necessarily a disadvantage as was shown by the addition of water to combinations of polyols and isocyanates. Most work was carried out on polyols containing secondary hydroxyl groups in order to obtain the most reliable balance between the water/isocyanate and hydroxyl/isocyanate reactions. When the water content rose to above 2% of the polyol content, the resistance to attack by caustic alkali was markedly improved, indicating an increase in crosslinking. The author stressed certain formulation guidelines for two-component polyurethanes, stating that it is the combined function of hydroxyl, isocyanate, and water that is of paramount importance and that specific resins require specific isocyanate equivalents. This may be in line with the hydroxyl value of the resin and the isocyanate value of the isocyanate, but it may also be necessary to allow for adventitious or atmospheric moisture. The constitutions of the isocyanates used and those of the hydroxyl-containing resins can be used to predict performance.

Polyesters have been devised with different degrees of branching and with different -OH content. Polyethers from diols and triols are limited to the same number of hydroxylic groups as in the polyhydric alcohol from which the polyether is derived, but molecular weight can be varied. The polyisocyanates have differing molecular complexities, differing isocyanate values, and differing physical forms. Thus a great number of possible combinations exist; generally the higher the hydroxyl content of a polyester or the greater the degree of branching, the harder and more chemically resistant will be the resultant films. Properties can also be controlled by -OH/-NCO ratios; the lower the isocyanate content, the more likely the flexibility is to increase and the resistance to weathering, solvents, and chemicals to decrease. Catalysts, pigments, and solvents can also profoundly influence the polymer structures and chemical linkages formed in a cured film. The relationships between these and the properties in aircraft finishes, e.g., phosphate ester resistance are not well known. In practice, however, it is possible to formulate durable two-component coatings consistent in resistance to aircraft fluids and in drying and pot life under controlled conditions.

In the U.K. these are mainly based on the Bayer Desmophen 650/Desmodur N systems and a seven-year life without repainting is expected of the glossy exterior finishes. Modifications with other polyesters or with polyethers may sometimes be desirable in order to achieve specific requirements, but this is usually accompanied by some loss in solvent resistance. The more highly branched polyesters give the most chemically and weather resistant films, slightly branched ones being used for flexibilizing purposes. Table 1-10 gives details of the principal Desmophens used in aircraft coatings.

Desmodur N was commercially introduced in the early 1960's as the first aliphatic polyisocyanate suitable for use in coatings. It contains biuret groups and is produced by the reaction of hexamethylene diisocyanate with water thus

$$3\ O{=}C{=}N{-}(CH_2)_6{-}N{=}C{=}O + H_2O \xrightarrow{-CO_2} O{=}C{=}N{-}(CH_2)_6{-}N\begin{cases} \overset{O}{\overset{\|}{C}}{-}\overset{H}{\overset{|}{N}}{-}(CH_2)_6{-}N{=}C{=}O \\ \underset{O}{\underset{\|}{C}}{-}\underset{H}{\underset{|}{N}}{-}(CH_2)_6{-}N{=}C{=}O \end{cases}$$

It is supplied as a 75% solution in equal parts by weight of xylene and ethoxy ethanol acetate in which form it contains 16.5% -CNO.

TABLE 1-10. Desmophens in Common Use in Aircraft Finishes

Resin	% OH	Solids	Description	Composition	
Desmophen 650	8.0	100%	Branched polyester	Trimethylolpropane — 1 mole	phthalic anhydride 1 mole
Desmophen 800	8.8	100%	Highly branched polyester	Trimethylolpropane or glycerine 4.1 moles	phthalic anhydride 0.5 mole
Desmophen 1100	6.5	100%	Branched polyester	1-4 butylene glycol 2 moles Hexanetriol 2 moles	Adipic acid 3 moles
Desmophen 1200	5.0	100%	Branched polyester	1,3 Butylene glycol 3 moles Trimethylolpropane 1 mole	Adipic acid 3 moles
Desmophen RD18	5.0	100%	Polyester modified with saturated fatty acids of low molecular weight		

Selection of solvents for the two-pack coatings is of importance. They should have the minimum practicable content of compounds capable of reacting with isocyanates; their isocyanate content should be determined and is expressed as the weight in grams of a product which will combine with 1 g equivalent weight of isocyanate. Commonly used solvents in urethane coatings such as ethoxy ethanol, methyl ethyl ketone, toluene, and butyl acetate may have isocyanate equivalents of 5000, 3800, >10,000 and 3000, respectively; a low boiling solvent mixture in the ratio of 4/4/1/1 may be used for a Desmophen 650 solution for use with Desmodur N by spray application. Higher boiling solvent mixtures may consist of 90-95% of ethoxy ethanol acetate with 5-10% of xylol. Other factors to be borne in mind in formulating solvent mixtures for Desmophen 650 are (a) Viscosity control is usually done by choice of solvents, since Desmophen 650 is incompatible with high molecular weight polymers. Methyl ethyl ketone gives the lowest and ethoxy ethanol acetate, methoxybutyl acetate, or diacetone alcohol the highest viscosity solutions at equal solids. (b) Methyl isobutyl ketone and butyl acetate are diluents. (c) Pot lives are influenced by solvent mixture selection.

Flow control agents may also be necessary; solutions of $\frac{1}{10}$ or $\frac{1}{20}$ sec viscosity grades of cellulose acetate butyrate in equal parts of ethoxy ethanol acetate and toluene are recommended for this purpose at a level to give 1-2% solid concentration based on the resin content. Small quantities of selected silicone oils or compatible urea-formaldehyde resins have been successful in this application.

Desmophen 650/Desmodur N combinations generally give optimum results at a 100% crosslinking ratio of -NCO/-OH. Higher isocyanate ratios tend to give harder films of somewhat improved fluid resistance. Small additions of RD.18 to Desmophen 650 for use with Desmodur N are useful in improving pigment wetting and giving slightly softer films of slightly lower solvent resistance. Durability of these combinations is however excellent and indeed the already excellent color retention may be improved. Since the reactivity of the aliphatic polyisocyanate is relatively low, such systems tend to be slow drying and the use of a catalyst is normally necessary to obtain acceptable drying times.

The catalysts of the isocyanate-hydroxyl reaction have been the subject of extensive research and a considerable literature because of their importance in the preparation of polyurethane foams. A screening method applicable to coatings technology was described by Britain and Gemeinhart (28) in which gel times at 70°C were recorded from a polyether triol reacted with tolylene diisocyanate at an -NCO/-OH ratio of 1/1 in the presence of 1% of catalyst on the weight of the polyether. Data were presented from screening tests on several hundred possible catalysts and also included a comparison of catalytic activity when m-xylylene diisocyanate and hexamethylene diisocyanate replaced tolylene diisocyanate. Many metallic compounds were found to be catalysts for the isocyanate-hydroxyl reaction; the catalysts tested could be classified into three general groups. The first group of tertiary amine catalysts had little effect on the relative reaction rates of the different isocyanates. The second group activated the aliphatic isocyanates more than tolylene diisocyanate so that the relative ratio of the reactions were approximately equal, and included stannous, lead, bismuth, and organotin compounds. A third group included zinc, cobalt, iron, stannic, antimony, and titanium compounds which caused the aliphatic isocyanates to react faster with the triol than tolylene diisocyanate. A similar technique was also used by Britain (29) to evaluate the effects of different types of catalyst on prepolymer stability in the presence of free tolylene diisocyanate.

There are many potential catalysts for accelerating the curing rates of two-pack polyurethane coatings, and synergistic effects are common. The preferred types are tertiary amines, lead, cobalt, and zinc octoates or naphthenates and dibutyl tin dilaurate. Harmful effects such as reduced potlives, yellowing or reduced durability, and water resistance can result from overaccelerating the cure, and the benefits of rapid drying can be much reduced in terms of protection. Zinc octoate is often used at a level of 0.1% to 0.2% on the total solids as is a blend of 29 parts by weight of 24% lead octoate with 1 part by weight dibutyl tin dilaurate. In the case of Desmophen 650/Desmodur N combinations, the use of Bayer's catalyst Desmorapid PP at a level of 0.2-0.5% on the total solids is recommended as not impairing durability.

Pigmentation is important both in respect of initial performance, e.g., flocculation, potlife, etc., and in the maintenance of good gloss and color retention on exposure. Desmophen 650 is not a good wetting medium and work is being carried out to improve it in this respect. Current Bayer recommendations are for the use of small additions of RD. 18 or if zinc octoate is used in the coating, this too may be added prior to grinding. Items such as "Antiterra U" (a salt of a high molecular weight acid ester of a long chain polyamine amide) may also be used. Bayer published information on suggested inorganic or organic pigments and their levels of use for optimum results, based on comprehensive weathering tests carried out in Central Europe at an angle of 45° facing south. The least suitable pigments which affect gloss retention in the full shade do so also on dilution with rutile titanium dioxide even down to the level of 1:50. The more easily wetted pigments tend to give better results. Matteing pigments may also be used, but will reduce durability and cause chalking. The well-known "stir-in" grades of fine silicas may be used at levels of approximately 10% on the binder solids but tendencies to settlement are likely. Talc may also be used as a matteing agent; another technique is to increase the pigmentation of the lacquer substantially to break the gloss before the addition of a "stir-in" in silica, but again this will result in reduced durability.

Glossy and matte finishes of the above types are required to conform to Specification DTD. 5580. Two schemes are possible, scheme one consisting of a two-pack chromated epoxy primer (DTD. 5567) at a film weight of 0.85 ± 0.15 oz/yd^2 followed by a finish at 1.25 ± 0.25 yd^2 as an intermediate coat. Drying times for overcoating primer or filler are 4 hr and the finishing coat is required to be hard dry in 6 hr. Simple mixing ratios of components are required and these are generally two parts by volume to one part by volume of curing agents. Mixtures should remain suitable for use for not less than 8 hr at a temperature of 65-70°F or 4 hr at a temperature of 90-95°F. Rigorous performance tests are applied to the systems after drying the finish for seven days at 18-21° C. These include toughness, hardness, adhesion, resistance to sea-water, to cold, hot, and hot, pyrolized synthetic lubricating oils, cold and hot mineral based hydraulic oils, cold and hot phosphate ester based hydraulic oils, to cold castor oil/glycol ether hydraulic oils, to methylated spirits, and to natural weathering. Fluid resistance tests are mainly carried out by immersion tests, the paint film being scribed to examine undercutting. However, smear tests are used in hot pyrolized ester lubricant and phosphate ester hydraulic oil tests, the test fluid being renewed at regular intervals during a test period of 1000 hr at 70 ± 2°C. After removal of the test fluid and cooling to room temperature, the test panel is subjected to a scratch test of 1500 g.

Although the specification does not make the use of an aliphatic polyisocyanate mandatory, the general requirements for color retention and performance indicate the use of such types. Other aliphatic prepolymers for use with polyesters to give films of good aircraft fluid resistance are reported though the long background evidence of Desmophen 650/Desmodur N combinations has established a considerable usage of these materials for aircraft applications. Such uses are for the Anglo-French "Jaguar," multirole fighter, and Anglo-French helicopters. Alternative two-pack materials based on aromatic polyisocyanates as well as polyurethane oils may find usages where long term durability or maximum resistance to aircraft contaminants is less important. A great range of components exists, including isocyanurates, e.g., Suprasecs 3150 and 3340 ex. Imperial Chemical Industries, among the vapor hazard-free types of isocyanates, and polyethers, polyesters, and a variety of copolymers among the hydroxyl-containing coreactants. By judicious blending, high gloss, quick drying chemically resistant finishes with many desirable properties may be produced at a price advantage over the aliphatic polyisocyanate cured types. Polyurethane oils of good through drying, abrasion resistance, color retention, and improved chemical resistance compared with the alkyds may also find aircraft applications, though their chalk resistance is normally low. Specification MIL-C-38412 for a gloss, general purpose exterior coating covers the requirements of both a one-component system and a two-component system in which the use of an aromatic polyisocyanate with UV absorbers is permissible. Guide formulations for the composition of the two-components systems are given and an equal volume mixing ratio is suggested. A high -NCO/-OH ratio is indicated, the hydroxyl numbers of the blended polyesters and the isocyanate percentage of the polyisocyanate being specified. Both types of coating are required to be quick drying ($2\frac{1}{2}$ hr tack-free, $4\frac{1}{2}$ hr hard dry) and the two-pack material should have a pot life of 10 hr at $75 \pm 2^{\circ}$F when the components are mixed. Alclad test panels are coated with wash primer (Mil-C-8514) followed by zinc chromate primer (Mil-P-8585 Type 1) before overcoating with finish. Test clauses include hardness, adhesion, recoatability, flexibility, water resistance, accelerated weathering, and salt spray tests as well as resistance to hydrocarbons and synthetic lubricating oil. The latter test requires immersion in the oil for 24 hr at $121 \pm 2^{\circ}$C without blistering, softening, or any other film failure. Specification Mil-C-27227A is for a two-component highly reflective white polyurethane gloss coating for use on exterior metal surfaces of aircraft but more specifically for the lower areas which may encounter an environment of high thermal flux. The medium composition is similar to that in Mil-C-38412; the coating should have a hard dry of 4 hr and the mixed components a pot life of 10 hr. Gloss and color are to be of a high order. Alclad panels are cleaned for test purposes and coated with zinc chromate primer (Mil-P-8585 Type 1). This is given 4 hr drying before overcoating with the finish which is given seven days of drying before tests are carried out for hydrocarbon resistance, synthetic fluid resistance, and water resistance. Immersion times are 48 hr at 75°F in hydrocarbon (TT-S-735), distilled water, and synthetic oil (Mil-7-7808). Slight softening may be disregarded but the films should show no blistering, wrinkling, discoloration, or other defects. Epoxy primer (Mil-P-27316) also performs satisfactorily with this finish. Specification Mil-C-38412 requires that the free diisocyanate in the mixed coating shall not exceed 1% when tested by the test method defined in U.S. Air Force Bulletin 535. Isocyanate vapor hazards are of importance in the application of aircraft coatings and this subject is dealt with later (see under "Application of Aircraft Finishes," p. 64).

As has been indicated, the versatility and range of materials available for formulating polyurethanes lends them to many applications, two of which are discussed briefly. One is for the erosion protection of nonmetallic materials such as fiberglass reinforced laminates used in radomes exposed to impact with rain, hail, and sand. This is a constant source of research and Booker and Nash (30) describe the facilities at the R.A.E. for the simulation of flight through rain. Elastomeric materials that absorb the impacting energy are used for coating radomes, multicoats being applied to give thicknesses of not less than 10×10^{-3} in. Rubbers have long been used for this purpose, specifications Mil-P-9503 and DTD.856 describing coatings of this type. Fyall (31) reviews some of the properties of polyurethanes as a class in comparison with neoprene rubbers for erosion protection. He describes work carried out under the sponsorship of the U.S. Air Force Materials laboratory [see also report by Schmitt, Jr. (32)]. Polyether polyols and isocyanates of various molecular weights and structures were reacted at various rates of -NCO to -OH groupings, the resulting polyurethane coatings being examined for physical properties and erosion resistance at 500 mph and 2-in./hr rainfall. Polyols of molecular weight 650, 1000, and 2000 were mixed in various proportions. Generally tensile strength and modulus decreased and elongation at break and abrasion resistance increased with increasing polyol molecular weight. Coatings with a polyol of molecular weight of 650 exhibited the best rain resistance; a ratio of 1.6/1 isocyanate/hydroxyl is favored. An 11×10^{-3} in. thick coating of such types withstood 85 min at 500 mph in 2 in./hr of rain. Formulations to protect against the dual erosion of rain and sand were developed but coats applied by brush or spray needed 24 hr curing between coats and a great many coats to build up the necessary thickness. Multicoat application is also necessary with rubber-based radome finishes; quicker processing is possible by the spray use of thixotropic polyesters cured by the cobalt/peroxide process to give thick films. Such coatings are normally used over a coat of epoxy primer.

The other example of the versatility of polyurethanes relates also to flexible types. Jointing compounds and sealants are widely used in the aircraft industry to provide an elastic seal in joints and cracks. They are a vital part of the common protection scheme, and many have been chromated to reduce corrosion between dissimilar metals; specification DTD.369 calls for a jointing compound of this type, barium chromate being used as the inhibitive pigment. The requirements could be met by the use of semi-oxidizing media in slowly evaporating solvents to allow for wet assembly. Present-day aircraft requirements are, however, vigorous in their demands on sealants, and elastomeric materials such as the polysulfides, fluorocarbon rubbers, silicones, vinyl plastisols, and polyurethanes have been examined and are used for this purpose. Saunders and Frisch (33) describe the potentialities of polyurethanes, quoting evaluations by Gamero and Le Fave (34) as to their general properties in relation to those of silicones, polysulfides, and vinyl compounds as evidence of their good electrical properties, wide range of mechanical properties available, and good solvent resistance and oxidation resistance.

Polysulfide sealants are particularly used as sealants for integral fuel tanks and pressurized cabins, as gaskets for removal parts, and as watershed fillets. By special preparation of the faying surfaces, various degrees of adhesion may be obtained. Primers are available to increase adhesion to a cleaned metal surface or a thin film of lubricating oil or parting agent may be applied to prevent adhesion. Preferential adhesion to all of the surfaces can be governed by the use of such techniques. The cured sealant has good resistance to aromatic fuel, petroleum, water, oil, and weathering ageing and has good properties over a temperature range of -75 to 110°C.

Stevens (35) describes the formulation of polysulfide sealants. Viton,* a rubber-based sealant, is valuable for use in integral fuel tanks at temperatures up to 230°C for prolonged periods and up to 315°C for short periods. Metal surfaces require careful cleaning before the application of special primers to give maximum adhesion of the Viton sealant. It has excellent resistance to fuel, water, alcohols, petroleum and synthetic lubricating oils, and petroleum-based hydraulic fluids and has good low-temperature flexibility.

Miller (36) emphasizes the importance of the inclusion of leachable chromates in elastomers to produce corrosion inhibitive sealants; he describes laboratory and field tests in evidence and recommends the use of a special aluminum pigmented topcoat to avoid depletion of soluble inhibitive ions.

7. Light Aircraft

Although the current high performance schemes approved for Service aircraft are suitable for light aircraft use, the special needs are more often for good appearance during service, the finishes with resistance to aircraft fluids and chemicals being reserved for turboprops, etc., and aircraft used for agricultural purposes. Indeed localized painting by, e.g., epoxy/amine finishes in areas subjected to oil drippings such as wheel covers may suffice,

---

*Trade-mark of E. I. Dupont de Nemours and Company.

other areas of the aircraft being painted with gloss and color retentive finishes. Nonyellowing finishes of good durability and gloss retention in white and a great many shades are required, the need for polishing during service being kept to a minimum. Rapid drying of the finishes is required by the constructor so vinyls, acrylics, quick-drying nonyellowing alkyds, nitrocellulose lacquers, and sometimes polyurethanes are in use as exterior finishes, selection being governed by the likely areas of operation and cost factors. Interior protection is afforded by a coat of a zinc chromate/alkyd primer of the "universal" type applied to anodized or Alocromed surfaces, wing interiors which are subject to condensation being protected by epoxy/polyamide primers. Alclad exterior surfaces are degreased with toluene and primed with an etch primer. Acrylic, vinyl, or alkyd finishes follow, though an additional primer coat of the "universal" type primer is often used to give better intercoat adhesion. The acrylic and alkyds finishes are of the types used in the DTD specifications, nonyellowing properties being important. Vinyl systems also show good color retention, durability and drying and such finishes based on blends of Vinylite resins VMCH and VYHH with chlorinated diphenyls as plasticizers have been much used.

1-5. MISCELLANEOUS MATERIALS AND PROCESSES

A. FLUORESCENT FINISHES

High-visibility coatings are required for certain applications, e.g., on trainer aircraft, and for this purpose finishes pigmented with powdered resins containing fluorescent dyes are used. As these dyes are soluble in powerful solvents such as ketones and alcohols, the nonvolatiles of the coatings are essentially hydrocarbon in type. The good color and color retention of acrylic media do not impair the desired reflectance, and by incorporating stabilizers such as substituted dihydroxy benzophenones into such media, protection from deterioration by ultraviolet light is possible. British practice is to use a red-orange color of dominant wavelength 615 ± 5 nm with a peak reflectance not less than 200% of that of magnesium oxide.

Components of the scheme for direct application to metal are a primer (etch, "universal," or epoxy) followed by a white pigmented synthetic resin vehicle, the red-orange fluorescent finishing coat, and an overlay varnish. Primers are omitted if the fluorescent system is to be applied over completed DTD finishing schemes, a feature of the white undercoat being its compatibility with many types of finishes. The overlay varnish can be based on an acrylic resin with pendant hydroxyl groups for reacting with an aliphatic isocyanate in circumstances where resistance to contamination with ester lubricants is required; it also contains stabilizers to ultraviolet light and does not impair the selective strippability of the systems if suitable strippers are used.

Such a system is tested for drying properties, color and peak reflectance, toughness, hardness and adhesion, resistance to synthetic lubricating oils, fastness to light, protection against artificial seawater, cold crack resistance, resistance to natural weathering, freedom from lead (where required), and keeping qualities of 12 months in temperature climates, or six months in tropical climates. Oil resistance of the system is such that a scratch test of 1000 g is resisted after 500 hr immersion of test panels in diester oil at 18-21°C; hot pyrolized ester lubricants resistance is such that after smearing test pieces with 0.1 ± 0.02 g of test fluid per 5 in.^{2} and heating them at 70° ± 2°C for seven days, a scratch test of 1000 g on the cooled or wiped panels is resisted. Light fastness is such that after exposure to a carbon arc for 2000 hr (Def-1053, Method No. 33), a film of the paint scheme shall not fall in peak reflectance below 160°% when compared with magnesium oxide. The natural weathering clause requires 12 months exposure in the open facing south at an angle of 45° to the horizontal, the panel being sprayed with artificial seawater three times a day. No cracking, blistering, flaking, or loss of adhesion is permitted.

Similar American specifications are MIL-P-4563 for a "permanent" type and MIL-P-21600 for a removable type. The latter uses a mineral spirits soluble polybutylmethacrylate as the medium for the color coat, and an aromatics-soluble unmodified acrylic polymer for the clear.

B. CHEMICAL MILLING

Much machining of metals, primarily to reduce the weight of steel and light alloy components, is avoided in the aircraft industry by the use of the chemical milling process. Localized areas of unwanted metal can be chemically removed from a workpiece on which a coating, resistant to the etchant, is present in the areas requiring protection. The process is applicable to aluminum, steel, magnesium and titanium alloys, and two variants, the "positive" and the "negative/positive," are in use.

In the latter, a strippable coating based on vinyl resins, is applied over the workpiece and given a short period of airdrying (approximately 2 hr). The desired pattern is then reproduced on the surface. After cutting through the outlines of the design with a knife to the surface of the workpiece, the strippable coating is peeled from those areas which do not require etching. A coat of a suitable etch primer is then applied overall at a dry film thickness of approximately 0.5×10^{-3} in. and given a short period of airdrying (1 hr) or forced drying (10 min at 100°C) before a resistant convertible coating is applied over the primer. This coating, e.g., an epoxy ester enamel is usually supplied in two shades for wet-on-wet application so that easy identification between coats is ensured. Adequate curing is then allowed, stoving being preferred (e.g., 30-45 min at 125°C), in such a manner that the remaining strippable coating now overcoated with the primer and resist can readily and cleanly be peeled from the areas to be etched. After the workpiece has been immersed in the chemical bath and the localized areas have been etched to a sufficient depth and undercut, the unwanted resistant enamel and primer are removed in a hot bath of noncorrosive stripper. After washing clean, it is pretreated for painting as specified by the design authority. This process, although involving more operations than in the positive process, requires somewhat lower coating film weights and allows latitude in the selection of suitable resists, the criteria being good chemical resistance, good adhesion, and good flow with freedom from pinholes if seepage of the etchant into undesired locations is to be avoided. Reject rates are low.

In the positive process, the maskant is applied to the workpiece and after curing the coating is scribed by cutting through it with a knife down to the surface of the metal. The maskant is then stripped away from the surface where etching is to take place and the workpiece immersed in the chemical bath for the desired period. After removal of the item from the bath, the maskant is peeled off the protected areas. Rubbery polymers are preferred for this usage, several coats being necessary to build up a film thickness of 8×10^{-3} - 10×10^{-3} in. As in the "negative/positive" process the maskant can be applied by dipping, spraying (conventional or airless), flowing, or by electrostatic spray (hot or cold). This type of maskant generally requires a minimum of 24 hr airdrying at room temperature for cure but may be stoved at 100° C for 30 min before exposure to etchants, suitable "flash-off" times being allowed to avoid solvent boiling in the coating. Retention of its strippability properties after stoving should be such as to enable the chemical milling process to be carried out satisfactorily within the next 14 days. A bonus from its airdrying application is its use also as a temporary protective and in such circumstances, clean peel would be expected after storage of the component for some months. A suitable peel strength would be 1.6-2.0 lb/in. Other desirable qualities would be tensile strength of 1800 psi, an elongation of 600%, and an etch factor of 1.0. The etch factor is the ratio of etching of depth to undercut and a radius in which the depth is equal to the undercut is generally preferable. The coating would be expected to have a high level of resistance to acids or alkalies depending on the metal to be milled; thus in the case of aluminum alloys, no breakdown would be expected after exposure to a 15% caustic soda solution at 80° C for up to 6 hr.

Translucency in these coatings offers several advantages. Thus marked grid patterns are allowed to show and Vidi gauge readings can be seen easily, permitting selective milling of high spots. Further, discolored regions will be visible through the maskant layer, particularly when the reaction products of the process are dark as in the case of sodium hydroxide attacking aluminum/copper alloys. A faulty area can thus soon be detected and the workpiece saved from spoiling by cutting out maskants in this area and replacing it by a spot-patching operation. U.S. Patent 3,227,589 describes translucent chemical milling maskants, an important feature of which is the inclusion of a filler in a rubbery polymer matrix. Styrene-butadiene, chloroprene, and acrylo-nitrile butadiene copolymers are quoted as rubbery polymers of suitable physical properties. In cured film form they are inherently translucent and will, on being cut, abraded, or otherwise disturbed undergo a reorientation of their molecular arrangements as a result of the mechanical stress. However, a film of pure rubbery polymer generally will not show a contrast between an undisturbed and a mechanically stressed portion. Thus, in the scribing process carried out on pure polymer films it would generally be the case that an irregularity of cut or some material removal at the boundary of a region later to be etched would be substantially undetectable to the naked eye. It is therefore important to obtain visibility of the scribe line and the patent describes criteria to achieve this. The refractive index of stressed areas should differ from that of unstressed areas, the unstressed areas of the film in its initially cured condition being referred to as in the film's first condition of visibility and the stressed areas as being in the film's second condition of visibility. Inclusion of a suitable filler with the appropriate matrix will lead to a noticeable change of color between the first condition of visibility and the second condition of visibility adjacent to the incision. If η_T is the refraction index of the matrix and η_ρ is the refraction index of the additives, when contacted by the matrix material, the system is translucent when Δ_η, which is defined as $\eta_\rho - \eta_r$, is less than 0.5 and the system tends too far toward the opaque if Δ_η is greater than 0.5. Examples of maskants using these design criteria and quoted by the patent are given in Table 1-11.

As in all aircraft coating applications, the preparation of the substrate is of great importance. The adhesion of the maskant in the positive process must be adequate to allow clean milling but not sufficiently great to present difficulty in manual peeling after its use as a protective. Removal of oil, etc., from the workpiece by trichlorethylene vapor followed by immersion in a nonetching solvent cleaner generally gives good results; in some circumstances, however, it may be necessary to provide a slightly etched surface by a bath treatment using, e.g., a chromic/sulfuric acid pickle.

C. REFLECTANT FINISHES

Finishes formulated to reflect or absorb radiation of specified wavelengths are important in the aerospace industry for applications such as temperature

TABLE 1-11. Translucent Milling Maskcants

	Example 1 parts by weight	Example 2 parts by weight
SBR 1012 (butadiene styrene copolymer ex Shell Chemical Co.)	100	—
Solid chloroprene elastomer	—	100
Calcium silicate precipitated anhydrous	30	20
Fibrous magnesium silicate	50	50
Hard clay-Dixie (aluminium silicate)	50	30
Stearic acid	2	—
Zinc oxide	5	5
Magnesium oxide	—	2
Sulfur (rubber grade)	1.5	
Mercapto benzo thiazole disulfide	2	
Diphenyl guanidine	0.5	
Thermosetting phenolic resin	25	25
Antioxidant such as hindered phenol		2
Accelerator (NA22; 2-mercapto imidazoline)		1
Toluene to form a solution containing 25% by weight of the above		Toluene or a blend of toluene and M.E.K. to form a solution containing 30% by weight of the above

control, increased substrate life, and camouflage purposes. Examples of noncamouflage usages are the antiflash coatings of the British V-bombers which were formulated on suitable white pigments in heat stable media to give resistance to nuclear flashes, and coatings for aerospace objects requiring prolonged resistance to solar radiations. Walker (37) describes work carried out in the U.K. A.E.A. to develop thermal control coatings on satellites, etc. Such coatings are exposed to a prelaunch environment comprising those of manufacturing, testing, and then assembly on the launch site with the many possibilities of contamination, etc., next to the ascent environment which includes further contamination possibilities combined with vibration and thermal shock, then to the space environment. The space environment is the most exacting and includes thermal irradiation, solar irradiation, penetrating radiation, physical impact, and high vacuum. A property not required in space is long-term corrosion resistance though this is required for storage; thermal control failure, however, is likely to lead to the malfunction of the spacecraft or satellite with rapid complete failure. This is because certain essential components of a satellite have only a narrow effective temperature range over which they can function. The necessary precise balance between absorbed and radiated heat can be achieved by surface coatings when thermal control surfaces of the passive type are incorporated. These coatings are characterized by the ratio of their solar absorptance (α_s) to their infrared emittance (E), and they can be used on solar reflector and flat absorber surfaces. The latter — surfaces which absorb the incident energy throughout the spectral range from ultraviolet to far infrared — do not represent a problem as many black matte paints or black anodized aluminum surfaces give the desired α_s/E ratio of 0.90. A solar reflector — a surface which reflects the incident solar energy while emitting infrared energy — is characterized by a low α_s/E ratio in the region of 0.12-0.24. Problems are inherent in producing a stable solar reflector, and Walker (37) describes work carried out to develop a suitable space-stable thermal control paint for this purpose. The properties required were adequate mechanical properties, reflectance to be greater than 70% of the total received radiant energy, high emittance in the infrared (over 90°), stability of both these properties within ±2% of the original values, and resistance to thermal shock (cycled between 80°C and -80°C).

Performance during ultraviolet radiation in vacuum was tested by exposure of specimens to a 1 kW mercury discharge lamp, whose intensity approximated four suns at the distance used, the test chamber being evacuated to a pressure of 1×10^{-5} Torr. After exposure, reflectance measurements were obtained in the solar spectral regions using an integrating sphere attachment on a Beckman DK2A spectrophotometer, readings being relative to freshly coated magnesium oxide, and thus the solar absorbance calculated. Disks of compressed pigments were first studied; disturbingly only zinc oxide and a few titanate pigments appeared to be of promise. These were incorporated into potassium silicate and silicone resin vehicles, and again promising results were obtained. However, during the operation of measuring reflectance after specimens had been removed from the vacuum, it was observed that some which were visibly yellowed became paler in air. Use was therefore made of an in situ R.A.E. test facility in which it was possible to irradiate test specimens in self-contained chambers kept under vacuum while reflectance was measured. Clear evidence of reversible degradation was obtained thus, e.g., barium titanate, while showing little or no degradation in the post irradiation tests, showed pronounced loss of reflectance over the entire range of wavelengths in the in situ tests. Zinc oxide showed this type of degradation in silicone but not in potassium silicate. By treating the zinc oxide in a mill with a dilute solution of potassium silicate and washing away the residual untreated silicate, a protected zinc oxide was prepared which did not suffer degradation in the silicone vehicle. The effect of pigment volume concentration (PVC) was also studied. The initial reflectance and ultraviolet stability both increased steadily with increase in reaching a practical maximum at 70%. The requirements therefore of suitable reflectance and stability to ultraviolet radiation in vacuum are met by zinc oxide/ silicone paints. Resistance to thermal shock was tested in a thermal cycling equipment in which samples were attached to thermocouples and cycled into and out of hot and cold shrouds at 80°C and -80°C, respectively, the test chamber being evacuated to a pressure of 10^{-2} Torr. A complete cycle normally took 20-30 min; however a highly reflective coating would take considerably longer. Another test was to heat specimens to 80°C in an oven and then to drop them into liquid nitrogen, repeating the cycle 10 times. The best of the coatings would withstand this test. It was in physical properties, however, that the coatings left much to be desired. The silicone resin based paints required improvements in toughness and adhesion, and the silicate based paints often showed surface cracking and crazing on cure, did not adhere well to substrates other than aluminum and did not have particularly good thermal shock properties. Similar American work on the influence of ultraviolet radiation on a white coating for aerospace vehicles is described by Brandenberg (38).

The use of specially formulated paints for camouflage purposes is well known and an extensive literature exists. Their purpose is to modify the reflectance of the coated object to simulate that of the immediate environment, thereby rendering the object indistinguishable from its surroundings. To be effective, camouflage must operate not only against direct visual observation, but against other means of detection, particularly aerial photography. This has led to much study of the use of surface coatings where radiations of longer wavelength than visible light are involved, particularly those in the near infrared region. Infrared radiation is defined as that region of the electromagnetic spectrum bounded at its lower wavelength limit by the extreme visible red and at its upper limit by the shortwave radio region. These radiations lead to an increase in temperature by their absorption, which can be detected and used as a quantitative measurement. Also the extension of the sensitivity of photographic emulsions beyond the extreme visible red has provided a further means of detecting and measuring infrared radiation. Although present-day emulsions are capable of detecting radiations lying well into the infrared region, many earlier materials were insensitive above 900 nm and accordingly it has become the practice to refer to the 700-900 nm waveband as the photographic infrared. Solar radiation is a strong source of infrared; Baker (39) states that the solar energy spectrum at sea level with clear sky conditions consists of approximately 5% ultraviolet, 45% visible, and 50% infrared radiation. As the relative energy is particularly high in the region of the photographic infrared, use is made of this component of sunlight in outdoor infrared photography. Furthermore, by virtue of their longer wavelength, these radiations are more capable of negotiating haze and light cloud than is visible light and thereby render the photographic operation less dependent upon good weather conditions.

Thus for camouflage purposes it is no longer merely necessary to ensure that a paint film exhibits similar visual reflectance characteristics (color) to those of the surroundings; a similar response to infrared radiation is equally important. Moreover, the applied film must possess adequate hiding power at these wavelengths in order that the characteristic reflectance properties of the uncamouflaged object are effectively masked. Furthermore, as the spectral region covered by photography can be varied to some extent by different film/filter combinations, a knowledge of the variation of reflectance properties with wavelength is necessary to ensure reasonable similarities in the response of the camouflaged object and its surroundings at all wavelengths within the photographic infrared waveband. Taylor (40) quotes approximate percentage reflectance limits in the 700-900 nm waveband for common natural and constructional materials (Table 1-12). As green vegetation, soil, and sand are the more important, it follows that the common camouflage shades are characteristically drab greens, browns, and grays whose reflectance properties are a reasonable simulation of possibly a mixed environment. Thus U.S. specifications for camouflage colors, e.g., TT-L-20 specify an infrared reflectance of not under 28% for olive drab, medium green, and sand and a maximum of 55% for olive drab and medium green. A high hiding power contrast ratio is also specified. As a camouflaged object may move to different environments, temporary camouflage paints, removable by aliphatic solvents, are available for quick changes of color schemes to adopt to the new surroundings and Specification MIL-P-6884 describes such types.

It is apparent from the above considerations that in formulating camouflage finishes for aircraft, a knowledge of the optical properties of pigments in the near infrared is essential with emphasis on the variation of these properties

TABLE 1-12. Approximate % Reflectance Limits — 700-900 nm Waveband

Material	% Reflectance
Green vegetation	50-70
Damp soil	10-15
Dry soil	15-20
Sand	30-40
Concrete	40-50
Building brick	30-40
Galvanized iron	15-20
Bituminous felting	10-15
Asbestos cement	30-40

with wavelength, their hiding power, and the relative contributions of absorption and reflectance processes to the overall hiding power of the system. It is also important that such factors should be examined in paint films and not on dry pigments, less because of the nature of the medium employed, but because such properties are dependent on the pigment concentration and the film thickness.

Infrared reflectance values are measured chiefly by recording spectrophotometers or photographic processes, the former being preferable where it is necessary to determine the variation of optical properties with wavelength. In such circumstances, the use of a Beckman DK2A spectrophotometer with an integrating sphere attachment is recommended, measurements being taken relative to freshly coated magnesium oxide. A portable reflectometer of relatively low cost and simple and quick to use has recently been developed by the Royal Radar Establishment for giving single reflectance values between the 700-900 nm waveband, the equipment being standardized by reference to tiles of known reflectance values. Such an instrument is valuable for quality control purposes, as is equipment described by Scofield (41).

Photographic methods are usual for obtaining a simple reflectance value for the photographic infrared band as a whole. The general principle consists of the recording on a sensitized emulsion, then measurement of the diffuse reflectance from the test pattern relative to a set of standards whose reflectance values are accurately known. The infrared source consists of one or more incandescent filament lamps mounted relative to the camera to avoid access of specularly reflected radiation. Visible light is excluded from the camera by using a filter with a transmission threshold of around 700 nm in front of the lens. A Wratten No. 89A filter used with Kodak Infrared Sheet Film is typical. The reflectance standards which are photographed alongside the test patterns usually consist of a series of neutral gray patterns covering as wide and as evenly spaced a reflectance range as possible, and their reflectances are accurately known from spectrophotometric measurements. After processing to the negative stage, comparisons are made between the image densities of the test patterns and reflectance standards by means of a densitometer. Positive prints can be produced for subsequent reference. The waveband covered in photographic methods is defined by the transmission characteristics of the filter and the spectral sensitivity of the emulsion. Thus Vesce (42) in studying pigment properties used a Wratten K2 filter with Eastman Super XX Panchromate film for the visual band (400-700 nm), a Wratten 89A filter with Eastman Spectroscopic Plate Type S for the borderline band (680-700 nm), a Wratten 89A filter with Kodak Infrared Sheet Film for the 680-860 nm band, and a Wratten No. 87 filter with the same film for the 740-860 nm band.

Taylor (40) has compiled infrared reflectance data from spectrophotometric measurements on dried paint films applied over white and black backgrounds.

He incorporated a large range of pigments into a long oil alkyd medium using a 10% pigment volume concentration. Films of 3×10^{-3} in. wet film thickness were applied over black and white checkerboards for which the infrared reflectances values were 3 and 90% respectively. After drying, spectrometric reflectance curves were plotted for each film over the black and white substrates covering the 700-900 nm waveband. Average reflectance values over black and white were calculated by measurement of the areas under the curves; from these, the contrast ratio of reflectance over black to that over white was calculated for use in comparing hiding powers. Reflectance values at a number of selected wavelengths are also quoted to indicate the manner in which the optical properties varied with wavelength. From this work he classifies the various pigments into three major optical groups and where they clearly fall between these into the relevant transitional groups. Those pigments with high reflectance values over both white and black have good hiding power, high diffuse reflectance, low absorption, and low transmission. Those having low reflectance values over both white and black have good hiding power, low diffuse reflectance, high absorption, and low transmission, whereas those having a high reflectance over white and a low reflectance over black have poor hiding power, low diffuse reflectance, low absorption, and high transmission. Such data are applied to a discussion on the requirements of camouflage paints, and in particular those intended to simulate green vegetation. The green pigment, chlorophyll, responsible for the characteristic shade of vegetation is also responsible for its surprisingly high infrared reflectance. Although this may vary somewhat from one type of vegetation to another, all values are of a relatively high order and the manner in which reflectance varies with wavelength follows a characteristic shape. From a peak in the green band, a dip is shown in the far visible red at 650-680 nm followed by a sharp rise in the region of the borderline band to a high and essentially constant value above 730-740 nm. Infrared reflecting paints can be classified into three groups, the E, O, and L groups according to the aspect of their reflectivity curves. The E and L pigments are the so-called "early and late risers" the term referring to the position of the rise of reflectance as compared with that of chlorophyll. In the early type the curve rises at shorter wavelengths than that for chlorophyll, whereas in the late type the rise occurs at longer wavelengths than chlorophyll. Most of the known green pigments are of the E and L types, so mixtures of yellow, blue, and orange or red pigments, suitably selected and blended from reflectance data, are used to obtain the green shades common to camouflage applications. The choice of the blue pigment presents the major difficulty. Most blue pigments absorb infrared strongly and their use in even small amounts can seriously reduce the reflectance level of an otherwise highly reflectant paint. Light fastness and tinctorial strength requirements are factors further reducing the scope of selection. The author in his work preferred to use a high-strength light-fast strongly absorbing blue, the reflectant properties of which were virtually independent of wavelength, and thus its use did not produce the late rising common to other light fast blues which would be detectable with a filter transmitting only above, say, 750 nm.

In matching reflectance curves of camouflage paints to those of vegetation, it often happens that the shape of the curve within the 700-900 nm waveband is similar to that required, but the overall level of reflectance is somewhat higher. The author draws attention to the use of small quantities of carbon black to reduce this level without at the same time modifying unduly the general shape of the curve. It is usual with these small additions that the visual shade is practically unchanged.

The use of infrared camouflage colors is also important to reduce the heating effects of solar radiation. Experience in India in World War II of camouflaged doping schemes showed that a reduction of approximately 20°C in surface temperature could be achieved by correct camouflage pigmentations.

Solar and heat reflecting paints are also important for use on rockets; U.S. Specification MIL-E-46061 makes use of cobalt titanate of good reflectivity in the 400-1100 nm waveband. A styrenated alkyd medium is used as vehicle.

D. HEAT RESISTANT FINISHES

Work on heat-resistant finishes commands much attention by the aerospace industry; Wright and Lee (43) have reviewed the search for thermally stable polymers. Restraints are imposed by the requirements of resistance to chemicals allied to good mechanical properties and corrosion protection to metals. Of the newer coatings polyamide-imides or polyimides possess many desirable properties in thin film applications; they possess good adhesion and flexibility on suitably pretreated metals and excellent heat and solvent resistance. As with other established high heat-resistant coatings, a high stoving schedule is necessary, 10 min at 275°C being typical. Air curing systems to resist temperatures in excess of 200°C such as the catalyzed silicones or butyl titanate-based zinc or aluminum-pigmented finishes are limited in usage by their susceptibility to fluid or mechanical damage before exposure to their operational environment and preferably require mechanical preparation of the surface. It is with the heat curing silicones that most success has been achieved.

The pure silicones contain methyl and phenyl groups attached to the silicon atoms and a few silicon-bonded hydroxyl groups which condense under the influence of heat to give a cured film. The inclusion of phenyl groups improves the heat resistance and compatibility with organic resins. The resins are supplied as 50-80% solutions in aromatic hydrocarbons, and have molecular weights in solution ranging from 1000 to 5000. Cure schedules are normally 1-4 hr at 230-250°C to give films which can be hard and brittle or soft and flexible. The resins can be usually blended to give any desired intermediate properties. Zinc and cobalt naphthenates at 0.1-0.5% metal on the resin solids accelerates cure and this can be further assisted by the inclusion of lead or iron salts at one-fifth or one-tenth of the catalyst concentration.

Though enamels properly formulated from pure silicones may be expected in most applications to have a useful life of several thousand hours at 250°C, their deficiencies in solvent resistance, etc., and the high curing temperatures needed have led to their modifications with organic resins for aeroengine applications. Thus a metallic pigmented silicone/ amino/epoxy resin system has given satisfactory resistance to all the corrosive materials met on jet engines and has continued to provide this protection at temperatures up to 600°C. Butler (44) describes developments from this system to service fused ceramic coatings, one of which has withstood 1200°C on a nickel alloy. Ceramic frits fusing between 500 and 900°C are incorporated into a silicone/amino resin/epoxy resin binder. Substrates should be chemically cleaned and ideally alumina blasting is desirable on metallic surfaces. The coatings can be applied by brush, dip, spray, or electrostatic spray. A stoving schedule of 2 hr at 190°C is recommended. Coating thicknesses of 1×10^{-3} - 2×10^{-3} in. after this treatment are flexible enough to withstand bending round a $\frac{1}{4}$ in. mandrel and have resistance to atmospheric and salt-water corrosion fuels, lubricants, coolants, etc. The coatings will air dry but in this state have no resistance to corrosion or contaminants.

In service, the coatings function in three overlapping phases:

1. As an organically bound stoving enamel up to approximately 400°C.
2. As a complex silicone oxide matrix above 400°C when the major portion of the organic binder is destroyed. This matrix is firmly adherent and with the aid of the inorganic pigment filling retains its full anticorrosive and chemical resistant properties.
3. As a glass firmly attached to the substrate. As the fusing temperature of the incorporated frits is reached, the frits flux out and combine with the silicone dioxide matrix to provide a coating having outstanding properties of resistance to heat, corrosion, and contaminants.

Full protection is given by the coating over the full range of its operating temperatures from below freezing to very high temperatures. The ultimate heat resistance afforded by an individual coating depends on the substrate employed; the coating should be selected carefully to match the heat expansion characteristics of the substrate on which it is applied. The test methods employed to assess the properties of the coatings are also described.

E. STRIPPABLE COATINGS

These are much used in the aircraft industry in a variety of roles. They may, for example, be used for protecting aircraft during storage or shipment, as temporary protectives on painted or unpainted metal surfaces and as maskants to protect solvent sensitive surfaces, e.g., acrylic moldings during spray operations. The main bulk of spray-applied coatings supplied to the aircraft industry for such purposes is based on vinyl resins. Their properties of toughness, flexibility, good barrier properties, rapid drying, noninflammability, and limited adhesion at normal operating temperatures render them eminently suitable for these applications. Solvent-based coatings on polyvinyl chloride-acetate copolymers or emulsions based on polyvinyl butyral are most used.

McKnight and Peacock (45) describe in detail the range and properties of vinyl chloride-acetate resins for solution coatings marked by Union Carbide Corporation, Chemicals and Plastics. The principal resins of interest in the U.K. for strippable coatings are shown in Table 1-13. Vinylite VYHH and VYNS are the more important of these for the high-service coatings for use in the U.K. aviation industry. Maximum solids are required in the solutions to reduce to a minimum the number of coats to give adequate protection; factors of importance are that the films should have satisfactory levels of tensile strength and flexibility, suitable "peelability," and satisfactory protective qualities. It is necessary also that these properties should be retained during long periods of exposure in various climatic conditions.

Solvent mixtures are important in achieving maximum solids. Ketones are the most suitable primary solvents for vinyl solution resins and give the higher resin concentration solutions without gelation. Methyl ethyl ketone is particularly effective, but other factors such as evaporation rates, cost, etc., may call for its use with slower evaporating ketones and with diluents, e.g., aromatic hydrocarbons. Vinylite VYNS gives strong, tough films of excellent water and chemical resistance. Thus although Vinylite VYHH is often used alone to give satisfactory strippable coatings, many combinations of properties may be achieved by judicious blending of VYHH with VYNS, with the latter as the minor component.

The use of plasticizer is very important and a considerable literature exists on the stress/strain relationships of vinyl films in which the type and content of plasticizer have been varied. Typical of such comprehensive data is that given by Reed and Harding (46). Solvent type plasticizers are preferred to promote their better retention in the film; they also minimize solvent retention. They should be used at a level to give nontacky films of good tensile strength and elongation at break and low brittle temperatures. Diisooctyl phthalate or di-*n*-octyl phthalate generally perform satisfactorily at levels of the order of about one-third of the vinyl resin content. Polymeric

TABLE 1-13. Vinyl Chloride-Acetate Resins for Solution Coatings[a]

Product	Approximate chemical composition, wt %			Specific gravity	Wt %	Solvent ratio	Viscosity, 25°C, cp	Properties and uses
	VCl	VA	Other					
VYHH	86	14	—	1.36	22	1:1 MIBK-toluene	375	Basic coatings resin for all coatings uses; good adhesion upon baking; good air-dry adhesion with addition of VMCH
VYHD	86	14	—	1.36	25	2:1 MIBK-toluene	175	For general coatings use; similar to VYHH, but has greater solubility; may be used when extreme toughness and durability of VYHH are not required
VYLF	88	12	—	1.37	30	1:1 MIBK-toluene	188	Uses similar to VYHD when even higher solids and greater gloss and build required, but with much less flexibility and toughness
VYNS	90	10	—	1.36	12	MEK	135	Like VYHH but less soluble and tolerates more plasticizer; has better heat stability

[a]Union Carbide Corporation, Chemicals and Plastics

plasticizers resistant to leaching by some aviation fuels are also often used; polypropylene sebacate is of this type.

Stabilizers, e.g. propylene oxide, for use in the presence of iron and ultraviolet absorbers may be included; pigmentation also has considerable influence in combatting the deteriorating influences of ultraviolet light. Aluminum powder is particularly valuable in this respect. Release agents to ensure easy strippability are sometimes used; these may be of a waxy or oily nature. Care should be exercised in their use, particularly in the case of silicones, to avoid subsequent surface contamination.

Application is generally by spray using wide aperture nozzles for high solids types. Flat sheets, however, can readily be coated by curtain coating methods.

Alternative types of water-based strippable coatings are available for use, e.g., on acrylic sheets and solvent-sensitive surfaces. Examples of polyvinyl butyral emulsions for this purpose are quoted by Chandhak and Agarwal (47); rubber latexes containing small quantities of water-soluble gums have also been used.

The current U.K. specification for strippable coatings is DTD. 5588 which covers colored coatings supplied at a spray viscosity. The principal requirements are a solid content greater than 20%, good storage properties, a surface dry time of 30 min at 18-21°C, a pressure test for blocking, an abrasion resistance test, a $\frac{1}{4}$-in. mandrel bend test at 0°C, a peel test on a 2-in.-wide strip under a load such that the film will not peel under a load of 100 g but shall peel under a load of 500 g, and tests for resistance to salt spray and accelerated weathering on light alloy substrates.

American specifications include MIL-B-6030, MIL-C-16555C, and MIL-L-20209. MIL-B-6030 in particular is for spray use on aircraft surfaces. The coating should contain aluminum and a minimum of 30% solids so that an 8×10^{-3} in. dry film can be deposited through a 0.07-in. nozzle in one coat without sagging. Five coats should give a through-dry film in 72 hr and should show no bridging on a concave surface after 144 hr accelerated weathering.

Its adhesion should be 0.5-2.0 lb/in. width before and after accelerated weathering. The ultimate elongation under similar conditions should be not less than 150%. Numerous other clauses define the protective properties and package stability required, the low temperature performance, and combustion requirements of the film and also govern the use of the coating on painted or doped surfaces as well as coated acrylic sheet.

F. PROTECTION OF MAGNESIUM ALLOYS

Special care is necessary in the preparation and painting of magnesium-rich alloys to secure adequate protection from corrosion. Process specification DTD. 911C outlines complete protective treatments of cleaning, chromate treatment, sealing and painting, taking into account both the operations necessary on the material in its semifinished form as a casting, forging, extrusion or sheet and those further operations required after all forming and machining operations have been completed. The intention is to ensure that the magnesium rich alloy surfaces, whether machining may have been carried out at some stage on them, have been suitably cleaned, then chromate treated by one of the methods to DTD. 911C followed by sealing with a sealant to DTD. 5562 in an application process to DTD. 935 prior to painting by an approved paint scheme incorporating chromate-rich primers.

The cleaning of castings, other than die castings not contaminated with sand, may be by steel or nonmetallic grit blasting, scurfing, chemical milling, or by acid pickling. If they are then machined all over, they are degreased by organic solvent treatment and immersion in a suitable alkaline solution prior to chromate treatment, sealing, and painting. If not to be machined all over, fluoride anodizing is employed for cleaning. The fluoride film left by this process is then removed by immersion of the parts for up to 15 min in a boiling 10-15% (w/v) solution of chromic acid followed by immersion either for 10 min in a boiling 5% (w/v) solution of caustic soda or for 5 min in a cold 15-20% (w/v) solution of hydrofluoric acid. Chromating and sealing then follow with painting if no machining is to be carried out. If this, however, is required then the machined areas are again degreased, chromated, and sealed to be followed by painting overall.

Die castings not contaminated with sand are given fluoride anodizing cleaning only, the next steps being removal of the fluoride followed by chromating, sealing, and painting. Any machining however will require further degreasing, followed by chromating, sealing, and painting.

Forgings not to be machined follow the sequence of cleaning by fluoride anodizing, removal of the magnesium fluoride film, chromating, sealing, etc.; but if they are to be machined all over, then the operations of solvent and alkaline degreasing followed by chromating, sealing, and painting are used. Extrusions and sheets too are degreased, cleaned by fluoride anodizing, followed by removal of the fluoride film, chromated, surface sealed, and painted.

Specification DTD. 911C gives appendices outlining the use, composition, and control of fluoride anodizing and chromate treatment baths. The method of cleaning magnesium alloys by the fluoride anodizing process given in British Patent 721,445 utilizes a high voltage (90-120 V) over a 15-25% w/v solution of ammonium bifluoride. Treatment is for 10-15 min at a temperature below 30° C, followed by washing.

Three types of chromate treatment are specified. The R.A.E. hot half-hour process requires a bath containing 3.0% w/v ammonium sulfate, 1.5%

w/v ammonium dichromate, 1.5% w/v potassium dichromate, and 0.25-0.43% v/v ammonia of S.G. 0.880. Chromating is carried out by immersion of the metal at the boil for 30 min followed by immediate washing in warm water (not above 50°C). A permitted variant — the de Havilland 167 bath — uses a solution of 3% ammonium dichromate, 1% ammonium sulfate, and 0.10% ammonia.

The acid chromate bath in DTD. 911C has a composition clause requiring that a solution containing 15% w/v of sodium or potassium dichromate and 20-25% v/v nitric acid of S.G. 1.42 be used. Immersion periods can vary from 10 sec to 2 min and the components are first drained before washing with water at a temperature of not more than 50°C.

The other permitted chromating bath is the chrome-manganese bath consisting of a solution of 10% w/v sodium dichromate, 5% w/v manganese sulfate, and 5% w/v magnesium sulfate. This bath may be operated for $1\frac{1}{2}$ hr at 20-30°C, 30 min at 50-60°C, 15 min at 70-80°C, or 10 min at boiling point to give adherent chromate films before washing free of unwanted solution.

After such pretreatments, the sealing operation to DTD. 935 is required. This employs three successive applications of the resin solution to DTD. 5562. In order to ensure even coating by dipping, the component should be held in a different attitude for each coating procedure. Preheating of the chromated component to a uniform temperature of 180-200°C for at least 10 min is required. It is then cooled to 60° ± 10°C and immediately dipped into the resin solution. Simple techniques using movement of the component are described to ensure thorough impregnation and drainage from recesses during or after a flash-off period of 15-30 min, tears and drips then being removed from low points with a palette knife or brush. Stoving for 15 min at not less than 180°C follows. After baking of the first coat, the component is again cooled to 60° ± 10°C, dipped a second time, given a drain time of 15-30 min, then baked at the approved temperature, the component being inverted from its former position of baking. A third dip-coat is then applied (after cooling of the component to 60°C) in a manner similar to that of the second coat, but a baking schedule of 45 min at not less than 180°C is employed, the component preferably being positioned at a different angle from the first and second bakings. Objectionable tears and drips are removed carefully with a sharp knife, a fine file, or glass paper normally after the first coat of resin. For convenience in this operation, the component is allowed to cool to room temperature, but it should then be gently warmed to 60°C before application of the second coat.

The thickness of coating achieved should be not less than 0.001 in.; a partly masked panel of magnesium alloy to DTD. 626 coated at the same time in the same bath and under the same application treatments may be used for determining film thickness by comparing micrometer readings of the masked and unmasked areas. Visual inspection for freedom from bubbles and pinholes is made, and the development of a golden yellow to dark brown color in the coating indicates adequate cure. A bright metal control panel coated and baked at the same time is used as an indicator if the magnesium component to be coated is dark, chromated metal.

A modified epoxy resin and an amino-resin curing agent fulfill the requirements of specification DTD. 5562 in which performance tests are described. Though the resin and hardener solutions separately are expected to have a storage life of twelve months in a temperate climate and six months in a tropical climate, the ready-for-use mixture should retain its properties for not less than three months in temperate climates if stored in a sealed container. A freshly prepared mixture of resin and hardener should stand for at least 24 hr before use. Bend, impact, and spalling tests are specified together with resistance to water, humidity, salt spray, trichlorethylene, and rechromating.

Solid contents and solvent mixtures are selected to give correct dipped film thicknesses and dip-tank stability. A typical solvent mixture would contain equal parts by volume of toluene, methyl ethyl ketone, ethanol, and diacetone alcohol and a solid content of 25-30% by weight is normal.

Very large components or those which for some reason it is not possible to coat by dipping may be coated by spraying. A full wet coat should be applied to the components, followed by heating and cooling as described in the dipping process. At least three coats are applied to give the requisite 0.001-in. film thickness. A suitable solvent mixture for spray application contains 70 parts by volume of diacetone alcohol, 20 parts of xylene, and 10 parts of ethyl acetate.

After final surface sealing parts are ready for painting to the requirements of DTD. 911C. Degreasing and light abrading may be necessary, and the application of one or more coats of primer prior to assembly is preferred, due regard being paid to tolerances on mating surfaces. Whatever the sequence, at least one coat of primer and one of finish should be applied overall after assembly. The paint scheme should consist of not less than one coat of chromated primer and not less than one coat of finish to an approved specification, the all-epoxy cold curing scheme to DTD. 5555 being preferred and stoving recommended. The primer should contain in the dried film not less than 15% by weight of a chromate pigment which should be either zinc chromate to BS389 Type 2, pure strontium chromate, or pure zinc monoxy chromate. The strontium chromate may contain up to 10% of pure barium chromate if desired. Freedom from mercury and lead is also a requirement and pigments to be used in conjunction with the chromate should be chosen from among the following: lithopone, barium sulfate, barium chromate, titanium dioxide, zinc oxide, kaolin, green oxide of chromium, magnesium carbonate, asbestine, talc, mica, silica, and aluminum stearate. They should be free from soluble salts.

Zinc tetroxy chromate as a pigment is acceptable only when in an etch primer. Use of this type of primer is recommended by DTD. 911C as part of paint systems to be used on previously sealed magnesium parts which may then contain inserts of aluminum alloys and/or cadmium plated metals. The minimum thickness of the total organic coating should be 0.004 in.

DTD. 911C describes also the local treatment of exposed metal areas on which the protective coating has been damaged during assembly. Ideally chromating, sealing, and repainting should be employed; but this may not be practicable in which case where there is no danger of the entrapment of treatment chemicals the exposed area should be first treated by either lightly swabbing it with the acid chromate bath solution with glass wool until a golden color is developed, followed by copious washing and then drying or by swabbing a solution containing 10% w/v selenious acid onto the metal with cotton wool until a permanent dark brown color develops; copious washing is followed by drying. Either method is also applicable for the repair of chromate films on magnesium but the selenious acid method should not be followed by an etch primer. The full painting scheme may then be used with not less than two coats of primer. If however, there is danger of entrapment of treatment chemicals, the bare metal should be given no treatment if it is in a mating surface but should be wet-assembled with jointing compounds, and over-painted after assembly with not less than one coat of primer and two coats of finish overall. Nonmating surfaces may be painted with an etch or epoxy primer followed by the coats comprising the main paint scheme for the parts. As the important stoved sealant has been omitted every effort should be made to protect the damaged areas with multiple coats of paint.

DTD. 911C also stresses the importance of jointing compounds, caulking compounds, and the appropriate application of sacrificial metal coatings to other materials which may contact the magnesium alloy as factors rigorously to be controlled during the assembly of components and parts.

A further section deals with the reclamation of corroded parts, the desired end result being that the magnesium-rich alloys should be satisfactorily chromate treated, surface sealed, and painted.

1-6. APPLICATION

All those concerned with aircraft finishes, be they aircraft manufacturers, government agencies, or airline operators, are fully aware of the need to ensure proper application of the complex coatings in use today if these are to achieve a long and reliable life. Comprehensive manuals and process specifications covering all details of the preparation of surfaces for painting, the adjustment of materials for the means of application, the use of equipment, the correct requirements of humidity and temperature to be observed during drying, safety requirements, quality and process control tests, etc., have been prepared. U.S. and British government specifications are typified by MIL-F-18264 (Application and Control of Organic Aircraft Finishes) and DTD. 902 (Application of Paint Materials to Metallic Surfaces); the principal constructors also apply rigid instructions and flow charts for painting processes. Gilder (48) describes the precautions taken during the repainting necessary in maintaining airline aircraft; the importance of paint and its proper application are appreciated from the experience gained from corrosion rectification over the years with polished, part painted, fully painted, and fully painted plus detail stage protected aircraft. The last method of finishing is considered the best.

Aircraft finishes are applied by all the well-known techniques of painting, e.g., hot-spray, airless spray, conventional or electrostatic spray, dipping, etc. Special care may be necessary to avoid droplet inhalation in the case of the much-used polyurethanes. The use of efficient respirators and protective clothing allied to good extraction from the areas containing the aircraft are recommended. Airless spraying techniques reduce the droplet concentration. Recommendations for the airless spraying of polyurethanes can be based on the use of the Grayco President or Nordson/Versa machines, for example. The former has a pump ratio of 28/1, an input pressure of 80 psi being recommended for use with a 163-413 tip and pattern width of 8 in.; a similar input pressure may be used with the Nordson, a No. 15 restrictor, a No. 15 screen, and a 0009/09 tip giving satisfactory results with material at a viscosity of 23 sec B.S. Cup B.4. Two single track coats may be applied allowing 10 min between coats.

Pressure rollering may also be used for the application of polyurethane finishes to avoid droplet hazards. The use of a polyurethane foam sleeve with a madapolam cover and an outermost nylon cover is recommended for application with an air pressure on the head of 4-5 lb/in. A rollering viscosity of 25 sec B.S. Cup B4 is normal, the roller head being fully filled while being held in a horizontal position. This eliminates airpockets which would give excessive bubbling. The roller is then evenly and completely wetted by rollering on a clean surface. Joints and rivet heads are brushed in immediately before roller coating; this is carried out at a fairly fast rate working to panel joints. A fairly wet coating is applied, the minimum amount of work on the surface to obtain good flow being desirable to avoid bubbles. Should these not all collapse it is possible to remove them by going over the surface lightly within 20 min of the application using a semidry roller. Two medium coats of finish are preferred to one heavy coat, overnight drying being allowed between coats. Large aircraft such as bombers are painted by this technique in the U.K.

The possibility of electropainting light alloys for aircraft use is an attractive one for laying down protective films on bolted assemblies and complicated fabrications where normal methods of application are unlikely to secure adequate paint penetration. Recent U.K. work sponsored by the Ministry of Technology shows some promise although the preferred use of chromate pigments with their tendency to settlement and the possibility of their reaction with the electropaint resin would indicate that stability problems might limit the application to small baths. Nevertheless, chromated electropaints laid down on Alocromed BSL72 at a film thickness of 1.3-1.8 mils for stoving 1 hr at 95-100°C have been developed to give satisfactory resistance to salt spraying by the ASTM B117-64 method for 1000 hours.

Chromate pretreatments on magnesium alloys generally interfere with the excellent lay-down characteristics of electropaints on magnesium alloys without pretreatment. It has been found that the chrome-manganese bath of DTD. 911C Appendix ii, however, gives good results and a pigmented electropaint requiring stoving for 30 min at 200°C is showing promising salt spray resistance at film thickness of 1-2 mils on magnesium alloys to DTD. 626B.

1-7. SURFACE CLEANING AND REMOVAL OF AIRCRAFT FINISHES

A. INTRODUCTION

We have already dealt with the application and formulation of aircraft finishes. Techniques for their cleaning and removal have in turn been profoundly affected by the increased requirements in resistance properties demanded of the coatings.

Many multicoat systems are susceptible to both chemical and solvent attack. The matrix of the film can be softened, swollen, and penetrated to the metal substrate by suitable groups of solvents, e.g., glycol ethers and chlorinated hydrocarbons. Also, the presence of unreacted cross-linking radicals, ester groups, and weak oxygen bridges constitute areas vulnerable to attack by organic acids or alkalies. Efficient paint strippers are, therefore, designed to attack the finishing schemes at the points of weakness in the minimum time, whereas, efficient aircraft cleaners are designed to have a negligible effect on the finishing scheme in spite of repeated applications over very long periods. The former requirement can be met by a blend of organic solvents containing a high proportion of chlorinated hydrocarbons and activated by organic acids or amines. The latter requirement can be met by an aqueous solution containing surfactants and with a pH > 9.5 for good cleaning power. However, both types of material, though dissimilar in composition and function, must satisfy the common requirements of being free from corrosive action on the materials of aircraft construction and, in particular, nonferrous metals.

Considerable attention has always been focused, in the United Kingdom, on the freedom from corrosive action of maintenance chemicals, and the development of standard techniques of evaluation has stemmed from military requirements. The following method has been standardized for both paint strippers and surface cleaners by the Ministry of Technology in the appropriate Aerospace Material Specifications.

B. PREPARATION OF METAL PANELS

Two sets of unused panels of the following metals, each 3 in. × 1 in. × 20 Standard Wire Gauge, pretreated or plated on faces and edges as described, shall be degreased in pure toluene, dried at 100-105°C for 30 min, cooled, and weighed to 0.0001 g.

1. Aluminum Alloy
   BS 2L70 acid chromate pickled as in Def. 1053, Method 2, para 5(b)(ii).

2. Magnesium Alloy
   Magnesium-Zinc-Zirconium alloy sheet BSL. 504 acid chromated as in DTD. 911C, Appendix ii, para. 2.

3. Copper
   BS. 899 C. 104 half-hard, freshly pickled as in specification DTD. 901F, Appendix I, paras. 12B.

4. Steel
   BS. 1449 Part 1B. CR3/FF freshly burnished as in specification BS. 3900, Part A.

5. Cadmium-Plated Steel
   BS. 1449 Part 1 B. CR3/FF cadmium plated to specification DTD. 904C.

Methods of immersion and time of exposure of the metals to the chemical concerned differ between paint stripper and aircraft cleaners, but the resultant performance required is identical. The freedom from corrosive action of the material shall be such that the metal panels shall not increase in weight by more than 1 mg nor lose more than 5 mg. There shall be no signs of corrosion such as pitting of the edges or surfaces, or formation of adherent deposits.

This stipulation relating to any maintenance chemical that can come into contact with metals of aircraft construction must be met to satisfy military requirements governed by the Ministry of Technology, commercial requirements of airline operators conforming to Air Registration Board standards, as well as the domestic requirements of each airframe constructor who may well adopt additional tests to safeguard any potential attack on specialized alloys which are not, in their opinion, adequately covered by satisfactory performance on the standard metal surfaces.

The developments in coating technology that have led to the introduction of highly resistant finishing schemes, based on, e.g., epoxies and polyurethanes, have produced the requirement for more aggressive strippers, but this increased resistance also allows a more aggressive cleaner to be used. However, this increased aggression is limited by the corrosive effect on metals as measured by the established standard method, and each product must be carefully examined in this context.

C. PAINT STRIPPERS

Paint stripping is required at four main stages in the life of an aircraft:

1. Spot rectification during airframe manufacture.
2. Maintenance repainting due to local operational damage between workshop overhaul periods.
3. Workshop repaints when progressive deterioration of the finishing scheme in service has lowered its protective and decorative value.
4. Major overhauls at a main base or at the airframe constructors workshops for structural checks after the designed number of hours has been flown. An important aspect of this type of overhaul is a physical examination of the metal surface of the skin, structural fabrications, and components such as ailerons, undercarriage mountings, and helicopter blades, as well as the detailed examination of each rivet head and sealant joint.

In practical terms, all these requirements can be met by an application of paint stripper, although this does not exclude the use of an immersion process for detachable detail such as cowlings, where advantage can be taken of more rigorous process control, greater flexibility in composition formulation, and increased speed of operation by raising the bath temperature.

The composition of an efficient, noncorrosive application paint stripper must be formulated around a large proportion of solvent that will effectively swell the polymer matrix. Such solvents should be composed of relatively small molecules to penetrate quickly and to reach the film/substrate interface to overcome the adhesive bond of the primer coat to the substrate. Minimal quantities of wetting agents assist penetration, as do organic acids which are particularly effective in breaking down adhesion. The most useful solvents are the chlorinated hydrocarbons, methylene dichloride being outstanding. Dimethyl formamide is also a powerful solvent. Other solvents may be used to increase the activity of the primary solvent; methanol and tetralin are frequently used in conjunction with methylene dichloride. A thickening agent is usually required to give a suitable application viscosity and cellulose ethers are commonly used. Since solvents composed of small molecules rapidly evaporate, a blanketing mechanism is required to retard evaporation, and by forming a skin toward the atmosphere, assist in preventing the draining of the stripper from vertical surfaces. This is generally achieved by incorporating a sufficient quantity of paraffin wax into the composition. Water rinsing of the surface is normal practice after the stripping action has ceased, and this can be allowed for, e.g., by using a water-soluble thickener and adjusting the concentration of wetting agents used. Small amounts of activators, e.g., amines, may also be added. Such a composition is described in Defence Specification 80-16/1:-

Dichloromethane	64.0%
Methanol	10.0%
Thickening agent (Celacol MM 100)†	3.0%
Wetting agent (Lubrol E *)	1.0%
Paraffin wax	1.0%
Cresylic acid (refined)	15.0%
Toluene-4-sulfonic acid	2.0%
Water	4.0%

This composition is of a reference fluid for comparing paint stripping power and establishes a minimum standard for capacity to lift and wrinkle paint films prepared under specified conditions. Such a composition does not comply with the freedom from corrosive action requirements already outlined and it is necessary to build into the composition sufficient inhibition to prevent attack on the standard metal surfaces.

Contemporary finishing schemes, however, have often been designed to resist aircraft fluids which are aggressive paint strippers by any standards. Glycol ether deicing fluids, phosphate ester hydraulic fluids, and certain proprietary fuels and lubricants are of this type; in order to provide paint surfaces resistant to attack by such fluids, aircraft constructors have substantially improved the adhesion of paint systems by the use of surface conversion processes and stoved primers; they also use finishing coats of enhanced resistance. As such schemes are designed to meet an operational life of many years in the face of spillage hazards, it is likely that effective strippers can only be maintained for such systems by the choice of more aggressive activators at the expense of reduced freedom from corrosive action.

---

†British Celanese.

*I.C.I., Ltd.

D. SURFACE CLEANERS

Surface cleaning is necessary to maintain the protective nature of the paint finishes, to prevent the accumulation of corrosive soils on unpainted areas, to preserve the decorative appearance of civil liveries, identification marks and hazard warnings, and to remove the superficial consequences of high altitude irradiation.

The soils can arise from three main sources:

1. Cyclic condensation can always give rise to accumulated water on external surfaces which can become mixed with surface dusts from ground activities to form inorganic clay-like structures which develop thin film adhesion forces to both painted and unpainted surfaces. Saline contamination can also occur in marine atmospheres.
2. From the engine compartment, cowlings can become coated in oils and fluids, and exhaust tracks on the aft fairings can be streaked with adherent carbonaceous films.
3. Apart from chalking residues generally associated with epoxy finishes, thin films of a light dust soil can be deposited on the aircraft skin by kinetically generated electrostatic forces which are only partially disturbed during high speed flights or by rain. Therefore, an effective cleaner must be capable of penetrating these soils, overcoming the forces of adhesion to the surface, and reducing the dislodged soils to a particle size that is mobile within the minimum of rinse water and does not become reattached during drain off with the formation of visual streaks.

The restrictions in its composition must be dictated by its freedom from corrosive action on metallic surfaces, and its attack on paints, jointing sealants, elastomeric grommets, and acrylic windows must be closely limited. There are two further strictures dictated by practical considerations which must be taken into account.

1. During rinsing the cleaning solution can enter structural compartments through vents or be trapped in butt joints and rivet heads. Once at rest the concentration of active chemicals could be increased by evaporation but, of more importance, could create a corrosive climate over extended periods of time.
2. The operational life of aircraft continues to be extended and the accumulative consequence of repeated cleaning must be taken into account.

Standard tests have been devised to meet these operational criteria for a free rinsing chemical cleaner. These are as follows.

1. Freedom from corrosive action. This test has already been detailed and is carried out by immersing the standard metal surfaces for 168 hr in both the concentrate cleaner and a 10:1 solution diluted with a standardized hard water to the following composition:

Calcium acetate dihydrate	0.40 ± 0.005 g
Magnesium sulfate heptahydrate	0.28 ± 0.005 g
Distilled water (boiled)	1 liter

2. An additional test is included in which aluminum panels to BS 2L 70 are prepared by drilling a $\frac{5}{16}$-in. diameter hole in the center before chromate pickling. The prepared panels are separately immersed in the concentrate cleaner and a 10% solution in standardized hard water to the bottom of the hole. The covered glass containers are tilted once daily for 5 days of every week so that the entire panel is momentarily immersed and the top half subsequently allowed to drain on regaining its original position. The test is continued for four weeks before the panels are washed, dried, and examined for visual signs of corrosion.
3. Freedom from damage to painted surfaces is assessed by appearance, bend and scratch tests after immersion of alkyd and polyurethane finishes in a 10% solution of the cleaner in standardized hard water for 20 min, rinsing, and allowing to dry for 24 hr before examination.
4. Effect of repeated cleaning. Cleaning of aluminum panels by the cleaner diluted to 25% in standardized hard water is carried out three times a week for ten weeks by comparison with water. Throughout this period the panels are exposed in the open facing south at an angle of 45°.
5. Freedom from crazing. A strip of clear polymethyl methacrylate $10 \times 1 \times \frac{1}{8}$ in. after annealing is stressed around a 10-in. mandrel to 3000 ft lb/in.^{2} at $20^{\circ} \pm 2^{\circ}$C for 24 hr after a liberal application of cleaner is made to the tension side of the strip. After rinsing the strip is examined in transmitted light for crazing.
6. Cleaning power is assessed as both a concentrate and 10% solution on an artificial soil 0.001 in. thick on aluminum. The soil consists of a film of equal parts of boiled linseed oil and engine oil heated to $99 \pm 1^{\circ}$C for 1 hr and allowed to cool for 30 min before use.

These tests are described in Ministry of Technology Aerospace Material Specification DTD. 5507B together with the following composition for a reference cleaning fluid to be used as a minimum standard for cleaning power:

Trisodium phosphate (dodecahydrate)	10.0%
Wetting agent (Lissapol N) (ICI)	2.0%
2-Ethoxy ethanol	6.0%
Distilled water	82.0%

Such a composition will behave efficiently toward oil-based soils but, under practical conditions will not easily remove light dust soils without mechanical assistance. More effective surfactants operating at a high pH value will improve cleaning power but are unlikely to give sufficient freedom from corrosive action to satisfy stringent tests on the many metals in aircraft use. Additionally, new materials of construction may well promote an added requirement in the United Kingdom that hydrogen embrittlement be evaluated in the testing of cleaners. However, the currently favored U.K. practice of the total painting of aircraft with highly chemical resistant finishing schemes may modify the present compromise of reduction in the corrosive tendencies of cleaners by buffer control and inhibition at the expense of their cleaning power.

Big changes can be foreseen in methods of cleaning. Light dry soils are peculiarly resistant to simple wetting dislodgement and some form of mechanical assistance is required. The enormous surface area of present day aircraft, increased labor costs, and the short time on the ground available for maintenance cleaning relative to the surface area involved, call for new techniques in cleaner application. One such is a foam-generating multilance application in which the whole of the external surface is covered in a foam cleaner (49). The foam is formulated to have a sufficiently stable dwell time to wet the surface before the kinetic energy of foam collapse overcomes the adhesion forces of stubborn light dust soils, allowing them to be rinsed away with any emulsified oils. Effluent problems are minimized by the choice of biodegradeable surfactants.

REFERENCES

1. J. S. Gourlay, J. Oil Colour Chemists' Assoc., 29, 311, 94 (1946).
2. H. Sutton and L. F. Le Brocq, J. Inst. Met., 57, 223 (1935).
3. H. G. Cole, R.A.E. Tech. Note Met. 185, No. V (1953).
4. L. F. Le Brocq and H. G. Cole, R.A.E. Tech. Note M.2584A, Sept. 1941.
5. R. J. Ledwith, J. Oil Colour Chemists' Assoc., 30, 330, 503 (1947).
6. L. R. Whiting and P. F. Wangner, U.S. Patent 2,525,107.
7. L. J. Coleman, J. Oil Colour Chemists' Assoc., 42, 1, 10 (1959).
8. J. L. Prosser, Paint Research Station Technical Paper No. 205, Sept. 1957, Research Association of British Paint Colour and Varnish Manufacturers.

9. D. A. Bayliss and D. C. Wall, J. Oil Colour Chemists' Assoc., 49, 770-804 (1966).

10. D. A. Bayliss and D. C. Wall, J. Oil Colour Chemists' Assoc., 51, 792-815 (1968).

11. H. G. Cole, R.A.E. Tech. Note Met. 209, Jan. 1955.

12. C. E. Hoey, J. Oil Colour Chemists' Assoc., 51, 9, 847 (1968).

13. G. B. Evans, AGARD Report 570, Pt. 1, 1968.

14. F. A. Champion, J. Oil Colour Chemists' Assoc., 41, 10, 730 (1958).

15. Imperial Chemical Industries Ltd. Metal Pretreatment Processes, Technical Data Sheets.

16. Anon Aircraft Painting, Ind. Finishing (Brit.), 21, 249, 33-5 (1969).

17. C. E. Hoey, Aircraft Eng., 39, 11, 465, 23 (1967).

18. W. Gotze, Jahr. Lachchemie 1959, quoted in C.I.B.A. Technical Bulletin TB.43.

19. D. Tuitt, Aircraft Eng., 39, 11, 465, 8 (1967).

20. C. E. Hoey, Paper presented at Air Force Materials Lab., 50th Anniversary Corrosion of Military Aerospace Equipment Tech., Conference, Colorado, 1967.

21. J. A. Scott, Aircraft Eng., 39, 11, 465, 23 (1967).

22. A. R. H. Tawn, 6th F.A.T.I.P.E.C. Congr., 323-6 (1962).

23. C. Barker and A. Lowe, J. Oil Colour Chemists' Assoc., 52, 10, 905 (1969).

24. J. H. Saunders and K. C. Frisch, Polyurethanes: Chemistry & Technology, Wiley (Interscience), New York, 1962 and 1964.

25. R. Hebermehl, 4th F.A.T.I.P.E.C. Congr., 85-93 (1957).

26. W. Berger, 6th F.A.T.I.P.E.C. Congr., 300 (1962).

27. A. Lowe, J. Oil Colour Chemists' Assoc., 46, 10, 820 (1963).

28. J. W. Britain and P. G. Germeinhardt, J. Appl. Polymer Sci., 4, 207 (1960).

29. J. W. Britain, Ind. Eng. Chem. Prod. Res. Develop., 1, 261 (1962).

30. J. D. Booker and M. J. B. Nash, R.A.E. Tech. Rept. 67245, 1967.

31. A. Fyall, Shell Aviation News, 377, 7 (1969).

32. G. F. Schmitt, Jr., U.S. Gov. Res. Dev. Rept., 68, 13, 86 (1968).

33. J. H. Saunders and J. C. Frisch, Polyurethanes: Chemistry & Technology, Wiley (Interscience), New York, Vol. II, 1964, p. 773.

34. R. Gamero and G. M. Le Fave, Paper, Symposium on Sealants and Sealing of Aircraft Missile and Electrical Components, Society of Aircraft Materials and Process Engineers, Oct. 28-30, Los Angeles, 1959.

35. W. H. Stevens, J. Oil Colour Chemists' Assoc., 42, 10, 663 (1959).

36. R. N. Miller, Soc. Aircraft Materials and Process Engineers J., April/May (1970).

37. P. Walker, Paint Technol., 34, 3, 388, 14 (1970).

38. W. M. Brandenberg, U.S. Gov. Res. & Dev. Rept., 69, 2, 87 (1969).

39. P. W. Baker, Environ. Eng. Quart., June, 1963, pp. 17-19.

40. D. Taylor, J. Oil Colour Chemists' Assoc., 41, 10, 707 (1958).

41. F. Scofield, Natl. Paint Var. Lac. Assoc. Sci. Sec. Circ., No. 638, March, 1942.

42. V. C. Vesce, Official Digest, 227, 217-64 (1943).

43. W. W. Wright and W. A. Lee, Progr. High Polymers, 2, 189-313 (1968).

44. C. Butler, J. Oil Colour Chemists' Assoc., 51, 2, 177 (1968).

45. W. H. McKnight and G. S. Peacock, Treatise on Coatings (R. R. Myers and J. S. Long, eds.), Vol. I, Pt. 2, Dekker, New York, 1968, p. 310.

46. M. C. Reed and J. Harding, Ind. Eng. Chem., 41, 675 (1949).

47. Y. M. Chaudhak and S. N. Agarwal, Paint Manuf., 40, 7, 35 (1970).

48. J. E. Gilder, Shell Aviation News, 376, 16 (1969).

49. Sunbeam Anticorrosives, Brit. Pat. 1,001,868; U.S. Pat. 3,231,134, Can. Pat. 723,861.

50. C. E. Hoey, J. Oil Colour Chemists' Assoc., 53, 1026 (1970).

51. D. D. Taft and A. H. Mohar, J. Paint Technol., 42, 550, 615 (1970).

Chapter 2

APPLIANCE FINISHES

E. Gustave Shur

Inmont Corporation
Clifton, New Jersey

2-1. INTRODUCTION

The industrial product designer is charged with the responsibility of endowing his creation with attractiveness as well as with function. When he has finished his design, a choice of external finish has to be made. In an increasing number of instances, the outer portion of the product may be fabricated of plastic or a noncorroding metal such as stainless steel which does not require a protective coating. In most instances, however, steel is used for its combination of strength with low cost. Here a suitable choice of coating must be made. In the case of plastic, coatings may be required to perform a decorative role since changing the inherent coloration of plastic parts may be more expensive than post coloration with an organic coating.

2-2. FUNCTIONS OF COATINGS

The generic term paint, is the most widely used word describing organic protective and decorative finishes. In addition to the basic functions of protection against weather, heat, chemicals, and other corrosive environments, paint also plays a secondary role in improving sanitation, in radiation control, antibiosis, and in controlling light reflectance, heat radiation, temperature, and acoustical properties.

2 - 3. HISTORY

Prior to World War 1, industrial coatings were based on natural oils and resins and on varnishes cooked from them. A limited number of pigments and dyes were available for coloration. Drying schedules were long: overnight for ambient temperatures or several hours if baked. The most durable coatings were black "japans" based on bitumens cooked with drying oils.

The home appliance industry began to develop in the 1920's, as electricity became widely available in the USA. The federally sponsored rural electrification programs of the 1930's, such as the Tennessee Valley Authority, played a vital role in this respect. The electric refrigerator was the first major appliance to make an impact on the quality of American home life in the twenties. The first small electrical appliance of numerical importance was the electric fan.

The types of coatings available at that time were fairly restricted. A great advance in varnish technology had taken place when the use of phenolic and modified phenolic resins with tung oil gave the paint industry fast-drying vehicles with excellent acid, alkali, and corrosion resistance. These made, and still make, excellent primers. After World War I, the Weizman bacteriological fermentation process for the production of acetone for smokeless powder, together with the availability of lower viscosity grades of nitrocellulose, made possible the introduction of fast drying lacquers. These revolutionized the paint industry. Other synthetic resins followed.

By the mid-twenties, lower viscosity grades of "cotton" were available and a real revolution in finishing took place. Slow drying varnishes gave way to lacquers which dried in minutes by evaporation of the solvents. These lacquers were modified with dewaxed damar gum and plasticizer and were widely used as topcoats for refrigerators.

Alkyd resins were first synthesized in the late twenties and were available commercially in the beginning of the thirties. An excellent history of the development of the alkyd resins is given by Kienle (1). The chemistry is also discussed by Kraft (2). Alkyd or "Glyptal" based finishes made high solids baking enamels suitable for use on appliances. They were superior in "build" to the nitrocellulose lacquers but required long bakes of one to two hours at temperatures of 200-250°C to cure properly.

From this point on, the type, number, and complexity of raw materials available for use in appliance finishes proliferated. Solution vinyls and urea resins became available about 1936, melamine in 1940, silicones in 1944, and epoxies in 1947. Each of these resin types contributed in some way to the achievement of accelerated rates of cure and improved resistance properties.

2-4. THE APPLIANCE FINISH MARKET

The appliance industry represents a large potential market for industrial product finishes. Banov (3) discusses the problems faced by the industry supplying finishes, including the appliance industry. Figures for shipments of household appliances are available from the office of Business Economics of the U.S. Department of Commerce. The value of these shipments doubled from $3.5 billion in 1960 to $6.7 billion in 1969. The potential for sales remains impressive since only 2% of American homes have dishwashers and 37% have clothes dryers. Despite the seeming saturation level reached with electric refrigerators, over six million units were sold in 1970, primarily as replacements.

Statistics for the volume of appliance finishes were last issued in 1967 (4), when $48.4 million worth of coatings were used. It may be estimated that $60 million worth of coatings were put on by the appliance industry in 1970. This contrasts with $40 million applied by the coil industry and $10 million by electrodeposition.

2-5. PRODUCTION AND APPLICATION METHODS

Production in America is almost completely conveyorized with the use of an overhead monorail to carry the ware through the following steps:

1. Four-to-six step metal degreasing and phosphate passivation of steel and small quantities of zinc and aluminum parts.
2. Paint application.
3. Baking operation in direct or indirect oil- or gas-fired or infrared ovens.
4. Cool off and storage conveyor line.

A. METAL PREPARATION

Metal preparation is by proprietary treatment of two general types. In the first, microcrystalline zinc phosphate is deposited for maximum corrosion resistance. The porosity of this deposit makes the use of a two-coat system almost mandatory for good holdout. The first coat helps fill the pores of the zinc phosphate layer.

Iron phosphate treatment is the second type used. The deposit is amorphous, thinner, more dense, and generally less corrosion resistant than the deposit of zinc phosphate treatment. It is widely used where one coat finishes are required. Holdout and flexibility are superior to the zinc phosphate pre-treatment. In both cases the surface is left acidic and care should be taken to pretest coating systems to ensure that there is no interference with the normal mechanism of cure. Because most thermosetting systems respond positively to acid catalysis, this is not normally a problem.

B. PAINT APPLICATION

The primary consideration in the application of appliance primers is the degree of protection required for protection against corrosion of the base metal during use. For appliances such as clothes washing machines and air conditioners, which may be used in cellars and in windows, the environment is considered highly corrosive. Primers for this equipment are usually applied by flow coating (5). This technique has largely replaced the older method of paint application by dip coating, which had been used where conveyor line speeds of ten ft/min or less could be tolerated. The complexity of mechanism and equipment maintenance were minimal, but even at these low line speeds a typical appliance tank required a capacity of about 2000 gal, and the attendant risks in paint inventory are quite obvious.

In flow coating, the size of the paint supply tank bears no relation to the line speed. Its maximum capacity is of the order of only 150 to 250 gal. A separate storage tank holds the solvent used to clean the system, to reduce viscosity to the operating range, and to maintain that viscosity during the working day. Most appliance operations are fully automated with a continuously recording viscometer regulating the input of additional reducing solvent, a heat exchanger controlling temperature, and recirculating filters maintaining cleanliness.

The paint is pumped from the reservoir to stationary or revolving risers from which it is emitted under pressure. In one case a series of nozzles is arranged to impinge the enamel on all areas of the parts. In another centrifugal force is used to assist in reaching difficultly accessible areas. Provision must be made for a solvent vapor chamber where the bubbles formed by impingement may break and where flow marks are minimized. Large volumes of solvent are used to maintain the solvent atmosphere and maintain viscosity. These constitute a considerable safety and pollution problem as well as a potential economic loss. Appearance leaves something to be desired. Film thicknesses vary from top to bottom and edge coating is thin. Although tears and fat edges may be minimized by proper hanging, by viscosity control, and by formulation, a supplemental reinforcing spray may be necessary.

The application of topcoat enamels by flow coating is feasible but is generally restricted to colors on appliances such as air conditioners where the finish is not readily inspected by the housewife.

At the present time there is considerable pressure in the appliance industry to change to the application of primers by electrophoresis. Among the potential advantages are elimination of runs and sags, elimination of solvent, and insurance of complete paint coverage on all areas, including holes, cutouts, raw edges, and internal boxed areas. Electrophoretic coating or electrodeposition is discussed at greater length in a separate chapter.

The electrodeposition of water based primers is threatening the future of solvent soluble epoxy basecoats, the main advantage being the deposition of a film of paint in areas difficult to coat by conventional means. The result is better protection against corrosion. The appliances most likely to be finished by electrocoating are those exposed to corrosive conditions in use such as air conditioners and laundry appliances.

Most appliance topcoats are currently applied by automatic electrostatic spray. Many companies supply equipment suitable for the application of topcoats using conventional compressed air atomization into a electrostatic field. The paint particles acquire a negative charge while the ware is grounded. Low air pressure is used to deliver the atomized particles to the field where high potentials of about 90,000 V take over and accelerate them to the parts.

The Ransburg Electrocoating Corporation supplies bell-shaped heads and disks which eliminate the need for compressed air. The finish is delivered to the head in an insulated tube from a gear-driven metering pump. The head is rotated at a speed of about 900 rpm. Centrifugal force atomizes the coating, providing the proper solvent and viscosity are used. More than one head may be used to improve paint particle distribution.

An advance in electrostatic spraying is the use of a flat disk in place of the bell-shaped heads. The disk may be stationary or angled, but in most cases it is reciprocated. One disk can do the work of several bell-shaped heads. Unpainted parts are moved on a bunched conveyor along the periphery of the circle whose center is the disk. Paint is pumped to the center of the charged disk where centrifugal force carries it spirally to the edge, from which the paint droplets are propelled. Two disks may be required to coat large parts such as refrigerators. Some manual control may be necessary since the volume of finish delivered to the heads is constant, while the conveyor moves.

Proper racking of parts to maintain approximately the same area to be coated is required, and this is a problem when there are different components on the same line. Spacing is also an important factor. As a result, provision for manual spray to reinforce areas not receiving the full amount of finish, is required.

Scholtz (6) reviews application methods and furnishes finishing costs for appliances in Table 2-1.

Application of finishes to flat plates by direct roller coating has been practiced in the can coating industry for half a century. Great impetus has been placed on the application of coatings by direct and reverse roller coaters on continuously moving strip steel and aluminum. The subject of coil coatings is discussed separately. Some use has been made of strip-coated metal in the appliance industry.

Maxwell (7) describes the use of precoated coil stock in the manufacture of freezer liners. The liner was designed to be made of prefinished 0.04-in. aluminum sheet which withstands blanking, forming, and assembly to a formed top and bottom. Both the ceiling and freezer bottom were made from the same material. In addition, a portable microwave oven with an outer case of precoated, textured sheet metal has been introduced.

TABLE 2-1. Cost of Finishing Appliances

	Primer 0.3 mil		Enamel 1.25 mil		
	Flow coating	Dipping	Manual spray	Steam and airless atomization	Electrostatic spray
Cost of coating material	$4.05	$4.05	$5.00	$4.00	$4.00
Amount of reducer (gal/gal finish)	2.0	1.0	0.25	0.125	0.75
Cost of reducer	$1.20	$0.44	$0.09	$0.04	$0.30
Total cost	$5.25	$4.49	$4.09	$4.04	$4.30
Theoretical coverage (ft^2/gal)	1890	1890	488	488	488
Cost/1000 ft^2 (100% efficiency)	$2.78	$2.38	$8.38	$8.26	$8.81
Cost efficiency of method	$3.09	$2.65	$16.76	$10.33	$9.27

The overall usage of coil coating is growing slowly. Technical problems exist:

1. Coatings with the requisite resistance properties generally have been deficient in formability;
2. There are bare edges which have to be coated or hidden by redesign;
3. Joining or welding prepainted parts is difficult;
4. Most appliance manufactures have large capital investments in existing facilities for post finishing prefabricated parts.

A comprehensive review of coating methods is given by Higgins (8).

Catalytic combustion of oven gases is used to reduce air pollutants. The waste gases provide heat energy which may be recycled and reduce fuel costs.

2-6. TYPICAL PAINT SYSTEMS

Refrigerators and home freezers generally use two-coat systems composed of varnish, alkyd, or modified epoxy-ether primer, baked, scuff sanded, and topcoated with an especially formulated spot resistant alkyd-amine or acrylic enamel. Epoxy-ether urea systems have been developed to replace vitreous enamel for interiors. Polyester and glass-reinforced plastic for interior cabinet shells have been used experimentally but require special primers and topcoats. Washing machines and clothes dryers require two-coat systems with specially formulated detergent resistant primers and acrylic or polyester topcoats. Epoxy-ether urea systems have been recommended to replace porcelain for interiors and tubs. Porcelain is still widely used for tops, lids, and tubs. Hot water heaters and kitchen cabinets generally use one coat of alkyd-amine enamel.

Porcelain enamels are most widely used on stoves, ranges, roasters, and ironers. Silicon-based enamels generally have poor grease resistance. In some cases specially formulated heat resistant oil free polyester amines are used on parts where direct exposure to high heat is not a factor. Electric-, gas-, and oil-fired space heaters use both one or two coats including primer and heat resistant alkyd-amine, generally a metallic containing aluminum pigmentation.

A. PAINTED STEEL VERSUS PLASTICS

Let us examine the exact range of physical properties of painted steel more critically, in comparison with plastic and porcelain. The primary deficiency of plastic parts lies in their strength — volume relationship. Much has been written describing and extolling the superior strength — weight properties of plastic materials versus their steel counterparts; however, the tremendous difference in specific gravity between plastic and steel means that there is a correspondingly great difference between the volume occupied by equal weights. Attempts have been made to increase the tensile strength and ductility of plastics by the incorporation of fibrous fillers. However, there remains no equivalence of function where a specific strength-volume relationship is required. Where this property is not a factor, the determining consideration is the economic one. Where large numbers of parts are involved, the fabrication of steel will usually be considerably more economical than the molding of plastic. However, many nonstrength, decorative parts of molded plastic have been used to replace coated steel. Organic coatings are used on plastics where a variety of colors is required and economy precludes molding these colors in situ. Enamels are also used to reclaim rejected plastic parts or further to decorate them.

Baked enamel coatings on steel of the type which we have been discussing, will equal or surpass the performance of most molded plastic articles provided the enamel has been supplied by a reputable vendor and has been formulated with a good understanding of the specific role to be performed by the coating and the term of durability required. The desired properties of physical strength, adherence, surface hardness, flexibility, ductility, and resistance to chemicals, salt, and water corrosion may most easily be obtained by a combination of carefully chosen primer and baking enamel.

B. PAINT VERSUS PORCELAIN

We come now to a critical comparative evaluation of the relative merits of paint versus porcelain; or organic versus inorganic coating over steel. The most important thing to remember about porcelain enamel is that it is vitreous in nature, that is to say it is composed of an inorganic glass which has been fused into a continuous film on a metallic substrate. The overall resistance properties vary with specific composition so that different formulations are recommended for alkali as opposed to acid resistance, etc. Generally speaking, the durability of vitreous enamel is considered superior to organic industrial finishes with the exception of flexibility and distensibility. Although partaking of the chemical resistance of glass, porcelain also has the intrinsic brittleness of glass.

This is an extremely important point to stress when considering the vitreous versus organic enamels for a specific end use, for if the normal requirements for the physical film properties of an organic coating could be altered to include acceptance of the brittle characteristics of vitreous enamels, then markedly greater resistance to attack by chemicals and other reagents, as opposed to that of organic films, would be possible. This is based on the well-known principle of high polymer chemistry that the greater the degree of crosslinking or thermosetting within the resin structure, the greater the inherent resistance properties but the poorer the innate flexibility. The paint chemist must work under a severe handicap in attempting to approach the ultimate in properties.

Other disadvantages of porcelain are well known. They require more expensive specially treated grades of steel usually of thicker gauge. They must be fired at temperatures of 1000 to 1500°F before fusion occurs. The "slips" must be dispersed by the user and are of limited storage stability. They are difficult to control for reproducibility of color during the firing process. They require a special cobalt primer.

Instances have been reported describing the feasibility of the direct application of vitreous enamels to chemically treated metals without the use of primer. Glasses have been made which may be fused at temperatures as low as 400°C. Experience with these enamels has indicated that they have a greater tendency to fracture on normal usage than the conventional porcelain finish. In addition, much of the normally high inherent resistance to corrosion of vitreous enamels has been lost in the process.

To recapitulate the differences, organic coatings are lower in cost, easier to apply, require lower temperatures of cure, have better physical properties of impact and flexibility and permit improved reproducibility of color.

2-7. FUNDAMENTAL PROPERTIES

An appliance finish should protect the underlying metal from water, salt, detergent, and other corrosive compounds acting at room to moderately elevated temperatures. Ideally, such a film should be completely impermeable to degrading materials, however, repeated experimentation has shown that this condition is never achieved by organic polymers.

A considerable effort has been made in studying the means by which surface active solutions penetrate films. In this connection ionic permeability has been studied by applying a voltage across a film and measuring the effect on its membrane potential and related osmotic phenomena.

To give a more detailed picture of what happens when a detergent molecule penetrates the film to the metal interface, surface diffusion measurements have been made by Carangelo (9) using radioisotope labeled surface active agents. In this way, information has been obtained which enables the investigator to determine whether the film has been displaced from the metal by the surface active agent.

Although all films are permeable to water and to ions in solution, there are great quantitative differences among them, depending upon the composition of the polymer backbone and the nature and extent of crosslinking present. Films selected for use as appliance finishes should be highly resistant to permeation by water and to osmosis. Water vapor diffusion may be studied in several ways. The Payne cup is widely used in the paint industry. Its development and use are given by Payne (10). An extremely sensitive method has been developed using radioactive tritium labeled water (11).

The mechanism of permeation may be explained somewhat as follows: Water molecules dissolve in the film on the side exposed to the vapor. An individual water molecule moves about in the film by jumping into intermolecular spaces in its immediate neighborhood. These "holes" are constantly forming and disappearing because of the random motion of segments of the long chain molecules. Because of the concentration gradient across the film, the net effect is a drift of the water molecules toward the dry side. The number of molecules transported across a given area is proportional to the concentration of water molecules and, therefore, to the solubility of water in the film. Permeation is thus dependent on both diffusion and solubility, both of which are also temperature dependent.

The diffusion of ions through organic films may be studied by measuring the drop in electrical resistance of especially purified water. A diffusion cell is used by Harvey (12) in which the film membrane under study separates a salt solution of 0.1 N electrolyte from the high resistance "conductivity" water. A potential is placed across the membrane and changes in conductivity measured on either side of the film. Glass (13) uses a radioactive tracer procedure to measure the diffusion of sodium chloride ions through free films.

Before undertaking any large scale experiments on appliance finishes, it is advisable to determine if the laboratory procedures and equipment used are sensitive and reproducible. Statistical analysis is required to maintain adequate confidence in the validity of test data. Among the factors known to affect test results are composition, oven temperature, oven position, and film thickness. Detergent composition, concentration, bath temperature, and the type of bath, i.e., static versus circulating, are factors affecting results in the determination of detergent resistance.

A. PAINT COVERAGE

The calculation of paint coverage efficiency has been reduced to simple arithmetical processes, using well established principles, by Shur (14). A gallon of paint will cover 1600 ft^2 at a film thickness of 1.0 mil if no volatile solvent is present. To obtain the figure of 1600 ft^2/gal, a simple calculation involving a conversion is required from the unit of volume per gallon, to that of square feet per mil.

In order to calculate the theoretical mileage one must know what fraction by volume of the gallon of paint is present as solid film former. Multiply 1600 by the fraction of solids present, to give theoretical coverage. Unfortunately most paint suppliers do not furnish this type of physical data. However, they do furnish physical constants such as weight per gallon and per cent nonvolatile (or solids) by weight which permit calculation of the fractional volume of solids.

B. FILM DENSITY MEASUREMENT

The density of films may be determined without stripping them from the metal substrate using the method described by Shur (14). Densities have been determined for both thermoplastic and thermosetting films at a number of typical curing schedules. The results indicate that where drying occurs by solvent evaporation, density is independent of cure cycle. Where further polymerization occurs, the density increases with increased time and temperature until full cure is reached.

It is also possible to run a theoretical coverage determination from the empirical film density determined as indicated by Shur.

When a thermosetting paint film is cured properly, some of the theoretical solids are lost by condensation and volatilization. Although the quantity of solids lost in this way will vary with the composition and the baking schedule, the average loss is from 5 to 10%.

C. COST OF MATERIALS

Knowing the actual and theoretical mileage one may proceed to calculate the cost of paint materials per square foot of painted ware. The total number of gallons of paint and of solvent are multiplied by their unit cost, added and divided by the number of square feet covered to compute the cost per square foot.

D. EFFECT OF PIGMENT-TO-VEHICLE RATIO

Let us examine the pigment-to-vehicle relationship of a white appliance finish. An extremely low volume of film is contributed by titanium dioxide pigment because of its low bulking value. The pigment is so dense that on a pound for pound basis, it provides less than one-fourth the film thickness contributed by the clear vehicle. Assume that one can make a series of enamels covering a range of pigment volume from 12.9 to 25.3% by volume of the total solids, where the latter is kept at 61% nonvolatile, by weight. If one considers only that a given dry film thickness of paint must be applied, in this case 1 mil, then the lower the pigment content the higher the coverage since the film density will be lower.

In one-coat appliance white applications, film thicknesses of 1.2 to 1.7 mils are usually specified for reasons of resistance performance. The formulator will use the minimum of pigment required for adequate visual hiding at the prescribed film thickness, so that the coverage may be kept at a maximum. This will usually fall in the range of 15 to 18% pigment of the total solids by volume. Many paint consumers fail to check dry film thicknesses and apply only enough to give visual hiding. This means that as the pigmentation is increased, the thickness of the film required for visual hiding decreases. A very thin film may suffice for visual hiding and, on this basis only, the use of higher pigmentation to get higher hiding results in greater apparent coverage. It is apparent, however, that resistance decreases as a paint film is made thinner.

2-8. APPLIANCE PRIMERS

The formulations given below show that the pigmentation of primers for appliance end uses is relatively simple compared with that of automotive primers. In many cases the primer is partially visible or actually used as a finish coat. In all cases it has to permit topcoating with white enamels without bleed or primer color "show-through." This generally prevents the use of conventional colored corrosion resistant primer pigments such as zinc chromate.

In 1964, Payne (15) felt that most of the requirements for appliance finishes could be met with alkyd-aminoplast combinations both for primers and topcoats. Epoxy-ether resins have been widely used as primer vehicles when modified with urea or phenolic resins, or when esterified with fatty acids. Shur (16) reviewed this technology. Epoxies have largely replaced the older varnish and alkyd primers in the appliance industry.

Dow (17) recommends an epoxy-dehydrated castor oil fatty acid ester modified with melamine for caustic and hot water resistance (see Table 2-2).

An epoxy-urea appliance primer features: (a) good color and color retention, (b) outstanding adhesion, and (c) excellent detergent resistance.

The primer in Table 2-3 will consistently pass 300 hr immersion in a 1% detergent solution at 74° C.

A similar formulation for an appliance primer is given by Union Carbide in Table 2-4.

TABLE 2-2. Epoxy Ester Appliance Primer

Resin	Dow epoxy resin 664
Fatty acid	Dehydrated castor oil
Coreactant	Melamine-formaldehyde

Vehicle preparation (closed kettle ester)

D.E.R. 664 epoxy resin	55%
Dehydrated castor oil fatty acid (1)	45%
	100%

Esterify at 260°C for approximately 4 hr and dilute to 55% solids with xylene.

Ester properties

Acid number	3
Color — Gardner	5
Viscosity — Gardner-Holdt	Z

Pigment dispersion	lb	gal
D.E.R. 664-D.C.O. ester (55%)	560.0	70.0
Titanium dioxide (2)	170.0	4.7
Lampblack (3)	4.5	.3
	734.5	75.0
Coreactant reduction		
Melamine resin (4)	218.0	25.0
Composite blend	952.5	100.0

Reduce to spray viscosity (15-20 sec on a #4 Ford Cup) with Varnish Makers' & Painters Naphtha and cure the coating for 30 min at 177°C.

Constants

Pounds/gallon	9.5
Pigment volume content	9.9%
Total solids	64%
Pigment/Binder ratio	29:71

TABLE 2-2. (Continued)

The following brands and materials were used in the above formulation as indicated by the corresponding numbers in parentheses. Equivalent brands and materials can be substituted and should give similar results.

(1) 9-11 Dehydrated castor oil fatty acids	The Baker Castor Oil Co.
(2) TiPure R-510	E. I. DuPont DeNemours & Co., Inc.
(3) Lampblack M-8416	Mineral Pigments Corp.
(4) Uformite MX-61	Rohm & Haas Co.

Ciba (18) reports improved hardness, flexibility, stain and alkali resistance, and reduced yellowing on overbake when their epoxy resin Araldite 7097 is crosslinked with hexamethoxymelamine instead of butylated urea formaldehyde resin. A formulation containing 25% hexamethoxymethyl melamine and 10% catalyst is recommended.

An appliance primer vehicle suitable for use on steel and galvanized steel is based by Higgins (19) on a polymeric polyol of styrene and allyl alcohol, epoxy resin, and phosphoric acid. Improved salt spray creep resistance, better flexibility, and less blistering are claimed.

Hornibrook (20) copolymerizes styrene, acrylic acid, and acrylic ester in aqueous dispersion. These latices are crosslinked with phenol, urea, or melamine formaldehyde condensation resins and are used as appliance primers.

2-9. APPLIANCE TOPCOATS

A. ALKYDS AND POLYESTERS

White finishes for metal objects such as washing machines, lighting fixtures, kitchen cabinets, and others, may be made with combinations of alkyd and amino resins (21). Faster cure may be obtained with styrenated alkyds with some sacrifice in color, toughness, and solvent resistance, or by using melamine formaldehyde resins. Urea resins are less expensive but not as resistant to moisture and alkalies. A triazine resin is recommended for resistance to hot soaps. Enamels having high amino resin contents and containing nondrying alkyds require primers such as the epoxies discussed

TABLE 2-3. Epoxy Urea Appliance Primer

Material	lb	gal
Ti-Pure R 110 (E. I. DuPont deNemours & Co., Inc.)	80.2	2.29
Lampblack-Blue Tone A (Monsanto Chemical Co.)	0.8	0.06
Binder		
EPON Resin 1007 (Shell Chemical)	226.7	23.74
Beckamine P-196 (Reichhold Chemicals, Inc.)	161.9	18.88
Solvents		
Cellosolve acetate (Union Carbide Chemicals Co.)	210.5	25.96
Xylene	210.5	29.07
	890.6	100.00

Formulation Constants

Nonvolatile content, % by weight	45.4
Pigment/binder, weight ratio	1/4
Pigment volume concentration, % by vol	6.6
EPON Resin 1007/Beckamine P 196, weight ratio (solids)	70/30
Weight/gallon, pounds	8.9

MANUFACTURING INSTRUCTIONS

Prepare an EPON Resin 1007 solution of such a concentration that the pigment and all of the EPON Resin 1007 will be included in the dispersion. Disperse this mixture on a three-roll mill. Let down with the urea resin and remaining solvents.

Note: Poor gloss and poor hiding may result if a portion of the EPON Resin is used in letting down the dispersion.

APPLICATION INSTRUCTIONS

For spray application, reduce to a #4 Ford cup viscosity of 20 sec with Cellosolve acetate/toluene, 1/1 by weight.

Recommended baking schedule 20 min at 196°C.

TABLE 2-4. Epoxy Urea Appliance Primer

Grind	
DuPont R900 titanium dioxide	80
Beckamine P196 (Reichold)	100
Nuodex Nuosperse 657	1
Monsanto lampblack 11	0.8
Cellosolve acetate	20
Let down	
Union Carbide EKS-2314 (50% N.V.)	454
Cellosolve acetate	77
Xylol	97
Beckamine P196	61
Union Carbide silicone S-10	1

above. It is interesting to study the literature and note the progression from alkyds to today's so-called "polyester" finishes.

Weaver, in a general review article (22) in 1958, reports the development and increasing acceptance of "oil-free" alkyds in white, nonyellowing appliance enamels. Hard, resistant coatings superior to conventional alkyds are reported by the California Research Corporation (23). Oil-free plasticizers for aminoplast resins are made. A typical formulation is based on 2 ethylhexoic acid, isophthalic acid, tris-methoxyethane, and propylene glycol. A polyester of p-tertiary-butyl benzoic acid, pelargonic acid, phthalic anhydride, glycerol, trimethylol ethane, and maleic anhydride, is acrylated by Allied Chemical (24) using p-tertiary-butyl perbenzoate as catalyst. The resulting acrylated oil-free polyester is clear, hard, and glossy when used in coatings. Emery 3393-D (25) is a dimer acid developed for use in "saturated polyester" or "oil-free" alkyd type coatings. Hydroxyl terminated polyesters crosslinked with amino-formaldehyde resins may be formulated with rapid cure to films with excellent hardness, adhesion, flexibility, heat stability, and chemical resistance. A typical polyester formulation and coating properties are noted in Table 2-5. The use of dimer acid to modify polyesters is described by Boylan (26). Dimer acid contains two carboxyl groups and 36 carbon atoms with a molecular weight of 565 and an equivalent weight of 283. It imparts flexibility and toughness.

TABLE 2-5. Polyester Formulation and Coating Properties

Polyester formulation	Equivalent ratio	Wt ratio
Emery 3393-D dimer acid	0.20	54.5
Phthalic anhydride	1.13	83.6
Maleic anhydride	0.67	32.8
Trimethylol propane	3.40	153.0
Xylene	—	20.0

Processing: Esterification occurs at 200-230°C in typical alkyd manufacturing equipment until the desired characteristics are obtained. The resin is then reduced to 60% NVM with 80/20 xylene/butanol.

%NVM	60
OH value (solids)	260
Acid value (solids)	15.8
Viscosity (G-H)	Z3-Z4

Coating formulation	Wt ratio
Emery 3393-D polyester resin (60% NVM)	21.4
Cymel 301[a]	0.7
Rutile TiO_2	16.4
Xylene	9.2
Butanol	2.3
%NVM	60
Polyester/melamine (solids)	95/5
Pigment/binder (solids)	1.2/1.0
Film thickness (dry)	1 mil
Cure schedule	30 min at 177°C
Substrate	Treated 0.025 in. aluminum[a]
Coating properties	
Hardness (pencil)	3H
Impact resistance (in.-lb)	
Forward	>30
Reverse	>30
Flexibility (cylindrical mandrel)	pass 1/8 in.
Gloss	
Initial	85
Overbake 2 min/288°C	83

TABLE 2-5. — Continued

Overbake 6 min/288°C	67
Chemical resistance (4 hr spot test/75°F)	
5% HCl	10
20% NaOH	10
Clorox bleach	10

Rating scale: 10 denotes no effect
0 denotes film disintegration

Variations in polyester formulation, type, and amount of amine resin, and pigmentation may be made to adjust specific coating properties.

[a] Americal Cyanamid Co.

Edwards et al. (27) make enamels with high hardness, gloss, adhesion, alkali, grease, and stain resistance from oil-free polyesters of trimethyl propane, neopentyl glycol, isophthalic, and adipic acids.

Isophthalic acid has been found to upgrade the properties of alkyds of varying composition (28).

Oil-free resins have shown an unusual combination of properties and were compared with a thermosetting acrylic vehicle and a commercial oil-free alkyd. The properties of the stain-resistant isophthalic, oil-free alkyd, are quite similar to the acrylic with the exception of better flexibility after heat aging. Formulations and properties are noted in Table 2-6.

An oil-free polyester resin of lower raw material cost has been formulated (29) by increasing the ratio of diol to triol. This resin has better salt spray, detergent, stain, and overbake resistance while retaining excellent physical properties. Application properties are reported superior to previously suggested formulations based on isophthalic acid. The resin and enamel formulations are noted in Table 2-7.

Roth et al. (30) patented coating compositions based on a diepoxide-containing oil-free isophthalic polyester blended with aminoplast resins. High detergent resistance, flexibility, toughness, and adhesion are claimed. Polyesters based on benzophenone tetracarboxylic acids (BPTCA) are synthesized by Horn and Van Strien (31). A typical resin has a composition of 1 mole BPTCA/ 2.25 propylene glycol/2.25 neopentyl glycol/2.25 adipic acid. The hydroxyl/1 carboxyl ratio is 1/1.5. The acid number is 58. The resin is neutralized with triethylamine, dissolved in water, and baked for 30 min at 182°C. Films are glossy, of 5H pencil hardness, and resist 80 in.-lb impact. The resin may also be cured in 30 min at 149°C with tolylene diisocyanate.

Emery Industries (32) described the use of pelargonic acid in producing alkyd resins having improved resistance to chemicals, light, and heat. Alkyds were prepared from glycerol and phthalic anhydride in which pelargonic was compared with mixed C-8 to C-10 acids and cocoanut oil. The pelargonic alkyd is reported to have better initial gloss and gloss retention on overbake, better impact resistance, with a harder film, better stain resistance, and superior chemical resistance. Naumann (33) describes the use of Versatic acid 911 in making oil-free alkyds. When cured with aminoplasts, the properties of chemical resistance are superior to alkyds. The polyesters are compatible with acrylamide interpolymers and thermosetting acrylics with pendant hydroxyl groups.

Hensley (34) crosslinks polyesters with hexamethoxymethylmelamine to achieve more elastic, flexible, thermosetting systems. An example of a commercial modified polyester designed for use with hexamethoxymethylmelamine, is Cyplex Resin 1473-5 (35). This resin has high nonvolatile content, hardness with flexibility, and impact resistance coupled with stain and detergent resistance. It can be recoated without sanding. Contamination of equipment with oil-modified alkyds or other resins should be avoided to prevent cratering and dimpling. A formulation for detergent-resistant high-gloss white enamel based on this resin is given in Table 2-8.

Finney (36) described a study of the effect of incremental changes in glycol composition on the properties of baking enamels. The glycols studied included neopentyl glycol (NPG), trimethylpentanediol, 1,4-cyclohexanedimethanol (CHDM), ethylene glycol, propylene glycol, and 1,3-butylene glycol. The resin system was an isophthalic adipic polyester. See Table 2-9 for compositions and average performance characteristics. The enamels were applied on 24 guage sheet treated with Bonderite 37 (37), cured for 30 min at 177°C, and evaluated by typical appliance test methods. Resins prepared at 30% excess hydroxyl produced films with better hardness, stain resistance, and color retention than those at 20% excess. Flexibility and impact resistance were better at the 20% level. Of extreme interest are the outstanding results obtained with a 75/25 blend of NPG/CHDM. Enamels based on this polyester had outstanding hardness, impact, solvent, stain, and detergent resistance.

Enamels with improved impact resistance are obtained by Simroc (38) by modifying alkyd aminoplast mixtures with an allyl alcohol-styrene-maleic anhydride adduct. Kapalko and Martin (39) patented oil-free polyester aminoplast compositions upgraded by modification with xylene formaldehyde resin reacted with a bisphenol A based polyepoxide (Methylon 75202). When cured for 20 min at 191°C, the modified enamel has superior detergent resistance.

Berger and Wynstra (40) have studied the use of esterdiol-204, 2,2-dimethyl-3 hydroxypropyl-2, 2-dimethyl-3 hydroxy-propionate in oil-free polyesters.

TABLE 2-6. Isophthalic Oil-Free Alkyd

Material	Charge wt., parts	Resin properties	
Trimethylolpropane	204	Acid number (solids)	7-8
Neopentyl glycol	317	Color, Gardner	1
Adipic acid	111	NVM, %	60
AMOCO IPA-95	505	Volatile	90/10:xylene/n-butoxy ethanol
	1137		
		Viscosity, Gardner-Holdt	Y-
Less water	-137		
Total	1000		

Resin processing

1. Charge all materials into kettle at room temperature. Fit with steam heated condenser. Heat slowly to 65-93°C to avoid charring any material.
2. At 65-93°C the charge should form a slurry. At this point begin agitation and inert gas sparge.
3. Heat to 177°C and hold until water of reaction slows.
4. Heat to 204°C and again hold until water of reaction slows.
5. Heat to 232°C and hold for V viscosity (60% NVM in 90/10:xylene/n-butoxy ethanol).
6. Reduce to 60% NVM in 90/10:xylene/n-butoxy ethanol.

TABLE 2-6. — Continued

Enamel formulation

Material	Parts by wt stain resistant	Formulation constants	
Titanium dioxide[a]	200	NVM, %	55
Oil-free alkyd	278	Pigment/binder ratio (solids)	0.9/1.0
Stain resistant type melamine resin[b]	101	Melamine/alkyd ratio (solids)	25/75
Butoxyethanol	15	Pigment dispersed on the 3-roll mill. Mill base at 200/90 parts by wt of pigment/alkyd (60% NVM)	
Xylene	174		
	768		

Physical properties (30 min at 149°C cure over CRS)

System	Film thickness, mils	Gloss		Impact (in.-lb)		Adhesion, cross-hatch	Bend, conical mandrel	Hardness		Abrasion Taber
		60°	Visual	Dir.	Rev.			Pencil	Sward	
Oil-free alkyd — stain resistant type	1.2	88	8	15	2	3	0	2H	62	100
Commercial control (oil-free alkyd)	1.2	88	8	15	2	1	0	2H	70	130
Commercial control (thermoset acrylic)	2.0	89	8	15	2	1	0	3H	60	80

Stain and solvent resistance (30 min at 149°C over CRS)

System	Film thickness, mils	Stain				Solvent	
		Mustard	Iodine	Ink	Lipstick	Xylene	Gasoline
Stain resistant type	1.5	10	9	7	10	10	10
Control (oil-free alkyd)	1.0	5	0	2	10	5	9
Control (thermoset acrylic)	2.0	10	5	7	10	10	10

Heat-aged properties (4 hr at 232°C over CRS)

System	Film thickness, mils	Gloss		Color		Dir.	Rev.	Adhesion, cross-hatch	Bend, conical mandrel	Pencil hardness
		60°	Loss	N	Visual					
Stain resistant type	1.6	60	(-18)	0.0284	9	25	2	8	20	6H
Control (oil-free alkyd)	1.3	76	(-12)	0.0086	10	10	2	0	0	H
Control (thermoset acrylic)	1.2	82	(-7)	0.0869	7	10	2	2	0	2H

TABLE 2.6 — Continued

Salt spray (250 hr over CRS)		
System	Rust creep from scribe, in.	
	Visual	Taped[c]
Oil-free alkyd — stain resistant type	2/16	5/16
Commercial control (oil-free alkyd)	10/16	13/16
Commercial control (thermoset acrylic)	10/16	14/16

[a]R-900 E. I. duPont de Nemours & Co.

[b]Cymel 248-8 American Cyanamid Co.

[c]Includes loss of adhesion away from scribe as well as rusting.

TABLE 2-7. Lower Cost Oil-Free Polyesters

Material Resin PE-100	Parts by weight
Neopentyl glycol	406
Trimethylolethane	80
IPA-95	542
Adipic acid	119
	1146
Less water	-146
	1000
Properties	
Acid number (solids)	3.3
Gardner-Holdt viscosity (60% NVM)	X-Y
Gardner color	1
Solvent	xylene
Excess hydroxyl (equivalent)	20%
Diol/triol equivalent ratio	4/1

Enamel formulation

Material	White
Resin PE-100 (60% NVM)	36.20
Rutile TiO_2	26.04
Melamine	13.20
Xylene	24.46
Catalyst	0.10
	100.00

TABLE 2.7. — Continued

Enamel	White (PE-100)	White commercial oil-free alkyd	White commercial acrylic
Mechanical properties			
Bake, min	30	30	30
Temperature, °C	177	177	177
Gloss, 60°	85	95	90
Pencil hardness	4H	2H	3H
Sward rocker hardness	48	50	54
Direct impact, in.-lb	50	80	10
Reverse impact, in.-lb	8	80	2
1/8 in. conical bend, % pass	60	100	40
Cross-cut adhesion, % pass	100	100	100
Mar resistance	exc.	exc.	exc.
Salt spray (ASTM B-117-64)			
Hours	672	672	408
Max. creep, in.	0	0	1/4
% score line rusted wider than 1/16 in.	0	0	40
Resistance properties			
Stain resistance (a qualitative rating with 10 = no effect)			
Iodine	5	6	4
Mustard	10	5	10
Ink	10	10	10

TABLE 2.7. — Continued

Enamel	White (PE-100)	White commercial oil-free alkyd	White commercial acrylic
Xylene resistance	soften	pass	pass
Detergent resistance (ASTM D 714-56)			
Hours	480	456	480
Blister rating (ASTM D 714-56)	8F	8D	9F
Overbake resistance			
16 hr at 177°C			
Gloss after overbake	72	67	82
Loss from initial gloss	13	28	8
Δb	-0.41	12.21	+0.28
Direct impact, in.-lb	10	10	10
Reverse impact, in.-lb	<2	<2	<2
1/8 in. conical bend, % pass	10	0	2
4 hr at 232°C			
Gloss after overbake	72	28	70
Loss from initial gloss	13	67	20
Δb	2.17	8.73	0.70
Direct impact, in.-lb	16	10	10
Reverse impact, in.-lb	<2	<2	<2
1/8 in. conical bend, % pass	30	10	20

TABLE 2-8. Detergent-Resistant High Gloss White Enamel Cyplex Resin 1473-5 (35)

Binder: CYPLEX Resin 1473-5/Epoxy resin—low molecular weight/CYMEL 301 hexamethoxymethylmelamine—75/10/15 solids

Catalyst: 1010 = 1% on total resin solids

Formula No. 595	lb
Unitane OR-580 titanium dioxide	295.1
Cyplex Resin 1473-5 (65%)	151.6
Xylol	12.2
Disperse on 3-roll mill and add:	
Cyplex Resin 1473-5 (65%)	228.0
Epoxy resin, low molecular weight,[a] 60% in MIBK	54.8
Cymel 301 hexamethoxymethylmelamine	49.3
General Electric Resin SR-82[b]	7.0
Dow Corning Paint Additive No. 1[c]	3.6
Xylol	12.4
Butanol	59.1
Methyl isobutyl ketone	101.8
Catalyst 1010	3.3
Cellosolve acetate	20.4
Isophorone	20.6
Pine oil	20.8
	1040.0
Viscosity (#4 Ford cup), sec	15
Wt/gal, lb	10.4
% Nonvolatiles, wt	60.8
vol	40.6
P/B in grind	300/100
in formula	90/100

TABLE 2-8. — Continued

Solvent mixture — hydrocarbon/alcohol/ketone/esters/pine oil/
= 40/15/35/5/5

Application: Spray on phosphate-treated metal

Baking schedule: 30 min at 149°C or equivalent

[a]Product of Shell Chemical Co., Epon 1001; Product of Ciba Corp., Araldite 6071 or equivalent.

[b]2% on total resin solids.

[c]1% on total resin solids.

This diol has complete substitution of the two carbon atoms beta to the primary hydroxyls, thereby blocking the most likely sites of decomposition. The oligomer found to be of most promise is based on 2.7 moles of Esterdiol-204, 2.0 moles of trimethylolpropane, and 4.0 moles of isophthalic acid. A typical white enamel formulation with properties is given in Table 2-10.

B. ACRYLIC

The impact of the chemical industry on the major appliance industry was examined in a special report by the editors of Chemical & Engineering News (41) in 1965. All appliances have reaped the advantages of the technological developments of the chemical industry resulting in new plastics, new insulating materials, and new finishes.

Extended use of the principle of copolymerization has resulted in great advances in the properties of appliance enamels. Monomers such as styrene, methyl methacrylate, etc., have been used to modify alkyds and epoxy ethers to produce very fast drying primers and enamels. Synthetic polymers may be made by copolymerization of the proper combination of unsaturated monomers, each one of which alone is capable of free radical polymerization. The properties of the resultant copolymer may usually be predicted from an average of the properties of the homopolymers, although there is generally a lowering of the softening point and an increase in solubility. The softening point, viscosity, tensile strength, and elastic modulus increase with an increase in degree of polymerization, while solubility decreases. Properties such as refractive index, hardness, electrical properties, color, and density are essentially independent of change in the degree of polymerization.

Functional groups such as carboxyl, hydroxyl, amido, etc., may be included in the backbone by proper choice of monomer and these may be crosslinked by an appropriate mechanism. Crosslinking tends to increase the effective molecular weight, the physical properties being dependent upon the nature of the main chain and the type and number of crosslinks. Polar groups occurring along the molecule of a high polymer exert strong attractive hydrogen bonding forces on molecules around them. This is reflected in higher tensile strengths, higher melting points, and a pronounced tendency

TABLE 2-9. Effect of Glycols on Performance of Oil-Free Polyesters

	Resin composition	
	20% excess hydroxyl	30% excess hydroxyl
Trimethylolpropane	3.23	5.70
Glycol	8.75	7.29
Isophthalic acid	6.00	6.00
Adipic acid	4.00	4.00

Glycol variations studied

	Molar %				
Neopentyl glycol (%)	100	75	50	25	0
Other glycol (%)	0	25	50	75	100

Prepared at both 20 and 30% excess hydroxyl levels.

Total resins prepared — 42.

Enamel composition

	Wt %	
Pigment (R-900 TiO_2 duPont)	40	
Vehicle (alkyd and amine)	60	
Alkyd		80
Amine		20

Total solids — 60%. Reduced with xylene/Ektasolve EB[a] to 17 sec (#4 Ford cup).

TABLE 2-9. — Continued

Average performance characteristics of neopentyl glycol resins

	Modified with		
	Cyclohexane-dimethanol	Ethylene glycol	Propylene glycol
Overbake gloss	Excellent	Excellent	Poor
Overbake color (Δb)	Average	Average	Poor
Hardness	5H	>6H	>6H
Solvent resistance	Excellent	Excellent	Excellent
Stain resistance	Excellent	Fair	Excellent
Detergent resistance	Excellent	Fair	Poor

[a]Eastman Chemical Products Inc.

toward crystallinity. A symmetrical structure without branches or side-chains permits closer molecular packing and greater intermolecular attraction. These attributes all affect important properties such as permeability, solubility, adhesion, flexibility, and chemical resistance. Acrylic ester resins are discussed by Allyn (42).

The technical, patent, and trade literature for those copolymers of direct application in appliance finish formulation is extensive.

C. CARBOXYL CROSSLINKED BY EPOXY

Segall and Cameron (43) report an acidic copolymer typically composed of 15-80 styrene/15-80 alkyl acrylate/5-10 acrylic acid. This is crosslinked with 15.55% epoxy and catalyzed with a quaternary ammonium hydroxide containing an alkyl group of 12-18 carbons, to a tough resistant film.

Devoe Raynolds Company (44) patented an acid containing copolymer such as one with a composition of 221 acrylic acid/196 propylene oxide/400 styrene/16 benzoyl peroxide/23 benzyl trimethyl-ammonium hydroxide. It may be crosslinked with isocyanates, epoxy, and aminoplast resins.

Aliphatic liquid epoxies are described by Applegath (45) as used to crosslink carboxyl functional acrylic copolymers. The films have good solvent and ultraviolet resistance.

TABLE 2-10. Flexible White Enamel

Grind	gal/100 gal	lb/100 gal	%, wt
MC-3314 Esterdiol 204 polyester[a]	54.87	496.00	42.51
"TiPure" R-900 TiO_2[c]	10.84	372.00	31.88
Cellosolve acetate	11.48	93.00	7.97
	Let-down		
Cellosolve acetate	11.48	93.00	7.97
"Cyzac" 1010[b]	0.17	1.24	0.11
"Cymel" 301[b]	11.16	111.60	9.56
	100.00	1166.85	100.00

Procedure

Charge ingredients into a ball mill or sand mill. Add let down materials in a mixing tank

Typical Properties — MC-3312

Pigment volume content	19.4
Coverage, 1 mil dry	896 ft^2/gal
Nonvolatile	71.03% by wt
Wt/gal	11.67 lb
Viscosity	148 sec (#4 Ford cup)
Gloss	over 94
Impact, direct, "Bonderite" 1000, unprimed	120 in.-lb
Impact, reverse, "Bonderite" 1000, unprimed	150 in.-lb
Bend conical mandrel, "Bonderite" 1000, unprimed	Excellent
Stain resistance	
Merthiolate	9+
Acetone	9
Ink	10 (no change)
Mustard	9+

TABLE 2-10. — Continued

MEK Test (50 rubs)	Excellent		
Taber abrasion CS-17 1000 g wt 1000 cycles	6 mg loss/100 cycles		
Bake schedule	30 min at 177°C or 5 min at 232°C		
Overbake	1 hr	246°C	No change
	16 hr	177°C	No change

[a] Eastman Chemical Products Inc.

[b] American Cyanamid

[c] DuPont

Berger, Jenson and Nicholson Ltd. (46) patented, a copolymer of 1080 styrene/360 ethyl acrylate/60 methacrylic acid, cured with epoxy and catalyzed with a quaternary imido-azolinium salt. Baked for 30 min at 177°C to high gloss, hard films, they are detergent, impact, and grease resistant.

Becalik and Gentles (47) describe a copolymer of methyl methacrylate, an unsaturated carboxylic acid, and a polyepoxy compound, plasticized with polypropylene glycol phthalate or succinate. The latter is claimed to provide improved flexibility, adhesion, and resistance to detergents. A terpolymer of styrene, ethyl acrylate, and maleic anhydride, crosslinked with epoxy and catalyzed with a quarternary ammonium salt was patented by Fang (48).

Vasta (49) has patented the use of octadecyltrimethyl ammonium tetraphenyl borate as a catalyst for curing an acid containing acrylic terpolymer with epoxy. The films are useful as appliance finishes.

Woodruff (50) cures an acrylic terpolymer with benzyldimethyl amine, in 20 min at 177°C to hard, glossy films. A typical composition is composed of 426 butyl methacrylate/708 vinyl toluene/77 methacrylic acid/29 allyl glycidyl ether.

Coatings resistant to household detergents and suitable for clear or pigmented appliance finishes are given by Rohm & Haas Co. (51). As an example, 45 styrene/45 methyl methacrylate/10 maleic anhydride, are copolymerized, crosslinked with epoxy, and catalyzed with benzyl dimethyl ammonium hydroxide. The baking cycle is 30 min at 150°C.

Gaske and Brown (52) patented a combination of 45 methyl methacrylate/45 ethyl acrylate/10 methacrylic acid, crosslinked with epoxide and catalyst.

The Dow Chemical Co. (53) describes a terpolymer of styrene/acrylic acid/octyl acrylate crosslinked with epoxy and catalyzed with dodecyltrimethyl ammonium chloride. The film cures in 30 min at 149°C to stain resistant, flexible, impact and detergent resistant coatings.

The DeSoto Co. (54) describes a copolymer of 20 2-ethylhexyl acrylate/ 42.5 methyl methacrylate/75 methacrylic acid/30 vinyl toluene, crosslinked with 21.27 parts of epoxy resin of epoxide equivalent weight 200, and 24.25 parts of butylated dimethylol urea. The bake is 20 min at 177°C to flexible, solvent, stain-resistant films which can be recoated.

Another variation of the carboxyl-containing copolymer is given by the American Marietta Co. (55). As an example, a copolymer of 5 acrylic acid/ 5 acrylonitrile/67.5 methyl methacrylate/22.5 styrene, is cured with 25% polyepoxide. Heppolette (56) claims that the amount of epoxide required to cure an acid-containing terpolymer may be reduced without impairing film hardness, by grafting on two or more per cent of a rigid polymer such as polystyrene.

A terpolymer of 1993 styrene/1328 ethyl acrylate/475 maleic anhydride is modified by Vasta (57) with epoxy resin, an epoxidized soya bean plasticizer, and catalyzed with a solution of octadecyl trimethyl ammonium acid phthalate.

Nix (58) patents a blend of two copolymers, 954 ethyl acrylate/636 methyl methacrylate/426 glycidyl methacrylate with 1080 ethyl acrylate/720 methyl methacrylate/216 methacrylic acid. Catalyzed with a quarternary mono-imidazoline, it is stable for 4 to 10 weeks of storage at 52°C. A copolymer of styrene, 2-hydroxymethyl 5-norbornene, modified with drying oil fatty acid and epoxy is claimed by Schefbauer (59). When baked for 30 min at 177°C, outstanding resistance to water, alkali, and dilute acids is achieved.

Lashua and Lee (60) patented a ternary interpolymer of styrene, an unsaturated monocarboxylic acid monomer, and an alkylene glycol ester of such an acid, crosslinked with epoxy. The cure is 30 min at 177°C to form coatings resistant to impact, staining, and detergent while retaining flexibility.

Fry (61) synthesizes detergent resistant compositions of acrylic acid containing polymers reacted with a polyglycidyl ether. The coatings are hard, tough, flexible, and light in color.

Hicks (62) patents a copolymer of 55 styrene/20 methyl acrylate/25 acrylic acid crosslinked by epoxy, where more than 20% acid monomer is used. When cured at 150°C for 30 min, the polymers give flexibility, hardness, and mar resistance together with good adhesion.

Carboxyl containing addition copolymers are cured with epoxy by Vasta (63) using a synergistic mixture of curing catalysts, the quaternary ammonium salt of Versatic 911 acid, and the quaternary ammonium salt of a dialkyl sulfosuccinate. The composition is package stable and cures in 30 min at 149°C. Excellent adhesion, gloss, hardness, and chemical resistance are claimed. Carboxyl containing acrylic copolymers are modified by Gaske and Brown (64) with epoxy and dicyandiamide catalyst. Cured for 20 min at 163°C, the films are high in gloss, hard, adherent, mar, impact, and chemical resistant.

Zimmerman and Lyons (65) prepare a copolymer of styrene and isooctyl acid maleate: 37.5 g of this are treated with 4.4 of butyl vinyl ether and modified with epoxy and trisbutoxymethyl melamine. Hard, resistant coatings are formed in a 30-min cure at 175°C.

Acrylic stoving finishes suitable for can linings or domestic appliances are based by the Chemische Werke Albert Co. (66) on acrylic copolymers with free carboxyl groups. These are partially esterified with a monoepoxy compound such as propylene oxide and cured on stoving with more epoxy and with an amino resin at 130-180°C.

D. CARBOXYL CROSSLINKED BY AMINOPLASTS

Allenby describes an acrylic copolymer of typical composition, 60 ethyl acrylate/35 styrene/5 methacrylamide, modified with Uformite F240 and acid catalyzed to a hard, resistant film (67).

Thermosetting copolymers suitable for appliance finishes are described by Gaylord (68). A copolymer based on 50 glycerol monoallyether/80 methyl methacrylate/120 butyl acrylate/10 methacrylic acid is blended with urea and melamine formaldehyde resins.

Rohm & Haas (69) has patented a copolymer of acrylic acid and acrylic ester modified with an aminoplast to form hard, solvent resistant coatings. A typical composition is composed of a copolymer of 35 methyl methacrylate, 55 ethyl acrylate, 10 methacrylic acid, modified with a butylated hydroxymethylolated benzoguanamine resin.

Enamels suitable for appliance use are made by DuPont (70) based on compositions of 53-85 styrene/5-35 ethyl acrylate/5-15 methacrylic acid crosslinked with urea formaldehyde resin.

The Berger, Jenson and Nicholson Co. (71) describes a terpolymer of methyl methacrylate, alkyl acrylate, and methacrylic acid useable for appliance finishes. It is cured with amine formaldehyde resin.

Petropoulos et al. (72) describe the synthesis of linear acrylic copolymers containing acrylic acid, methylolacrylamide, beta-hydroxyethyl methacrylate, or methacrylic acid. These monomers introduce pendant reactive groups onto the chain which may be crosslinked with amine formaldehyde resins.

The resulting properties may be correlated with chemical composition and structure.

Gaylord (73) describes the grafting of epoxy resin onto an acrylic copolymer containing the norbornene nucleus. When blended with butylated melamine, hard, flexible films with good gloss and resistance to solvents and detergents are formed. To a copolymer of 90 styrene/14 methyl methacrylate, 54 2-hydroxymethyl-5-norbornene/22 butyl acrylate/13 methyl methacrylate/13 methacrylic acid, are grafted. When crosslinked with a butylated triazine formaldehyde resin, Gaylord (74) claims that hard, glossy, flexible films with excellent resistance to soaps and detergent are formed.

Vasta (75) takes copolymers of acrylic and methacrylic acid, treats them with 1,2 butylene oxide, and mixes with aminoplast resins to heat cure as a coating composition.

Tough, flexible, detergent resistant coatings are made by Gaylord (76) by blending a terpolymer of ethyl hexyl acrylate/glycidyl methacrylate/ hydroxymethyl norbornene with a butylated triazine formaldehyde resin.

Gaylord (77) patents blends of acid containing acrylic terpolymer with amine aldehyde resin and oil-free polyester. Good cure is obtained in 30 min at 177°C to glossy, hard films with good color and resistance to soaps and solvents.

Rohm & Haas (78) copolymerizes 36.1 methyl methacrylate/36.1 styrene/ 14.8 ethyl acrylate/11 betahydroxypropyl methacrylate/2 itaconic acid and heats together with a melamine formaldehyde resin for complete compatibility. Levantin (79) patents this composition, curing for 30 min at 149°C to form films with excellent adhesion and detergent resistance.

Costanza and Waters (80) copolymerize styrene, butyl acrylate, 2-ethylhexyl acrylate, propylene glycol monoacrylate, and acrylic acid. The copolymers are cured with amine formaldehyde resins and acid catalysts. Styrene containing acrylics have better solvent and detergent resistance while the all acrylics have better exterior durability.

O'Donnell and Suen (81) describe a copolymer of 30 acrylic acid/85 butyl acrylate/85 isobutylene, dissolved in ammonia and crosslinked with hexamethoxylmethyl melamine.

Fry et al. (82) claim an interpolymer of styrene, ethyl acrylate, methyl methacrylate, acrylonitrile, and acrylic acid, crosslinked with hexamethylol melamine hexamethyl ether.

Hard, flexible, and yet impact resistant enamels are made by the DeSoto Company (83) by copolymerizing acrylamide with 2-ethyl hexyl acrylate, ethyl acrylate, styrene, methacrylic acid, and an hydroxyl containing monomer. Examples of the latter are allyl glycerol ether, monoallyl ether of trimethylol propane, 2-hydroxyethyl methacrylate, or 2-hydroxymethyl-5-

norbornene. The copolymers are crosslinked with benzoguanamine formaldehyde resin and oil-modified alkyds. Butoxymethylolation improves impact resistance.

The PPG Industries (84) patent a ternary blend of hydroxyl and carboxyl containing acrylic addition copolymers with amine aldehyde resin and an alkyd.

Fry (85) crosslinks an acrylic copolymer containing both hydroxyl and carboxyl with hexakis, using p-toluene sulfonic acid as catalyst and castor oil as a plasticizer.

Vasta (86) formulates thermosetting copolymers with pendant substituted amide radicals. For example, 25 styrene/28 ethyl acrylate/47 acrylamide are copolymerized, esterified with Cardura E, and crosslinked with melamine formaldehyde resin for 30 min at 149°C. The finishes are hard, glossy, and heat resistant. A copolymer of 110 styrene/40 butyl acrylate/50 methacrylic acid/180 Cardura E is made by Cox and Swann (87). This may be cured with aminoplasts, epoxy, or isocyanates. Coatings resistant to alkali, detergents, stains, grease, vapors, and heat are made by Vasta (88). A typical formulation is composed of 7.46 phthalic anhydride/39.3 methyl methacrylate/10.65 methacrylic acid/42.58 Cardura E. The copolymer is cured for 30 min at 149°C with melamine formaldehyde.

Miranda (89) synthesizes acid containing copolymers modified with amino-hydroxy compounds. 87.5 styrene/137.5 2-ethylhexyl acrylate/25 acrylic acid are compolymerized. Tris (hydroxymethyl) aminomethane is added. The resultant copolymer may be cured in 30 min at 149°C with melamine resins to give hardness greater than 8H, with adhesion to metal and resistance to alkali. 2 amino-2 methyl propanol and 2 amino-2 methyl-1, 3, propanediol may also be used.

In a British patent (90), Union Carbide describes the cure of copolymers of acrylic acid with ethyl acrylate using hexamethylol melamine hexamethyl ether. Cured for 30 min at 149°C, the films are hard and resistant to impact, ammonia, and gasoline.

Gilkes (91) coreacts 148 parts phthalic anhydride with 120 of glycerol at 240°C to an acid value of 6. Maleic anhydride (330 part) and 74 part Cardura E† are added and the mixture further reacted at 150°C to an acid number of 8. A copolymer of 238 styrene/50 butyl acrylate/12 methacrylic acid is made by free radical polymerization. The polyester and acrylic polymer are blended in a 3:1 ratio and crosslinked with melamine formaldehyde to hard, flexible, adherent films which cure in 30 min at 130°C. Cardura E is

---

†An alkyd resin based on glycidyl esters of C_{12}-C_{14} branched mono carboxylic acids.

heated with acrylic acid by Hall and Price (92) to give an hydroxylated ester which may then be copolymerized with butyl acrylate. The copolymer is crosslinked with melamine formaldehyde and cured for 30 min at 120°C to hard, glossy, chemical resistant films.

Nyquist (93)describes the synthesis of thermosetting acrylics based on 2-hydroxyethyl acrylate and hydroxypropyl acrylate. Acrylic copolymers containing methylol ether, hydroxyl or carboxyl groups, are crosslinked with aminoplast resins by Rohm & Haas (94). The enamels are useful for washing machines, since they resist boiling water and detergents.

E. METHYLOLATED ACRYLAMIDE

An acrylic copolymer with a typical formulation of 40 styrene/45 ethyl acrylate/15 acrylamide is given by Vogel and Bittle (95). This is methylolated with two moles of formaldehyde per amino group. When heat cured, this copolymer forms hard, flexible, heat stable, and corrosion resistant films. The copolymers may be modified with vinyl and epoxy. When formulated into appliance enamels, excellent resistant to salt, fog, humidity, and detergent, and food staining are claimed. Christenson and Vogel (96) describe an interpolymer of maleic linseed esters with a copolymer of acrylamide and styrene.

Christenson and Hart (97) describe copolymers of acrylamide. These are treated with formaldehyde in butanol-containing acid catalyst. The films crosslink to hard, tough, chemically resistant films.

Vogel and Bittle (98) report baked finishes based on acrylamide interpolymers, some of which may be modified with other types of resins and a wide variety of properties attained including appliance finish end uses. An acrylamide copolymer of 10 acrylamide/65 ethyl acrylate/25 styrene is methylolated with formaldehyde by Lynch (99). The resin is refluxed for 1 hr. at 130° C with an alkyd resin to improve gloss.

Gaylord (100) has also patented a copolymer of 180 acrylamide/900 styrene/84 ethyl acrylate/36 methacrylic acid/44 Epon 1001/38 of 40% formaldehyde. A second copolymer of butyl acrylate/methyl methacrylate/ norbornene is blended with the first and cured for 30 min at 177°C to hard glossy, flexible films.

In a modification of prior patents, Gaylord, (101) teaches the formulation of coatings with good gloss, color, resistance to salt spray and detergents made by blending an alkoxyalkylated acrylamide-epoxy condensation product, with polyester and acrylic copolymer.

Vogel and Bittle (102) make a copolymer of 15 acrylamide/40 styrene/45 ethyl acrylate in 100% butanol using cumene hydroperoxide and p-tertiary dodecyl mercaptan as catalyst and chain transfer agent. This is then methylolated with 10.5 butyl formcel and treated with 0.13 maleic anhydride. Ninety three parts of this copolymer are modified with seven of tetrabutoxy titanate and cured for 20 min at 82°C to form finishes suitable for appliances such as washing machines and ranges. A methylolated acrylamide copolymer is mixed by Shell (103) with an alkyd based on the glycidyl esters of C12-14 branched monocarboxylic acids (Cardura E), epoxy resin, and melamine formaldehyde to a hard, resistant coating with high gloss.

Sekmakas (104) patented a blend of ethylenically unsaturated polyester with methylolated acrylamide copolymers.

Christenson and Shahade (105) copolymerize 39 styrene/44 ethyl acrylate/ 15 acrylamide/2 acrylic acid, methylolate with formaldehyde and blend with Shell's X-450 styrene-allyl alcohol copolymer. The coating has good recoat adhesion, hardness, resistance to kitchen stains, detergent, solvents, and marring.

Sekmakas (108) also teaches the copolymerization of acrylamide or other amide of an unsaturated carboxylic acid, with unsaturated epoxy resin, styrene, and ethyl acrylate. It is methylolated with formaldehyde, catalyzed with phosphoric acid, and cured in 20 min at 163° C. The films are hard, adherent to metals, flexible, glossy, and resistant. In U.S. 3,230,275, such methylolated acrylamide copolymers are modified with alkyd and triazine resin.

Flegenheimer (109) makes an interpolymer of dehydrogenated castor oil, styrene, and methylolated acrylamide. Applied on metal and cured 20 min at 81°C, it develops resistance to Tide, solutions of detergents, and impact resistance.

Michelotti et al. (110) make flexible thermosetting coatings by blending methylolated acrylamide interpolymers with epoxide free, hydroxyl containing esters of epoxide resins. When cured at 177°C excellent properties are claimed.

Haskell et al. (111) patent coating compositions based on blends of a copolymer containing an alkylolated acrylamide and at least one other ethylenically unsaturated monomer, with a copolymer of an ethylenically unsaturated carboxylic acid and at least one other ethylenically unsaturated monomer.

The Nobel-Bozel Co. (112) patents a copolymer of 3.4 acrylamidoglycolic acid/4 acrylic acid/88 ethyl acrylate/15 acrylamide/78 styrene, which is methylolated with formaldehyde and modified with Epon 1001. When cured at 130°C, hard, flexible, shock resistant enamels are made.

In 1966, Gaylord (113) patents blends of copolymers of 2-hydroxymethyl 5-norbornene/butyl acrylate/methacrylic acid with copolymers of alkylolated acrylamide. When pigmented and cured on steel for 30 min at 177°C, the thermosetting compositions are suitable as finishes for household appliances.

In a patent disclosure (114), Rutherford teaches an interpolymer of aldehyde modified amides and the polyallyl ethers of polyhydric alcohols such as the diallyl ether of trimethylol propane, with a methylolated acrylamide copolymer. Excellent flexibility, impact, and adhesion are claimed. The use of mercaptans as chain terminators reduces mechanical properties.

Beck, Koller and Co. (115) patents enamels based on copolymers of styrene, ethyl acrylate, butyl acrylate, methacrylic acid, and acrylamide. The copolymer is methylolated with formaldehyde and treated with propylene oxide. When baked for 30 min at 150°C, the films are adherent and exhibit high resistance to solvents.

Sekmakas (116) teaches coating compositions based on coreacted blends of acrylic copolymers. For example, 480 vinyl toluene/360 ethyl acrylate/100 propylene oxide is coreacted with a polyester based on dehydrated castor oil fatty acids, crotonic, isophthalic, phthalic, and azelaic acids, with glycerol. In another example, a copolymer containing methylolated acrylamide is copolymerized in the presence of a polyester. Typical cures are 15 min at 177°C. Excellent adhesion, pencil hardnesses of 2-3 H, intercoat adhesion, and resistance to solvent and impact result.

Sekmakas and Stanel (117) copolymerize unsaturated polyesters based on fumaric acid, with acrylamide. This is then methylolated. The copolymer is cured with melamine formaldehyde resin for 20 min at 163°C to hard, impact resistant films.

Rauch-Puntigam and Tulasz (118) copolymerize a maleic-ethylene glycol ester with styrene, butyl acrylate, and butyl acrylamide. The copolymer is methylolated and butylated, and it cures in 30 min at 150°C to hard, elastic, glossy, chemical-resistant coatings.

An interesting thermosetting acrylic is reported by Rauch-Puntigam (119). Hexamethoxymethyl melamine (195 parts) is heated with 35 parts acrylic acid in the presence of 0.035 parts hydroquinine, 11 parts montmorillonite, and activated alumina catalyst, in 240 parts of toluene, to an acid value of 7. Methanol is azeotroped off and this is then copolymerized with butyl acrylate and styrene. Hard, lustrous, resistant films are obtained using curing schedules of 30 min at 150°C.

Sekmakas (120) claims that superiority of performance over commercial methylolated acrylamide copolymers is achieved by reacting an unsaturated hydroxy functional polyester, with an hydroxy and carboxyl containing acrylic copolymer. This is crosslinked with hexamethoxymethyl melamine.

F. EPOXY MODIFIED ACRYLICS

Glycidyl methacrylate is copolymerized with styrene or ethyl acrylate by Simma (121). These copolymers may be cured with epoxy curing agents such as amines and diacids.

Abbotson et al. (122) report the copolymerization of methyl methacrylate with glycidyl methacrylate. The copolymer is modified with plasticizer and cured with typical epoxy catalysts to form resistant surface coatings.

Two British patents by Nelan (123) describe an interreactive system of two copolymers containing carboxyl and oxirane groups, respectively. An example is given of a 20 styrene/9.6 methyl methacrylate/14.4 ethyl acrylate/3.8 acrylic acid copolymer blended with a similar one containing glycidyl methacrylate. The cure is for 30 min at 150°C and the enamels are recommended for appliances. Acrylamide may also be included in the polymer.

Ravve and Khamis (124) copolymerize 65 styrene/20 ethyl hexyl acrylate/11 glycidyl methacrylate. This resin may be crosslinked with citric acid. Flexible, smooth, adherent, chemically inert films are made by curing for 12 min at 1 9°C.

G. AMIDO MODIFIED ACRYLICS

Melamed (125) describes an acrylamide copolymer of 39 methyl methacrylate/1'acrylamide with 1% maleic acid and 10% epoxide.

Souder and Melamed (126) copolymerize methyl methacrylate with acrylamide and crosslink with N,N' bis (methoxy), N,N' ethylene urea, using a maleic acid salt as catalyst.

Wilhelm et al. (127) copolymerize N-2-oxyhexyl methacrylamide, cthyl hexyl acrylate, acrylic acid, 2 butene-1, diol, styrene, and 4-hydroxybutyl acrylate. The cure is 30 min at 149°C to glossy enamels, resistant to shock and dilute alkali.

Christenson et al. (128) patent copolymers of acrylamide, styrene, ethyl acrylate, methacrylic acid, maleic anhydride, and the styrene-allyl alcohol-maleic anhydride adduct.

A thermosetting acrylic is made by Ashjian (129) by reacting an epoxy resin with acrylamide and then copolymerizing with a carboxyl containing acrylic copolymer. Urea, melamine formaldehyde, or polyamide resins are used to cure to hard, chemical resistant films.

Hard glossy films are obtained by Daniel (130) using a water soluble acrylic, the ammonium salt of a copolymer of acrylic acid, and acrylic ester. This is cured by modification with a water solution of the polymethyl ether of a polyhydroxymethyl melamine.

Glossy coatings are made from aqueous acrylic copolymers by Canadian Industries Ltd. (131). Ethyl acrylate, acrylamide, and itaconic acid are copolymerized in isopropanol, the water and alcohol azeotroped off, and the resin solubilized by adding 27% ammonium hydroxide. This resin is cured with epoxy and melamine formaldehyde.

Brown et al. (132) patent an adherent and rust-resistant aqueous film based on an emulsion of methyl methacrylate/ethyl acrylate/methacrylamide cross-linked with a water-soluble methylolated melamine formaldehyde resin.

Tough, durable, flexible, salt spray resistant finishes based on thermosetting, water-based acrylic resins of high molecular weight, are described by Allyn (133).

A copolymer of ethyl acrylate and acrylic acid in ethanol, is hydroxyethylated by Canadian Industries Ltd. (134) with ethylene oxide, solubilized with ammonia water and cured in 30 min at 150°C to hard, glossy adherent, acetone and water resistant coatings.

Aronoff et al. (135) patents a latex of 42.5 ethyl acrylate/10 2-hydroxymethyl 5-norbornene/10 glycidyl methacrylate/2.5 methacrylic acid refluxed with sodium lauryl sulfate in 300 water. A mixture of 127.5 ethyl acrylate and 7.5/methacrylic acid is added dropwise. The latex achieves 96% conversion at 38.5% nonvolatile and may be blended with aminoplasts. Baked films are hard, glossy, water and solvent resistant, and useful for metal appliances.

A copolymer of 85 butyl acrylate/15 acrylic acid or 39.2 styrene/54.1 ethyl methacrylate/3.1 acrylic acid/3.6 beta-hydroxyethyl methacrylate, is dissolved by Koral et al. (136) in ammoniated water, blended with hexakis, and cured for 30 min at 149 C, forming films with excellent gloss, hardness, and resistance properties.

H. ISOCYANATE MODIFIED

BASF (137) patent copolymers of acrylic esters and hydroxyl containing monomers mixed with isocyanates to give tough, alkali resistant coatings. As an example, 20 isobutyl acrylate/5 acrylonitrile/6-hydroxy 5-methyl bicyclohept-2-ene/1 trimethylol propane/3 TDI is cited. Pot-life is about 24 hr.

Finishes resistant to detergent, water, sodium hydroxide, and acetic acid are patented by Farbenfabriken Bayer (138). An acrylic copolymer is based on ethyl acrylate, 2-hydroxypropylmethacrylate, beta-isocyanatoethyl methacrylate, acetoxime, and styrene. Films are cured in 30 min at 150°C. N-Butoxymethacrylamide may be used in place of beta-isocyanatoethyl methacrylate.

I. HYDROXYL-CARBOXYL CONTAINING

Beardon (139) copolymerizes 23 acrylic acid/26 hydroxyethyl acrylate/17 ethyl acrylate. This type of hydroxyl and carboxyl containing copolymer is said to cure in 30 min at 177°C to films with excellent flexibility, impact and solvent resistance. One to thirty per cent of castor oil is used by Union Carbide (140) to plasticize hydroxyl and carboxyl containing acrylic copolymers.

J. ADDITIVES

Kapalko and Bittle (141) claim that an enamel based on an acrylamide copolymer may be corrected of bubbles and solvent popping by adding about 10% polyethylene on the acrylic resin, in the form of a 10% gel of polyethylene in xylene.

2-10. REVIEW ARTICLES

Kelly et al. (142) review thermosetting vinyl and acrylic copolymers. Carboxyl containing acrylics react with melamine formaldehyde resins when heated for 30 min at 120°C. Epoxy or amido copolymers require the addition of acid catalysts before crosslinking with melamine. Hydroxyl substituted copolymers may be made of a glycidyl methacrylate copolymer treated with diethyl amine or an acrylic acid copolymer by refluxing with butylene oxide. They cure with melamine formaldehyde resins.

Brown and Miranda (143) review the physical and chemical properties of acrylic polymers prepared by free radical mechanisms. Thermosetting polymers containing acid functionality are crosslinked with epoxy. Those containing hydroxy ethyl methacrylate are crosslinked with melamine. Ethyl and 2-ethyl hexyl acrylates increase elongation and decrease tensile strength. Epoxy/acid terpolymer blends in hydroxyl containing solvents, lose oxirane oxygen with time.

The properties and design of acrylic resins are discussed by Mercurio (144). The properties of a carboxyl containing acrylic cured with epoxy, are compared with a 70 nondrying alkyd/30 triazine formaldehyde resin blend. The former is harder, more flexible, and more resistant to stains and detergent. Fischer (144 a) reviews the composition, properties and uses of thermosetting acrylic resins in coatings. The chemistry of hexakis (methoxymethyl) melamine is discussed by Koral and Petropoulas (145). Carboxyl containing acrylic copolymers are cured with hexakis without acid catalyst but hydroxyl containing polymers require external catalysts such as p-toluene sulfonic acid. Hexakis confers greatly improved flexibility at equivalent hardness as compared with conventional polymeric butylated melamine formaldehyde resins

This characteristic is ascribed to the low tendency of hexakis to condense with itself.

A review of the physical nature of thermosetting acrylic resins, their mechanisms of cure, and the degree of crosslinking achieved, is given by King (146). In an article entitled Acrylic Resins for Industrial Chemical Coatings, Wampner (147) compares various alkyd amino, epoxy amino, urethane, and unsaturated polyester enamels for typical appliance finish properties. Fifteen commercial acrylics were also included and the data are presented to indicate cost-property relationships. A review of the process of free radical addition polymerization of acrylic monomers is given by Klein (148). The mechanisms of chain initiation, propogation, and termination are described. Monomer reactivity ratios, the effect of alkyl chain length on glass transition temperatures, the design of thermoplastic and crosslinking polymers, and their solubility behavior are discussed.

2-11. PROPRIETARY SYSTEMS

Proprietary acrylic resins 100 and Acrylic 120 were evaluated by Union Carbide (149) using General Electric Co.'s specifications. Acrylic 120 modified with Uformite MM47 provided the best overall performance in this work. Union Carbide Formula Suggestion MP-3230 is typical. See Table 2-11.

Rohm & Haas Co. (150) present data and properties of their proprietary thermosetting acrylic resins—Acryloids. These are subdivided into two groups according to the mechanism of their cure. The hydroxyl functional types are cured with aminoplast resins while the carboxyl types are crosslinked with epoxy resins. Acryloid AT-50 and 51 are characterized by very hard, glossy, stain- and detergent-resistant films. Acryloid AT-56 requires the addition of an amino resin and gives intermediate hardness with excellent adhesion and flexibility. A typical suggestion is given in Table 2-12. Properties may be found in Table 2-13. Acryloid AT 51 is similar to AT 50 in composition and performance but is lower in cost.

The Acryloid AT-70 series is composed of thermosetting acrylic resins with carboxyl functionality which require an epoxy resin for crosslinking. These cure to tough coatings having excellent flexibility and resistance properties. A suggested formulation for an appliance white baking enamel is shown in Table 2-14.

Experimental resin QR-463A (151) is a carboxyl functional thermosetting acrylic resin designed for appliance finishes. A typical formulation and film properties are shown in Tables 2-16 and 2-17.

TABLE 2-11. White Appliance Enamel

	gal/100 gal	lb/100 gal	%, wt
TiO_2, OR600	7.43	254.25	24.47
Nuosperse 657	0.11	0.89	0.08
EKS-2002	9.86	89.73	8.63
Uformite MM47	14.07	119.64	11.51
MIBK	11.41	83.75	8.06
Acrylic 120	13.38	119.64	11.51
	Let-down		
Acrylic 120	31.25	279.38	26.89
Cyzac 1010	0.24	1.79	0.17
MIBK	12.22	89.73	8.63
	100.00	1038.85	100.00

Procedure

Grind the raw materials in a pebble mill 18 hr. Add ingredients in let-down with agitation.

OR600 American Cyanamid

Neosperse 657 Nuodex Division

EKS-2002 Union Carbide

Uformite MM47 Rohm & Haas

Cyzac 1010 American Cyanamid

Acrylic 120 Union Carbide

Properties

Viscosity #4 Ford 85 sec

Ohms resistivity Ransburg Tester #234 180K

Thinned to 16 sec butanol

Ohms resistivity 40K

Gloss 60° 90

60° 70

TABLE 2-11. — Continued

Applied over Bonderite 1000 and epoxy primer	
Impact direct	110
Reverse	20
Salt spray 1000 hr, excellent	
Pigment volume content, 16.45	
Mil feet 724 ft^2/gal at 1 mil	
N.V. 61.0%	
Solids vol 45.18%	

2-12. MISCELLANEOUS RESINS

Bakelite Vinyl solution resins VERR and VMCA (152) are epoxy and carboxy functional, low molecular weight, reactive solution resins. Coatings based on these show an unusual combination of properties such as formability and chemical resistance. Although the spray solids at the gun are about one-third higher than Vinylite VYHD and VMCC solutions, the primary use of these coatings is for coil coating stock to be postformed into appliance parts (see Table 2-18).

Dietrich et al. (153) describes the properties of polyvinyl fluoride and formulations based on polyvinyl fluoride. The principles of PVF dispersions are discussed and their use in various industries including the appliance industry is reported.

2-13. SILICONE RESINS

Dow Corning (154) recommends the use of silicone modified oil-free polyester as an appliance finish where exposure to high temperature and a chemical environment are factors to be considered (see Table 2-19).

2-14. TESTING OF APPLIANCE FINISHES

Zahn (155) discusses in some detail the modern way of evaluating finishes. Although written 15 years ago when the concept of weighted numerical scoring

TABLE 2-12. Acryloid AT-50 White Baking Enamel Modified with Epoxy Resin

Roller mill grind	lb	gal
Titanium dioxide	259.5	7.45
Acryloid AT-50 (50% solids)[2] (add toners as desired)	259.5	31.10
Mix with		
Acryloid AT-50(1) (50% solids)	279.2	34.30
Epoxy resin (50% in Cellosolve acetate)[b]	95.5	10.65
Xylene	81.5	11.25
Solvesso 150	21.5	2.90
Cellosolve acetate	23.2	2.86
Raybo 3 (antisilk agent)[c]	0.5	0.07
	1020.4	100.59

Physical constants of enamel

Wt/gal		10.2 lb
Total solids		56.6%
Pigment		45.0%
Vehicle		55.0%
Acryloid AT-50	85.0%	
Epoxy resin	15.0%	

Note that the reducing solvents used in Formulations 183 and 184 differ from the solvent blend in which Acryloid AT-50 is supplied.

[a] Rohm & Haas.

[b] Epon 1001 or other compatible epoxy resins.

[c] Raybo Chemical Company, Huntington, West Virginia.

TABLE 2-13. Properties of White Enamels

Property	Formulation 183 (epoxy modified)	Formulation 184 (unmodified)	Uformite HX-61 baking enamel
Initial solids, %	56.6	55.3	62.8
Initial viscosity (Ford #4 cup)	60 in.	60 in.	60 in.
Spray solids, %	48.4	47.0	54.0
Spray viscosity	22 in.	22 in.	21 in.
Bake schedule	149°C-30 min	149°C-30 min	149°C-30 min
Hardness (high values better)			
Tukon (Knoop)	27.2	26.7	12.1
Pencil	4H	4H	HB
Adhesion (high values better)			
Microknife "H"—Bonderite steel	30.6	26.7	23.8
Knife adhesion—Bonderite steel	Very good	Good	Very good
60° Gloss	93	88	92
Color—(low values better)			
Original (toned)	8.1	7.2	8.1
Overbake 177°C—16 hr	16.5	7.2	11.5
Print resistance			
2 psi—1/2 hr at 82°C	No print	No print	Light print
Cold-rolled steel			
Flexibility—1/2 in. mandrel			
Bonderite steel	Very fine cracking	Very fine cracking	Very fine cracking

TABLE 2-13.—Continued

Stain-spot tests			
Mustard — 30 min	No stain	No stain	Light stain
Ink — 30 min	No stain	No stain	Light stain
Iodine — 5 min	No stain	No stain	Light stain
Lard oil/oleic acid overnight	No stain	No stain	Light stain
Resistance to Tide			
20 hr — 1% Tide at 74°C			
ASTM Blister Ratings[b]			
Unprimed Bonderite	Excellent few — 6	Medium dense — 4	Poor medium dense — 1
Primed Bonderite steel[c]	Very few — 9	Very few — 8	Few — 4
Cellosolve acetate resistance — pencil hardness			
Before immersion	4H	4H	HB
After 60 min immersion	H	H	Partially dissolved

[a]Vehicle was 30:70 Uformite MX-61: Duraplex ND-77B with equal parts titanium dioxide.

[b]ASTM blister ratings. A rating of 10 means no blistering, a rating of 0 means very large blisters, with intermediate ratings judged by ASTM photo standards.

[c]Primer was a commercial gray washing machine primer of the epoxy ester type. The same primer was used for the primed panels in all of the tables in these notes.

of appliance finishes was new, the basic tenets still apply. Each property is scored numerically, multiplied by a weighting factor dependent upon the specific appliance involved, and a total grade compiled. Minimum values are required in each paint property to prevent excessively high scores in any one test from unbalancing the total of properties desired. This approach is now widely used outside the appliance industry as well as wherever paint specifications are utilized. A typical test specification follows in Table 2-20.

TABLE 2-14. Acryloid AT-70 White Baking Enamel

Roller mill grind	lb	gal
Rutile titanium dioxide (Titanox RA or Ti-Pure R-900)	274.0	7.81
Acryloid AT-70 (50% solids) (add toners as desired)	182.6	22.00
Mix with		
Acryloid AT-70 (50% solids)	218.4	26.23
Epon 1001 (50% solids in Cellosolve acetate)	267.0	29.70
Xylene	77.8	10.96
Cellosolve acetate	26.3	3.29
Raybo 3 (antisilk agent)[a]	0.6	0.08
	1046.7	100.07

Physical constants of enamel

Wt/gal		10.5 lb
Total solids		58.0%
Viscosity (#4 Ford cup)		60 in.
Spray solids		48.7%
Spray viscosity #4 Ford cup		21 in.
Pigment		45%
Vehicle		55%
Acryloid AT-70	60%	
Epon 1001[b]	40%	

Application properties

The application properties of Acryloid AT-70 were tested in our laboratory using the white baking enamel Formulation 272. The enamel was reduced with a mixture of xylene and Cellosolve acetate in a 3/1 ratio to spray viscosity (19 in. to 21 in. in a #4 Ford cup). The recommended minimum baking schedule of 30 min at 177°C was used.

[a]Raybo Chemical Co., Huntington, West Virginia. This agent is recommended to give better surface appearance and smoothness.

[b]Shell Chemical Co.

TABLE 2-15. Film Properties of White Enamel Formulation Made with Acryloid AT-70 (Baking Schedule — 30 min at 177°C)

Tukon hardness	16.2
Pencil hardness	2H
Solvent resistance, glass	
Cellosolve acetate, 15 min	6B
Cellosolve acetate, 60 min	6B
Xylol, 15 min	5B
Stain resistance, CRS	
Mustard, 30 min	No stain
Ink, 30 min	Trace
Print resistance, CRS	
80°C, 30 min, 2 psi	Light print
Tide resistance, Bonderite 1000[a]	
1% Tide, 74°C, 200 hr	Few — 6
Optical properties, CRS (2 coats)	
Original 60° gloss	95.5
Gloss after 16 hr at 177°C	95.4
Original color	9.2
Color after 16 hr at 177°C	13.0
Microknife adhesion, CRS "H" value	22.8
Mandrel flexibility[b]; 1/2 in., 1/4 in., 1/8 in.	
Cold rolled steel	0-0-0
Bonderite 1000	0-0-0
Reverse impact, in.-lb	
Cold-rolled steel	22
Bonderite 1000	15

TABLE 2-15. — Continued

Direct impact, in.-lb	
Cold-rolled steel	35+
Bonderite 1000	50+

[a]ASTM Blister Rating. A rating of 10 means no blistering, a rating of 0 means very large blisters, with intermediate ratings judged by ASTM photo standards.

[b]0 = no cracks; 9 = delamination.

The results of testing of a current high quality appliance white baking enamel are listed in Table 2-21. This type of enamel would be suitable for use on home laundry, dishwashers, refrigerators, and freezers.

2-15. EMPHASIS ON COST REDUCTION

The most prominent recent trend has not been in improvement of quality, but in a search for methods which reduce cost. These include:

1. Complete elimination of manual spray.
2. Reduction of overspray loss by means of hot spray.
3. Reclaim of sludge enamel.
4. Catalytic fume combustion of waste solvent and oven fumes.
5. Roller coating in the flat and post forming into ware.
6. Flow coat application of enamels.
7. Electrodeposition of primers.

2-16. WATER SYSTEMS

From the standpoint of safety, a critical, industry-wide reevaluation of water systems for industrial use is under way. Several serious industrial accidents occurring in the course of use of solvent based coatings and the

TABLE 2-16. White Baking Enamel Formulation

	lb	gal
Rutile titanium dioxide (Titanox RA or Ti-Pure R-900)	220.0	6.3
Experimental resin (QR-463A (50% solids)[a] (add toners as desired)	220.0	26.1
Mix with		
Experimental resin QR-463A (50% solids)	184.0	21.8
Epon 828[b]	68.0	7.0
Cellosolve acetate	72.1	8.9
Xylene	215.0	29.7
Raybo 3 (antisilk agent)[c]	0.5	0.07

Physical constants of white baking enamel

Total solids		52.0%
Viscosity (#4 Ford cup)		61 sec
Spray solids		43.5%
Spray viscosity (#4 Ford cup)		20 sec
Pigment		45%
Vehicle		55%
Experimental Resin QR-463A	75%	
Epon 828	25%	

Application

Reducc to the spray viscosity indicated with xylol/Cellosolve acetate: 3/1 and bake at 177° for 30 min or 191°C for 20 min.

Stability

Enamels based on Experimental Resin QR-463A made according to the White Baking Enamel Formulation show excellent viscosity and stability properties after heat aging for 160 hr at 60°C. Room temperature enamel stability results have also demonstrated excellent viscosity and stability properties.

[a]Rohm & Haas Co. [b]Shell Chemical Company.

[c]Raybo Chemical Company, Huntingdon, West Virginia. This agent is recommended to give better surface appearance and smoothness.

TABLE 2-17. Properties of White Enamel in Table 2-16.

Baking Schedule — 30 min at 177°C	
Tukon hardness	22.3
Adhesion — Knife	
Cold-rolled steel	Excellent
Bonderite 1000	Excellent
Gloss 60°	97
Gloss retention 177°C — 16 hr	97
Color (two coats)	6.9
177°C — 16 hr	8.2
Reverse impact	
Cold-rolled steel	Less than 2
Bonderite 1000	Less than 2
Flexibility (1/8 in.)	
Cold-rolled steel	No cracking
Bonderite 1000	Very fine cracking
Print resistance	
82°C — 30 min — 2 psi	None
Solvent resistance	
Initial pencil hardness	4H
Pencil hardness after 15 min in xylol	3H
Tide resistance (74°C 1%)	
Blistering at 500 hr on Bonderite 1000	Few — medium 4[a]
Syndet resistance	
Blistering at 500 hr on Bonderite 1000	Few — medium 6[a]

[a]ASTM Blister Ratings. 10 = no blistering. 0 = very large blisters, with intermediate ranges judged by ASTM standards.

TABLE 2-18. White Solution Vinyl Metal Coating

	lb/100 gal	gal/100 gal	% by wt
Formula			
Bakelite Vinyl Resin VERR[a]	122.7	10.68	12.54
Bakelite Vinyl Resin VMCA[a]	122.7	10.98	12.54
TiO_2 Rutile, nonchalking[b]	163.4	4.75	16.71
Antimony oxide	17.9	0.38	1.83
Flexol Plasticizer 10-10[a]	63.8	7.93	6.52
Cyclohexanone[a]	307.6	38.88	31.45
"Solvesso" 100[c]	30.9	4.22	3.16
Standard thinner #325[d]	130.2	19.88	13.32
Tin stabilizer[e]	2.5	0.28	0.26
Stannous octoate	1.2	0.15	0.12
Lecithin grinding aid[f]	2.9	0.37	0.30
Union Carbide Silicone L-45 1% in MIBK[a]	4.9	0.73	0.50
Bakelite Epoxy Resin ERL-2774[a]	7.4	0.77	0.75
	978.1	100.00	100.00

Procedure

Resins were dissolved in solvents, plasticizer, grinding aid, catalyst, Silicone L-45, and stabilizer. Solution was charged to a pebble mill with pigments and ground to 7+ Hegman reading.

Typical paint properties

Viscosity	1440 cps
Wt/gal	9.80 lb
Pigment volume content	13.9%
Gloss (60° Gardner) cured 90 sec at 500°F on steel	90
Nonvolatile by wt	52.1%
Nonvolatile by vol	37.0%
Pigment/resin ratio by wt	75/100

[a]Union Carbide Corporation.
[b]Such as "Ti-Pure" R-960 or equivalent.
[c]Humble Oil Co.
[d]Standard Oil of California.
[e]Such as "Advanstab" T-17-M or equivalent.
[f]Such as American Lecithin "Lexinol" AC-1 or equivalent.

TABLE 2-19. Preparation of Dow Corning XR-6-0031 Resin

I. Equipment

Both the organic polyester and the silicone modified polyester are prepared in a 3-neck balloon flask equipped with a thermometer, nitrogen inlet, agitator, water cooled condenser, and volatile trap. The flask is heated by a Variac controlled electric mantle.

II. Formulation

A. Polyester

	Moles	g	%
Trimethylolethane (Celanese)	2.67	320	30.3
Trimethylolpropane (Celanese)	1.33	179	17.0
Isophthalic acid (Amoco)	3.00	558	52.7
		1057	100.0

Solvent: Cellosolve acetate

B. Silicone polyester

Polyester solids	50 parts
Dow Corning Z-6188 intermediate	56 parts[a]

Processing solids: 65% (assume silicone reacts completely)

Catalyst: Tetraisopropyl titanate 0.5% based on silicone

Solvent: Cellosolve acetate butyl alcohol (3% of total silicone polyester solids)

III. Procedure

A. Polyester

1. Charge the TME, TMP, 60% of the IPA, and a small amount of water to avoid charring.
2. Heat to 205°C (clear solution) and cool to 190°C before the addition of the remaining IPA.
3. Heat to 230°C and hold for at least one hour.
4. Cool to 170°C and add the solvent.

Table 2-19. — Continued

B. Silicone polyester

1. Heat the polyester to 120°C and add premixed; catalyst, solvent for 65% processing solids, and half of the Dow Corning Z-6188 intermediate. This mixture is stable.
2. Heat to 120°C and add the second half of Dow Corning Z-6188 intermediate. Heat to 130°C and hold at that temperature.
3. The resin solution will start to body after about 50% of the theoretical methanol has been removed. The viscosity will increase rapidly and the reaction should be stopped in order that the final viscosity is between 100-400 cps at 50% solids. The reaction is stopped by the addition of solvent.
4. Cool to 100°C before the butyl alcohol is added.

IV. Silicone Polyester Properties

% solids (3 hr at 135°C)	50
Specific gravity	1.07
Viscosity, cps	100-400
Acid number	2-4

[a]Dow Corning Z-6188 intermediate is 89% silicon when completely reacted.

ecological problem of air pollution have spurred research and development in this direction. Unfortunately, those water-reducible coatings which have been made suffer from deficiencies in resistance properties, are excessive in cost or both.

2-17. THE CHALLENGE TO ORGANIC COATINGS TECHNOLOGY IN THE APPLIANCE INDUSTRY

Those of us who are in the industrial finishes business must face the challenge of the following problems:

1. Performance - increasing the durability of our finishes to approach the resistance properties of vitreous enamels without loss of toughness and flexibility.

TABLE 2-20. Appliance Finish Specifications for Washing Machines

Property	Normal min level of performance	Min score acceptable		Max value possible
Pencil hardness	2H		6.5	10
Pencil adhesion	2H		4.5	7
Humidity, % blister (1000 hr)	0	8.0	19.64	20
Softening	0	8.0		
Color	ΔE = 0.5	1.80		
Gloss	ΔG = 2%	1.84		
5% Salt fog-creep 15° from vertical (1000 hr)	1/16 in.	2.11	4.36	5
TFC	0	2.25		
$1\frac{1}{2}$% Detergent, % blister	5	10.27	15.31	18
74°C (240 hr): Color	ΔE = 2	2.16		
Gloss	ΔG = 5	2.88		
1000 cycles abrasion				
Taber, CS-10, 1000 g	100 mg		3.75	5
Color stability (24 hr) Atlas Fadeometer	ΔE-1.23		6.78	9
Grease-softening	2 pencils		4.23	5
50% Oleic acid/cottonseed oil				
Stain— Color	ΔE = 1.5	1.75	3.60	5
Gloss	ΔG = 6.5	1.85		
Impact — Direct	24 in.-lb		2.32	5
Reverse	2 in.-lb			
(Steel Kitchen Cabinet Institute)				
Flexibility — conical Mandrel	1/4 in.		4.26	6
Overbake — color	ΔE = 1	1.20	4.11	5
(Bake time gloss temp.)	ΔG = 5	1.20		
Double bake — flexibility	9/16 in.	0.79		
impact	12 in.-lb avg	0.92		
Total			79.36	100

TABLE 2-21. Properties of High Quality Appliance Baking White Enamel

	Specification	High quality white enamel
Film thickness	1.2-1.5 mils	1.3-1.6 mils
Gloss	75% min (60° meter)	100%
Pencil hardness	F min	2H (3H, 4H and 5H pencil penetrate the film, but do not remove film from panel)
Pencil adhesion	HB min	H
Intercoat adhesion	HB min	F
Humidity — 30 days		
Blistering	5% max	None
Softening	To HB min	H
Adhesion loss	To B min	H
Color change	ΔE = 0.7 max	0.1
Gloss change	4% max	4%
24 hr recovery	F min hardness	2H
	HB min adhesion	H
Salt fog — 5% for 21 days		
Creepage	1/8 in. max	1/16 in.
TFC	10% max	None
(Through film corrosion)		
Detergent[a] — 1½%		
10 days at 74°C		
Blistering	10% max	10%
Softening	To B min	2H
Adhesion loss	To 2B min	F
Color change	ΔE = 1.0 max	0.8
24 hr recovery	HB min hardness	2H
	B min adhesion	F

TABLE 2-21. — Continued

Conical mandrel	$\frac{1}{2}$ in. cracking from apex of cone (max)	No cracking
Impact		
Cracking of film	Direct — 20 in.-lb min	56 in.-lb — no cracking
	Reverse — 6 in.-lb min	20 in.-lb
Heat resistance		
Overbake		
Color change	$\Delta E = 1.0$ max	0.4
Flex. change	$\frac{3}{4}$ in. cracking — max	No cracking
Impact change	Direct — 15 in.-lb min	56 in.-lb — no cracking
	Reverse — 4 in.-lb min	20 in.-lb
Heat resistance aging 7 days at 160°F		
Color change	$\Delta E = 0.5$ max	0.35
Gloss change	5% max	None
Flex. change	$\frac{3}{4}$ in. cracking max	No cracking
Impact	Direct — 15 in.-lb min	56 in.-lb — no cracking
	Reverse — 4 in.-lb min	20 in.-lb
Taber abrasion		
200 cycles		
Weight loss	0.208 mg/cycle-max	0.058
Grease resistance		
Oleic/cottonseed 24 hr at 49°C		
Softening	To HB min	2H
Adhesion loss	To B min	H
Color change	$\Delta E = 1.0$ max	0.25
Gloss change	5% max	No change

TABLE 2-21. — Continued

24 hr recovery	F min hardness	2H
	HB min adhesion	H
Lard/Crisco 48 hr at 49°C		
Softening	None	None
Adhesion loss	None	None
Color change	ΔE = 1.0 max	0.2
Gloss change	5% max	No change
Stain resistance		
Color change	ΔE = 1.0 max (except mustard which is checked visually)	
Tomato juice	ΔE = 0.25	
Lemon juice	ΔE = 0.15	
Grape juice	ΔE = 0.30	
Mustard	Slight to moderate stain	
Ultraviolet exposure		
200 hr — color change	ΔE = 1.0 max	0.5

[a]The detergent solution is a solution of the following:

Sodium pyrophosphate (ACS Grade)	51.0%
Sodium sulfate (ACS Grade)	16.0
Sodium alkyl aryl sulfonate purified (National Aniline's Nacconal NRSF)	23.0
Sodium metasilicate (ACS Grade)	8.5
Sodium carbonate (ACS Grade)	1.5
	100.0%w

2. Economy
   a. Lowering the raw material costs of organic coatings, and
   b. Lowering the cost of application by reducing paint losses in application, by speeding up the rate of cure at present baking schedules, by lowering the temperature required for cure, or by a combination of all three.

3. Pollution
   a. Eliminating the necessity for the use of solvents in application. Failing this, to replace organic, flammable, and/or toxic solvents with water.
   b. Catalytically combusting all oven reaction products to carbon dioxide and water.

REFERENCES

1. R. H. Kienle, Ind. Eng. Chem., 41, 726 (1949).
2. W. M. Kraft et al., Treatise on Coatings (R. R. Myers and J. S. Long, eds.), Part I, Vol. I, Dekker, New York, 1967, Chap. 3.
3. A. Banov, Am. Paint J., 54(43), 4 (1970).
4. U.S. Dept. Comm., Census of Manufacture, Paint and Allied Products, Wash. D.C., 1969.
5. E. A. Zahn, Flow Coating, Research Press, Dayton, Ohio, 1954.
6. H. H Scholz, Interchem. Rev., 15, 4, 110 (Winter 1956-57).
7. R. D. Maxwell, Prod. Finishing, 34(9), 83 (1970).
8. D. G. Higgins, Encyclopedia of Polymer Science and Technology, Vol. 3, Wiley, 1965, p. 165.
9. S. Carangelo, unpublished paper, Interchemical Corp., 1956.
10. H. F. Payne, Offic. Digest. Fed. Soc. of Paint Technol., 159, 297 (1936).
11. Shell Chemical Company, unpublished communication, 1963.
12. E. N. Harvey, Jr., unpublished paper, Interchemical Corp., 1955.
13. A. L. Glass, J. Paint Technol., 42, 449 (1970).
14. E. G. Shur, Interchem. Rev., 14, No. 3, (Autumn, 1955).
15. H. F. Payne, Encyclopedia of Chemical Technology, Vol 5, Wiley (Interscience), New York, 1964.
16. E. G. Shur, Modern Plastics, 33(8), 174 (1956).
17. Dow Chemical Company, Midland, Mich., Formulation 35-0-21 Epoxy Ester Appliance Primer.
18. Ciba Products Co., Technical Service Notes 508.
19. W. A. Higgins, (to Lubrizol), U.S. Pat. 3,133,838 (1964).
20. W. J. Hornibrook, (to DuPont), U.S. Pat. 2,918,391 (1959).

21. H. F. Payne, Organic Coating Technology, Vol. II, Wiley, New York, 1961.

22. J. C. Weaver, Paint Varnish Prod., 48, No. 10, 33 (1958).

23. California Research Corp., Brit. Pat. 821,176 (1959).

24. Allied Chemical Corp., Brit. Pat. 957,367 (1964).

25. Emery Industries, "3393-D Dimer Acid" Development Data Sheet No. 130.

26. J. B. Boylan, Paint Varnish Prod., 56, 71 (1966).

27. D. L. Edwards, et al. Oil Free Alkyds for Enamels Modified with Neopentyl Glycol," Paint Varnish Prod., 56, 44 (1966).

28. Amoco Chemical Corp., Tech. Bulletin, IPA-95.

29. Amoco Chemical Corp., "Polyester Coating Resin Based on IPA," Tech. Bulletin, IP-29.

30. H. T. Roth, C. H. Lamendola, and N. J. Kennedy, (to Interchemical) U.S. Pat. 3,173,971 (1965).

31. J. E. Horn, and R. E. Van Strien, (to Standard Oil, Indiana) U.S. Pat. 3,213,063 (1965).

32. Emery Industries, Inc., "Pelargonic Acid in Baking Alkyds," Technical Bulletin No. 406B, 1966.

33. W. C. Naumann, "Alkyd Intermediates," Paint Varnish Prod., 56, 68 (1966).

34. W. L. Hensley, Paint Varnish Prod., 56, 68 (1966).

35. American Cyanamid Co., "Cyplex Resin 1473-5," Technical Bulletin CRI 8 2A.

36. D. C. Finney, Paint Varnish Prod., 59 (9), 27-31 (1969).

37. Parker Rust Proof Division of Hooker Chemical Corp.

38. C. V. Simroc, (to Pittsburgh Plate Glass) U.S. Pat. 3,320,975 (1968).

39. E. J. Kapalko, and R. A. Martin, (to PPG Industries), U.S. 3, 192,256 (1970).

40. D. M. Berger and J. Wynstra, "High Performance Esterdiol Based Oligomers for Coatings," Western Coating Societies Symposium, 1970, Federation of Societies for Paint Technology, Philadephia, Pa.

41. C & EN Special Report, "Chemicals and the Major Appliances," Chemical and Engineering News, 74-86, July 26, 1965.

42. G. Allyn, Treatise on Coatings (R. R. Myers and J. S. Long, eds.), Dekker, Vol. 1, Pt. 1, 1967, p. 35.

43. G. H. Segall and J. L. Cameron, (to Canadian Industries Ltd.) Can. Pat. 534,261; U.S. Pat. 2,798,861 (1956).

44. Devoe Raynolds, Brit. Pat. 831,056 (1960).

45. D. D. Applegath, Ind. Eng. Chem., 53 (1961).

46. Berger, Jenson and Nicholson Ltd., Brit Pat. 882,113 (1961).

47. A. J. Becalik, and R. P. Gentles, (to Canadian Industries Ltd.) Brit. Pat. 911,050 (1962).

48. J. C. Fang (to DuPont), U.S. Pat. 3,048,553 (1962).

49. J. A. Vasta (to DuPont), U.S. Pat. 3,065,195 (1962).

50. H. C. Woodruff, U.S. Pat. 3,052,659 (1962).

51. Rohm & Haas Co., Brit. Pat. 940,695 (1963).

52. J. E. Gaske and W. H. Brown (to DeSoto), Brit. Pat. 943,217 (1963).

53. Dow Chemical Co., Brit. Pat. 930,035 (1963).

54. DeSoto Co., Brit. Pat. 925,856 (1963).

55. American Marietta, Brit. Pat. 932,250 (1963).

56. R. L. Heppolette (to Canadian Industries Ltd.), Can. Pat. 658,199 (1963).

57. J. A. Vasta (DuPont), U.S. Pat. 3,154,398 (1964).

58. V. G. Nix, (to Sherwin-Williams), U.S. Pat. 3,234,157 (1966).

59. R. J. Schefbauer (to Interchem), U.S. Pat. 3,234,157 (1966).

60. S. C. Lashua and L. H. Lee (to Dow), U.S. Pat. 3,235,528 (1966).

61. J. S. Fry (to Union Carbide), U.S. Pat. 3,008,914 (1967).

62. D. D. Hicks (to Celanese), U.S. Pat. 3,305,601 (1967).

63. J. A. Vasta (to DuPont), U.S. Pat. 3,347,951 (1967).

64. J. E. Gaske and W. H. Brown (to DeSoto), U.S. Pat. 3,301,80 (1967).

65. R. L. Zimmerman and C. E. Lyons (to Dow), U.S. Pat. 3,342,896 (1967).

66. Chemische Werke Albert, Ger. Pat. 1,544,713 (1970).

67. O. C. N. Allenby (to Canadian Industries Ltd.), Can. Pat. 567,165 (1958).

68. N. G. Gaylord (to Interchem), U.S. Pat. 2,853,463 (1958).

69. Rohm & Haas Co., Brit. Pat. 831,898 (1960).

70. E. I. DuPont Brit. Pat. 849,066 (1960).

71. Berger, Jenson, and Nicholson, Brit. Pat. 882,113 (1961).

72. J. C. Petropoulos, C. Frazier, and L. E. Caldwell, Ind. Eng. Chem., 53, 466 (1961); Offic. Digest Federation Soc. Paint Technol., 33, 719 (1961).

73. N. G. Gaylord (to Interchem), U.S. Pat. 3,028,359 (1962).

74. N. G. Gaylord (to Interchem), U.S. Pat. 3,035,008 (1962).

75. J. A. Vasta (to DuPont), Belg. Pat. 634,310 (1963).

76. N. G. Gaylord (to Interchem), U.S. Pat. 3,083,190 (1963).

77. N. G. Gaylord (to Interchem), U.S. Pat. 3,098,853 (1963).

78. Rohm & Haas, Brit. Pat. 986,921 (1965).

79. A. M. Levantin (to Rohm & Haas), U.S. Pat. 3,198,850 (1965).

80. J. R. Costanza and E. E. Waters, Offic. Dig. Federation Soc. Paint Technol., 37, 424 (1965).

81. D. V. O'Donnell and T. J. Suen (to American Cyanamid), U.S. Pat. 3,214,488 (1965).

82. J. S. Fry, G. S. Peacock, J. W. Hagen, and G. A. Senior, Jr., (to Union Carbide), U.S. Pat. 3,267,174 (1966).

83. DeSoto Chemical Coatings, Fr. Pat. 1,462,560 (1966).

84. Pittsburgh Plate Glass Co., Brit. Pat. 1,041,425 (1966).

85. J. S. Fry (to Union Carbide),Fr. Pat. 1,503,303 (1967).

86. J. A. Vasta (to DuPont), U.S. Pat. 3,312,646 (1967).

87. P. D. Cox and G. Swann (to Beck, Koller), Brit. Pat. 1,060,711 (1967).

88. J. A. Vasta (to DuPont), Fr. Pat. 1,466,823 (1967).

89. T. J. Miranda "Oxazoline Modified Thermosetting Acrylics," J. Paint Technol., 39, 40 (1967).

90. Union Carbide, Brit. Pat. 1,105,561 (1968).

91. K. B. Gilkes (to B. P. Chemicals), Brit. Pat. 1,145,203 (1969).

92. R. W. Hall and T. I. Price, Brit. Pat. 1,146,474 (1969).

93. E. B. Nyquist (to Dow), Am. Paint J., 54, 18 (1969).

94. Rohm & Haas Co., Brit. Pat. 1,178,903 (1970).

95. H. A. Vogel and H. G. Bittle, (to Pittsburgh Plate Glass), U.S. Pat. 2,870,116; 2,870,117 (1959).

96. R. M. Christenson and H. A. Vogel, (to Pittsburgh Plate Glass), U.S. Pat. 2,919,254 (1959).

97. R. M. Christenson and D. P. Hart, Ind. Eng. Chem., 53, 459; Offic. Digest Federation Soc. Paint Technol., 33, 684 (1961).

98. H. A. Vogel, and H. G. Bittle, Ind. Eng. Chem., 53, 461 (1961); Offic. Digest Federation Soc. Paint Technol., 33, 699 (1961).

99. C. L. Lynch, (to Pittsburgh Plate Glass) Brit. Pat. 881,498 (1961).

100. N. G. Gaylord (to Interchemical) U.S. Pat. 3,060,144 (1962).

101. N. G. Gaylord (to Interchemical) U.S. Pat. 3,067,776, 3,062,783 (1962)

102. H. A. Vogel and H. G. Bittle (to P.P.G.) U.S. Pat. 3,151,101 (1964).

103. Shell International, Belg. Pat. 635,871 (1964).

104. K. Sekmakas (to DeSoto) U.S. Pat. 3,163,615 (1964).

105. R. M. Christenson and F. S. Shahade (to P.P.G.), U.S. Pat. 3,118,852 (1964).

106. D. P. Hart and P. M. Chairge (to P.P.G.) U.S. Pat. 3,118,853 (1964).

107. DeSoto Company, Brit. Pat. 998,205 (1965).

108. K. Sekmakas (to DeSoto) U.S. Pat. 3,222,309, 3,222,321 (1965), U.S. Pat. 3,230,275, 1965.

109. H. H. Flegenheimer (to Devoe & Raynolds), Belg. Pat. 659,208 (1965).

110. F. W. Michelotti, A. W. Pucknat, and B. H. Siverman (to Interchemical U.S. Pat. 3,242,111 (1966).

111. E. C. Haskell, F. W. Michelotti, H. Burnell, and E. G. Shur (to Interchemical) U.S. Pat. 3,249,564 (1966).

112. . . Nobel-Bozel, Fr. Pat. 1,458,364 (1966).

113. N. G. Gaylord (to Interchemical) U.S. Pat. 3,249,657 (1966).

114. W. F. Rutherford (to Freeman Chemical) U.S. Pat. 3,298,1978 (1967).

115. Beck, Koller & Co., Brit. Pat. 1,093,367 (1967).

116. K. Sekmakas (to DeSoto) Fr. Pat. 1,485,575 (1967).

117. K. Sekmakas and R. F. Stancl (to DeSoto) U.S. Pat. 3,399,153 (1968).

118. H. Rauch-Puntigam and H. Tulasz, Ger. Pat. 1,904,579 (1969).

119. Rauch-Puntigam, Austrian Pat. 271,880 (1969).

120. K. Sekmakas (to DeSoto) U.S. 3,457,324 (1969).

121. J.A. Simma, Am. Chem. Soc., Div. Paints, Plastics, Pigments, 19 No. (2), (1959).

122. W. Abbotson, D. H. Coffey, F. K. Durbury, and R. Hurd, Brit. Pat. 809,257, 809,658 (1959).

123. N. Nelan (to C.I.L.), Brit. Pat. 1,083,135, 1,083,136 (1967).

124. A. Ravve and J. T. Khamis (to Continental Oil) U.S. Pat. 3,306,883 (1967).

125. S. Melamed (to Rohm & Haas) U.S. Pat. 2,992,132 (1961).

126. L. C. Souder and S. Melamed (to Rohm & Haas) U.S. Pat. 2,955,055 (1960).

127. H. Wilhelm, M. Marx, A. Vlachos, and G. Faulhaber (to Badische) Ger. Pat. 1,234,897 (1967).

128. R. M. Christenson, K. R. Gosselink, and S. Porter, Jr. (to P.P.G.) U.S. Pat. 3,362,844 (1968).

129. H. Ashjian (to Mobil Oil) U.S. Pat. 3,456,036 (1969).

130. J. H. Daniel, Jr. (to Cyanamid) U.S. Pat. 2,906,724 (1959).

131. Canadian Industries Ltd., Fr. Pat. 1,287,278 (1962).

132. G. L. Brown, R. E. Harren, B. J. Kine, and E. E. Wormser (to Rohm & Haas) U.S. Pat. 3,033,811 (1962).

133. G. Allyn, "Water Reducible Acrylic Resins in Industrial Finishes," Paint Varnish Prod., 53, 66 (1963).

134. Canadian Industries Ltd., Brit. 940,766 (1963).

135. E. J. Aronoff (to Interchemical) U.S. Pat. 3,083,171, 1963.

136. J. M. Koral, G. E. Bruner, and J. H. Daniel, Jr., (American Cyanamid), U.S. Pat. 3,218,280 (1965).

137. Badische Anilin Soda Fabrik, Belg. Pat. 663,496 (1965).

138. Farbenfabriken Bayer, Fr. Pat. 1,551,838 (1968).

139. C. R. Beardon (to Dow) U.S. Pat. 3,311,583 (1967).

140. Union Carbide, Brit. Pat. 1,177,929 (1970).

141. E. J. Kapalko and H. G. Bittle, U.S. Pat. 3,011,993 (1957).

142. D. P. Kelly, G. J. H. Melrose, and D. H. Solomon, J. Appl. Polymer Sci., 7, 1991 (1963).

143. W. H. Brown, and T. J. Miranda, "Chemistry of Acrylic Solution Polymers," Offic. Digest, 36, 92 (1964).

144. A. Mercurio, "Thermoplastic and Thermosetting Acrylics," Spring Lecture Series, Chicagoland Paint Industry, February, 1964.

144a. A. E. Fischer, Paint Oil Colour J., 48, 99 (1965).

145. J. N. Koral, A. E. Fischer, and J. C. Petropoulas, Paint, Oil, Colour J. 48, 999 (1965). J. Paint Technol., 38, 600 (1966).

146. R. J. King, J. Oil Color Chem Assoc., 52, 1075 (1969).

147. H. L. Wampner, J. Oil Color Chem. Assoc., 52, 30 (1969).

148. D. H. Klein, J. Paint Technol., 42, 545, 335, June, 1970.

149. Union Carbide Corp. Chemicals and Plastics, "Thermosetting Acrylic Coatings: Refrigerator Finish and Primer Evaluation," Union Carbide Corp., Coatings Materials, "Bakelite Acrylic Resins 120," Bull. F41990, 1968.

150. Rohm & Haas Co., Technical Bulletin C-170, Sept., 1969. "Acryloid Thermosetting Acrylic Resins."

151. Rohm & Haas Company, "Experimental Resin QR-463A," Technical Bulletin RD-26.

152. Union Carbide Corp., Coating Materials, "Reactive Vinyl Resins for Solution Coatings," Technical Bulletin F-42268A.

153. J. J. Dietrich, T. E. Hedge, and H. E. Kucsma, Paint Varnish Prod. 56, 75 (1966).

154. Dow Corning Corp., Technical Bulletin, "Preparation of XR-6-003 Resin."

155. E. G. Zahn, Scientific Paint Evaluation, Research Press, Dayton, Ohio, 1955.

Chapter 3

AUTOMOTIVE COATINGS

Bruce N. McBane

PPG Industries, Inc.
Coatings and Resins Division
Springdale, Pennsylvania

3-1. THE SYSTEM

Any commercial automotive coating is a composite of chemical products which interact with each other, usually in the form of a sequence of films at the respective surfaces or interfaces. The combination of substrate, substrate surface treatment, undercoats or primers, and topcoats is referred to as a system. Even though in some systems some of the components mentioned are not present (self-priming topcoats, for instance), each of their respective functions must always be performed in a satisfactory system.

Adequacy of performance of a coatings system in service is judged by observing its efficiency and longevity in accomplishing the protective or decorative function for which it was intended. The importance of the automotive market for coatings technology is indicated by the existence of about 90 million operating vehicles currently in the United States. To the degree that measurements may be applied to performance, therefore, the better the subject is understood and handled by the formulator with problems.

A. PERFORMANCE PARAMETERS

1. Appearance

The most important single performance parameter considered in judging automotive coatings is excellence of appearance. The ultimate standards for appearance of an automobile are somewhat subjectively established by stylists who strive to bring an image of beauty of line and color into harmony with utility and unique effect. A high quality of performance of the coating, indeed of the entire automobile, tends to be assumed by the ultimate purchaser. By contrast, the appearance of the machine, in large part controlled by the appearance of the coating, is obvious by inspection and compromises are not tolerated by ever-critical future owners.

"Appearance" for the automotive formulator can usually be resolved into smoothness and uniformity of gloss, color, and pattern, if metallic flake is a feature of the pigmentation. Appearance begins with eliminating substrate profile, a function which usually falls to formulation and handling of undercoats. To the degree that smoothness is a function of the topcoat, formulators are concerned with selection of resinous components that atomize easily for efficient spray application and whose atomized droplets coalesce completely into continuous, level films. It is desirable in the event that several vehicle components are required, that they be compatible enough not only to form single-phase films, but that at least one of the polymeric constituents has a surface energy that wets pigment particles to effectively deflocculate them and prevent their interference with image reflection from film surfaces. High-gloss topcoats for automotive use, therefore, are specified to be measured in terms of sharpness of image, percentage of light reflected at an angle nearly normal to the film surface, 20° usually, rather than by the more conventional 60° glossmeter.

2. Mechanical Properties

The strength of the system is measured frequently and usefully by testing adhesion to substrates, intercoat adhesion of system components, and retention of film integrity under deformation by bending, stretching, sharp impact, or stress developed by sudden thermal shock. Film hardness, which will not necessarily be low for good flexibility, is also normally specified for the system. Earlier chapters in this treatise (1, 2) deal in detail with measurement of adhesion and mechanical properties. Our discussion in this chapter, therefore, is very specific. It should be noted that while the following sections of the chapter develop relationships between mechanical properties and coating composition, demonstration of excellence or failure mechanically is also a function of film thickness, proper and complete film formation from liquid ingredients, and the provision in the film structure for absorption and harmless dissipation of stress energy.

3. Durability and Other Forms of Environmental Resistance

There is a strong relationship between mechanical strength, first discussed, and durability, since in many cases durability is simply a matter of retaining a high order of mechanical performance during exposure to the environment of use of the automobile. For instance, blister resistance or resistance to gloss loss during exposure to condensation of humidity can frequently be promoted by developing stronger adhesion. Good film gloss is likewise promoted by polymers that are good pigment wetters. However, if during environmental exposure there should be chemical degradation of the polar groups that promote adhesion, or if hydrolysis of the polymer sacrifices its ability to continue to coat pigment particles or maintain film integrity, or if peeling, blistering, chalking, gloss loss, or checking and cracking occur, there will be a likelihood of field complaints.

4. Economics — Raw Material Selection and Product Applicability

It is usually a simple matter for the formulator who knows his raw material, processing, and labor costs to figure the cost of manufacturing a coating. While such costing is obviously essential to remaining in profitable business, the most successful formulator will consider a much broader view which is implied in using the term "coating economics." There are options from the viewpoint of the automotive manufacturer that allow for several coatings to present a range in economic value. The purchaser of the coating writes a performance specification for a given end use. The optional products which meet a common specification may have similar component raw material costs. Presumably, the formulator who achieves the high value level and succeeds in making his customer aware of it will be favored in business award

Several typical formulator-controlled economic options are cited as simple examples. Weight per cent total solids can be increased by using a low-density thinner combination that maintains requisite compatibility relationships with the resinous portion of the vehicle. Such a maneuver is useful in meeting some automotive coating specifications because per cent volume solids, a truer measure of film-forming capability, remains unaffected.

Formulating or color matching with complementary colors produces light absorption similar to that of black pigment and is frequently practiced to obtain opacity without involving light-scattering pigments. One situation where intense light scattering is not tolerable is the pigmentation for metallized coatings, because geometrical metamerism (mismatch due to change in viewing angle) would result. While colors obtained by complementary combinations are chiefly of interest for their unique chromaticity, the concept may have economic value in cases which have limited tolerance for black as a light-absorbing pigment because the level of total pigmentation can be reduced.

One of the most frequently measured economic variables in automotive factories is "mileage," or coating cost per unit of production. There are a variety of properties whose balance affects mileage, but one of the most important is efficiency of utilization of sprayed coatings. It is not unusual to expect that 50% or more of the liquid passing through spray guns will be wasted by dissipation into ventilating air, water wash curtains, or elsewhere in the application other than on the ware surface. Both the manufacturers of spray equipment and coatings formulators have been diligent in efforts to improve mileage by increasing spray efficiency.

Equipment for hot pneumatic or airless (hydraulic) atomization and electrostatic attraction of charged, atomized particles to grounded ware surface have all been important aids to increased utilization of sprayed coating. There are also, clearly, variations in efficiency of spray deposition as the character of the coating is varied. One useful principle is to maximize the liquid-atomizing air ratio (maximum liquid and minimum air) that produces atomization uniformly fine enough to yield a smoothly coalesced film. Since less energy is normally required to atomize low-viscosity liquids than those with greater rheological structure, the use of heated liquids (hot spray) serves to increase the liquid-air ratio to benefit efficient spraying. It is shown later in this chapter that polymerization control to produce narrow molecular weight distribution poly (methyl methacrylate) leads to acrylic lacquers that atomize with enough greater ease to facilitate mileage appreciation when sprayed (see Fig. 3-5).

B. ELEMENTS OF THE SYSTEM

1. Substrates for Automotive Coatings

Until recently, the surfaces to be coated in automotive finishing have been cold-rolled, mild steel reinforced by galvanizing for corrosion resistance at critical sites, such as rocker panels under doors, or made from zinc die castings used for grilles and fender extensions. Aluminum, stainless steel, and nickel-chrome plated parts were attached usually without coating, frequently by conductive clips which promoted bimetallic corrosion.

More recently there has been a trend toward the use of rigid plastics, such as injection-molded acrylonitrile-butadiene-styrene (ABS), polysulfones, polycarbonates, polyformaldehyde resins, and nylon, as materials of construction for exterior attachments. Plated metal bumpers and metal side moldings are increasingly being replaced by nonrigid, energy-absorbing, solid elastomers, high-density foam moldings, or extrusions. Modified poly [ethylene-propylene-dicyclopentadiene (EPD)], or polyurethane-type polymers are regularly employed for impact or scuff-resistant areas of automotive interiors and exteriors. Rigid injection moldings of modified polyphenylene oxide, noted for dimensional stability during post-form baking, or the older phenol-formaldehyde or modified epoxies, may be encountered as substrates in interior finishing problems.

Finally, one takes note of intensive activity to extend the use of molded plastics, particularly fiber-reinforced, thermosetting resins as materials of construction for automotive body shells.

The significance of this recital of the variation in automotive substrates lies in their universal need for coatings not only for appealing styling, but for protection against service environments. The observation is occasionally made that since plastics do not corrode, they need not be painted. Practice indicates, however, that degradation of uncoated plastics exposed to weather creates the same need for coating that exists with commercial metals. The problems of color control and the expense of pigmenting the entire cross-sectional structure of plastics of construction further mitigates against the strong appeal of color modifying the plastic composition to eliminate its need for a coating.

2. Substrate Pretreatment

Both metallic and nonmetallic substrates require attention to some type of preparation for organic coating application. Pretreatments are designed to promote coating adhesion, and further, in the systems that involve metals as substrates, to inhibit corrosion.

Metal pretreatment will normally be concerned with modifying the surface of steel or zinc galvanize, in which three to seven-step alkali cleaning, acid etch, and chromate rinse operations are utilized. The purpose of the acid etch is not, as popularly supposed, to roughen the surface, but to achieve chemical passivation of the metal surfaces by involving them in a reaction which deposits an adherent, uniform, and relatively fine-grained crystal structure as a new surface. The surface composition found most useful in combining fine-grain structure required for optimizing profile, with the crystal composition needed for corrosion inhibition, is a hydrated zinc iron phosphate (3). The deposit is dense enough to inhibit moisture diffusion to the metal and is usually in the range of 2 to 3μ thickness. Due to its crystalline nature, the pretreatment reaction product is somewhat friable. Thicker deposits, therefore, are not encouraged because of their tendency to fracture and lose adhesion when the metal is impacted or deformed. The crystal layer does serve as a basis for mechanical adhesion of subsequent coatings, but more importantly, adhesion can also be implemented by using its hydrated structure to hydrogen bond with pendant functionality in properly designed metal primers (Fig. 3-1).

Surface preparation of plastics may vary from simple hot water or solvent washing to remove soil and mold release agents, to gentle abrading, or to chemical pretreatment with reagents, or occasionally brief radiation with ultraviolet. The latter treatment might be expected to be effective, because enhanced coating adhesion has been observed when weathered plastic is coated.

3. Undercoat Contribution to System Performance

Undercoats are conventionally differentiated according to their main function into:

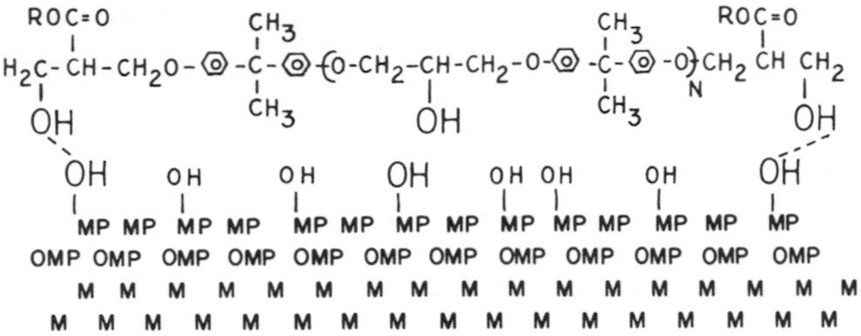

```
ROC=O          CH3                                  CH3       ROC=O
   |            |                                    |          |
H2C-CH-CH2O-◎-C-◎-(O-CH2-CH-CH2-O-◎-C-◎-O)-CH2 CH CH2
   |            |              |                     |       N      |
  OH           CH3            OH                    CH3            OH

  OH   OH    OH    OH    OH OH    OH    OH
   |    |     |     |     |  |     |     |
 MP MP MP MP  MP MP MP MP MP MP MP  MP MP  MP
OMP OMP OMP OMP OMP OMP OMP OMP OMP OMP OMP OMP
  M  M  M  M  M  M  M  M  M  M  M  M  M  M  M  M  M
M  M  M  M  M  M  M  M  M  M  M  M  M  M  M  M  M
```

R= OIL ACID
M= METAL
OMP= PHOSPHATED METAL OXIDE
MP= METAL PRETREATMENT

FIG. 3-1. Epoxy ester adhesion to metal.

a. Primers

Primers are used principally to ensure adhesion of the coating system to substrates and, occasionally in those situations in which plastic surfaces are coated, to serve as barriers to the migration of low-molecular-weight species from the plastic to the coating's surface. Even near monolayers of such migrating fractions, which may be amines, sulfur, waxes, or oligomers, could promote discoloration or interfere with the intended cure of crosslinkable topcoats.

b. Surfacers

These may combine the functional role described for primers with the role of obscuring substrate profile. Total elimination of a pronounced profile requires that the surface be sanded after drying; to ease that operation, surfacers are usually highly filled (short of critical pigment volume concentration) with inert-type pigments.

c. Sealers

Sealers occasionally referred to as midcoats or tiecoats, are low-level pigmented undercoats designed to enhance intercoat adhesion of primer and topcoat, and have been recommended in systems demanding prêmium appearance to reduce porosity of primers or surfacers.

Referring to Fig. 3-1, where development of coating adhesion by bonding is illustrated, one is correct in concluding that many of the important performance requirements of the system are a function of the creation of strong adhesion at the several interfaces. Zisman (4) and Huntsberger (5) are among many who have studied mechanisms that act to produce adhesion. Huntsberger comments that "a reasonable approach to the problem is to ask not why certain pairs of materials adhere to one another, but rather to ask why any two materials should not adhere to each other. . . . If we can assume that calculated values for dispersion force interactions are reasonable approximations, we are led at once to the conclusion that materials not adhering strongly to each other must not exhibit intimate contact at the interface." Attractive forces acting across an interface may be very large, but they diminish very rapidly as the planes of the interfaces are separated by very small distances. It is for this reason, therefore, that if good adhesion is to be obtained between the interfaces, maximum contact of the interfaces must be made.

In the conventional cases discussed here, i.e., of liquid coatings applied to solid substrates, "maximum contact" means that the liquid must wet the solid surface so completely that areas of the two interfaces are essentially equal. Moreover, since in film formation the liquid coating not only shrinks by losing volatiles, but frequently changes its chemical nature by curing or crosslinking, there must be concern by the formulator that the intimate contact between film and substrate be maintained throughout the transition of the coating from liquid to solid. It is quite likely, therefore, that modifications of a coating by incorporation of unreacted chemical groups into films is just as important to the physical attainment of adhesion by good substrate surface wetting as it is to the chemical development of adhesion by interfacial bonding.

The problem of obtaining completely wetted substrate upon application of primer assumes further importance in the matter of good adhesion when one considers the minute roughness of the crystalline substrate surface produced by the pretreatment. Surface roughness or "tooth" is intuitively considered to be beneficial to adhesion because the actual surface area of a given plane is increased by the degree of roughness, but the benefit only accrues if wetting of the total profile is obtained. Huntsberger's data (5) confirm that increased fluidity of a polymeric coating induced by baking the applied material, or plasticizing the polymer, does indeed serve to reduce the number of interfacial discontinuities and increase adhesion.

It is useful to note that much of the consideration given to understanding undercoat adhesion to substrates can be extended to the intercoat adhesion of undercoats and topcoats.

More subtly dependent on adhesive forces than the obvious coating resistance to delamination or peeling from substrates, are such other performance properties as resistance to condensed humidity, blistering, retention of film integrity when confronted with thermal shock, or ability to dissipate impact energy without chipping or cracking. Automotive system chipping resistance, for example, is tested in the S.A.E. J-40 Gravelometer, and Bender (6) has established a relationship between T_g (the glass transition temperature of polymers) in the primer vehicle solids and chip resistance. Relaxation of the stiffness of the primer vehicle by lowering its T_g, allows induced stresses (such as chip-producing impacts) to be relieved more easily, thus avoiding film discontinuities. The usual primer vehicles are heat convertible to crosslinked structures, and T_g is controlled by the crosslink density produced upon curing or by varying the concentration of hard polymer segments, such as phthalic anhydride in polyesters. While the suggested mechanism for minimizing chipping, i.e., impact energy dissipation, clearly has merit and experimental support, the concurrent role of the lower T_g polymer in developing increased adhesion must also have considerable potency. Chipping has also, for instance, been substantially reduced by the incorporation of adhesion-promoting chemical groups into the polymers used either for metal or plastic primer vehicles or sealer vehicles.

A related function of automotive primers for metals is the inhibition of corrosion. The corrosion reactions of steel and zinc are oxidation-reduction and, of course, involve a flow of electrons in an electromotive cell.

$$Fe \longrightarrow Fe^{2+} + 2e \qquad \text{anodic reaction} \tag{1}$$

$$\tfrac{1}{2}O_2 + H_2O + 2e \longrightarrow 2OH^- \tag{2}$$

$$2H^+ + 2e \longrightarrow H_2 \tag{3}$$

(Reactions (2) and (3): cathodic reactions)

As in the control of any chemical reaction, the principles of the law of mass action can be invoked to inhibit corrosion. The system component next to the metal, metal pretreatment, has the opportunity to exert the greatest influence on chemical inertness of the metal, and the primer, being next in line geographically in the system, is assigned much of the responsibility for protection. Application of the mass action law to corrosion inhibitive primer formulation takes the form of:

1. Attempting to insulate the anode electrically.

2. Reversing Reactions (1) and (2) above by supplying from an external source a flood of electrons or hydroxyl ions.

3. Passivating those local areas of steel which may be exposed through the pretreatment layer by accidental scoring or fracture of the crystalline deposit, by some chemical action of the primer.

In the design and construction of metal automobile bodies, engineers have learned to anticipate the areas most likely to be attacked by corrosion. Strategically located drains are installed to prevent accumulation of electrolyte picked up during service, and most usefully employed is anodically reacting sacrificial galvanize in constructing critical box section members. Epoxy esters, for example, are popularly used as vehicles for inhibiting corrosion by insulating the pretreated metal because of their strong adhesion to the metal surface and their longevity in the presence of an alkaline environment [see Reaction (2)]. The epoxy condensation resins used in making the oil-acid-modified esters are polyethers with typically strong resistance to saponification. They are usually crosslinked with alkylated urea or melamine formaldehyde resins, which introduce additional ether linkages upon their reaction with a portion of the pendant hydroxyl in the epoxy ester. Further saponification resistance may be obtained by blending modest concentrations of solutions of unesterified epoxy condensation resin into the epoxy ester primer vehicle.

The use of corrosion-inhibitive pigmentation in automotive primers is a positive aid in the battle against corrosion, but is limited to products such as dip primers that can be applied without risking ingestion or skin contact by workmen in the painting operation. Mildly soluble chromates such as calcium, strontium, or zinc salts are believed to passivate the iron surface by formation of thin, tightly held oxides. They are, however, strongly irritating to human skin or mucous membranes and are therefore avoided in spraying applications or in primers that are sanded before topcoating.

It was observed with corrosion reaction (2) that hydroxyl ions are generated as corrosion proceeds. Were it not for the removal of hydroxyl ion from the electrolytic corrosion cell by precipitation of hydrated ferrous hydroxide (the forerunner to rust), the cell would become so polarized that the reaction would eventually cease. Experiment has proved that corrosion does not proceed well in a basic environment. If basic pigments, zinc oxide for instance, can be tolerated in the primer pigmentation, it is apparent that another technique for inhibiting corrosion hy primer formulation is at hand. Zinc oxide does, of course, tend to form high-viscosity soaps with acid vehicles, and before introducing basic pigments, precautions must be taken to minimize vehicle acidity and monitor product package stability.

Fontana (7) has warned against an additional pitfall associated with incorrect use of corrosion-inhibiting chromates. "The addition of chromates to stifle the anodic process may have deleterious effects if added to a solution of the wrong pH. At pH's below about 3, chromates reduce to Cr^{3+} and at higher pH the chromates reduce to chromium oxides or hydroxides according to the reactions:

$$6e^- + Cr_2O_7^{2-} + 7H^+ \longrightarrow 2Cr^{3+} + 7OH^- \quad \text{(acid range)}$$

$$6e^- + Cr_2O_7^{2-} + 4H^+ \longrightarrow Cr_2O_3 + 4OH^- \quad \text{(less acid)}$$

Producing Cr^{3+} provides an additional reduction reaction and accelerates corrosion; producing Cr_2O_3 stifles further anodic reactions, raises the potential, and reduces the corrosion."

Undercoats which offer outstanding support to system appearance will resist replication of substrate profile and retain topcoat gloss by allowing minimal migration of topcoat vehicle into the undercoat film. Undercoats that are proficient in obscuring substrate profile are said to have good "filling," while minimal absorption of topcoat vehicle is called "holdout."

An appearance of filling could be achieved by a primer that simply bridged the high spots of an irregular profile, but that product would be expected to fail all tests dependent on good adhesion. Plainly, the primer that combines appearance and performance must assume the profile of the substrate at the lower interface and be nearly planar at the upper surface. If one envisions a saw-tooth profile at the interface, it is not unusual to find that wet films on such a profile are nearly planar, but unfortunately, upon drying, the profile reappears in the new surface. The problem is found to be largely a matter of shrinkage during primer film formation. Because good filling primers are usually sprayed, and substantial amounts of volatile solvents are employed in the application, the thicker portions of the film between high points in profile undergo more shrinkage than the thinner film over the high points. Original finish automotive primers are usually baked to accomplish chemical cure, and since further differential shrinkage occurs during curing, the tendency to replicate substrate profile is aggravated.

It has been mentioned that premium appearance is achieved by formulating the primer to allow sanding of the dry film. By careful sanding, one may arrive at a planar surface, and in the event that very marked profile is encountered in the substrate, sanding the primer may be the only approach to good appearance. Automobile factories, however, find the sanding of primer to be expensive in terms of manpower and production speed; there is obviously a preference for filling reasonably smooth substrates without sanding. Moreover, the sanding operation is carried out on a water or solvent wet surface which must be meticulously cleaned before topcoating. If the sanded bodies are water washed to remove sanding sludge, they must be blown dry with clean air to minimize evaporation of water spots on the primed surface. Water droplet evaporation inevitably leaves behind small aggregates of previously dissolved salts or mineral which inevitably lead to film blistering in condensing humidity. While good blow-off technique solves the water-spot problem on primers, a more fundamental solution once more is the elimination of sanding.

Filling without sanding has been profitably attacked by formulating to minimize primer shrinkage during cure. A high percent volume solids combined with a chemical cure reaction that evolves the minimum volume by-product is probably a satisfactory answer in most of the cases.

"Holdout" is a function of effectively reducing cured primer film porosity to topcoat vehicle. This is one of the few problems in automotive coating formulating for which the obvious approach is the successful one. The selection of fine particle size primer pigments, good pigment dispersion techniques and the minimal use of fibrous shaped pigments, along with minimizing pigment volume concentration consistent with primer film performance, will usually provide a satisfying response.

There is a widespread attitude that good levels of intercoat adhesion between primers and topcoats depend on a certain degree of solubility of the primer surface in the topcoat so that the two films are said to be "knit together" or "diffused into each other." There is a scientific basis for this attitude in that topcoats that wet primer films very well will in most instances have similar thermodynamically described free energies and strong intercoat adhesion. The description just given might also be reasonably applied to favorable interaction between organic solvent and solute. The problem as it related to appearance of the system, however, is that formulators following the "knitting" principle have been decoyed into preparation of weak interfaces by diluting the surface polymer with low-molecular-weight ingredients, or undercuring their primer in the interests of topcoat solubility. The net result is not only mechanical or environmental failure at the interface, but also poor topcoat appearance due to distortion of primer film swollen by the attack of topcoat solvents. There may well be valid cases of strong intercoat adhesion due to interdiffusional activity, but frequently these cases reduce to self-adhesion in which performance has benefitted by the useful device of formulating similar compositions on both sides of the interface. Our emphasis is simply that there should be no need to sacrifice holdout at interfaces to obtain intercoat adhesion.

Fiber-reinforced polyester substrates sometimes offer a similar problem in profile replication in that if the polyester is undercured when the surface is exposed to coating solvent or subjected to humidity or water immersion testing, the resin swells differentially with respect to the fiber. The resulting profile is accordingly the reverse of the original, but, of course, objectionable. Post-curing such substrates prior to coating may be helpful.

4. Topcoat Contribution to System Performance

It should by now have become apparent that while the automotive coating system will be judged by the casual observer in terms of the visible appearance afforded by topcoats, the successful functioning and attractive appearance of the topcoat is very largely dependent on prior components of the coating system. The contributions to system performance that are uniquely topcoat function are accordingly relatively few.

Probably the most distinguishing characteristic of an automotive topcoat is its pigmentation, which provides color and opacity to the coating and, by reason of its opacifying activity, may offer valuable protection to the resinous binder against destructive wavelengths in sunshine. Reciprocal protection is afforded to the pigment by the binder which, in its preferred activity, coats the surface of the pigment aggregate intimately with a relatively chemically inert, ultraviolet-transparent, resinous layer capable of tying the composite into an integral film.

Traditional styling of automotive topcoats has dictated that they be glossy to the degree that the more desirable dry films are described as having a wet look. Such appearance is achieved by choice of polymers, pigment dispersions, and formulating techniques that promote a thin layer of nearly clear vehicle at the surface of the dry film by resin reflow during film formation. The formation of a resin-rich film surface is a fairly natural part of film formation from low-viscosity, relatively low-molecular-weight polymers, which are fluid enough to be deposited at the surface by evaporating solvent and are then crosslinked by oxidation or a polyfunctional resinous component present as a crosslinking agent. When a baking period is utilized to produce rapid chemical cure, the reduction in polymer viscosity at the elevated temperature can usually be expected to induce the required resinous reflow by convection.

Alternatively, it has recently become an industry practice with original finish topcoats that form films from high-molecular-weight polymers by simple solvent evaporation (no crosslinking), to use baking temperatures in the vicinity of 150°C to induce high gloss and film smoothness by resin reflow. We see here a reversal of the traditional handling of automotive enamels and lacquers, the latter being the evaporative dry film formers. Lacquers were originally introduced because of their drying speeds at temperatures of 38 to 71° C. The composite vehicles of those early days were rich in plasticized cellulose nitrate, which is subject to reflow with difficulty even at elevated temperatures. The dried films were low in gloss until subjected to extensive polishing, and the attendant friction produced a plasticizer-rich surface whose soft luster was associated with high quality, referred to as "rubbed effect." The current lacquers (using modified acrylic polymers) are heat reflowed at temperatures higher than normally used for the chemically cured enamels, and highly glossy films are produced without the labor of polishing.

Concern for highway safety has been responsible for the suggestion that automotive topcoats should not be glossy, but preferably have surfaces that offer matte or highly diffuse reflectivity. Such films have low initial gloss and must maintain a uniform diffusivity of incident light during exposure to weather, the polishing activity of cleaning operation, and the spillage or collection from contaminated atmosphere of oily residues that are gloss producing. Satisfactory low-gloss topcoats must also be chalk resistant and highly mar resistant. There are amorphous, nonabrasive, mineral pigments which satisfy these requirements to a degree when used as flatting agents, but relatively transparent high-molecular-weight organic materials, insoluble in the coating, are more favored because of their mar resistance. Powdered polypropylene and dispersion-grade vinyls have yielded encouraging properties in this currently limited market.

The key to the utility of a topcoat seems to begin with formation of films with a very high degree of continuity, or integrity, and chemical inertness, and to end with retaining these properties throughout its service history. The specific problems that formulators and their raw material suppliers are solving for the automotive industry are illustrative.

Polymeric binders and pigment. surfaces with a high degree of chemical inertness minimize the effect of combined chemical attack of moisture, oxygen, and atmospheric contaminants under the influence of solar radiation and heat. Inorganic protective layers are deposited on titanium pigments and the light-sensitive chromate, molybdate, and Prussian blue-type pigments. Metallic flake is increasingly passivated by surface treatment, and polymeric encapsulation of pigments has been favorably reported (8).

While the mechanical properties of coating systems are highly dependent on undercoats, much cooperation is expected from the topcoat in yielding to spray atomization and coalescence of atomized particles to produce film smoothness, as well as high enough adhesiveness and tensile strength to resist blistering or dulling in condensing humidity and cracking under thermal shock. Automotive performance specifications are placing emphasis on retesting exposed film specimens for humidity resistance and thermal cycle resistance as well as for gloss retention.

Studies of the diffusion of moisture vapor into films (9, 10) have been related to their practical performance. The conclusion has been drawn that diffusion rates are probably more significant than amounts of water diffused and that if integrity and protection of the system are to be long lived, the diffusivity of the primer and topcoat should be similar.

Finally, consideration should be given to the problem of humidity etching and water spotting of automotive topcoats, which is proportional in severity to the degree of thermoplasticity of the film and is, therefore, encountered more usually with lacquers than with enamels. Automobiles that are stored in atmospheres subject to a dewpoint will be covered with droplets of condensed water which are still present when sunshine later warms the surface. The film is softened by plasticization of water and by the sun's heat as the water is evaporated. The water droplet pulls the softened film into a series of concentric rings as it diminishes in size and the film surface eventually is roughened to the degree that it begins to display diffuse rather than mirror reflectance. Severe exposures of susceptible films can produce a badly weathered appearance on a new car in a few weeks. The adoption by the auto producers of higher heat reflow temperatures for original finish lacquers has made it feasible for formulators to raise the glass temperatures (T_g) of the coatings to solve the problem. Figure 3-2 shows the variation in T_g as homopolymer methyl methacrylate is copolymerized with increasing amounts of ethyl hexyl acrylate (11).

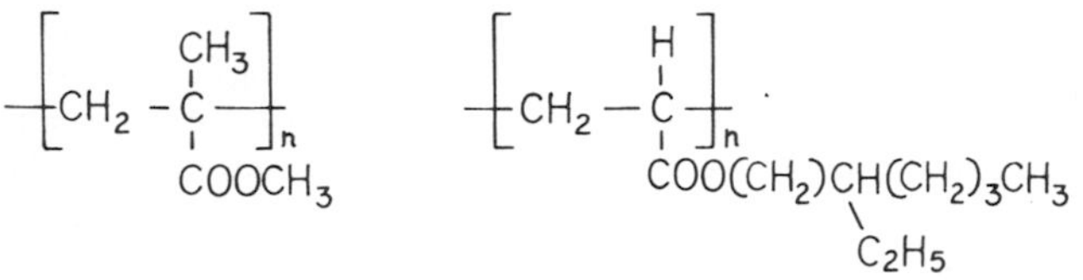

$$\left[-CH_2-\overset{CH_3}{\underset{COOCH_3}{C}}-\right]_n \qquad \left[-CH_2-\overset{H}{\underset{COO(CH_2)CH(CH_2)_3CH_3}{C}}-\right]_n$$

(with C_2H_5 branch on the CH)

% MMA	% EHA	Tg
100	0	221 °F
95	5	187 °F
90	10	160 °F
85	15	131 °F
80	20	106 °F
0	100	- 121 °F

342 °F (Approximatety Linear)

FIG. 3-2. Structure and T_g range of copolymers of varying ratio of "hard" to "soft" monomer.

5. Thinners and Reducers

The presence of the volatile portion of coatings is motivated principally by a need to diminish the interaction of high-molecular-weight polymer molecules by dilution. Viscosity must be reduced sufficiently to enable the forces available in simple application technique (air atomization for spray, surface tension for dipping) to spread films of controllable dimensions. With the advent of some understanding of the role of surface-free energy in spreading a film on a substrate, there is also a constructive use of volatiles in assisting substrate wetting, polymer compatibility, coating package stability, and film formation.

On the other hand, evidence that release of organic vapors into the atmosphere contributes to environmental pollution, and the growing body of restrictive legislation, have initiated vigorous reformulation activity aimed at a substantial reduction in organic volatile content of coatings. Correspondingly, therefore, each of the specific discussions of coatings formulations which follow has dealt with the subject of minimal organic emissions.

Even in those cases of reduced organic volatile content, it is not unusual to find 5 to 30% of the coating weight lost as volatiles during application as a result of a combination of the presence of volatile reaction products of film formation, volatile organic coalescing agents, or low-molecular-weight reactive components of which a portion evaporates before chemical incorporation into the film. Those problems in film continuity or appearance which stem from volatile release by the film still prevail. Particular reference is made to the disruption of film continuity occasioned by the sudden release of solvent bubbles from semirigid films, which takes the form of crater-shaped holes in the film called "solvent popping." The craters are so pronounced that even polymers or formulations with excellent heat reflow cannot obscure the deformities in the film. Efforts must be made via formulation to allow such volatiles either to escape while the coating is liquid, or for their release to be slow enough for nondisruptive diffusion through the film.

A great forward step in the understanding of good coating formulation has been articulated by those who have been active in the application of the solvent parameter concept whose basis was laid down as early as 1916 by Hildebrand in the thermodynamics of the solubilities of nonelectrolytes (12, 17). The concept takes account of the attractive forces between solvent molecules that cause the compound to exist at room temperature and pressure as a liquid. Hansen (15) has paid particular attention to the dispersion forces, polar forces, and hydrogen bonding forces of the volatile liquids used to dissolve or dilute polymers, expressing the solubility parameter as a vector in space, a resultant of the indicated attraction force components. If the magnitude of each of the components is identified along an axis in three-dimensional space, each of the common solvents can be located as to solubility parameter in the volume defined by the units of the three axes. If then, the solubility of a given polymer is determined (as soluble, insoluble, or borderline, using dilute mixtures of polymer) in a series of mixtures with individual liquids, it will be found that a volume of solvency can be defined for the polymer in terms of the attractive forces for the liquids. Liquids having a three-dimensional solubility parameter within the volume of solvency are compatible with the polymer, and the closer they are located toward the center of the volume of solvency, the better their solvency will be found to be. Molecules with similar attractive forces have difficulty distinguishing between themselves and can associate readily (compatibly), whereas other molecules with widely different vectors are excluded from this association even to the point of immiscibility if the vectors are sufficiently different. The "vector-in-a-volume" concept has particular value for the formulator of automotive coatings who will normally be required to use a rather complex mixture of solvents to control evaporation rate as well as a mixture of polymers to obtain optimal film properties. If a borderline solvent (with respect to the volume of solvency) is considered to be desirable to the coating's performance, it can usually be made more tolerable by mixing with another borderline liquid lying on a diagonal through the volume. Obviously, if one thus chooses a mixture of nonsolvents to become the solvent for a polymer, the component evaporation rates from the coating solids need to be similar to avoid polymer precipitation as volatiles evaporate from films.

Even in the cases of essentially solventless coatings, data accumulated by mapping solubility parameters for a variety of polymers can be useful in predicting polymer:polymer or plasticizer:polymer compatibility. It will be unusual for solubility maps of different solutes to coincide exactly, but in those areas of overlap (assuming the same solvents have been used to define the volume of solvency), one can expect the polymers to be miscible.

A concept widely used for automotive coatings by which solvent evaporation rate is put to work to control film thickness is credited to McMaster of General Motors Corporation (18) and identified as "hi-lo" reducer formulation. The concept was developed to aid in applying lacquers based on cellulose nitrate, which was traditionally formulated at relatively low solids content to obtain spray atomizable dilutions. Consequently, the multiple coats required to deposit film thicknesses needed for protective coatings posed problems in economy and production speed. In formulating high-low solvent blends, a mixture of fast-evaporating and slow-evaporating volatiles is used with the expectation that most of the fast-evaporating fraction will volatilize during the air atomization of the coating. The fast-evaporating fraction may consist of less effective solvents than the slow-evaporating portion, which should be retained by the applied coating film long enough to induce leveling and coalescence of atomized paint particles.

Advantages of high-low solvent blending are associated with greater freedom to operate the spray equipment in an efficient manner. Not only are smoother films obtained by wet spraying without danger of sags, but because of the easy volatility of the "fast" portion of the thinner, the coating may be sprayed with the minimum air pressure that produces atomization. The lower air pressures produce a narrower spray fan and enough less overspray, that those using high-low formulation claim reductions in the number of coats required to obtain necessary film thickness.

3-2. MECHANISMS OF FILM FORMATION UTILIZED FOR AUTOMOTIVE COATINGS

Films designed for automotive coatings are formed by four general mechanisms. The high-molecular-weight polymers, of which thermoplastics consist, may be prepared in a prior polymerization step, mixed with solvents to form a solution, or stably suspended to form dispersions, but in either case the film formation is initiated by evaporation of the volatile component. A stable formulation of such a thermoplastic film former dictates a nonreactive, linear (two-dimensional) polymer structure, which is resoluble in selected solvents. The term "thermoplastic," which originally implied softening by heat, must be used with caution in categorizing a film, since softening is a relative property displayed by many organic films not generated by the evaporating process. "Lacquer" and "thermoplastic" are terms used almost interchangeably in discussing automotive coatings.

A second important film-forming mechanism is, by contrast, called thermosetting and involves a chemical crosslinking, often as a result of condensation reactions. Characteristic of formulation of thermosettable coatings is the presence of reactive groups (conventially hydroxyl) pendant from a relatively low-molecular-weight polymer, together with a crosslinking agent, also polyfunctional, such as urea or melamine resin. It is important to remember that the condensation reactions will evolve a low-molecular-weight by-product such as water or an alcohol.

Addition polymerization describes a third mechanism of film formation in which small but reactive molecules are caused by appropriate initiation to react with each other, forming useful film molecules. Vinyl polymerization or the urethane reaction are examples.

Finally, a fourth type of reaction is started by oxidation at the site of unsaturation, such as occurs in drying or semidrying oils.

A. FILM FORMING BY SOLVENT EVAPORATION

Automotive thermoplastics or lacquers were originally devised to expedite production-line painting in which coatings used vehicles based on plasticized cellulose nitrate. In the late 1950s the need for more prolonged gloss retention of topcoats forced those formulators who wished to participate in the automotive thermoplastic market to convert compositions to plasticized solution of acrylic polymers.

All thermoplastic solution vehicles share a problem inherent in their nature. Because strong films derived from high-molecular-weight polymers are required, and because the viscous solutions of the high-molecular polymers formed prior to application require extensive dilution to render them sprayable, the content of solids at the gun is relatively low. It follows that formulating ingenuity should be directed toward obtaining extraordinary film build with low solids solutions, or toward so manipulating the high-molecular-weight polymer that it can be spray applied at high solids without encountering high-solution viscosities. Of course, spray equipment engineers have been concerned enough to design mechanical aids such as hot spray or airless spray, but the formulator's contribution might, for example, take the role of devising a high-low spray reducer (see Sec. 3-1, B, 5), or preparing an organosol or dispersion coating (as in Sec. 3-3, A, 4).

Despite the difficulty of obtaining film build with the typical thermoplastic coating, a lacquer has points of advantage, as indicated by the fact that for the past 25 years about half the automotive topcoat used in the United States has been thermoplastic. Figure 3-3, with its simplified structure, demonstrates the linear character of a typical acrylic thermoplastic polymer. The structural linearity, which allows resolubility of the finished coating, is

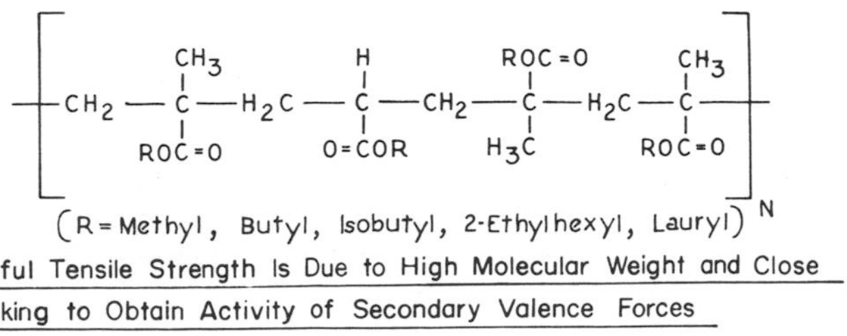

$$\left[-CH_2-\underset{ROC=O}{\overset{CH_3}{C}}-H_2C-\underset{O=COR}{\overset{H}{C}}-CH_2-\underset{H_3C}{\overset{ROC=O}{C}}-H_2C-\underset{ROC=O}{\overset{CH_3}{C}}- \right]_N$$

(R = Methyl, Butyl, Isobutyl, 2-Ethylhexyl, Lauryl)

Useful Tensile Strength Is Due to High Molecular Weight and Close Packing to Obtain Activity of Secondary Valence Forces

FIG. 3-3. Thermoplastic linear acrylic polymer structure.

useful in achieving a special type of factory production-line repair of scratches or mars occurring on painted auto bodies during final assembly operations. Badly disfigured auto bodies may be totally repainted before assembly; in other cases, "panel repairs" are made in which an entire door, hood, or deck lid may be repainted at the end of the assembly line. Much more common, however, with thermoplastics at the end of the assembly, is a "spot repair" which is a repaint of the very immediate, local area surrounding a blemish. The ability to make spot repairs is valuable because by far the majority of blemishes found by inspection after assembly are small. Considerable time is saved in minimal masking of surrounding areas or otherwise preparing for the repair as compared to repainting an entire panel. After the mar or blemish has been removed by local sanding, a fast-drying primer is applied to protect exposed metal and the repair area is sprayed with another coat of thermoplastic, which may consist of a portion of the same coating used for the original body finish. Occasionally, a special repair thermoplastic, more suitable than the original finish for the low-temperature drying conditions at the end of assembly, is formulated for factory repairs. Frequently, if the original finish thermoplastic supplied to the factory is used for repairing, a special composition reducer will be used to facilitate the operation. In any event, the original finish thermoplastic is softened by the repair composition applied. The original and repair melt together under the influence of solvent contained in the repair, a final application of "mist-thinner," either 100% volatile or a very low-solids solution of clear vehicle, is applied to obscure the edge of the repair, and the operation is concluded, after a short drying period, by a light polishing operation. If all factors involved in the spot repair have operated as per design, the repair is very rapidly concluded and is not visible even to close inspection.

At this point, brief discussion on handling problems that can arise when repairs do not follow the optimal path just described is in order. It must be understood that at the end of an assembly line an automobile body contains temperature-sensitive parts such as glass, rubber, plastics, and fabrics. Paint-repair baking schedules are therefore restricted to the time and temperature which can be tolerated by those materials of the construction which are easily deformed or discolored by heat. Spot repairs, while in many cases are force dried by local application of heat lamps, must be made with formulations that approximate not only the appearance, but also the long-term performance of the original finish. This means that repair pigmentation and vehicle quality must be virtually identical with the original finish, for small deviations in repair quality show up after minimal weathering of the automobile as a "bull's-eye" of faded, dulled, or even cracked coating.

It has been stated that an advantage of the thermoplastic coating is the ease of conducting spot repairs based on the resolubility of the original finish vehicle. Obviously, in a coating which must resist attack or spilled gasoline, alcohols or glycols in windshield washes, or radiator antifreeze, and the mineral oil or aliphatic volatiles present in waxes and polishes, easy resolubility can be overdone (19). Particularly damaging is the aromatic (toluene) content of high-test gasolines which not only softens or distorts the acrylic film, but also leaches plasticizer from the film, leaving it prone to failure by weather-induced cracking. Even so, a balance point of calculated resolubility of original film is essential to a spot repair, for if the original thermoplastic topcoat is formulated exclusively to maximize its insolubility, rather than redissolving the original film, the repair coat may only soften and swell it. The original film, if incorrectly formulated, may be in a state of stress from the reflow baking operation due to differential cooling rates of the metal substrate and film, which also have substantially different coefficients of heat expansion. Now, a repair operation that only swells the original film in its stressed state can provoke severe cracking, crazing, or even lifting of the original film in the repair area. Section 3-3, A should be studied for identification of formulation compositions that are intended to inhibit such disasters.

Further advantages of the thermoplastic topcoat have to do with the fact that in film forming by solvent evaporation the ultimate film composition consists fairly precisely of the additive composition of the blended components. This fact provides the formulator with the opportunity to characterize his components individually and readily determine such compositional adjustments and compromises as are required to achieve, or at least approach, a desired end in film performance. One outstanding example can be cited. Acrylic lacquers have been famed for their versatility in satisfying styling requirements of metallic pigmentation. Metallic coatings gain much of their distinctive appearance not only from the reflective glitter of the metallic flake, but also from their propensity to offer continually changing nuances of color as the viewing angle changes. The widely variant contour of the automobile surface and the mobility of the surface with respect to the observer's eye provide the occasion for change in the viewing angle. This effect is known by the somewhat vulgar but descriptive term of "flop."

Color flop is a function of two pigmentation properties — a relatively high degree of coating transparency requiring that presence of light-scattering pigments be minimized, and orientation of most of the flake particles in the same plane. This combination of circumstances provides that light reflected from within the film, where characteristic, selective wavelength absorption produces color, will arrive at the observer's eye in a nondiffuse manner. The oriented reflectors, the metallic flakes, reflect light in nearly parallel rays. Since these reflectors are ideally immobilized by the coating binder, as the angle of reflection changes by variation in surface contour with respect to the observer, the color of the surface appears to change as the depth of film penetration of the reflected beam changes.

Metal flake for pigment use, when at rest within the film, orients with its principal surface area parallel to the film surface. Shrinkage of the film due to solvent evaporation promotes this desirable orientation, but solvent streaming to the surface to evaporate tips the flake and tends to combat orientation by randomizing the flake position. The structure of the high-molecular-weight polymer in lacquers, combined with the shrinkage of the low-solids coating, assists the flake to retain its orientation. The lacquers for automotive use, both cellulose nitrate and acrylic types, have accordingly found great favor with stylists because of the pronounced color flop they are able to generate.

The use of lacquers as sealers or midcoats under lacquer topcoats is fairly common as a part of the original finish system, but in automotive technology, the formulation of lacquer primers is restricted to field refinishing and is discussed in that section of the chapter (Sec. 3-6, C).

B. FILM FORMING BY CHEMICAL CROSSLINKING

Chemical crosslinking of films is employed broadly in formulating both automotive topcoats and undercoats. The thermoset films that result have their appeal chiefly for two types of properties. Because the structural complexity needed for high-level performance of films in antagonistic environments is developed by chemical reactions that take place after application, enamels can be applied at relatively high-solids content without the complications of high viscosity. Furthermore, the crosslinked structure produces films that are highly solvent resistant, as well as being resistant to dimensional distortion in humidity, and somewhat more impermeable than thermoplastics to penetration of staining fallout from atmospheric pollution.

As a result of their higher solids content and low-application viscosity, pattern control and color flop of metallic pigmentations in thermosetting vehicles are less versatile than are possible with thermoplastics. Formulators working with thermosetting topcoats must resort to special modifications to encourage orientation of metallic flake.

The factory repair operation involving thermosettable topcoats (enamels) is certainly more critical than that described for thermoplastics. Spot repairs are almost never attempted at the end of the line because the edge of the repair remains too obvious if it occurs in the center area of a section. Panel repairs are therefore the ordinary approach with enamels. The special case of formulation for repair-in-process enamels is discussed in Sec. 3-3,B.

Original finish enamels are formulated to be cured at schedules involving temperatures between 120 and 150° C. Since an assembled automobile has heat-sensitive components that are distorted by such schedules, the original finish enamel is modified for lower temperature cure repair work by the addition of cure catalysts. The chemistry of most topcoat cures involves condensation reaction between alkoxy groups in melamine formaldehyde resin and hydroxyl sites on the polymer (see Fig. 3-4). That reaction is highly responsive to acid catalysis. Repair baking schedules are usually found in the range of 82 to 91°C for 20 to 30 min.

Formulators have employed a variety of catalysts such as p-toluene sulfonic acid, maleic anhydride, phthalic anhydride, citric acid, cyclamic acid, phosphoric acid, and acid phosphate esters. The selection of the catalyst is usually moderated by its solubility in the resin system and by untoward side effects such as production of color bodies, a tendency to remain active after the cure to cause film embrittlement, and failure of the cured repair to have adequate intercoat adhesion. While local sanding of blemishes in the original enamel film is necessary to remove them, it is not normally the practice to sand the entire panel being recoated as a means of achieving repair coat adhesion to the original film. The original finish baking schedule is restricted, therefore, within limitations of good film formation, to facilitate repair intercoat adhesion to the unsanded, full-gloss original.

Enhanced gloss of automotive topcoats by a polishing operation is not unusual for thermoplastic formulations because the local heating of the film by friction of polishing produces a degree of clear resin reflow to the surface of the film. Polishing operations involve hand labor and are therefore too expensive for general use as a production practice. It has been observed, however, that a polishable topcoat lends itself conveniently to end-of-line repairability. Very small blemishes or shallow scratches can be lightly sanded, with the film being restored to full gloss by polishing if it has sufficient heat reflow capability. Acrylic enamels are almost unique among thermosetting compositions in that their crosslink density can be controlled by composition and selection of cure schedule to allow acceptable polishing properties after cure.

Undercoats for original finish automotive systems are almost universally formulated with thermosetting vehicle systems, partially because of the chemical protection afforded by such resins as oil-modified epoxy esters,

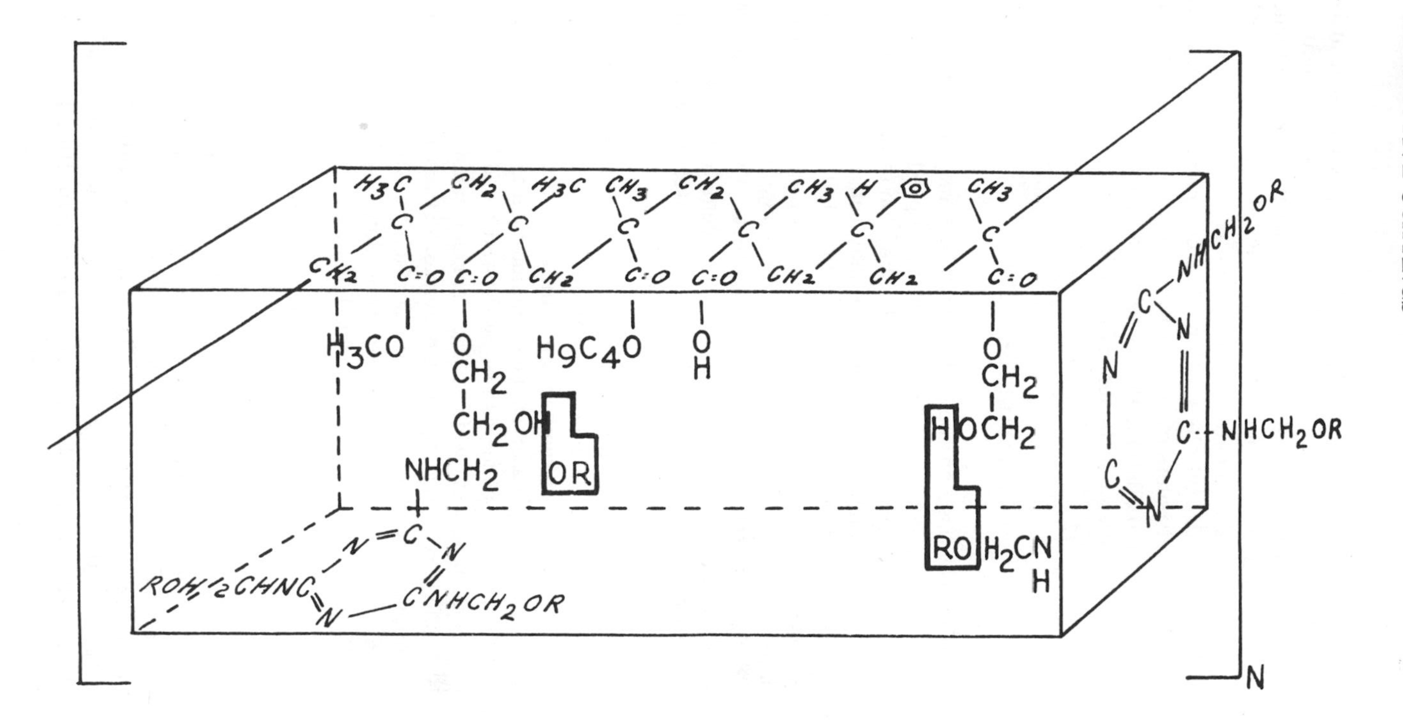

FIG. 3-4. Thermosetting acrylic polymer crosslinked with melamine formaldehyde resin.

partly because of mechanical strength of the crosslinked structure, and finally because of their contribution to system appearance due to their efficiency in obscuring substrate pattern and resisting the swelling tendency of topcoat solvents. It should be noted that when semidrying oil derivatives, such as soya acids, are used to esterify epoxy resins, the films cure both by being crosslinked at hydroxyl sites with melamine or urea formaldehyde resins, and by oxidation at points of unsaturation. If metallic soaps are added as driers to initiate oxidation, their choice and concentration is critical to avoid post-cure embrittlement of the film.

C. COALESCENCE

Almost every film that finds an application in an automotive coating system is formed as a result of some sort of coalescence. Dipped film based on a solution polymer would seem to be a practical exception to the above statement, but even here, elimination of the volatile may require some progressive flow of the residual film in order that its thickness and ultimate performance be uniform. Certainly, in the case of objects coated by spraying solutions, emulsions, dispersions, or powders, an important stage in film formation is the flowing of discrete particles into an integral sheet. The action of boundary layers of naturally occurring or added surfactants, vehicle components of limited compatibility, or vehicle fractions of widely differing molecular weights must be overcome to produce an integral film.

When the interfaces of particles to be coalesced are of identical composition so that they have equal surface tension, coalescence is normally spontaneous, particularly when the surface over which they must spread to form a continuous film has greater surface tension than the surface tension of the liquid coating. Coalescence can be facilitated by adding very small amounts of flow-control agents, which reduce surface tension of the liquid coating. Poly(2-ethylhexylacrylate) is an example. Silicone oils are used occasionally in minute quantities to equalize surface tension at interfaces. While coalescence may be promoted, experimentation may be required to find a type of silicone oil that does not inhibit leveling.

Coalescence is of particular importance in those instances (emulsion or dispersion vehicles) when two-phase vehicles are used in the formulation. The continuous phase of the vehicle is frequently largely made up of a volatile liquid (water, for instance), some of which must remain in the film long enough for polymer aggregates to fuse. Shrinkage of the continuous phase caused by evaporation brings the polymer particles into contact. Fusion of particles is accomplished because their composition has been chosen in such fashion that hard monomers are sufficiently softened by comonomers that deformation of spherical aggregates into films is promoted (see Fig. 3-2).

Nonvolatile plasticizers, or slowly volatile solvents referred to as coalescents, may be present in the continuous phase to swell the aggregate after loss of faster-evaporating diluent, thus ensuring film coalescence. Baking the two-phase vehicle is a useful operation to ensure particle coalescence, since polymers are softened by heat, but two precautions must be observed. If the continuous phase of an emulsion or dispersion is sharply reduced by inadequate baking, prolonged air dry prior to baking, or absorption by porous substrate, the distance between polymer aggregates can be too great for coalescence, and mud-cracking is found instead of the continuous film that is desired. Furthermore, if the polymer is of the reactive or thermosettable type, the formulator must employ retarders in the form of a solvent which evaporates slowly enough to provide time for coalescence before crosslinking occurs and blocks further flow of the vehicle.

Coalescence is particularly critical when films are to be formed from powder application. The formulator is limited in type and concentration of flow-promoting coalescents that may be introduced. Premature softening of the particles will cause them to stick together, thus inhibiting application, while hard particles that are deformed with difficulty into films will require excessive baking temperatures or produce a very rough film surface.

3-3. SPECIFIC TOPCOAT FORMULATION

A. THERMOPLASTIC TOPCOATS FOR ORIGINAL AUTOMOTIVE FINISHING

Acrylic thermoplastics began to be widely used as the principal component of original topcoat thermoplastics for automobiles in 1950, and, with a succession of modifications, have continued to be used into the 1970s. They replaced formulations based on plasticized cellulose nitrate vehicles, and it is not surprising to find that the original acrylic lacquers tended to approach the result of simple substitution of a portion of the cellulose nitrate with a linear acrylic polymer. The formulations that have evolved, however, are the result of recognizing properties unique to acrylic polymers and capitalizing on these properties to solve problems in the automotive finishing industry.

Figure 3-3 illustrates the chemical structure of a typical thermoplastic acrylic polymer. The property that led to the rapid adoption of acrylics as replacements for cellulose nitrate was the high order of ultraviolet light transmission of acrylate and methacrylate ester polymers, and the resultant resistance to photodegradation on exposure to weather. Coatings formulation and application, therefore, must be accomplished in a manner that retains the bulk of the advantage originally associated with acrylics. The additional modifying raw materials with which we are concerned are pigments and pigment dispersants, cellulosic modifiers, plasticizers, solvents, diluents, and ultraviolet screening agents. A typical formulation for an acrylic lacquer, based on 100 lb of film-forming components, is shown in Table 3-1. The spray application of the acrylic topcoat described in Table 3-1 is shown in Table 3-2.

Excellent pigment dispersions for acrylic lacquers are prepared by premixing dry pigments with a portion of the polymer solution and passing the mixture through a sand mill or steel ball attritor. Special copolymers may be preferred for dispersing pigments, because the polar groups introduced by co-monomers, such as methacrylic acid, hydroxyethyl methacrylate, or aminoethyl methacrylate, can be counted upon for easy wetting of pigment surfaces and the deflocculation of pigment particle aggregate. The concentration of polar grinding vehicles is minimized in lacquer formulation since their reaction is not normally a part of the film-forming reaction. Such polar groups as are introduced are likely to remain as undesirably active reaction sites in the film during attack by the environment.

Acrylic polymers for thermoplastic coatings are designed in part for their mechanical strength, which is, of course, a function of their high molecular weight. The structure of these polymers as a portion of the film of a protective coating, is the same structure that exists in the coating material as applied, indeed the same structure that was formed in the polymer reaction kettle. Strength is manifested principally in the ability of the polymer to maintain film integrity for a lifetime measured in years. It is true, of course, that other performance areas such as softening point and reflow temperature are critically important, but these properties are controlled by a combination of comonomers and modifying resins.

1. Selection of Polymers

Tables 3-3, 3-4, and 3-5 have been reproduced from descriptive literature prepared by a commercial acrylic polymer manufacturer. They describe a range of acrylic compositions and polymer properties that might typically be used by an automotive lacquer formulator.

Note that solution polymers 6010, 6011, and 6012 are all described as methyl methacrylate polymers in Table 3-3, and that while there is some correspondence between resin properties in solution and bead form (6010 and 2010 for instance), bead polymers 2008 and 2041 are included to extend the molecular weight range. The higher the inherent viscosity, the higher the molecular weight. Polymer solution 6012, with its polar modification, could be used as a pigment dispersant, and 6011 as the principal lacquer polymer.

TABLE 3-1. Package Composition: Typical Thermoplastic Acrylic Automotive Topcoat (Pastel)[a]

Material[b]	Material (lb)	Material (gal)	Film forming (lb)	Film forming (gal)
Elvacite #6012[c]	54.6	6.80	21.9	2.13
Cellosolve acetate[d]	35.7	4.46	—	—
Toluene	71.4	10.20	—	—
Rutile titanium dioxide pig. plus trace tinting	163.0	4.64	163.0	4.64
Grind to Hegeman 7+ high-density ceramic media attritor				
Plasticizer				
TMPD-DB[e] or paraplex G-30[f]	47.3	5.44	47.3	5.44
Cellulosic EAB-381-2[g]	47.3	4.86	47.3	4.86
Elvacite #6011[c]	310.0	38.74	124.0	11.74
Cellosolve acetate[d]	14.0	1.73	—	—
Acetone	84.5	12.83	—	—
Toluene	72.7	10.30	—	—
	900.5	100.00	403.5	28.81

[a] Total solids (w/w) 45%
Total solids (vol/vol) 28.8%
P/B (w/w) 1/1.47
% pig. of film (w/w) 40
% pig. of film (vol/vol) 16
% vehicle solids (w/w) 32.5

[b] Compositional ratios: resin — acrylic/plasticizer/cellulosic 60/20/20
solvent — acetone/toluene/cellosolve acet. 30/60/10

[c] E.I. Du Pont de Nemours & Co., Inc.

[d] Union Carbide.

[e] Velsicol Chemical Corporation.

[f] Rohm & Haas Co.

[g] Eastman Chemicals Products, Inc.

TABLE 3-2. Spray Application of Acrylic Topcoat Described in Table 3-1[a]

Materials	lb	gal
Acetone	1258	190
Toluene	2305	320
Xylene	720	100
i-Propyl alcohol	922	140
Cellosolve acetate	2000	250
	7205	1000

[a]Reduce package composition of Table 3-3 with an equal volume of spray reducer made by blending the above ingredients. Recommended reduction produces approximately 15% vehicle solids, the property which is usually the limiting factor in controlling sprayability of the coating. More thorough atomization and smoother, wetter films are obtained by increasing the reduction to use 1.25 to 1.50 vol reducer per vol of package lacquer.

Bearing in mind the performance restrictions that control ease of atomization and polymer reflow as opposed to cold cycle resistance, opportunity is afforded to choose a molecular weight range that is optimal for the end use. Within limits of compatibility, coating properties such as film hardness, flexibility, and resolubility can be modified by blending homopolymers and selected copolymers. The compatability limits are sometimes surprisingly narrow and the formulator who blends resins is warned to keep a check of compatibility by examining the clarity of unpigmented, thoroughly dried, reasonably thick films applied to clean glass. Borderline compatibility revealed by hazy films of blended resins may signal later problems of package stability on pigmented film gloss.

The relatively high Tukon hardness values of 18 and 19 for resins 6012 and 6011 suggests immediately that for obtaining a desirable level of film reflow after spray application, plasticizer modification in the range of 25 to 40 parts per 100 parts of polymer solids would be required. Alternatively, plasticizer concentrations can be decreased by choosing a softer polymer (6013 or 6024 copolymers, 13 Tukon, or 6014, 4 Tukon) to replace 6011, or if compatibility allows, to blend with 6011. Number 6024, for

TABLE 3-3. Elvacite Acrylic Resins — Commercially Available Grades[a]

Elvacite (grade number)	Polymer type	Typical inherent viscosity[b]	Molecular weight	Principal characteristics and uses
				Bead polymers
2008	Methyl methacrylate	0.20	Low	Lowest molecular weight grade. Used in lacquers and gravure inks requiring low viscosity at high solids. Highest tack temperature.
2009	Methyl methacrylate	0.45	Medium	A more soluble, softer version of Elvacite 2010. Shows unusual chip resistance for a hard polymer.
2010	Methyl methacrylate	0.45	Medium	General-purpose grade for lacquer coatings such as barrier topcoatings for vinyl fabric. Highest tensile and hardness.
2013	Methyl/n-butyl methacrylate copolymer	0.20	Low	Fast-dissolving, low-viscosity grade with quick solvent release for use in industrial lacquers, aerosols, inks, and coatings for plastics. Softer than the methyl methacrylate polymers.
2041	Methyl methacrylate	1.20	Very high	The highest molecular weight grade. Used in barrier topcoating of vinyl fabric for dielectric heat sealability, maximum abrasion resistance, block resistance, and slip.
2042	Ethyl methacrylate	0.91	Very high	Tough, alcohol-tolerant, broadly compatible grade for use in abrasion-resistant coatings such as high-gloss clear lacquers for decals and outdoor signs. Slightly softer than Elvacite 2013.
2043	Ethyl methacrylate	0.20	Low	Fast-dissolving, low-viscosity resin with alcohol solubility for flexographic inks. Most compatible grade. Has excellent pigment wetting ability and broad solubility for use in lacquers for solvent-sensitive substrates. Equal in hardness to Elvacite 2042.

TABLE 3-3. — Continued

Elvacite (grade number)	Polymer type	Typical inherent viscosity[b]	Molecular weight	Principal characteristics and uses
2044	n-Butyl methacrylate	0.53	High	Softest, most flexible bead grade. Useful in adhesives for smooth plastic films and aluminum, and in silk screen inks. Will plasticize and improve adhesion of harder butyl grades (Elvacite 2045 and 2046) and nitrocellulose. Improves outdoor durability of vinyl chloride resins in pigmented lacquers.
2045	Isobutyl methacrylate	0.66	High	Hardest of the weak-solvent-soluble butyl grades. Useful in seamless flooring, in topcoats for vacuum metallized plastics, and in lacquers for other solvent-sensitive substrates. Softer than Elvacite 2042 and 2043.
2046	n-Butyl/isobutyl methacrylate, 50/50 copolymer	0.61	High	Medium-hardness butyl grade. Soluble (like Elvacite 2044 and 2045) in mineral spirits, VM & P naphtha, and some alcohols. Useful in aluminum finishes, ceramic coatings, and aerosol snows.
				Solution polymers
6010	Methyl methacrylate	0.45	Medium	Solution in methyl ethyl ketone (40% solids by weight). Resin properties correspond to bead grade 2010.
6011	Methyl methacrylate	0.34	Medium	Solution in 30/70 weight ratio of acetone/toluene (40% solids by weight) for convenience in lacquer formulation. Useful for automotive exterior finishes and barrier coatings for plastics.

6012	Methyl methacrylate	0.38	Medium	Solution in 30/70 weight ratio of acetone/toluene (40% solids by weight). Carboxylated modification of Elvacite 6011 for improved pigment wetting and adhesion. Used in exterior automotive top-coat lacquers.
6013	Methyl/n-butyl methacrylate copolymer	0.20	Low	Solution in toluene (40% solids by weight). Resin properties correspond to bead grade 2013. Softer than Elvacite 6010, 6011, and 6012.
6014	Methyl methacrylate copolymer	0.45	Medium	Solution in 80/20 weight ratio of toluene/isopropanol (40% solids by weight). Softest and most flexible of the Elvacite copolymer resins. Good pigment wetting. Useful for clear, translucent or pigmented metal lacquers, e.g., for aircraft. Also useful for "Electrofax" copy paper coatings and seamless flooring.
6015	Methyl/n-butyl methacrylate copolymer	0.25	Low	Solution in methyl ethyl ketone (45% solids by weight). Useful for specialty plastic coatings. Corresponds to Elvacite 6016 in resin properties.
6016	Methyl/n-butyl methacrylate copolymer	0.25	Low	Solution in toluene (40% solids by weight). General-purpose lacquer grade for metal, plastic, wood, paper and cementitious substrates. Clears withstand stress cracking on metal in outdoor exposures. Useful for seamless flooring and walls. Softer than Elvacite 6013.
6017	n-Butyl methacrylate	0.53	High	Solution in methyl ethyl ketone (50% solids by weight). Forms tough, flexible films with high elongation. Shows excellent adhesion to most substrates including nonporous surfaces. Resin properties correspond to bead grade 2044.

TABLE 3-3. — Continued

Elvacite (grade number)	Polymer type	Typical inherent viscosity[b]	Molecular weight	Principal characteristics and uses
6024	Methyl/n-butyl methacrylate copolymer	0.20	Low	Solution in toluene (40% solids by weight). Resin properties similar to bead grade 2013 but with greater adhesion and gloss.
6026	Methyl/n-butyl methacrylate copolymer	0.25	Low	Solution in toluene (50% solids by weight). High solids permit dilution to meet air pollution rules. Resin properties correspond to Elvacite 6016.
6036	Methyl/n-butyl methacrylate copolymer	0.25	Low	Solution in toluene (50% solids by weight). High solids provide formulating latitude including dilution to pass Rule 66. Useful for cementitious coatings. Intermediate in hardness between Elvacite 6016 and 6013.

[a]Courtesy E. I. Du Pont DeNemours, Du Pont Elvacite acrylic resin.

[b]Inherent viscosity of a solution containing 0.25 g polymer in 50 ml chloroform, measured at 25°C using a No. 50 Cannon-Fenske Viscometer.

instance, is reported to be compatible with 6012 and 6011 up to equal parts. Film properties of blended resins are not necessarily arithmetic averages of those properties obtained with individual components. The innovative formulator may therefore look for blends that develop and reinforce desirable properties of components and suppress those that are less desirable.

2. Role of Plasticizers

Plasticizers were originally incorporated into thermoplastic coatings to soften such naturally hard polymers as cellulose nitrate in the interest of increasing their flexibility. Acrylic polymers, however, where flexibility can be extensively modified synthetically by monomer selection (see Fig. 3-2), depends on plasticizers to function more as nonvolatile solvents in promoting heat reflow leading to film leveling and gloss. Phthalate diesters, butyl benzyl phthalate for instance, have been found to be excellent solvents for acrylic polymers. However, at the reflow temperatures (150°C or greater) used for thermoplastic original finish today, it would not be unusual for as much as one-third of the butyl benzyl phthalate plasticizer to be volatilized during the reflow schedule. Accordingly, nonvolatile, non-drying polyesters are favored for their permanence in the film. Phthalic derivatives are broadly compatible, but glycol or glycerol esters of aliphatic acids, sebacic, adipic, or azelaic are used for special formulations.

Good compatibility of the plasticizer with the polymer is important for several reasons. Not only is solvent activity of plasticizer on the polymer essential for the lubricating effect, which allows plasticized polymer molecules to slide over each other under mechanical stress or impact, but also incompatible plasticizers tend to be squeezed to the film surfaces producing a "spew" or greasy layer which may be unsightly, or worse, will collect dirt, discolor in high humidity, or initiate gloss loss during exterior exposure. Exuding plasticizer can also accumulate at the lower surface of the film at the interface with substrates, adversely affecting adhesion. Another property which is partially a function of compatibility is resistance of plasticizer to extraction by water and gasoline. Even plasticizers which are mildly soluble in gasoline, for instance, tend to resist extraction from the film if they are preferentially soluble in the polymer. Classification of plasticizers by the solubility parameter concept can, therefore, be handled in the same manner, and can be as beneficial to the formulator as solvent parameter classifications of thinners.

A full listing of available plasticizers, complete with properties and suppliers, is readily available (20).

3. Cellulosic-modifying Resins for Thermoplastic Reflow Topcoats

Since current automotive top coats are purchased for heat reflow application, it is important that any modifying cellulosic as well as the acrylic

TABLE 3-4. Typical Properties of Elvacite Bead Polymers[a]

	Methyl methacrylate resins				Methacrylate copolymer resin	Ethyl methacrylate resins		Butyl methacrylate resins		
Elvacite grade	2008	2009	2010	2041	2013	2042	2043	2044	2045	2046
Density, lb resin/gal resin[b]	9.87	9.54	9.98	9.96	9.60	9.23	9.51	8.86	9.08	9.04
Bulking value, gal resin/lb resin[b]	0.1013	0.1048	0.1002	0.1004	0.1042	0.1083	0.1051	0.1129	0.1101	0.1106
Specific gravity, 25°C/25°C[c]	1.19	1.15	1.20	1.20	1.15	1.11	1.14	1.07	1.09	1.09
Tukon hardness, Knoop No.[d]	18	17	19	19	13	11	11	<1	8	4
Tack temperature, °C	96	68	88	90	77	57	50	40	63	52
Acid number[e]	9	0	0	0	5	0	9	0	0	0
Tensile strength (22.8°C, 50% RH), psi[f]	4000	9500	10,500	7800	1600	5400	1000	500	3600	2100

Elongation at break (22.8°C, 50% RH), %[f]	0.5	3.5	3.5	1	0.5	25	<1	300	1	175

[a]Courtesy E. I. Du Pont De Nemours, Du Pont Elvacite acrylic resin.

[b]Calculated from density (ASTM D 1475) of 20% solutions of bead grades in methyl ethyl ketone or of solution grades as supplied.

[c]Calculated from density, using water = 8.32 lb/gal.

[d]Measured on 1/16 in. pressed sheet.

[e]Milligrams potassium hydroxide per gram polymer.

[f]Determined using compression-molded samples.

[g]Brookfield Model LVF, 60 rpm; #2 spindle for viscosities up to 400 cp; 400-1650 cp, #3 spindle; above 1650 cp, #4 spindle.

TABLE 3-5. Typical Properties of Elvacite Solution Polymers[a]

	Methyl methacrylate resins			Methacrylate copolymer resins							Butyl methacrylate resin
Elvacite grade	6010	6011	6012	6013	6014	6015	6016	6024	6026	6036	6017
Solution as supplied											
Solids, % by wt	40	40	40	40	40	45	40	40	50	50	50
Density, 25°C, lb/gal	7.78	8.01	8.03	8.05	7.90	7.90	7.95	8.05	8.14	8.22	7.69
Bulking value (calc'd), gal resin/lb solution	0.0401	0.0363	0.0360	0.0417	0.0421	0.0480	0.0430	0.0417	0.0553	0.0542	0.0554
Viscosity at 25°C, Brookfield, cp[g]	5000	2400	2760	300	1000	300	300	700	2500	3900	3000
Gardner	Z5-Z6	Z-Z1	Z1-Z2	L-M	V	L-M	L-M	U-V	Z1-Z2	Z2-Z3	Z1-Z2
Flash point, Tag open cup, °C (ASTM D 1310)	3	-4	-4	5.5	4.4	1.6	4.4	1	15	16	4.4
Contained resin											
Density (calc'd), lb resin/gal resin[b]	10.26	10.34	10.36	9.63	9.51	9.38	9.34	9.63	9.23	9.41	9.03
Specific gravity, 25°/25°C[c]	1.23	1.24	1.25	1.16	1.14	1.13	1.12	1.16	1.11	1.13	1.08

Tukon hardness, Knoop No[d]	19	19	18	13	4	7	7	13	7	11	<1
Tack temperature, °C	88	98	100	73	63	60	60	73	60	67	40
Acid number[e]	0	0	13	7	13	3	3	13	3	3	0
Tensile strength (23°C, 50% RH), psi[f]	10,500	—	—	1600	3500	1000	1000	1600	1000	1000	500
Elongation at break (23°C, 50% RH), %[f]	3.5	—	—	0.5	18	0.5	0.5	0.5	0.5	0.5	300

[a]Courtesy E. I. Du Pont De Nemours, Du Pont Elvacite acrylic resin.

[b]Calculated from density (ASTM D 1475) of 20% solutions of bead grades in methyl ethyl ketone or of solution grades as supplied.

[c]Calculated from density, using water = 8.32 lb/gal.

[d]Measured on 1/16 in. pressed sheet.

[e]Milligrams potassium hydroxide per gram polymer.

[f]Determined using compression-molded samples.

[g]Brookfield Model LVF, 60 rpm; #2 spindle for viscosities up to 400 cp; 400-1650 cp, #3 spindle; above 1650 cp, #4 spindle.

polymers be selected to contribute to heat reflow. The mixed ester, cellulose acetate butyrate, is the preferred modifying cellulosic for acrylic lacquers, its choice over other cellulose derivatives being based not only on its contribution to heat reflow, but also on its compatibility with acrylic polymers and on its durability during exterior exposure. Cellulosic modification of acrylic lacquers was initiated and probably continues to be used because acrylic polymers, so modified, release solvents more rapidly than films of the neat polymer. Faster solvent release from films enhances pattern control of metal flake pigment and reduces the opportunity for solvent "popping" when thick films are baked. Furthermore, the modified acrylic is more easily atomized during spraying so that less dilution and lower air pressure is needed than would be true for the same thermoplastic acrylic polymer unmodified.

Cellulose acetate butyrate is sold in a range of butyrate levels. For automotive coatings, the useful butyrate level varies between 38 and 55%, the higher butyryl esters usually selected for most compatible blending with methyl methacrylate homopolymers, the lower butyryl esters more normally being used with copolymer resins. The higher butyryl esters as acrylic modifiers have better exterior durability than the cellulose acetate butyrate with lower butyryl content (21).

Originally, the use of cellulose acetate butyrate to replace cellulose nitrate as modifying resin in acrylic lacquers was adopted to facilitate heat reflow in a topcoating formulation called "bake-sand-bake." After a preliminary bake at a schedule in the vicinity of 15 min at 107°C to evaporate the bulk of its solvent, film irregularities such as application-derived orange peel, or subsurface profile replications, are removed by sanding. The sandpaper chosen must be #500 grit or finer, and the sanding paper must be lubricated with a nonsolvent, such as mineral spirits, to obtain uniform abrasion. If the film has a composition that is swollen by mineral spirits, water is occasionally used to advantage as a sanding lubricant. The sanded surface is thoroughly cleaned and then subjected to a reflow schedule (20 to 30 min at 150°C) to smooth the fine sand scratches and restore gloss without polishing.

The bake-sand-bake schedule has not been used in very recent years because formulations and baking schedules provide film gloss and smoothness without resorting to the expense of sanding.

4. Heat-reflowable Thermoplastic Organosols—Acrylic Dispersion Lacquers

Experimentation begun in the 1960s (22, 23) matured in the 1970-1971 production of automotive thermoplastic acrylic finishes aimed at circumventing certain problems of the lacquer applications caused by the structure and high-molecular-weight of the polymers. It has been observed with aqueous emulsions for instance, that their viscosity is essentially that of their continuous phase. High-molecular-weight polymer dispersions in organic nonsolvents accordingly, as compared to "solutions" of the same polymer, can be made to have very low viscosities at fairly high total solids content. Their application by air atomization is as feasible as the more dilute solutions, and because application solids level is higher and fluid delivery rates of dispersions can frequently be made greater than solutions, film build is expedited so that two coats per auto body suffice where three or four coats were needed with solution formulations. While aqueous emulsions (particularly those designed for baking during film formation) have been modified to have fair gloss, the dispersion polymers whose suspensions are stabilized in a nonsolvent, organic, continuous phase may surpass solution formulations in developing film gloss. Strong solvents are incorporated as coalescents, or filming aids, and the emulsifying agents common to aqueous suspensions are supplanted by "stabilizers" which become an integral part of the final film.

Modifying plasticizers for dispersion lacquers must be carefully selected for their compatibility with the nonsolvent continuous phase or, if incompatible, stably dispersed in the nonsolvent for blending with the disperse polymer, a difficult maneuver. Cellulosic modifications are infrequently used in dispersion lacquers, again because of their difficult compatibility with the nonsolvent continuous phase present in the package and spray stages of the liquid coating.

Once the application of the thermoplastic acrylic organosol or dispersion coating is accomplished, heat-reflow schedules and ultimate film performance are similar to those encountered with commercial solution lacquers. The same gamut of testing procedures is used with both these types of automotive topcoats.

5. Ultraviolet Screening Agents

In view of evidence that film degradation reactions are energized by ultraviolet radiation, it is natural that there has been an extensive search for effective ultraviolet-screening agents (24-26). Many pigmentations that effectively scatter or absorb visible light, titanium dioxide and carbon black for instance, are effective enough in protecting polymers against ultraviolet that addition of screening agents to coatings so pigmented has not been very useful. Clear coatings and the semitransparent coatings with metallic pigmentations are more likely prospects for utility of screening chemicals.

Highly useful ultraviolet screening agents absorb strongly in the ultraviolet region of incident wavelengths while transmitting visible light without color change. Absorbed energy must be dissipated harmlessly, preferably as heat. Screening agents must themselves be stable to ultraviolet except for formation of equilibrium rearrangements by which absorption and dissipations take place. They must also be nonvolatile at film-forming, heat-reflow temperatures and, while highly compatible with polymers they are intended to protect, they must be insoluble enough to resist extraction from films by elements of the environment of use.

Three general types of screening agents in commerical use are: (a) benzophenones, example, 4-dodecyloxy-2-hydroxybenzophenone (Eastman Inhibitor DOBP), (b) benzotriazoles, example, 2-(2' hydroxyphenyl benzotriazole (similar to Geigy Tinuvin 328), and (c) salicylic esters, o-nitrophenol derivatives, substituted acrylonitriles, and some triazines. This latter group of compounds is seldom used in automotive topcoats becuase of discoloration that occurs during exposure. It has been suggested that a rearrangement of their structure occurs by resonance to produce yellow-toned hydroxybenzophenones.

B. THERMOSETTING TOPCOATS FOR ORIGINAL AUTOMOTIVE FINISHING

Mechanisms of crosslinking film formation by thermosetting have been presented in Sec. 3-2,B and Fig. 3-3. The utility of applying relatively low-molecular-weight polymers at maximum sprayable solids and continuing to build molecular weight after application by condensation or addition reactions was also presented.

1. Selection of Polymers

The modern automotive enamel topcoat in the United States is almost exclusively derived from an acrylic polymer containing hydroxyl functionality. The hydroxyl groups react under acid conditions with melamine formaldehyde polymer whose reactive functionality consists of a mixture of:

1. amino hydrogen $-N{\angle}H$.
2. methylol groups from reaction of formaldehyde with amino hydrogen $-N{\angle}CH_2OH$.
3. alkoxy groups $-N{\angle}CH_2OR$ ($R{=}CH_3-$, $\begin{matrix}CH_3\\CH_3\end{matrix}{>}CH_2-$, C_4H_9- , etc.).

The reactivity of the melamine resin functionality decreases as one proceed from 1 through 3, while the compatibility and viscosity stability when mixed with acrylic polymer increases as measured by increasing mineral-thinner tolerance. The concentrations of reactive functionality present in acrylic polymers and melamine resins for automotive enamels is such that weight ratios ranging from 70/30 to 60/40, respectively, of the resin solids will usually optimize crosslink density, chemical resistance, and film appearance.

Table 3-6 shows the composition of a typical pastel automotive enamel. Tables 3-7 and 3-8, respectively, list available raw material intermediates as supplied by the Rohm and Haas Company.

The acrylic polymers for thermosetting automotive enamels are frequently interpolymers in which, in addition to the hydroxy acrylic esters used to introduce crosslinking sites (hydroxy ethyl acrylate or methacrylate for instance), other monomers are employed to design a backbone that fulfills the complete range of required topcoat performance. In contrast to the practice earlier described of modifying the polymer by blending with plasticizers, the enamel flexibility and reflow character are better controlled by selecting as some portion of the monomers those which have low glass transition temperatures. The alkyl acrylates are convenient for this purpose and their effectiveness increases in proportion to the alkyl chain length. To ensure a film of useful hardness, its glass transition temperature must be higher than the temperature of the anticipated environment of use for which purpose the more rigid methacrylates and styrene are applicable as yet another portion of the polymer backbone. Concentrations of each of the components of the interpolymer must be controlled within fairly close limits, particularly that of the styrene, which contributes not only hardness, but also lower cost and higher pigmented film gloss than the alkyl methacrylates. Styrene, because of its aromatic character, is often suspect in coatings requiring maximum durability, since in high concentration it contributes to early gloss loss during exposure to sunshine relatively rich in ultraviolet wavelengths. One study (27) of the optimum monomer ratio reports that an interpolymer of approximately equal parts of methacrylate esters, acrylic esters, and styrene is preferred for film gloss retention.

It is not unusual that a low weight percentage ($\frac{1}{2}$ to 5%) acrylic or methacrylic acid would be included in the interpolymer to facilitate the curing reaction with melamine resin during the film bake.

2. Enamel Solvents

Solvents for acrylic enamels can be selected on the usual basis of polymer compatibility, contribution to atomizing ease and film flow, and desired release from the film so that cratering is not encountered during film flow. It should be noted that 3% by volume of n-butyl alcohol is shown in the formula of Table 3-4 to provide package viscosity stability during enamel storage; the balance is largely aromatic hydrocarbon. Should it be a requirement that the formulation conform with air pollution legislation, which in some cases restricts aromatic hydrocarbon content, blends of aliphatic hydrocarbons with esters or straight chain ketones can be substituted for xylene. In selecting such alternate volatile combinations, it is usually necessary to approximate the viscosity reduction curve of the aromatic

TABLE 3-6. Package Composition: Typical Thermosetting Acrylic Enamel (Pastel)[a]

Material[b]	Material (lb)	Material (gal)	Film forming	
			lb	gal
Acryloid AT - 56[a]	55.7	6.86	28	2.90
Xylene	111.2	15.45		
Rutile titanium dioxide pigment plus tinting colors	205.0	5.88	205	5.88
Grind to Hegeman 7+ in high-density ceramic attritor; reduce with agitation using				
Acryloid AT - 56[c]	347.0	42.75	173	18.12
Uformite MM - 47[c]	180.0	21.10	108	10.40
Butanol	19.0	3.00		
Sovesso 100[d]	35.7	4.96		
	953.6	100.00	514	37.30

[a]

Total solids (w/w)	54%
Total solids (vol/vol)	37.3%
P/B (w/w)	1/1.51
% pig. of film (w/w)	40%
% pig. of film (vol/vol)	16%
% vehicle solids (w/w)	41.3%

[b]Application directions: reduce package enamel to spray with 15% by volume of xylene or a 1:1 mixture of xylene: Solvesso 100, apply in three wet coats to a dry film of 2 mils, using an air dry flash period of 2 min between coats, and bake 30 min at 120–130°C. For factory repair add 3% by volume of reduced enamel of a 25% by weight solution of alkyl acid phosphate in xylene, spray, and bake repair coat 20 min at 82–93°C. Prepare new batches of repair catalyzed enamel at least every 24 hr to ensure repair film gloss at least equal to original finish gloss.

[c]Rohm & Haas Co.

[d]Humble Oil Co.

TABLE 3-7. Physical Properties of Uformite Resins

Grade	Type	Percent solids ±2%	Solvent	Acid number, solids basis	Specific gravity	lb/gal	Color	Viscosity, Gardner-Holdt at 25°C	Mineral thinner tolerance[b]
Uformite F-200E	Urea	50	Xylol-butanol (1:1)	6-9	1.01	8.4	Colorless and clear	W-Z	11-18
Uformite F-210	Urea	50	Xylol-butanol (1:1)	4-7	1.01	8.4	Colorless and clear to slightly hazy	W-Z	15-20[c]
Uformite F-222	Urea	50	Xylol-butanol (2:3)	14-18	1.01	8.4	Colorless and clear to slightly hazy	T-W	20-30
Uformite F-226E	Urea	50	Capryl alcohol-butanol	4-7	0.98	8.2	Clear light yellow	W-Z	45-90[c]
Uformite F-233	Urea	50	Xylol-butanol (1:1.5)	3-6	1.01	8.4	Colorless and clear	K-P	21-35[c]
Uformite F-240	Urea	60	Xylol-butanol (1:1.5)	3-6	1.02	8.5	Colorless and clear	K-P	50-75
Uformite F-240N	Urea	60	High-flash naphtha	2-5	1.03	8.6	Colorless and clear	Z_1-Z_5	20-30

TABLE 3-7. — Continued

Grade	Type	Percent solids ± 2%	Solvent	Acid number, solids basis	Specific gravity	lb/gal	Color	Viscosity, Gardner-Holdt at 25°C	Mineral thinner tolerance[b]
Uformite MM-46	Melamine	60	Xylol-butanol (1:1)	0-1	1.01	8.4	Colorless and clear	I-L	50-90[c]
Uformite MM-47	Melamine	60	Xylol-butanol (1:1)	0-1	1.02	8.5	Colorless and clear	P-T	35-65
Uformite MM-55	Melamine	50	Xylol-butanol (1:4)	0-1.5	0.97	8.1	Colorless and clear	F-K	40-70
Uformite MM-55HV	Melamine	50	Xylol-butanol (1:9)	0-1.5	0.98	8.2	Colorless and clear to slightly hazy	T-X	45-85
Uformite MM-57	Melamine	50	Xylol-butanol (1:9)	8-12	0.98	8.2	Colorless and clear	H-K	35-65
Uformite MX-61	Triazine	60	Xylol-butanol (1:1)	0-1	1.04	8.7	Colorless and clear	G-M	50-130

Uformite QR-336[d]	Triazine	100	None	0–1	1.10	9.2	Colorless and slightly opalescent	T–Y[e]	20–30
Uformite M-311	Modified triazine	50	Xylol	4–7	1.01	8.4	200 max.[f]	Q–T	7–15

[a]Courtesy of Rohm & Haas Co., Synthetic Resins for Coatings, Bulletin C-163.

[b]Cubic centimeters of mineral thinner tolerated by 10 g of resin solution.

[c]Cubic centimeters of iso-octane tolerated by 10 g of resin solution.

[d]Also available at 60% solids in Solvesso 150.

[e]Measured at 65% solids in solvent mixture of 1.3 parts butanol and 2.0 parts xylol.

[f]APHA Color.

TABLE 3-8. Physical Characteristics of Thermosetting Acryloid Resins[a]

Resin constants	Hydroxyl-functional types			Carboxyl-functional types		
	Acryloid AT-50	Acryloid AT-51	Acryloid AT-56	Acryloid AT-70	Acryloid AT-71	Acryloid AT-75
Percent solids	50	50	50	50	50	50
Solvent[b]	X/B/MC	X/B	X/B	X/CA	S-150/CA	S-150/CA
	60/22/18	78/22	90/10	75/25	75/25	75/25
Acid equivalent[c]	—	—	—	825	825	825
Viscosity (cP at 25°C)	625-1755	1070-2265	450-700	1200-2500	2700-3600	2700-3600
Color (APHA)	50 max	50 max	50 max	50 max	50 max	50 max
Wt/gal	8.4	8.4	8.1	8.3	8.4	8.4

[a]Courtesy of Rohm & Haas Company, Acryloid Thermosetting Acrylic Resins, Bulletin C-170.

[b]X is xylol, MC is methyl cellosolve, S-150 is solvesso 150, B is n-butanol, CA is cellosolve acetate.

[c]Acid equivalent is the number of grams of resin solids containing one gram equivalent of acid.

hydrocarbon for the acrylic polymer-melamine resin mixture over the range of resin solids that will be encountered in storing and applying the enamel, and to balance evaporation rates of the components. Not only should the liquid mixture evaporate from the film at a rate satisfactory for flow and uniform film thickness, but also solvent and diluent fractions should be chosen to evaporate at similar rates. Solvent activity of the evaporating volatile mixture should be ensured by examining solvency of the final 5 to 10% of the evaporating mixture to protect compatibility of vehicle components in the film-forming stage.

3. Heat-reflowable Acrylic Enamels

Influenced by the freedom from "orange-peel" of heat-reflowable lacquers, and the relative ease of accomplishing factory repairs with lacquer, the formulation of heat-reflowable enamels was reported in 1967 (21), and the product has had some commercial application. The principle which must be observed if the heat-reflowable enamel is useful on production lines, is to obtain wet films which remain highly thermoplastic in presence of enough heat to evaporate solvent, and which then crosslink completely at conventional baking schedules. Polymers and melamine-formaldehyde resins are prepared or selected for heat-reflow enamels such that a threshold with respect to time and temperature delays curing during a preliminary bake of 10 to 15 min at 82°C. The film should then be firm enough to permit light sanding with very fine sandpaper (500 to 600 grade grit) to remove blemishes, or if desired "orange-peel," from highly visible areas of the automobile body. If such deep sanding has been required in local areas that film thickness has been undesirably reduced, additional spray application of enamel may be made in these areas without masking adjacent areas. In a successful formulation, spray dust from this repair work will be accepted by the adjacent, less conspicuous, areas and melted into the original finishing without lines of demarcation remaining to identify repair work. The film is then cured by further baking at a schedule recommended to secure total crosslinking.

In order to facilitate adequate film hardening during the preliminary bake, it is useful to modify the formulation with 15 to 20% of resin solids using a cellulose acetate butyrate resin selected for compatibility with melamine resin and acrylic polymer. EAB-551-0.2, developed and sold by Eastman Chemical Products, Inc., Kingsport, Tennessee, is such a cellulosic ester. The high butyryl, low acetyl ratio provides desired compatibility; the viscosity of its solutions is low enough for minimal sacrifice of spray solids, and other effects on film properties, such as reflow and ultimate resistance to humidity and weathering, are acceptable. Even the most compatible cellulosic, however, has dubious miscibility with acrylic polymers containing more than 6% styrene. The polymers of formulation example in Table 3-6 have been chosen to accommodate cellulosic modification if desired, although it would be wise in preparing such an enamel also to modify the package solvents by replacing 5 to 8 gal of the hydrocarbon with ethoxyethyl acetate. Obviously, "snap-cure" melamine resins should be avoided in heat-reflow or "bake-sand-bake" enamels, and the maximum time/temperature relationships established for the preliminary bake to ensure compliance with specific production time conditions.

The heat-reflowable enamel (also referred to in the industry as "repair-in-process"), even though used on occasion to ease factory repair problems, will still be subject to minor assembly damage requiring end-of-line repair. Conventional catalysis with alkyl acid phosphate and a bake of 20 min at 82 to 93°C continue to be a feasibility.

4. High-solids Enamels

The advantages of high-solids enamels range around rapid deposition of thick films, or (especially in the case of automotive primers) reduced replication of substrate profile due to minimized shrinkage during film formation. Air pollution legislation has added new impetus to formulation of topcoats which minimize organic emissions from application and baking facilities. The formulating problem centers around selection of intermediates which have low enough viscosities for spray application, but which still can be expected to form high-strength films by subsequent crosslinking.

Most conventional enamels are stable enough to respond to heating to temperatures of 60 to 71°C as a means of reducing viscosity. Hot spray equipment is currently available to handle such applications, and will be valuable for applying, with no solvent additions, those coatings whose viscosities allow them to be pourable at solids of 70 to 80%.

An approach to high-solids, low-viscosity enamels that has received wide attention is that of replacing conventional solvent with compounds that have polymer solvent activity at storage and application temperatures, but which at some elevated temperature react with a polymeric component to become part of the ultimate film solids. Two examples of this "reactive solvent" approach, which has been principally experimental in scope, are cited. Acrylic monomeric esters are useful diluents for acrylic polymer solutions. They can undergo addition polymerization if the polymer-monomer mixture is applied to a thermoplastic primer containing polymerization initiator which can be extracted by the monomer. Alternatively, a large excess of ethylene glycol can be used in preparing a glycol adipate polyester, for instance. The glycol acts as a viscosity-reducing agent for the polyester and, after coating application, is crosslinked along with the polyester by reaction with melamine resin in the formulation. Each of these approaches has the same inherent defect, namely that film-curing temperatures must be approached gradually in order to contain the "reactive solvents" within the coating for time periods long enough to enable partial reaction to reduce their volatility. The formulator is forced to depend on polymer present in the liquid coating to depress volatility of monomeric components during application.

A more widely used formulation for high-solids liquid coatings is the two-package type in which low-molecular-weight polyester polyols or acrylic polymers containing pendant hydroxyl groups are blended with an isocyanate adduct or prepolymer containing unreacted isocyanate derived from aliphatic or alicyclic diisocyanates. Note that only the nonaromatic isocyanate derivatives are acceptable for finishes, such as automotive topcoats from which light stable color and exterior durability are expected. Depending on type and concentration of catalyst used to catalyze the urethane reaction, and the degree to which dilution with organic volatiles is tolerable, the mixture will have a sprayable potlife at room temperature varying between 15 and 20 min and 8 to 10 hr.

An example of two component isocyanate adduct-polyester urethane with light-fast properties is the coating prepared by following recommendations of Farbenfabriken Bayer and their U.S. affiliate, Mobay Chemical Company. Chemical equivalent weights of each component are measured and Desmodur N (aliphatic polyisocyanate adduct) is blended with Desmophen 650 (polymeric polyol) in amounts that yield ideally an NCO to OH ratio of 1.1 to 1. For instance:

	Total solids	NCO equiv wt
Desmodur N (Mobay Chemical Co.)	75%	250
	Total solids	OH equiv
Desmophen 650 (Mobay Chemical Co.)	100%	225

Combined weight ratio	Desmodur N	275 wt units
(1.1:1 isocyanate/hydroxyl)	Desmophen 650	250 wt units

Pigments may be dispersed in the Desmophen 650 and stored preparatory to blending. Silicone oils such as DC-200 (Dow-Corning) in amounts of as little as 0.1% (or less) of the total binder solids (after the combination) may be added to the pigmented base to improve film flow. Zinc naphthenate or octoate or dibutyl tin dilaurate (Union Carbide Corp. Niax D-22) may also be added to the pigmented base in anticipation of the combination with Desmodur N to accelerate film formation. Anhydrous ketones and esters such as ethyl acetate or ethylene glycol ether acetate may be used to thin the combination vehicle to sprayable solids to obtain good atomization during spray application.

Nationwide legislation for control of organic emissions to the atmosphere is tending to follow that established by California in which volatiles are classified by structure in terms of their photochemical reactivity. Allowable emission quantities are defined as pounds per day/per production unit of nonphotochemically reactive or "exempt" compounds. An average automotive production line might produce 60 painted bodies per hour for two shifts of eight hours each, making a total of approximately 1000 finished units per day. If one assumes current actual consumption of 1.75 gal of 30% spray solids enamel per body unit, there will be approximately 10,000 lb organic vapor emitted per day. Should it be desired to reduce the emission to 1000 lb per day (1500 lb/day would represent 85% reduction, a probable limit in California), the application solids must be raised to 80% to maintain the same film thickness at the same deposition efficiency.

While it is potentially feasible to make spray application of special formulations at such high solids, or to engineer solvent recovery systems that incinerate or recycle volatilized organic compounds, the consideration of water-thinned or dry powder coatings becomes increasingly attractive. In both the latter cases mentioned, film formation by thermosetting reactions is favored. It has usually been the history of thermoplastic aqueous emulsions that their high molecular weight inhibits the development of full film gloss. Similarly, thermoplastic powders which must be fused into film integrity tend to suffer in appearance due to incomplete heat flow of the high-molecular-weight polymers required for mechanical strength.

Aqueous solutions of amine soaps of carboxylic acid-modified polymers can be sprayed to smooth films and conventionally crosslinked with hexamethoxymethyl melamine resins, which are also water thinnable (33). Water-soluble, organic coalescents are required to facilitate film formation, and solubilizing amine as well as by-product methanol are driven off as the film cures. It is plain then, that water-thinned enamel is by no means free of organic emissions, but to encourage their use, legislative action in some control areas exempts water-thinned coatings that contain a maximum of 20% nonphotochemically reactive organic solvent in the total volatile.

5. Powder Coatings for Automobiles

If one should seek the coating material for automobiles that produces the minimum quantity of organic emission during application and cure, thermoplastic or thermosetting powder would probably qualify. Other compelling advantages that powders may have when compared with liquids could be adequate film thickness in one coat and, by means of recovery and recycling of powder not used, a high rate of deposition efficiency. As of 1972, automotive powder coatings used for production finishing are restricted to wheels and some small parts, but materials and techniques for finishing of bodies is an objective which is being actively pursued.

It is expected that powder application for automobiles will use an electrostatic device, probably automated spraying. To a considerable degree, the composition and physical nature of the powder coating is controlled by the operating principles of proposed application equipment. Powder particle size usually ranges from 30 to 70 μ diameter with the shape varying from roughly spherical to crystalline. Air-supported powders flow easily at low pressures through hoses and are given negative or positive static charges at the tip of a powder spray gun. Ware surfaces (auto bodies) are grounded and rapidly attract and retain a substantial share of charged powder particles sprayed into their vicinity. Powder coated objects then move into baking chambers where the loose powder coating is sintered into an integral film. Baking temperatures of 177 to 205°C are usually required for good film integrity and smoothness (28). Substrate adhesion is generated sufficient to satisfy environmental resistance and mechanical tests currently required of films from liquid coatings.

Powder applications in the coatings industry have traditionally been associated with use of fluidized beds which have the advantage of almost 100% deposition efficiency; however, they also have the disadvantage that preheated surfaces are required to obtain initial adhesion of loose powder to the substrate, and while the film thickness of the coating can be somewhat controlled by preheating temperature employed, thicknesses of 4 to 6 mils (excessive for auto parts) are considered minimal for the process. Electrostatic spraying permits application of powders to cold surfaces with film thickness control to yield the desired 1.5 to 3 mils for automotive coatings. Facile color change from one automobile body to another, when a succession of color changes is required in production finishing, will require drastic design and scheduling changes of current coating application areas in automobile factories.

Currently, powder coating formulation is based on melt mixing of ingredients in a screw-type extruder which simultaneously blends resinous ingredients intimately and disperses pigments. Time and temperature of mixing are meticulously controlled when thermosettable polymers are involved to minimize crosslinking in the extruder. The extruded product is finally dry ground and classified to obtain the desired particle size and size distribution. For automotive topcoats, thermoplastic and thermosetting polyesters and acrylics are the leading vehicle candidates, and pilot applications in automobile factories are being conducted (29).

3-4. EVALUATION AND TESTING OF AUTOMOTIVE COATING SYSTEMS

Automotive coating formulators are often confronted by questioning as to which of the two general types of topcoats, thermosetting or thermoplastic, is "better." A review of the performance specifications (30-32) which all automotive topcoats must meet to qualify as original production finishes will reveal very little difference in quality. Discussions in earlier sections reveal substantial differences in the chemistry of polymer preparation and film formation of lacquers and enamels which in turn are reflected significantly in respective application techniques and schedules. Accordingly, the different types of coatings require the formulating and product development chemist to resort to a wide variety of compositional adjustments to achieve specified performance levels. While the intricacy and ingenuity called forth to arrive at specified standards while remaining competitive may therefore vary widely, the commercial end products are remarkably similar in performance. A high degree of sensitivity on the part of formulator and customer alike to the constant upgrading of system performance demands that improvements made in one system should quickly be made available in competing systems. Because of the understandable emphasis on and constant concern for system performance, reliable evaluative procedures almost always require total automotive coating system testing, no matter how localized the variable may appear to be.

An abbreviated review of testing procedures with comments relative to reasons for failure and normal corrections is therefore in order (note that automotive coating specifications normally deal only generally with composition, and very specifically with performance).

TESTING PROCEDURE FOR TOPCOATS IN THE COATING SYSTEM

1. Package Properties

As a means of ensuring good money value, the purchaser of coatings material normally specifies a narrow range of package viscosities which are to be achieved simultaneously with minimum weight and volume solids. Film build per application operation (e.g., per spray coat) is similarly of economic value, since it is directly proportional to the amount of labor and production-line space required. Therefore, a critical requirement usually found in the specification is minimum application solids or maximum allowable thinning solvent for achieving application viscosity. The study of solubility parameter relationships between polymers and solvents can be invaluable in optimizing the solids/solvent/viscosity ratios.

The stability of the packaged coating must take into account a variety of concerns. First, it is important that viscosity/solids ratios for optimum application be maintained. This may be particularly troublesome when a thermosetting coating is subjected to prolonged storage in a warm warehouse. A small degree of crosslinking or association by secondary bonding of functional groups can increase the volume of reducing solvent required, causing sacrifice of film build or leveling of a sprayed film. If the functionality involved is intended ultimately for curing by condensation reaction under mildly acid condition, stabilizing manipulations could involve reduction of concentration of the acid component, replacement of several percent hydrocarbon with an alcohol, or introduction of a basic condensation inhibitor such as 0.1 to 0.5% of total resin as triethylamine. Stability testing usually involves an attempt to accelerate the development of problems by storing the package material for periods varying between several hours to several weeks at temperatures between 50 and 60°C. Effective criticism of the degree of viscosity stability should include a combination of viscosity monitoring as the test proceeds, together with periodic application to observe the degree to which initial appearance properties are maintained.

Other features which are assumed when reference is made to package stability, and which are also therefore monitored during the stability test, are pigment suspension and ease with which settlement is reincorporated, pigment dispersion stability (flocculation resistance) as judged by constancy of dry film color match, resistance to gassing or gelation of pigment, vehicle and moisture interaction, or resistance to skinning if oxidizable components are present that may react with air in the package.

2. Film Appearance

System appearance is usually described and specified in terms of topcoat gloss and leveling. Glossiness is evaluated subjectively and qualitatively in terms of visual "sharpness of image" when the observer, using a reference standard, makes visual comparisons of film haze or its lack, which is seen as degree of "wet look" in the dry film. Glossiness is also quantified for written specifications by measuring percentage reflectance of incident light from the dry film surface. A fairly close correlation in the ratings of visual and instrumental sharpness of image determinations is achieved when both the eye of the visual observer and the geometry of light path in the instrument is arranged to provide an angle of incidence and reflection that is nearly normal to the film surface. Accordingly, a 20° glossmeter is used to judge glossiness of automotive topcoats. This means that specular reflection from the film when the incident light beam is angled 20° from a perpendicular to the film surface, is collected by a photocell connected to a galvanometer, which is in turn calibrated to read percent reflectance relative to some glossy standard such as smooth glass. There are a few compositions, such as that of a film heavily pigmented with flake aluminum, which are so highly reflective that glossmeters produce readings that conflict with visual appearance. The alert evaluator will understand and make allowance for these apparent aberrations of his hardware.

A coating property closely allied to appearance is dry film opacity, which is most accurately characterized by the total reflectivity ratio of the film applied at a standard thickness over contrasting color backgrounds such as black versus gray or white. This determination is referred to as contrast ratio with perfect opacity, of course, producing a contrast ratio of unity.

The judging of color match to a standard is a matter of great complexity for automotive topcoats, which must provide nearly perfect spectral duplication of the standard, because of the variety of light sources and viewing angles which will be used in even a casual inspection. The first requirement for a spectral color match is pigmentation qualitatively and quantitatively equal to that of the standard. In the United States, however, a very large proportion of topcoat colors include metal flake as a portion of the pigment. It is essential that flake orientation be similar in batch and standard to avoid geometrical metamerism, the term used to define failure to achieve color match at all viewing angles of batch and standard. Flake orientation is a function of several variables, of which application technique is certainly important and must, therefore, be defined. The most effective method of securing defined and duplicable application is to use a programmed mechanical spray device. Another of the important variables is rate of solvent evaporation, which enters strongly into the rate of viscosity increase of the applied coating. Color control, therefore, must be held in mind by the formulator as he adjusts his coating composition for leveling or resistance to sagging by manipulating evaporation rate of volatiles.

Finally, as the formulator judges appearance of a coating, he is once again impressed by the interaction of system components. Undercoats, for instance, must be smoothly applied and present reasonably dense, nonporous surfaces to the topcoats, for the rough surface of an undercoat will amost certainly be replicated in the topcoat, and a porous undercoat film may contribute strongly to gloss deficiency of an otherwise excellent topcoat.

3. Film Hardness

Film hardness usually must be qualified to differentiate between surface hardness or uniform cross section hardness which may be a function of crosslink density in thermosetting coatings or degree of solvent release by thermoplastics. Surface hardness, important to mar resistance, has a strong bearing on readings obtained by pendulum-type hardness testing instruments, of which the Sward rocker is fairly typical. There is some trend to place more emphasis on hardness values determined by modified penetrometers such as the Tukon tester which measures resistance of the film to penetration by a very hard, weighted wedge (hardness reported in knoop units). Film hardness is a property which, except as a means for measuring uniformity of production, has a limited significance when taken by itself in assessing the coating. Performance areas normally associated with hardness, such as efficiency of film formation, flexibility, and intercoat adhesion, may be more effectively influenced and controlled by working through other properties.

4. Adhesion Flexibility

The end use intended for a recommended coating system should properly control the test procedures employed in its evaluation; adhesion flexibility testing of automotive systems is a case in point. Adhesion and flexibility are properties so closely allied that it is convenient to test them with the same procedure. Substrate adhesion and resistance to delamination of system components (intercoat adhesion) can usually be tested simultaneously by a crosshatch or "vaccination" test in which a very sharp blade is used to cut through an applied film to the substrate in a series of parallel lines, usually as close as 1/8 in. A further series of cuts is made at right angles to the first, and the resultant squares of isolated film are tested for adhesion by applying transparent adhesive tape to the area with standard pressure followed by removal of the tape by a sudden tug. Examination of both the tape and the film system remaining on the substrate will indicate whether failure has occurred and will show the part of the system requiring corrective adjustment. Crumbling of the film under the sharp knife blade is indicative of a brittle film; aggravations of the test procedure are frequently devised to further explore interaction of flexibility and adhesion in which the test coating is chilled, overcured, subjected to humidity, or exterior exposure prior to testing. The adhesion test is sometimes quantified by attaching one face plate of an Instron tensile test machine to the uncoated side of a substrate specimen, and the other face plate to the coating surface by special adhesives that do not attack or weaken the film. Energy required to produce delamination can then be tabulated for a test series of controlled variables.

A test widely employed by the automotive industry has been developed by the Society of Automotive Engineers, and is described in their Technical Report, SAE-J400. It is called the Gravelometer, and provides for controlled chipping of a coating system with standard screened natural road gravel by throwing a measured quantity of the gravel against the specimen with an air blast to simulate chip-producing incidents to automotive finishes in service. Coating system adhesion may also be seen to strongly influence other flexibility tests such as those that deform the substrate, falling dart impact or mandrel bend, for instance, so that adjustments for adhesion improvement are likely to be immediately obvious in flexibility testing.

5. Environmental Resistance

Condensing humidity from the environment is a normally encountered and strongly destructive hazard for organic coatings, causing dimensional change, gloss loss, and eventual blistering in the absence of strong system adhesion. Condensation on thermally conductive test surfaces is easily produced by exposing the coated side of a test panel to a controlled temperature steam chamber saturated with moisture vapor at 38 to 65°C and the uncoated side of the panel to room temperature of 21 to 27°C. The use of the higher humidity temperatures may be expected to induce severe failure when mediocre formulations are exposed in such a device for 24 to 72 hr. If gloss, adhesion, film integrity, and color match to standard are maintained for 72 hr of exposure to condensing humidity at 60°C, with no blistering, the coating formulator may be reasonably assured of freedom from field complaints due to natural humidity.

Thermoplastic films thought to be subject to the water spotting surface or etching problem discussed in Sec. 3-1, B, 4 may feasibly be tested for resistance to this surface distortion caused by water droplets which evaporate from sun-warmed films. Since the problem is believed to be a function of shrinkage of water droplets lying on the surface, a useful reagent in the test is 3 to 5% aqueous egg albumen. Small puddles of the reagent are placed at close intervals along the length of a test panel and allowed to dry at room temperature. The proteinaceous albumen develops strong adhesion to the coating film as the water of the solution evaporates. The test panel is then placed on a heater which develops a temperature gradient rising from 49°C at one end of the test specimen to 71°C at the other end. The albumen spots shrink and crack as they are warmed and will accordingly produce distortion of the coating surface if it is softened to mobility in the heat of the temperature gradient, which can be calibrated by using a contact pyrometer. Experience in correlating etch resistance in this test with practical exposure in warm and humid environments, such as the Gulf Coast in the United States, suggests that satisfactory coating should resist etching on the gradient bar at temperatures below 65°C.

Corrosion resistance of coating systems is reasonably well characterized by using the equipment described in ASTM test No. B 117-64 (34). Those coatings which resist corrosion creepage from scribe marks cut through the film into a steel substrate when exposure to 5% aqueous salt spray at 100°F has proceeded for 250 hr are regarded as corrosion resistant in the field. Most of the premium automotive coatings systems, however, are formulated to withstand 500 hr without developing under-film corrosion at the scribe mark of greater than 1/16 in.

Atmosphere fallout resistance tests are also usefully conducted under controlled laboratory conditions for film staining due to aqueous contact with fly-ash, dilute battery acid, antifreeze solutions, and gasoline. Intermittent, repeated exposure to gasoline spills in the area of the fill pipe can be insidious for plasticized thermoplastics due to plasticizer extraction which leads eventually to film cracking. Resistance to plasticizer extractability by gasoline, therefore, is an important consideration in evaluating a prospective finish.

In service, mechanical failure of the coating system evidenced by film cracking due to the differential coefficient of expansion of film and substrate was discussed in the earlier formulation sections. The problem is particularly severe when thermoplastic polymers of inadequate molecular weight are used, or when thermosetting coatings are insufficiently cross-linked by poor formulation or insufficient heat during the curing schedule. Dispersion or powder coatings which have incompletely coalesced may also be identified by a thermal cracking test. Accordingly, exaggerated or accelerated testing involves working with film thicknesses 50 to 100% greater than recommended, slightly inadequate bake schedules, and very sudden temperature changes of a magnitude greater than expected in service.

System test panels complete with repair areas are cycled between 16 and 18 hr each of 38°C condensing humidity and -11°C cold chest. The warm to cold transition should be as sudden as possible, with test panels transferred in small groups to allow for recovery of the freezer. Test panels removed from the freezer are allowed enough time at room temperature to inspect the surface for cracking. One complete transition from warm humidity through the freezer is regarded as a cycle, and 15 to 20 cycles without cracking failure are thought to indicate acceptable mechanical performance. Since poor coatings will undergo rapid mechanical weakening by degradation due to weathering, it is the usual practice to repeat thermal cycling of new formulations after three months of tropical exposure.

6. Weather Resistance — Exposure Testing

It is apparent that automotive coating systems must be highly weather resistant for long periods of time. At least 18 months in a tropical environment is required with reasonably satisfactory retention of gloss and complete film integrity. Five years or longer of normal temperate zone exposure will be required to equal the severe tropical test employed. It is not surprising, therefore, that much effort has been expended in exploring techniques for the acceleration of the film degradation or failure that occurs upon exposure to weather. Artificial weathering devices have been used for years with varying degrees of success in establishing a correlation with natural weathering, but none has been fully accepted by the automotive builder as completely reliable. Natural weathering presents a complex of variables involving the reactants of film degradation - atmospheric components including oxygen, ozone, and moisture vapor, together with heat and radiant energy from sunshine. Even those devices that induce a dew point as part of the artificial weathering cycle, while frequently producing some form of failure in as short a time period as several hundred hours, are found to offer comparative data, among a variety of coatings, of a different order of severity than occurs naturally. Artificial weathering has an advisory value to the practiced formulator of automotive topcoats that is limited to the elimination of prospective coatings, or compositional and processing variables which are responsible for gross failures at early exposure periods.

A greater utility than that available from artificial weathering has been accomplished by the discovery of the most effective techniques for presenting a coating specimen to natural weather. An angle of exposure that places the coating surface nearly normal to the incident radiation from the sun is essential to duplicating degradation of the coating on horizontal surfaces of an automobile. This is the "5° exposure" mentioned prominently in tabulations of weathering data.

A useful device that has recently come into prominence in exposure technology is the "black box." This is an arrangement for mounting flat, metal-coated exposure specimens at 5° from horizontal over a body of air trapped in an enclosure beneath the test panels. The enclosure, painted black, serves to insulate the back side of test specimens, so that during periods when the face side is receiving solar radiation, the surfaces are maintained, depending on reflectivity of the surface, at temperatures -7°C to 4°C greater than the ambient. During the night, the black box and the test panels, which constitute its cover, efficiently radiate heat energy to bring the surface temperature below the ambient, thus inducing a dew point which saturates the coating with moisture. The black box combination of tropical exposure, horizontal angle, high daylight temperature, and nighttime wetness has been credited with accelerating evidence of exposure failure by about 50% compared to similar exposures without the black box.

3-5. AUTOMOTIVE UNDERCOATS

A. SPRAYING AND DIPPING PRIMERS

The functions of primers, surfacers, and sealers have been discussed in connection with the various features of system performance, and the general chemical functionalities have been related to obtaining specific performance features such as appearance and corrosion resistance. It remains to offer examples of these compositions as currently used by the automotive original finish production lines.

A typical pigmentation for a spray primer might consist of:

50% by wt of barium sulfate

15% by wt of red iron oxide

25% by wt of talc

10% by wt of specific modifying pigments

Barium sulfate is a dense, high specific gravity pigment which is strongly resistant to chemical attack and very insoluble in water. Two grades are available, a low-cost, naturally occurring mineral known as barytes, and a synthetic, chemically precipitated pigment, blanc fixe, which normally costs about three times more than the natural mineral. Barium sulfate is valuable in primer pigmentations because it offers minimal interference with topcoat gloss and is therefore said to have good "holdout." As with all natural minerals, barytes tend to vary somewhat in degree of purity and in particle size. For a small cost premium it is possible to obtain a classified barytes which is distinctly advantageous over most of the natural mineral in terms of easier dispersion in the primer vehicle and more complete long-term suspension in the vehicle during storage, or circulation after the primer is thinned for application. Only in the instance of formulating for maximum ease of dispersion or maximum topcoat gloss holdout is it necessary to resort to using the precipitated grades.

Iron oxides are usually employed as the opacifying agents in primers, although there is also a substantial demand for gray primers, for which the red oxide is simply replaced with titanium dioxide and lamp or furnace black carbon pigments in ratios and amounts required for desired color and opacity. Formulators must take into account particle size and distribution of the iron oxide pigment they use, for color uniformity, ease of dispersion, and suspension properties can again be critical. They must also be alert, however, to the grade of insoluble impurities, such as silica, that are apt to be present in natural pigments, or water-soluble impurities that may occur with synthetic sources which affect moisture sensitivity of films.

Talcs are by nature needle-shaped crystals which are valuable in assisting the easy reincorporation of primer pigmentation that has settled after long-term package storage. Such pigments, however, do tend to protrude slightly from film surfaces, adversely affecting topcoat gloss unless the cured primer film is thoroughly wet sanded. Talc suppliers offer a micronizing or dry grinding operation that provides a fine particle grade which is highly desirable for obtaining the good properties of talc in primer pigmentation with minimal problems.

Specific modifying pigments in automotive primers will vary widely depending on vehicle type, application technique to be used in depositing the primer, and the rigor of performance requirements of the coating system specification. Corrosion-inhibiting pigments may constitute a large proportion of modifying pigmentation, and can vary from highly insoluble lead chromate to more soluble (therefore more effective, but also more hygienically hazardous) chromates, or to the much less physiologically reactive, but more expensive, zinc molybdates. Included in specific modifying pigments are a variety of montmorillonite clays (bentonites) which are highly useful in spray primers for inhibiting pigment settlement during package storage. These clays absorb moisture and alcohols from the solvents used in the pigment dispersion phase of manufacture, particularly when a mild temperature increase is allowed during the grind. The rheological structure produced by the absorption of the polar solvent serves to stabilize resin-wet pigment against hard settling without seriously interfering with leveling of the applied coating. The bentonite clays will usually be used at the rate of 1% of the total pigment in the average primer when pigment to binder ratio by weight is 1.5:1, and whose pigment volume concentration will be 30% of the total solids volume. Using approximately 20% of the total binder solids in the grind and controlling dwell time in a continuous sand mill a fineness reading of 7 can be obtained on a Hegeman grind gauge with the pigmentation described above.

A prior paper in this treatise (35) describes soya-dimer acid-epoxy esters which are highly suitable as vehicles for the primers under discussion, particularly when modified with urea or melamine formaldehyde resins. Modification with crosslinking agents such as the nitrogen resins mentioned is usually restricted to the minimum that produces requisite hardness and humidity resistance. This is because of the interest in optimizing those properties along with flexibility as measured by resistance to chipping-type impact and film fracture due to sharp substrate deformation. The modifying amino resin concentration will conventionally be found to be 3 to 10% of the total resin solids. Higher concentrations would add to chemical resistance, but in addition to sacrificing flexibility, intercoat adhesion between unsanded primer and topcoat would suffer.

Soya-dimer acid-epoxy ester primers of the type described are also suitable for dip tank operations, particularly if further modified with 0.25 to 0.50% by volume of slowly volatile antioxidant, such as cresylic acid, used to maintain viscosity stability. Stability monitoring of dip tank compositions is particularly important with automotive coatings because a tank large enough to immerse an entire automobile body traveling on a production line, will usually contain several thousand gallons of coating. Failure of the coating in such dip tanks is usually the result of excessive aeration, poor temperature control, and failure to replace evaporated solvent and antioxidant. Aeration due to turbulence of entrance and withdrawal of ware from the tank may not be avoidable, but if the other two factors are controlled, considerable aeration is tolerable with reasonable formulations. An operating dip tank should be sampled daily to determine viscosity and total solids corrections required. Frequently, the sample should also be subjected to accelerated aeration in the laboratory during which it will proabbly be observed that simple restoration of evaporated solvent does not restore the coating to the desired viscosity. A curve should be plotted showing rate of viscosity increase at standard total solids until the sample eventually gels. The slope of this curve would be expected to remain low if replacement of used coating keeps the dip tank reasonably fresh. If in the laboratory test, however, the slope of the curve is seen to increase rapidly toward gelation, antioxidant additions should be made, additions of more active solvents considered, or the tank drained for replacement with fresh coating, into which the unstable material formerly in the tank may usually be introduced slowly for work-off.

The urea- or melamine-modified soya-dimer acid-epoxy ester primer at dry film thicknesses of 0.75 to 1.2 mil should be baked at a schedule of 20 to 30 min at 163°C to achieve the cure required for optimal topcoat holdout, corrosion, cycle, and humidity resistance, as well as mechanical properties.

B. SEALERS

Sealers are midcoats between two surfaces and, in general use, their function in a coating system is to serve as a barrier against bleeding pigment or resin migration. Specifically with automotive systems it is more usual to find that sealers are used as tie coats to overcome weaknesses in intercoat adhesion. Amino functional thermoplastic acrylic copolymers are sometimes employed as sealer components. Epoxy resins and heat reactive phenolics, or copolymers containing 2-sulfoethylmethacrylate, also have polar functionality that not only promotes wetting and spreading in specific systems, but also becomes effective in producing adhesion through secondary valence bonding (36).

Plueddemann, in discussing the use of reactive silanes as adhesion promoters to hydrophilic surfaces (37), suggests an intriguing area for further study in the development of tie coats. He starts with the demonstration that wetting of a solid by a liquid occurs readily when the solid surface has a higher surface energy (measured by critical surface tension) than the liquid. He proposes that modification of the solid surface by coating it with a high-surface-energy film of reactive silane coupling agent will simplify the application of topcoats and enhance performance of the system due to superior intercoat adhesion.

C. PRIMING BY ELECTRODEPOSITION

A substantial portion of current automotive body priming is done by the electrodeposition process, and as production lines are remodelled and new plants built, the process promises to be extended even further. Numerous publications (38-47), reporting research and development in the field of electrodeposition, agree that an efficiently operable basic formulation involves pigmentation of water-thinnable, amino- or alkali-solubilized polycarboxylic acid resins. In water solution, the solubilized oil, alkyd, polyester, or acrylic polymer ionizes somewhat, so that if electrodes are immersed in the solution and a potential gradient (this may vary from 100 to 450 V in current practice, depending on the polymer) impressed by a source, the charged particles begin to migrate to the electrodes. In the case of the amine-solubilized polycarboxylic acid resin, the film-forming or polymeric particles are negatively charged and are attracted to the anode which is made the ware to be coated. At the anode, the negative coating particles release their charge to the circuit and in doing so revert to the physical and chemical water-insoluble state in which they existed prior to their solubilization. An evenly distributed layer of precipitated resin accumulates around the entire surface of the anode, eventually insulating the anode so that unless voltage is increased, current flow and deposition cease when the limiting film thickness is reached. This feature, self-limiting film thickness, is important in achieving deposition on the interior of box sections or closed channels in which deposition occurs more slowly than on the exterior surface. Coating activity continues on the inner surface, therefore, without building excessively on the outer. Quantitative tests have been devised for measuring the capability of a coating to deposit on these areas shielded from the cathode, and are reported as the "throwing power" of the coating material.

The electrically precipitated resin particles coalesce almost immediately to form a semipermeable film from which almost all of the water is driven by electro-endosmosis. The film, as withdrawn from the bath, therefore contains 85 to 95% solids, and even in the uncured state, is firm and sufficiently adherent to permit rinsing with water to remove the thin layer of undeposited coatings "dragged out" with the deposited film. Another very important feature of the high-solids deposition is that the continuous coating permanently resides on the interior of box sections and channels, since there is little or no occasion for condensed water from the coating to wash the protective film away from the interior walls of such parts during baking. Films deposited vary in practice between 0.5 and 1.0 mil in thickness, depending on the chemical nature of the coating, dip tank solids, immersion time, bath temperature, and, of course, the voltage applied.

During the electrodeposition of the coating at the anode, it is important to understand that the cationic portion of the paint solids is also being discharged at the cathode, regenerating the solubilizing agent. The accumulation of amine or alkali at the cathode would soon render the electrodeposition bath inoperable if it were not controlled. One control that is used is to separate the cathode from the main part of the tank by either semipermeable or cation selective membranes, which concentrate free amine or alkali formed in the cathode compartment. The free solubilizing agent is removed from the cathode compartment by flushing with water at a rate that removes the base but still maintains adequate conductivity of the compartment. A second control procedure involves replenishing the tank solids with unsolubilized or partially solubilized acidic polymer at frequent intervals. This feed material must be formulated to take into account differences in composition of the deposited film and the primer charged to the tank. Ideally formulated primer contains a solubilized resin that is itself an excellent emulsifying agent, and if a second resin such as amino-formaldehyde crosslinking agent or an epoxy ester is present, it will be entrained thoroughly enough into the total emulsion to electrodeposit the same proportions present in the tank. The same situation is also expected of the pigmentation. In practice, however, close monitoring of the tank composition is required to provide a compensatory feed that restores the tank to the desired ratio of components in the event one of the resins or pigments has deposited disproportionately.

A valuable adjunct to operating electrodeposition tanks is the ultrafiltration unit (46) which:

1. Contributes to operating economy by providing a means to recover coating drag-out.

2. Reduces or eliminates water pollution occasioned by rinse water disposal.

3. Offers a method for maintaining coating performance at the high level associated with a freshly filled electrodeposition tank.

When ultrafiltration is used to facilitate electrodeposition priming, mixtures of water-diluted polymer solution and polymer-pigment colloidal suspension are passed over the surface of a special synthetic, semipermeable membrance which may be in form a sheet, a tube, or arranged into hollow fiber bundles.

Water and some of the smaller solute molecules pass through the membrane under a relatively low pressure head, 10 to 100 psi. To avoid plugging the membrane with the macro-molecule retentate, fluid flow velocity of about 10 ft/sec over the filter surface is maintained. A standard ultrafiltration unit associated with the electrodeposition tank used to prime automobile bodies will contain 240 ft^2 of membrane and is capable of removing 2000 gal of water per day from the tank or from rinse water. Rinse water accumulated from clearing the electrodeposition primed body of dragged-out primer that has not drained back to the tank, will contain solids concentrations in the vicinity of 0.1%. Since the dip tank could not tolerate constant dilution by the low-solids rinse water, it was formerly disposed through settling ponds. With ultrafiltration, rinse can be concentrated sufficiently to return the solids to the tank with the filtrate recirculated for further rinsing.

Anodic electrodeposition has a tendency to produce some dissolution of the iron/zinc phosphate crystals resulting from the steel surface passivating chemical pretreatments. There has, therefore, been motivation to develop coatings for cationic deposition, and primers of this type are known. These cationic coatings should have not only better corrosion resistance, but should also avoid the film staining induced by iron ions in the bath.

The following typical examples of anionic electrodeposition formulations have kindly been made available by Amoco Chemicals Corporation (see Tables 3-9, 3-10, and 3-11).

3-6. AUTOMOBILE REFINISH SYSTEMS

Coatings formulated for automobile refinishing may carry specific instructions for their use, but frequently they are applied under circumstances that are far different from those described in the recommendations. The total coating system for refinish work is just as important a consideration as it is for original or factory finishing. However, because of the variation in surface, from bare metal to factory-primed new parts to weathered original finish topcoats, both thermoplastic and thermosetting, and the variation in refinish shop equipment and operator technique, the key word for coating system components is versatility.

A. METAL PREPARATION

An automobile that is to be refinished usually requires extensive preparation before the painting operation can begin. All surface imperfections in the old coating system must be eliminated. Imperfections commonly handled are chipped film or rusting around brackets or moldings where the original coating has been fractured. Where possible, moldings should be removed and blemishes sanded out of the original finish, starting with coarse paper and completing the sanding by removing scratches with fine paper. An area considerably larger than the blemish is sanded with base metal exposed at the center of the area and the cross-sectional edges of the "sand-through" sloped in an almost imperceptibly gradual level to the surface of the old finish. The practice is called "feather edging." As soon as possible after completing the sanding, base metal exposed should be pretreated by scrubbing with one of the diluted phosphoric acid preparations available. There are a number of these metal preparation solutions containing phosphoric acid and wetting agents which may be diluted with water and applied to metal surfaces with very clean cloths. Proper dilution with water is essential to control the thickness of the iron phosphate crystal layer formed, and since thin layers are preferable, the acid wet metal is wiped dry rather than allowing reaction of all applied acid.

TABLE 3-9. Formulation of Water-soluble Resin 3823 EC[a]

Materials	Charge weight	Mole ratio	Resin properties	
AMOCO TMA[b]	240	3	NVM, %	87
Neopentyl glycol	346	8	Acid number (solids)	35-37
AMOCO IPA-95[c]	138	2	Gardner-Holdt viscosity	Z_8-
Tall oil fatty acid[d]	356	3	Gardner color	5-7
Total	1080		Cure, sec @ 200°C	70-75
Water off	-80		Volatile	butoxyethanol
Yield	1000			

Resin processing

1. Charge neopentyl glycol, IPA-95, and 59% (wt) of the TMA to a reaction kettle equipped with an agitator, inert gas sparge, and a steam-heated partial condenser.
2. Slowly heat ingredients to 232-238°C. Begin agitation at about 168°C. Maintaining a maximum overhead temperature of 102°C, hold for an acid number of less than 10.
3. Charge tall oil fatty acid, maintain heat at 221-227°C, and hold for less than 10 acid number and viscosity at 80% NVM in butoxyethanol of T + (Gardner-Holdt).
4. Cool to 210°C and add remaining TMA. Hold at 204-210°C for 42-44 acid number. Cool to 170°C, hold for 35-37 acid number, and reduce to 87% NVM with butoxyethanol.
5. Filter.

[a]Courtesy of Amoco Chemicals Corporation, Bulletin TMA-103a.

[b]Trimellitic anhydride.

[c]Isophthalic acid, Amoco Chemicals Corporation.

[d]Acintol FA-1, Arizona Chemical Company.

TABLE 3-10. Primer and Enamel Formulations[a]

Materials[b]	Red iron oxide primer	Orange enamel	Yellow enamel	Green enamel	Pastel green enamel	Pastel blue enamel	Gray enamel	Black enamel
TMA Resin 3823EC	244	366.4	366.4	366.4	366.4	366.4	366.4	426.9
Butoxyethanol	26	42.3	42.3	42.3	42.3	42.3	42.3	49.3
Amino crosslinking agent[c]	38	56.2	56.2	56.2	56.2	56.2	56.2	65.5
Triethylamine	12	16.5	16.5	16.5	16.5	16.5	16.5	21.0
Water	400	700.0	700.0	700.0	700.0	700.0	700.0	700.0
Pigments								
Red iron oxide[d]	200	—	—	—	—	—	—	—
Pyrazine orange[e]	—	55.0	—	—	—	—	—	—
Cadmium yellow[f]	—	—	55.0	—	—	—	—	—
Chromium oxide green[g]	—	—	—	75.0	—	—	—	—
Phthalocyanine blue[h]	—	—	—	—	—	0.8	—	—
Phthalocyanine green[i]	—	—	—	—	0.8	—	—	—
Titanium dioxide[j]	—	20.0	20.0	—	74.3	74.3	74.3	—
Carbon black[k]	—	—	—	—	—	—	0.8	13.1
Grind base	920	1256.4	1256.4	1256.4	1256.5	1256.5	1256.5	1275.8
Water dilution for electrocoating	3580	3243.6	3243.6	3243.6	3243.5	3243.5	3243.5	3224.2
Total	4500	4500.0	4500.0	4500.0	4500.0	4500.0	4500.0	4500.0

Solution properties								
Hegman grind	7	7+	7+	7+	7+	7+	7+	7+
Initial pH	7.6-7.8	7.4-7.8	7.4-7.8	7.4-7.8	7.4-7.8	7.4-7.8	7.4-7.8	7.4-7.8
Plating NVM	10%	10%	10%	10%	10%	10%	10%	10%
Pigment/binder ratio	0.8/1.0	0.2/1.0	0.2/1.0	0.2/1.0	0.2/1.0	0.2/1.0	0.2/1.0	0.03/1.0
Melamine/resin ratio	15/85	15/85	15/85	15/85	15/85	15/85	15/85	15/85

[a] Courtesy of Amoco Chemicals Corporation, Bulletin TMA-103a.

[b] Quantities in parts by weight.

[c] Cymel XM-1116 — American Cyanamid Company.

[d] R-3200 — Pfizer Minerals, Pigments and Metals.

[e] 2917 Orange — Harshaw, Pigments & Dye Div.

[f] X-2826 — Imperial Colors & Chemicals.

[g] G-6099 — Pfizer Minerals, Pigments and Metals.

[h] Monastral Blue BT — E. I. du Pont de Nemours & Company.

[i] A-4433 — Imperial Colors & Chemicals.

[j] Ti-Pure R-900 — E. I. du Pont de Nemours & Company.

[k] Mogul A — Cabot Corporation.

TABLE 3-11. Plate Out Characteristics and Physical Performance of Paints[a]

	Primer, red iron oxide		Enamels						
			Orange	Yellow	Green	Pastel blue	Pastel green	Gray	Black
Plating NVM	← 10% →								
Bath temperature	← 80°F →								
Substrate	crs[b]	← Bonderite 37 steel[c] →							
Gassing voltage	160	160	100	100	100	120	120	120	80
Plating voltage	140	140	80	80	80	100	100	100	60
Film thickness, mils	0.6	0.6	0.7	1.0	0.7	0.9	0.9	0.7	0.5
Throwing power, 60 sec, %	54	—	20	22	22	20	20	22	15
Throwing power, 120 sec, %	64	—	25	28	30	28	30	28	27
Panel appearance	← Smooth gloss →								
Coulombic yields, mg/C	34.0	32.5	24.1	23.3	24.5	23.5	23.2	23.4	18.2
Bath stability, 82 ± 2°F	← 60–65 days →								
Cure cycle	← 20 min @ 350° →								
Hardness, sward	26	22	44	40	34	46	42	42	42
pencil	3H	3H	F	H	H	H	H	2H	H
Crosscut adhesion, % pass	← 100 →								
$\frac{1}{8}$ in. Conical bend, % pass	← 100 →								

Impact, direct, in lb		>80	>80	20	>80	>80	>80	>80	>80	>80
reverse, in lb		>80	50	2	>80	>80	50	60	>80	>80
Salt spray resistance,[d]	200 hr	fail	←	←	←	pass	→	→	→	→
	250 hr	—	pass ←	←	←	fail	→	→	→	→
	350 hr	—	pass	—	—	—	—	—	—	

[a]Courtesy of Amoco Chemicals Corporation, Bulletin TMA-103a.

[b]Cold rolled steel.

[c]Bonderite 37 steel panels from Parker Rust Proof.

[d]ASTM B-117-64.

ENAMEL FORMULATIONS

The ingredients of each system were dispersed in a pebble mill to a #7 Hegman grind. The pH was adjusted to 7.5-8.0, then the paints were diluted to 10% NVM in distilled water for application. The paints were formulated to maximize bath stability; thus plate out characteristics such as gassing voltage and throwing power have been somewhat sacrificed. Similarly, the low pigment to binder ratios that contribute to hydrolytic stability also dictate the use of titanium dioxide in some formulations to achieve good hiding power. In applications where plating parameters are more important than batch stability these paints can be reformulated to provide a different balance of properties.

B. BODY FILLERS

Automobile bodies which require a refinish operation will frequently have been exposed to road use and weather for periods of time great enough to require filling of dents, welded areas, or holes in the metal. If dents in sound metal are shallow, they can be filled by putty knife application of glazing compounds. These glazes are made by heavily pigmenting a fast-drying resin system such as a plasticized mixture of shellac and nitrocellulose. Pigmentation will consist principally of talc, clay, barytes, and color pigments that are wet with the composite vehicle during high shear mixing. Fineness of grind is not a critical property of the glaze, since it will be coated later with sanding surfacer. A small amount of solvent is used in formulating a glazing compound to enhance spreading, but any raw material (such as volatile solvent) which leads to shrinkage of the applied filler during drying will contribute to poor filling of the dent. Refinish operators overfill a dent or rough metal area with glazing compound; when shrinkage during drying occurs, there is still an excess of glaze present which can be removed by filing and sanding. It is apparent, therefore, that a high percentage of volume solids will be an advantageous objective in formulating a glaze, for much sanding can be avoided if the contour of the nonshrinking fill can be brought into conformance with adjacent areas while the glaze is fresh and soft.

Recently, there has been a tendency to replace the single-package glazing compound with a two-component package designed for blending the components immediately prior to use. Volatiles present are monomers, which chemically combine with the solid filler so that hardening occurs with almost no shrinkage. This filler consists of conventional pigment in unsaturated polyester which is "thinned" with styrene as a reactive diluent. Peroxide "catalyst" or hardener is added just prior to spreading, which must be done during a relatively short pot life. In some cases, an aluminum pigment is used to prepare body filler that simulates the hot solder used to fill weld seams in original body building. These catalyzed body fillers are particularly useful for filling holes in metal, since they can be applied to fiberglass cloth which then becomes a patch that accepts the coating system.

C. REFINISH PRIMER SURFACERS

It will have been observed that the refinish system is emerging as a step-by-step analog of the original or factory finish applied to automobiles on production lines. The important ingredient making this possible is formulator ingenuity in choosing or modifying chemical raw materials such that each of the analogous coatings forms films at air temperatures. To varying degrees, these films approach the performance of those of the production lines, where the capability for heat-driven reactions exists. The comparison of original finish and refinish is inevitable since both systems may frequently be observed on the same automobile and any marked difference becomes the source of complaints.

Speed of dry, good appearance, and ultimately, good bonding to substrates are requisites of the refinish primer surfacer. These properties are usually achieved with highly modified nitrocellulose lacquers (48). A conventional primer pigmentation (iron oxide 55%, zinc chromate 10%, talc, lithopone, and clay 10% each, barytes 5%) could typically be used in a vehicle which is divided between 0.5 sec resin soluble grade nitrocellulose and a compatible alkyd resin in a ratio of 45/55, respectively. The alkyd may vary between a composition comprising a short oil length, pure coconut oil, to a semioxidizable short oil, alkyd modified with maleinized ester gum to prevent the oxidizable component from causing lifting when topcoated with products rich in active solvent. A stable package should be obtained with pigment-binder ratios of 2/1 to 1.5/1 weight and 55 to 60% total solids. A typical solvent blend for this package would be 25% toluene, 15% xylene, 20% ethyl or isopropyl alcohol, 30% butyl or isobutyl acetate, and 10% methyl ethyl ketone.

Good atomization when sprayed applications are made will require at least 1:1 reductions with lacquer reducers, and for very smooth films requiring minimal sanding, the reduction may be increased to 1.5 (primer to reducer) at the expense of some film thickness and substrate filling.

Dry film thicknesses of 1 to 1.5 mil should be sprayed (2 to 3 coats), after which, given normal ventilation and temperatures between 21 and 27°C, the primer will be dry enough to wet sand (use rubber gloves to guard against skin irritation by zinc chromate) in 30 min. Sanding sludge is thoroughly removed by water washing, and as soon as the surface has been wiped dry with clean lint-free cloths or blown dry with clean air, it may be topcoated. Allowing the wash water to evaporate from the primed body may lead to topcoat failure because of local residues of soluble salts at sites of water droplet evaporation.

Using a pigmentation similar to that described for the sanding lacquer just reviewed, alkyd surfacers are also formulated using alkyds modified with oxidizable or semioxidizable oil (linseed or soya), but these surfacers require overnight drying before they demonstrate optimum performance and are usually restricted to those occasions when enamel topcoats are used, since there is a hazard of their being swollen into lifting if the topcoat contains lacquer-type reducers. A particularly insidious type of failure occurs when alkyd surfacers are applied over an old lacquer film and topcoated with lacquer so that alkyd is sandwiched between two lacquer films. Even if the alkyd film is not lifted by topcoat solvents, it is permeable enough so that the basecoat lacquer can absorb considerable amounts of the topcoat solvent. The differential drying speed of the three-layer system accordingly can cause dimensional differences in the respective layers that are great enough to cause "splitting" or fissuring of the topcoat, which will obviously shrink faster than the other layers which retain solvent longer. The alkyd modified lacquer primer surfacer is much more generally used, therefore, not only because of its greater drying speed, but also because of its greater versatility of performance in conjunction with the great variety of other film formers with which it must interact.

D. REFINISH TOPCOATS

1. Acrylic Lacquers

Automotive refinish topcoats are expected to perform in a manner similar to that specified for original finish topcoats. Of the various topcoat compositions commercially available that rapidly form films at air temperature, it is probable that all formulators working in the refinish market will agree that acrylic lacquers make the nearest approach to realizing the expectation.

It is not unusual for original finish acrylic lacquer to double as an air-dry refinish topcoat. Accordingly, the formulation displayed in Sec. 3-3,A (Table 3-1) can be used as a basic reference. If one were to criticize the performance of that composition for refinish, it would be in the area of its appearance. The original finish topcoat was intended to be heat reflowed to develop unpolished smoothness and film gloss. As an air-dry coating, the gloss will be in the low eggshell range, and unless spraying solids is substantially reduced, there will be a considerable degree of "orange-peel" texture. Both defects are overcome by a wheel compounding and polishing operation which, in cases of severe dry spray, can be preceded in critical areas by light sanding with fine grit. The natural desire for better appearance with a minimum of laborious and costly polishing leads to formulation variations strictly designed for air drying. The cellulosic component (EAB -381-2, Eastman Chemical Products, Inc.), present in the ratio of 20% of the vehicle solids, may be reduced by as much as one half in favor of polymer, or changed to a more soluble type such as EAB-531-1 (Eastman Chemical Products, Inc.). The plasticizer may be changed to a compound which is a better solvent for the polymer now that danger of loss by volatilization in a baking oven has been removed. Butyl benzyl phthalate (Santicizer #160, The Monsanto Company) is commonly used. The temptation to improve initial film appearance by resorting to lower-molecular-weight polymers, which unquestionably flow to smoother films, should be resisted unless the formulator is dealing with an end use in which weather resistance is not essential. Film cracking, blistering (49, 50), and early gloss loss by vehicle erosion are hazards associated with the lower-molecular-weight resins when they are used in acrylic lacquers for outdoor service. [See gel permeation chromatogram (GPC) polymer analysis, Fig. 3-5.]

The consideration of reducing thinner to prepare acrylic lacquer package compositions for spraying is important to a successful end use of the coatings. A single reducer was recommended in Table 3-2 for use with original finishing acrylic lacquer topcoat which can be associated with standard production line climatic conditions and a predictable and uniform surface. Neither of these favorable conditions is likely to occur in a refinishing operation except by pure coincidence. One additional degree of freedom which the refinish formulator can therefore supply to those who use his product is a small variety of reducing thinners to modify application properties to be consistent with air temperature and quality of the surface being refinished.

Two frequent variations in refinish reducers from so-called standard compositions are low-penetration and retarder or reflow types. The low-penetration reducer is a balance of volatile compounds calculated to have barely sufficient compatability with the polymer, even at the end of the evaporation period, to avoid precipitation or phase separation. At the same time, it is expected that its compatability with aged and thoroughly dried films of some prior coating will be poor enough to resist crazing or lifting that prior coating. Its formulation is based on extensive use of fast

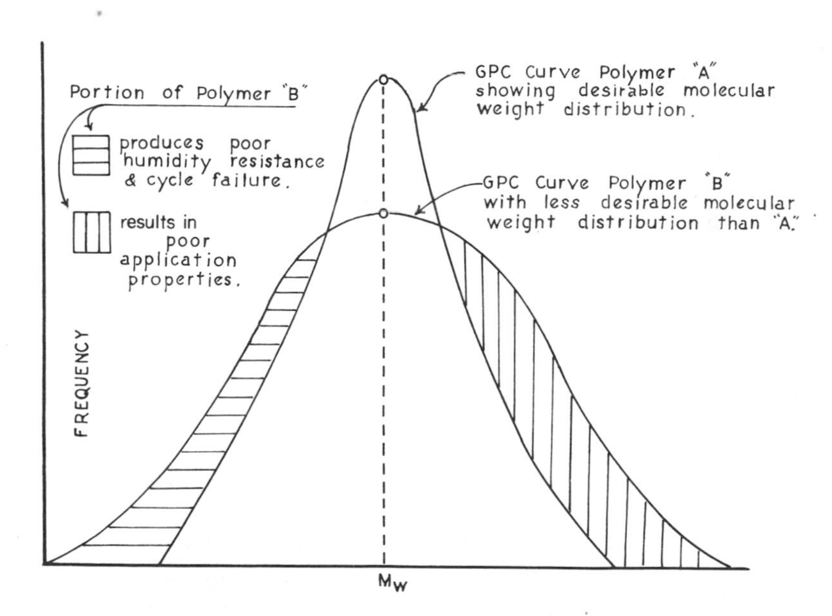

FIG. 3-5. Effects of molecular weight distribution.

evaporation of those components, such as acetone, which assist air atomization, but which might be expected to lift or craze a substrate. The experience is that such active, fast-evaporating solvents as acetone can be expected to volatilize during atomization, and if there is limited retention of a high concentration of active solvent in contact with a sensitive substrate, the substrate may escape being seriously swollen.

Retarders are simply active solvents that are evaporated slowly enough in exceptionally warm application environments to be retained in the applied coating long enough to promote desired flow and leveling. Retarders are particularly useful for applications in humid atmospheres which may cause moisture condensation on wet coating surfaces locally cooled by rapidly evaporating solvent. An example of a modification in the composition of the "standard" reducer of Table 3-2 that would constitute retarder action, therefore, might take the form of increasing the percentage of cellosolve acetate (Union Carbide Corp.) by 10 to 15%.

2. Refinish Enamels—Unmodified Alkyd Type

In its simplest form, a refinish enamel is formulated by combining lightfast pigmentations with a polymer that yields glossy, weather-resistant films in an air-drying schedule. A typical specification for satisfactory film-forming rates by air drying suggests that the coating become dry enough to be dust-free in 15 to 20 min after spray application, and tack-free under gentle pressure of finger tips or a paper test surface in 4 hr. Frequently, a small repair must be made before the automobile may be returned to the owner. If a blemish is removed by sanding and further application of paint, masking with adhesive tape and paper is required to protect surrounding coating areas. This need gives rise to an additional requirement in the drying specification which describes thoroughness of film cross-section drying in terms of time required for resistance to being imprinted by the masking tape.

In the case of enamels which ultimately form films by oxidative cross-linking, resinous vehicles are selected which are hard enough to meet the "dust-free" requirements merely by evaporation of the bulk of the volatile from the applied film. In the special case discussed below of the fast-drying acrylic or styrene-modified alkyds for automotive refinish, the polymer is usually hard enough prior to oxidation to meet the "tack-free" drying requirements as well.

A typical alkyd for the standard, or 4-hour, tack-free enamel, is Beckosol 1307-50 (Reichold Chemicals Inc.), which is a medium oil length soya alkyd, described (51) as having 41% soya bean oil, 42% phthalic anhydride, with a Gardner-Holdt viscosity of N to P at 49 to 51% total solids in xylol. This alkyd can be used for the preparation of pigment dispersions in a sand mill or attritor, as well as for the let-down vehicle that constitutes the major portion of the enamel's resin solids. Pigmentations of automotive refinishes are almost always scrupulously chosen to be as nearly as possible identical with those of the original finish standard, in order that true spectral color matches are provided for repair work when only a portion of the original finish is replaced.

A critical problem in establishing a refinish enamel formulation concerns the decision as to the type and concentration of drier to be used. Driers are the metal soaps that are added to an oxidizable oil-modified alkyd to accelerate the rate of conversion at air temperatures of the liquid coating to a dry film. Driers are, therefore, frequently referred to as "catalysts," although their precise chemical functioning is still a matter of some disagreement.

Refinish enamels of the unmodified alkyd type are applied at film thicknesses that are the maximum allowable for uniformity of drying through their cross-section because of the need to achieve the best possible appearance through filling of substrate sand scratches. Accordingly, drier concentration normally used in refinish enamels is somewhat greater than that usually recommended for general industrial air-dry alkyd-type enamels. Based on resin solids, a blend of the respective alkyd-soluble soaps would be used to supply 0.10 to 0.15% cobalt, 0.10 to 0.15% manganese, and 1.0 to 1.5% lead calculated as metal. Special variations of this blend will frequently occur to deal with specific problems. For instance, cobalt concentration may be increased slightly and manganese removed in preparing white or pastel colors because of the brown tone imparted by manganese. Calcium, which is almost totally inactive as a drier but has a very pale color, may be used to satisfy the absorption tendency that some pigment surfaces have for cobalt driers. The package storage stability of an enamel as it concerns its ability to dry on schedule after long periods of shelf aging is thereby enhanced. Zirconium complexes, which have relatively little dry-promoting activity in their own right, make a particularly happy combination with cobalt as a replacement for lead. The zirconium-cobalt combination may be found to be more soluble in alkyds than cobalt-lead, and not only produces greater drying activity, but eliminates film haze from any situation caused by precipitation of reaction products of lead drier blended with some alkyds. Zinc is occasionally used to inhibit surface drying of oxidative films, thus solving film wrinkling problems by ensuring a more uniform cross-sectional cure.

It will be noted that the alkyd selected for the alkyd refinish enamel is modified with soybean oil, which is substantially lower in unsaturation than linseed or dehydrated castor oil. With the soy-modified alkyd, there is not only less color contribution to very pastel pigmentations, but better control of the uniformity of cross-sectional drying. When the soy alkyd is used, one encounters less difficulty with the wrinkling of thick films in hot, humid weather, but there is also less etching of fresh dried films exposed to water droplets from rain or dew.

Solvents for alkyd-type refinish enamels are selected for compatibility with the polymer, for viscosity control of the enamel, and for control of the leveling and drying speed of the films. Compatibility of solvents with medium-oil-content alkyds is not the critical situation it frequently is with acrylic lacquers. Accordingly, hydrocarbon mixtures of aromatic (xylene) and aliphatic (high-flash naphthas) are common. Refinish enamel packages, because of the merchandising approach employed, are apt to be subject to prolonged shelf storage. Package viscosities ranging from 75 to 100 sec #4 Ford cup are sought, therefore, to aid in long-term pigment suspension. An enamel of the package viscosity indicated will contain 40 to 50% total solids, and require the addition of about 25% of its volume as a spray reducer to accomplish easy spray application. This reducer might typically be composed of 20% by volume toluene, 30% xylene, 35% VM&P naphtha, and 15% of a slower evaporating aromatic type (Solvesso #150, Esso Standard Oil) for good flow and leveling.

3. Refinish Enamels — Modified Alkyd Type

The usual purpose in making any substantial modification of alkyds for automotive refinish enamels is to increase the air-drying speed of the enamel. For this purpose, alkyds have been modified during their preparation with phenolic or occasionally with natural resin, such as rosin derivatives (such modifiers typically can be expected to degrade durability), or with styrene or acrylic monomers which can react with unsaturation in the oil that is present. Of course, to the degree permitted by compatibility, finished polymers can also be blended, but the degree of modification is frequently more limited than is the case when the modifier is one of the reactants. Styrenated alkyds were tested early in the attempts to develop faster drying refinish enamels, and refinements have appeared that perform respectably for rapid air-dry or low-temperature-bake general industrial schedules. The weather resistance of air-dry enamels based on styrenated alkyds has not been adequate, however, for automotive refinishing, and the use of acrylic monomers to replace substantial amounts of the styrene has led to preferred enamel vehicles.

An example of formulating with a modified alkyd for fast-drying refinish enamel is that achieved by blending Amberlac 292X (Rohm and Haas Co., Philadelphia, Pa.) with thermoplastic acrylic polymer B-99 [Rohm and Haas Co. (52, 53)].

Amberlac 292X, a castor oil-phthalic alkyd, 50% solids in xylene, is modified in such a way as to be compatible with acrylic polymer B-99 supplied at 50% solids in a 70/30 ratio of xylene and toluene. Cobalt drier is recommended at the rate of 0.01 to 0.05% metal based on the Amberlac 292X solid resin content, and the two resins are blended at a ratio of 75/25 alkyd to acrylic. Tack-free times for enamels as short as 10 min have been measured when such a blended vehicle is used.

There are other advantages to the acrylic-modified alkyd enamels in addition to rapid air drying. Their ability to resist gasoline spillage and water spotting within the first day of their existence as films has made them popular with the refinish paint shops which are thereby enabled to enjoy a much more rapid clearance of freshly painted ware from their shops to outside storage. The rapid air-dry film has its potential disadvantage also, particularly in the case of those compositions with enough unsaturation to oxidize rapidly to insolubility in the solvents utilized by the formulator.

Frequently it is necessary to make a small repair in the refinish job that requires recoating the freshly dried film with additional liquid coating. The successful formulator will control the rate of oxidative drying with a volatile antioxidant such as ketoxime to provide that the base refinish film does not become insoluble in the repair solvent for the first 8 to 10 hr of its life. Sufficient time is thereby afforded for recoating without the lifting that occurs when partial oxidation of the base film has had the opportunity to take place. There is little doubt that fast air-dry formulations that have some capacity for oxidation of films during early stages of dry will almost universally exhibit a "sensitive" period sometime in the drying process when they cannot be recoated without lifting. The formulator's task is seen, not so much as an attempt to eliminate this sensitive period, but as an attempt to accommodate its timing to coincide with the minimal inconvenience of the painter. For instance, sensitive periods that prevail between the eighth hour of drying and the twentieth, will coincide with the 12 to 15 nighttime hours when the shop is closed anyway. If the painter can schedule his topcoating for early in his day, he could do repair touchups later the same day, and still release the car the same day it was painted or, at the latest, on the following day.

Paint additives, whose recommendation consists of modifying refinish enamels with 2 to 6% (based on resin solids) of melamine resin, may be used to convert most air-dry refinish enamels to baking enamels. Formulators for automotive refinishing will capitalize on the capability of the melamine crosslinking reaction for those end use applications which have heat schedules of 82 to 91°C available, to recommend such additives. Immediate film hardness and mar resistance, coupled with long-range extension of weather resistance, make such coatings attractive to the large refinish operation or to the general industrial coatings market.

3-7. ELASTOMERIC COATINGS

A variety of elastomeric plastics have been used in recent years on the exterior of automobiles for absorption and dissipation of impact energy.

Front and rear bumpers and bumper assemblies, bump moldings along the belt line of the body, air conditioner and carburetor air scoops, and wheel covers, all from molded urethane, have been popular. The construction of these parts has sometimes involved molded casting of high-density, microcellular urethane foam, while at other times solid elastomers have been injection molded using EPDM rubber, urethanes, or other modified distensible plastics. The molded objects must be coated for use on automotive exteriors because many of the plastics involved have poor weather resistance and require the protection afforded by the coating. The coating, on the other hand, fulfills a difficult role because not only must it have weather resistance in the normal sense that it retains original color, gloss, and substrate adhesion, but it must also retain a substantial portion of its original mechanical properties. Of valuable mechanical properties in the coating, a high percentage elongation with a high tensile strength and nearly linear elastic modulus, are prized. These properties allow the coating to distend without fracture and recover without wrinkling from distortion as the elastomeric substrate is distorted by the high-energy impacts of even low-speed collisions of automobile bodies.

Some of the cast moldings, such as automobile bumpers, contain steel reinforcements which are placed in the mold cavity prior to pouring the liquid mixture, whose reaction forms the foam. The metal insert restricts the compressive distortion that occurs on impact and an elastomeric coating with 50 to 80% elongation, and tensile strength at break of 3500 to 4000 psi will usually be satisfactory.

The fatigue resistance to dynamic flexing need not be great with castings containing metal inserts. As unsupported elastomeric plastic parts are designed, coatings' requirements become more stringent, requiring 150 to 200% elongation without reducing tensile strength, and resistance to dynamic flexing at low temperatures (-11°C).

The breakthrough in urethane chemistry that permits formulation of urethane coatings for automotive topcoats is the commercial availability of aliphatic or alicyclic diisocyanates. Examples of such monomers are: hexamethylene diisocyanate (Mobay HX), methylene di (p-cyclohexyl isocyanate), for example, Du Pont Hylene "W", trimethyl hexamethylene diisocyanate, or isophorone diisocyanate. Urethane resins prepared with aliphatic diisocyanates are referred to as "light stable" in view of their resistance to discoloration and gloss loss during exterior exposure.

Coatings suitable for the end uses described may be either thermoplastic or thermosetting. The urethanes, which have alternating hard and soft domains resembling block copolymers, are particularly adaptable for meeting the mechanical requirements both of substrate and of coating. The crosslinked or thermoset enamel films are formed from the reaction between a light-stable urethane resin, which has hydroxyl functionality due to the use of a small excess of polyol in its manufacture, and a melamine-formaldehyde resin of either butylated or methylated types (54). Curing follows the route of other melamine resin crosslinked enamels discussed earlier, and within bounds of mechanical film properties required, may be catalyzed with soluble acids (p-toluene sulfonic or acid phosphate esters) to occur at schedules of 20 to 30 min at 120°C to as low as 93°C. Dry films of 1.5 mil are pigmented to follow the body color standards with similar opacity and initial gloss.

The elastomeric enamel described should be thinned for spray application with a mixture of xylene, butyl alcohol, and cellosolve acetate (Union Carbide Corp.) in a ratio by volume of 75/10/15 to 25 to 30% total solids.

Elastomeric lacquer coatings for automotive parts are formulated with high-molecular-weight, light-stable urethane resins that are essentially inert in that almost no unreacted monomer functionality remains. The high-molecular-weight structure produces a high-viscosity resin solution (approx Z_6 at 25% solids), and correspondingly low-sprayable solids (approx 10% at room temperature). Once again, pigmentation follows patterns established by body colors, but with the thermoplastic formulation, even baked film gloss is low enough with pigmented coatings to require a final application of clear elastomeric lacquer to achieve body gloss standards.

Mechanically, the essentially linear structure of the urethane lacquers permits a much higher elongation (400 to 600%) at break than the crosslinked enamel structure. Of course, it also forms films at lower bake temperatures (60 to 71°C) or even at room temperature. In the latter case of room temperature drying, enough time must be allowed before testing or packing for nearly complete evaporation of solvent, Depending on film thickness and presence of slow ketones such as cyclohexanone added as reflow agents, dry time for handling may be one to three hours, and dry time before environmental testing should be three days. Lacquer urethanes, therefore, tend to be restricted in their commercial use to such substrates as fabrics which receive extreme flexural stress and to automotive refinish operations which are conducted at ambient temperatures.

Further commercial considerations governing choice between thermosetting and thermoplastic urethanes concern their comparisons as barrier coats for sealing substrates. Some cast- or injection-molded substrates contain traces of amines, other catalysts, or unreacted monomers which, if allowed to migrate to the coating's surface, cause discoloration during interior or exterior exposure. It has been the history of the observation of total systems that the barrier activity of coatings for plastic substrates has been rather strongly in favor of the thermosetting compositions.

REFERENCES

1. A. F. Lewis and L. J. Forrestal, "Adhesion of Coatings," in Treatise on Coatings (R. Myers and J. S. Long, eds.), Vol. 2, Part I, Dekker, New York, 1969.
2. P. E. Pierce, "Mechanical Properties of Coatings," in Treatise on Coatings (R. Myers and J. S. Long, eds.), Vol. 2, Part I, Dekker, New York, 1969.
3. A. Neuhaus and M. Gebhardt, "Epitaxy and Corrosion Resistance of Inorganic Protective Layers on Metals," in Interface Conversion for Polymer Coatings (P. Weiss and G. D. Cheever, eds.), American Elsevier, New York, 1968.
4. W. A. Zisman, Influence of constitution on adhesion, Ind. Eng. Chem., 55 (10), 18-38 (1963).
5. J. R. Huntsberger, Mechanisms of adhesion, J. Paint Technol., 39 (507), 199 (1967).
6. H. S. Bender, The mechanical properties of films and their relation to paint chipping, J. Appl. Polymer Sci., 13, 1253-1264 (1969).
7. M. G. Fontana, Perspectives on corrosion, Corrosion, 27, No. 4, 129-145 (1971).
8. G. H. Cox, D. W. J. Osmond, M. W. Skinner, and C. H. Young (to Imperial Chemical Industries), U.S. Pat. 3,393,162 (1968).
9. T. D. Hoy and G. G. Schurr, J. Paint Technol., 43 (556), 63-72 (1971).
10. B. N. McBane, J. Paint Technol., 42 (551), 730-735 (1970).
11. Rohm & Haas Co., "Acrylic Glass Temperature Analyzer," Form SP-222 Philadelphia, Pa., March, 1965.
12. J. Hildebrand, J. Am. Chem. Soc., 38, 1452 (1916).
13. H. Burrell, Offic. Digest Fed. Soc. Paint Technol., 27 (369), 726 (1955).
14. J. Crowley, G. Teague, and J. Lowe, J. Paint Technol., 38 (496), 269 (1966).
15. C. M. Hansen, J. Paint Technol., 39 (505), 104 (1967).
16. L. H. Lee, J. Paint Technol., 42 (545), 365 (1970).
17. K. L. Hoy, J. Paint Technol., 42 (541), 76 (1970).
18. Hercules Incorporated, Bulletin M-299A, "High Film Build Lacquers Based on Hi-Lo Thinners," Polymers Department, Coating News, Wilmington, Del.

19. B. McBane, Automotive Coatings, Chemical Specialties Manufacturers Association, Inc., "Proceedings of the 56th Mid-Year Meeting," 50 East 41st Street, New York, N.Y., 1970.

20. "Plasticizers (Properties and Suppliers)," Modern Plastics Encyclopedia for 1970-71, McGraw Hill, New York, p. 856.

21. Eastman Chemical Products, Inc., Formulator's Notes, No. 3.3g, "Cellulose Acetate Butyrate Thermally Reflowable Finishing Systems," Kingsport, Tenn., Oct. 15, 1967.

22. D. W. J. Osmond, F. A. Waite, and D. J. Walbridge, British Pat. 1,174,391 ICI (1969).

23. M. Fryd (to E. I. DuPont De Nemours and Co., Inc.), U.S. Pat. 3,405,087 (1968).

24. E. C. Rothstein, Paint Varnish Prod., 58 (2), 39-43 (1968).

25. F. Golemba and J. E. Guillet, J. Paint Technol., 41 (532), 315-320 (1969).

26. Z. Raciszewski and D. H. Mullins, J. Appl. Polymer Sci., 14 (4), 967-977 (1970).

27. N. B. Graham, F. R. Crowne, and D. E. McAlpine, Offic. Digest Fed. Soc. Paint Technol., 37 (489), 1228-1250 (1965).

28. D. R. Savage, "Plastic Coatings via Dry Powder Formulation," Society of Plastics Engineers, Inc., Technical Papers 22nd Annual Tech. Conf., Stamford, Conn., 1966.

29. Proceedings of the First North American Conference on Powder Coatings, Maclean-Hunter Ltd., Toronto, Ontario, Canada, 1971.

30. Ford Motor Co., Specification M-32-J-100A, Revised, Dearborn, Mich., 1967.

31. Chrysler Corp., Specification MS-PA10-1, Revised, Detroit, Mich., 1967.

32. General Motors Corporation, Manufacturing Development Dept., Test Procedures and Specifications for Coating Materials, Warren, Michigan, 1972.

33. Monsanto, Publication No. 6288, "Resimene Melamines, Cross Linkers for Chemical Coatings, Textiles, Inks," St. Louis, Mo., 1971, p. 19.

34. ASTM Standards, American Society for Testing and Materials, 1971.

35. J. B. Boylan, "Epoxy Ester (Dimer Acids in Surface Coatings)," Treatise on Coatings (R. Myers and J. S. Long, eds.), Vol. I, Part III, Dekker, New York, 1972, p. 13.

36. Dow Chemical U.S.A., unpublished data, Conference: The Changing Technology of Coatings, Midland, Mich., Oct. 1971.

37. E. P. Plueddemann, Treatise on Coatings (R. R. Myers and J. S. Long, eds.), Vol. I, Part III, Dekker, New York, 1972, chap. 9.

38. S. Gloyer, D. Hart, and R. Cutforth, Offic. Digest Fed. Soc. Paint Technol., 37 (481), 113-128 (1965).

39. S. R. Finn and C. C. Mell, J. Oil Colour Chemists' Assoc., 47 (3), 219-45 (1964).

40. G. Burnside and G. Brewer, Paint Varnish Prod., 60 (4), 46 (1970).

41. M. R. Rifi, "Water Borne Coatings from Bakelite Polycyclol 1222 Modified with Bakelite Phenolic Resins," Technical Service Report, Bound Brook, New Jersey, 1972.

42. Amoco Chemicals Corp., Technical Memorandum, "Experimental Electrocoat Primer from Resin Based on TMA-IPA-95-HBPA," Chicago, Ill., 1965.

43. J. M. DeVittorio, Reconstituting Electrocoating Baths, U.S. Pat. 3,499,828 (1970).

44. Imperial Chemical Industries, "Coating Compositions for Electrodeposition," British Pat. 1,211,203 (1970).

45. O. W. Huggard (to Mobil Oil Corporation), U.S. Pat. 3,598,775 (1971).

46. L. LeBras and R. Christenson, J. Paint Technol., 44 (566), 63-69 (1972).

47. L. LeBras, Electrodeposition theory and mechanism, J. Paint Technol., 38 (493), 85-90 (1966).

48. Eastman Chemical Products, Inc., Formulator's Notes No. 3.5, Kingsport, Tenn., Nov. 1957.

49. E. I. DuPont de Nemours and Co., "Du Pont Elvacite Acrylic Resin, Properties and Uses," Du Pont Polymer Products, Bulletin A-76941, Wilmington, Del., Aug., 1971.

50. N. H. Frick, "The Effect of Molecular Weight Distribution on the Cold Crack Thermal Cycle Resistance of Automotive Lacquers," Applied Polymer Symposia No. 10 (C. A. Lucchesi, ed.), Wiley (Interscience), New York, 1969, p. 23-34.

51. H. P. Preuss, Synthetic Resins in Coatings, Properties of Alkyd Resins, Chart 9, Noyes Development Corporation, New York, 1965, p. 78.

52. Rohm & Haas Co., "Synthetic Resins for Coatings, Duraplex and Amberlac," Bulletin C-160, Philadelphia, Pa., Feb., 1966.

53. Rohm & Haas Co., "Acryloid B-99," Bulletin C-291, Philadelphia, Pa., Feb., 1970.

54. M. Kaplan, Advances in Urethane Science and Technology (K. C. Frisch and S. L. Reegen, eds.), Vol. 1, Technomic Publishing, Stamford, Conn., 1971, p. 133.

Chapter 4

COIL COATING — THE CONTINUOUS COATING OF METAL STRIP

Donald K. Lutes, Sr.

Technical Manager, Coil Coatings
The Sherwin-Williams Company
Chicago, Illinois

4-1. INTRODUCTION

The continuous coil coating concept dates as far back as 1936. The prototype for our modern coil coating lines was actually developed in 1936.

The continuous roller coating line idea was conceived by Joseph E. Hunter who, with his associates, engineered a workable unit. By 1937 the Hunter Engineering Company started to produce a continuous roller coating line for the venetian blind industry.

Today the modern coil coating lines still embody many of the principles engineered into the first Hunter line. The early Hunter line was capable of cleaning, treating, and coating 0.008 in. thick metal strip of 2 in. width. To a degree, the basic design of the early continuous coating line parallels today's equipment. Chronologically, from stage to stage, it consisted of a coil payoff reel, tension rolls, metal cleaning and treatment section, paint roller coater, bake oven, cooling section, and coil rewind unit.

The early Hunter line was capable of applying paint to both sides of the metal strip simultaneously. Maximum coil weight was limited to approximately 100 pounds and the coating, or line speed, was 50 ft/min.

The advancement of continuous roller coating lines was somewhat retarded during the war years, 1941 to 1945. Immediately following the war, interest was revived, and by 1946 several new venetian blind manufacturers appeared. These concerns, for the most part, installed continuous horizontal roller coating lines of the Hunter type. A few venetian blind manufacturers employed vertical roller coating units.

Venetian blind roller coating equipment improved considerably by 1948. By this time several other equipment manufacturers had entered the field. The newer coating units became more sophisticated and versatile, providing higher quality coated strip, higher production rates, and greater ease of operation. As the equipment improved, it was possible to coat strips of greater width. Within a few years the width of coated metal strip increased from 2 to 5 to 8 to 12 in. This significant breakthrough opened new markets for continuous coated steel or aluminum strip. Some of the first products to undergo this transition were actually accessories to the venetian blind unit. This included bottom rails, headbox, facia boards, end brackets, etc. It was not long before other products were being formed and fabricated from prepainted strip; among the first were sections for awnings, canopies, and aluminum residential siding.

By 1949 the continuous coil coating concept had really taken hold. Essential a new industry had been born. Manufacturers and fabricators of various products and commodities began to conceive the advantages and possibilities of this new approach. From this time on, significant and steady progress was being made in coil coating. Twenty years of additional technological progress on the part of equipment manufacturers, as well as metal cleaning and treatment concerns, the paint industry, coil coating concerns, and the metal fabricators, has resulted in today's ultramodern coil coating lines.

4-2. THE MODERN COIL COATING LINE

The large-volume coil coating line today bears little resemblance to the early lines. Modern coating lines are massive, highly automated units, capable of coating metal strip five feet wide. Even wider lines are being engineered at the present time. Through the years, improvements have been made in each component or section of the continuous operation - payoff reel, strip stitcher or splicer, accumulators, tension system, metal treatment stage, coater, oven, strip cooling equipment, and rewind unit. (See Figs. 4-1 and 4-2.) The modern lines are heavy duty units designed to handle both aluminum and heavy gauge steel. The production capacity of the modern line is high, since the average line speed is 200 to 250 ft/min. Several lines are operating at 300 ft/min and speeds of 400 to 500 ft/min are anticipated in the near future.

The large modern coil coating line requires considerable plant space, since the total length from payoff reel to rewind can range between 600 and 900 ft. There are a variety of coating line designs and configurations in the field. Some lines are single coating units, others are tandem or dual units adapted to applying two-coat paint systems. There are straight-run, one-level lines, and there are also double-tiered lines to conserve plant space. In the latter type line, the metal treatment section, or, in the case of a tandem coater, the primer application section, occupies the upper level.

Several modern coil coating lines have other innovations or modifications built into them to provide greater versatility. Some lines, for example, include film laminating equipment so that calendered vinyl, acrylic, fluorocarbon or similar films can be applied to metal strip with an adhesive. In-line slitters to enable the strip to be cut to designated widths before recoiling are also quite common. Rotary embossing rolls are often installed in a line to impart various embossed patterns in either the vinyl laminate or plastisol-type films. Specialized printing and overlay equipment is available to achieve attractive color combinations, wood-grain patterns, and high styling designs. The lines are designed so that any of the above ancillary equipment can be easily disengaged or bypassed if it is not required for a given operation.

In order to achieve an efficient operational level, many coil coating lines are equipped with elaborate control devices and electronic equipment. Automatic strip tracking guides maintain good linear tracking of the metal strip through the entire cycle. The metal cleaning and treatment section is equipped with automatic sensitizing units to control proper solution temperature, pH, and solution concentration. This assures adequate deposition of the conversion coating to the metal substrate. Sensitive automatic temperature controls assure consistent and efficient operation of the baking oven. In-line electronic instrumentation to determine and monitor the dry film thickness of the coating can be installed as an additional quality control measure. A

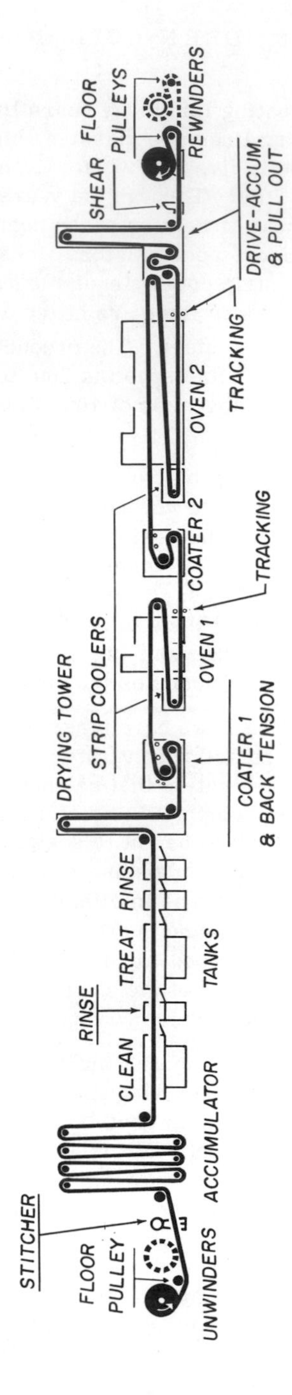

FIG. 4-1. Elements of a coated coil line.

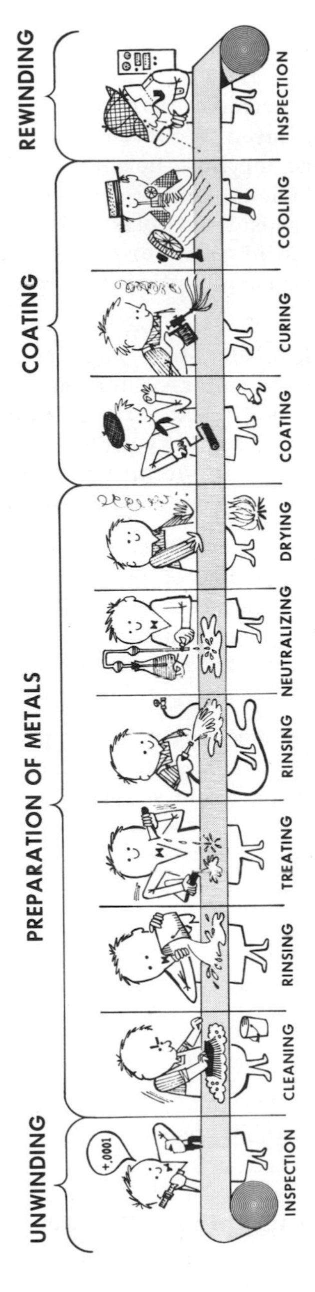

FIG. 4-2. A typical "Tandem" coil coating line.

similar unit can be utilized to monitor the color of the coated strip before it is recoiled.

The modern coil coating line operating at full capacity can utilize large volumes of paint. The typical 60 in. width line, for example, can utilize approximately two 50-gal drums of paint per hour, or 2400 gal in a three-shift day. This figure is based on conventional coatings such as solution-vinyls, amine alkyds, and acrylics, normally applied as a 1-mil dry film. With coatings of high volumetric solids that are often applied at higher film thicknesses, such as organosols and plastisols, the total gallons utilized for the same period of time could double or even triple. Since paint consumption is high, some modern coil coating plants provide for bulk paint delivery and bulk storage.

In order to assure a clean, dirt-free painting operation, most modern lines have a room built around the roller-coater unit. The room is usually pressurized and all intake air is filtered. The room can also be air conditioned for constant temperature and humidity control. The strip passes into and exits from the room through wall slots. Paint handling is generally minimized in the roller-coating area by having an automatic paint circulating system.

The installation of a current model maximum coating width line (66 to 72 in.) can involve a total investment of $850,000 to $2,000,000 depending on design and ancillary equipment desired. Many such lines have been installed and the number will certainly continue to grow. We must also remember that many large coating firms have two or more complete lines in a given plant area.

Several of the major coil coating firms also process their own metal, for example, in the continuous casting and rolling of aluminum. A good number of firms also have roll-forming and stamping equipment, modern quality control laboratories, and all the other components to form an integrated operation. There are actually three types of coil coaters. The categories are:

1. Strictly Captive Coater — Coating strip for their own products. Example:—awnings, siding, rain-carrying equipment, building sheet, etc.

2. Toll Coater - Coating strip for the general market, metal fabricators, or secondary user.

3. Toll and Captive Coater - Coating for their captive products and also for the general market.

The majority of coil coating firms, at the time of this writing, fall into categories 2 and 3.

4-3. ADVANTAGES OF COIL COATING

A layman who is not familiar or close to the coil coating industry may ask —why has this method of coating become so popular? Why has the continuous strip coating process enjoyed such rapid growth? Are there advantages? Yes, there are several advantages to prepainting metal by this method.

The most significant advantage is probably passed on to the metal fabricator or the manufacturer of the numerous commodities on today's market. Prior to the advent of coil coating, many metal products were prefabricated, cleaned, and treated, and then painted by spray, dip, or flow-coat methods. Each manufacturer had to equip his plant to handle all these operations. In most instances these operations were separate steps often involving numerous departments within the plant. Operating in this fashion required considerable interplant coordination, additional handling equipment, and, of course, additional personnel.

Now, by installing a coil coating line, all the above operations can be simplified and consolidated in one department. This often conserves plant space and eliminates a great deal of piece work and intermediate handling of the product. It very often can reduce the personnel requirements in the total operation. Many small manufacturers and fabricators have eliminated their metal cleaning and painting operation entirely and now purchase coated strip from a toll coater. This approach can often save money.

The application of paint by roller coating will also effect a savings by minimizing paint loss. Actually there is no paint loss in a roller coating operation. In contrast even with the most efficient spray-painting operation, manual or automatic, paint losses through overspray can be substantial.

Another definite advantage of roller coating is the exceptional uniformity achieved in applying the finish. The film thickness of the coating can be controlled easily and consistently. The mechanized and automated process provides a finish that is uniform in a day-to-day operation. This ultimately results in greater protection of the end product, as well as a better looking product.

Proper cleaning and treatment of the metal substrate is essential to achieve good performance of the finish. This is particularly important for those products destined for exterior exposure such as residential siding, awnings, canopies, mobile home sheathing, and other building components. In the area of metal cleaning and treatment, the modern coil coating line has definite advantages. The cleaning and treatment phase is easily controlled thereby producing, as in the painting operation that follows, uniform results. Paint is applied to the strip immediately after the cleaning, treating, and drying stage, before the surface ages or becomes contaminated in any way. This contributes to good adhesion, formability, corrosion resistance, and field performance.

Coil coating, or utilizing precoated strip, can actually simplify the quality control operation in a plant. The quality control is largely directed to the coil of stock. Once the coil stock has met the established standards or performance parameters, it can be stamped, drawn, or roll-formed into the end product. Often subsequent inspection of each unit is not really necessary. Applying paint to individual items by spraying, dipping, or flow coating, on the other hand, usually necessitates closer inspection of each unit.

Extremely high production capacity is another advantage of coil coating. The recent development of wider coating lines capable of higher line speeds has materially increased the output. This has increased to the point where it far exceeds the production volume of most spray lines. For example, a modern 48 in. width coil coating line operating at a speed of 180 ft/min can coat 24 tons of 18-guage steel per hour. A 60 in. width line operating at a speed of 250 ft/min can coat 75,000 sq ft of strip per hour. Based on a three-shift day, this amounts to 1,800,000 sq ft of finished strip.

The foregoing cites the major advantages of coil coating over other methods of painting. Some of the minor advantages include smaller paint inventories, a much cleaner operation, and the sales and market value in having a better, more appealing painted product.

4-4. COIL COATING FINISHES

The coil coating industry and the market it serves helped advance the technology of coatings as much as the coating industry helped advance the coil coating concept. In common vernacular, this definitely was a two-way street. As new markets were found for coated strip involving a variety of end uses, performance of the coatings has to be upgraded. The major requisite of a coil coating, of course, is film flexibility. It must be able to withstand fabrication into the final product. This often involves extreme and unique configurations depending upon the final commodity. Through the years the forming requirements have become more severe. Many of the commodities post-formed from painted strip have sharp bends, tight radii, deep draws, or crimped interlocking seams. This challenge was met head on by the paint industry, and as a result, new coatings were developed.

A second important requisite of coil finishes is exterior durability, or weatherability. The list of items made from coated stock reveals a marked increase in exterior products such as residential siding, awnings, commercial building panels, rain-carrying equipment, metal roof shingles, and other building components. It is estimated today that 65 to 75% of coil coated stock is destined for exterior service. This embraces two large markets; building components (including all types—siding, awnings, rain-carrying equipment, etc.) and mobile home and trailer sheathing.

In view of the above two major requirements, the paint industry developed flexible and durable coatings encompassing several generic classes. The most prominent types used in recent years are: (a) amine alkyds, (b) vinyl alkyds, (c) solution vinyls, (d) thermoset acrylics, (e) organosols, (f) plastisols, and (g) polyester.

The rapid advancement of paint technology has introduced several new coatings to the market place during the last five years. Most of these are used commercially at the present time. Among the recent developments are: (a) silicone polyesters, (b) silicone acrylics, (c) silicone alkyds, (d) polyesters, and (e) fluorocarbon coatings.

It is to be noted that many of the new coatings represent combinations of major resin classes. In a sense they are hybrids designed to combine the outstanding properties of the basic components.

A. DEFINITIONS

A description of the basic chemical components and the general formula composition will provide a better understanding of the coatings mentioned above. Although the composition may vary slightly from one coating supplier to another, within any generic type the following definitions are reasonable. They are taken from the National Coil Coaters Association Technical Bulletin No. IV, 3 (1).

1. Amine Alkyd

Amine alkyd finishes are also called alkyd/amino types. An alkyd resin is the oil or fatty acid modified condensation product of principally phthalic anhydride or other polybasic acids and polyols (polyhydric alcohols). An amine or amino resin is the alkylated reaction product of urea or melamine and formaldehyde. The alkyd is cross-linked by the amino resin to form a thermosetting system on baking. Supplied in solution form.

2. Polyester

Polyester resins are the condensation products of polybasic acids and diols (dihydric alcohols) and are sometimes called oil-free alkyds. They may be modified by the addition reaction of other monomers such as styrenes or acrylic esters. Polyester resins are generally cross-linked with amino resins to form thermosetting systems on baking. Supplied in solution form.

3. Vinyl Alkyd

Vinyl alkyd coatings are based on the blend of a copolymer, or more commonly a terpolymer of vinyl chloride, vinyl acetate, and vinyl alcohol, and a compatible alkyd resin. These are thermoplastic coatings but may approach thermosetting types by the incorporation of amino resins. The amino resin will not only cross-link the alkyd, but will also react with the hydroxyl portion of the vinyl terpolymer. Supplied in solution form.

4. Solution Vinyl

Solution vinyl coatings are based on terpolymers or copolymers of vinyl chloride, vinyl acetate, or other monomers, usually high in vinyl chloride, together with solvents, plasticizers, and stabilizers. These coatings cure by solvent release and are thermoplastic. When reactive groups are present, cross-linking may be accomplished with amino resins, etc. Supplied in solution form.

5. Solution Epoxy

Solution epoxy coatings are based on epoxy resins which are usually the reaction product of epichlorohydrin and bisphenol. These resins are usually cross-linked with amino or phenolic resins to form thermosetting systems on baking. Supplied in solution form.

6. Epoxy Ester

An epoxy ester coating is based on a resin prepared by esterifying an epoxy resin with fatty acids. These esters are generally cross-linked with an amino resin to form thermosetting systems on baking. Supplied in solution form.

7. Thermoset Acrylic

Thermoset acrylic coatings are based on resins prepared from acrylic and methacrylic esters, acrylic and methacrylic acids, and/or styrene. They contain one or more functional groups such as amide, hydroxy, or carboxy. They form thermosetting resins when cross-linked with themselves, amino or epoxy resins. Supplied in solution form.

8. Silicone Alkyd

Silicone alkyd finishes are based either on a reaction product between an organo-siloxane intermediate with an alkyd resin, or a cold blend of a silicone resin and a compatible alkyd resin. These resins are generally cross-linked with amino resins to form thermosetting systems on baking. Supplied in solution form.

9. Silicone Polyester

Silicone polyester finishes are a reaction product between organo-siloxane intermediates and a suitable polyester resin intermediate typically characterized by the absence of drying oils. These resins are generally cross-linked with amino resins to form thermosetting systems on baking. Supplied in solution form.

10. Silicone Acrylics

A silicone acrylic finish can be either the product of the reaction of an organo-siloxane intermediate with acrylic monomers or a cold blend of an organo-siloxane with acrylic polymers. They form thermosetting systems on baking in the same manner as straight thermoset acrylics. Supplied in solution form.

11. Polyvinyl Fluoride

Polyvinyl fluoride coatings are based on fluorocarbon resins made by the polymerization of vinyl fluoride monomer. These resins are usually formulated into coatings by dispersion of finely divided particles in dispersants and diluents.

12. Polyvinylidene Fluoride

Polyvinylidene fluoride coatings are based on fluorocarbon resins made by polymerization of vinylidene fluoride monomer. These resins are usually formulated into coatings by dispersion of finely divided particles in dispersants and diluents.

13. Organosol

Organosol coatings are dispersions of finely divided polyvinyl chloride resin particles in plasticizers and volatile dispersants and diluents. Additives may be incorporated to improve adhesion, gloss, etc. During the baking process the solid resin particles are solvated by the plasticizer and dispersants to fuse to a continuous film.

14. Plastisol

Plastisol coatings are dispersions of finely divided vinyl resin in plasticizers. During the baking process the finely divided resin particles are solvated by the plasticiser and fuse to a continuous film.

15. Polyvinyl Chloride Laminate Film and Sheeting

Vinyl films and sheeting are composed of polyvinyl chloride resins, plasticizers, stabilizers, extenders, and pigments. These films and sheeting are laminated to metal substrates with an adhesive.

16. Polyvinyl Fluoride Laminate Films

Films made from polyvinyl fluoride, clear or pigmented. These films are laminated to metal substrates with an adhesive.

17. Acrylic Laminate Films

Plastic films, clear or pigmented, made from polymerized acrylic monomers. They are laminated to metal substrates by an adhesive.

Basically, all the major classes of coatings used through the years have provided good performance. For example, amine alkyd-type coatings have been used since the inception of coil coating. Fifteen to twenty years of exterior exposure on aluminum siding and aluminum awnings has indicated excellent performance with amine alkyd coatings.

The demand for greater film flexibility to withstand more severe fabrication requirements prompted the industry to move to vinyl-modified alkyds, solution vinyls, organosols, and plastisols. The acrylic-type finishes, with moderate forming characteristics, made a major breakthrough in about 1959. The major incentive in advancing coating technology was to provide more flexible and durable coatings for the coil coating industry. The trend today is in the same direction. The coil coating industry needs new coatings which will provide exceptional film flexibility and improved exterior durability. In the realm of durability the greatest emphasis is on providing improved fade and chalk resistance.

Considerable information on the major properties and advantages of the various coatings is readily available through both the coil coating and the paint industry. Suffice it to say that all the various coatings have certain advantages which make them adaptable to coating coil stock for specific end products. Certainly, the logical approach to the coil coater is to consider the ultimate end use or final product to be made from the coated strip, and then select the finish accordingly. Some of the important points to be considered are:

1. Exterior or interior use.
2. Type fabrication—roll form, brake formed, stamped, drawn, or seamed.
3. Severity of fabrication—bend radii, depth of draw, etc.
4. If for exterior use—type performance expected.
5. Specific end product involved.
6. Is the product to be subsequently overpainted, stencilled, printed, or decaled?

The coil coater very often solicits the assistance of a reputable coating supplier in an effort to select the proper finish for a given product. This approach is highly recommended since it will assure the very best coating recommendation for a given product. This cooperative effort will very often avoid field problems and costly adjustments because of complaints.

In addition to the direct services rendered by a coating supplier, a coil coating firm can also keep up to date by participating in the National Coil Coaters Association. The N. C. C. A. was founded in 1962. This young, dynamic Association has been very successful and of great value to the coil coating industry. The N. C. C. A. holds two conferences each year. The conference includes a business meeting, committee meetings, and a technical session during which several papers are presented on various subjects of interest to the membership. Some of the major topics presented are coating equipment advances, equipment modifications, ovens, metal cleaning and conversion coatings, advanced coating technology, quality control, test procedures, and marketing. Worldwide interest in coil coating has also enticed several international firms to join the N. C. C. A. This in itself certainly reflects the enthusiasm and vitality of the coil coating industry in the world today.

The N. C. C. A. Technical Committee has also compiled useful information comparing the film performance properties of major type coatings. The information is available in Table 4-1. The table rates 17 major generic coating classes for 32 characteristics properties. It is used as a helpful guide in selecting the proper finish for a given end use.

The N. C. C. A. "Coil Coating Finishes - Comparative Properties and Performance Chart" is perhaps the most comprehensive coil coatings technical guide available. This chart is under constant review by the N. C. C. A. technical committee in an effort to keep it up to date and in

Table 4-1. Coil Coating Finishes—

Generic coating type	Amine alkyd	Polyester (oil free)	Vinyl alkyd	Solution vinyl	Organosol	Plastisol	Straight epoxy
Ease of application	1	2	2	1	1	2	3
Average cure rate (based on 60 sec)	2	2	2	1	1	1	3
Minimum metal temperature for curing efficiency, °C	149	149	149	149	177	177	177
Color and gloss retention (double time in oven)	3	1	3	3	3	3	3
Thermoplastic	No	No	Yes	Yes	Yes	Yes	No
Thermosetting	Yes	Yes	No	No	No	No	Yes
Film hardness	2	1	2	2	2	3	1
Film adhesion	2	2	2	1	1	1	1
Film flexibility	3	2	2	1	1	1	3
Color and gloss retention at elevated temp. (on end product)	2	2	2	3	3	3	3
Color retention on aging film (storage)	2	2	2	2	2	2	3
Mar resistance (fingernail test)	2	1	2	2	3	3	1
Ability to achieve high gloss (above 85 units 60°)	1	1	2	1	4	4	1
Ability to fabricate after aging	4	2	2	1	1	1	3
Adaptability to embossing (with metal)	4	3	3	2	1	1	4
Embossing (film only)	4	4	4	4	4	1	4
Metal marking resistance	2	2	2	2	3	3	1

Comparative Properties and Performance Chart[a,b]

Epoxy ester	Thermoset acrylic	Silicone alkyd	Silicone acrylic	Silicone polyester	Polyvinyl fluoride	Polyvinylidene fluoride	Polyvinyl fluoride film laminate	Polyvinyl chloride film laminate	Acrylic film laminate
2	1	2	2	2	2	2	3	3	3
2	2	3	3	3	3	3	1	1	1
149	190	204	232	232	232	232	Bonding adhesive 149	Bonding adhesive 177	Bonding adhesive 149
3	1	1	1	1	2	2	2	3	2
No	No	No	No	No	Yes	Yes	Yes	Yes	No
Yes	Yes	Yes	Yes	Yes	No	No	No	No	Yes
2	2	2	2	2	2	2	2	3	2
1	1	2	2	2	2	2	2	1	1
3	2	3	3	3	1	1	1	1	1
3	2	1	1	1	2	2	2	3	2
3	1	1	1	1	1	1	1	2	1
2	1	2	2	2	3	3	2	3	2
2	1	1	1	1	3	4	4	4	2
3	2	3	3	3	1	1	1	1	1
3	3	4	4	4	1	1	1	1	1
4	4	4	4	4	4	4	1[d]	1[d]	1[d]
2	2	2	2	2	3	3	2	3	2

Table 4.1.—

Generic coating type	Amine alkyd	Polyester (oil free)	Vinyl alkyd	Solution vinyl	Organosol	Plastisol	Straight epoxy
Resistance to pressure mottling in coil	2	2	2	2	2	3	1
Abrasion resistance	3	2	2	2	1	1	2
Stain resistance (household agents and foodstuffs)	3	2	3	2	2	2	1
Salt spray resistance	3	2	2	1	1	1	1
Humidity resistance	2	2	2	1	1	1	1
Grease and oil resistance	2	2	2	1	1	1	1
Solvent resistance (aliphatic hydrocarbon)	2	2	2	2	1	1	1
Solvent resistance (aromatic hydrocarbon)	2	2	3	4	3	3	1
Solvent resistance (ketone or oxygenated types)	3	3	3	4	4	4	1
General chemical resistance — acids, alkali, etc. (spot tests)	3	3	3	2	1	1	1
General corrosion resistance (industrial atmospheres)	2	2	2	2	1	1	1
Exterior durability (pigmented)	2	2	3	2	1	1	4
Exterior durability (clear films)	3	3	3	4	4	4	4
Fade resistance in color (properly pigmented)	3	2	3	2	2	2	4
Unit finishing cost range	L	M	M	MH	L	L[c]	H

Continued

Epoxy ester	Thermoset acrylic	Silicone alkyd	Silicone acrylic	Silicone polyester	Polyvinyl fluoride	Polyvinylidene fluoride	Polyvinyl fluoride film laminate	Polyvinyl chloride film laminate	Acrylic film laminate
2	2	2	2	2	2	2	2	3	2
2	2	2	2	2	1	1	1	1	2
2	1	2	2	2	1	1	1	2	2
1	1	2	2	2	1	1	1	1	1
1	1	2	2	2	1	1	1	1	1
2	1	2	2	2	1	1	1	1	1
1	1	2	2	2	1	1	1	1	1
1	2	2	2	2	1	1	1	3	3
2	2	2	2	2	1	1	1	4	4
2	2	2	2	2	2	2	1	1	1
2	2	2	2	2	2	2	1	1	1
4	2	1	1	1	1	1	1	1	1
4	2	2	2	2	1	1	3	4	2
4	2	1	1	1	1	1	1	2	2
MH	M	VH	VH	VH	VH	VH	VH	VH	H

[a]Coating material cost per square foot (cents)

L — Low, 0.65 or less at 1 mil dry film
M — Medium, 0.65 to 0.85 at 1 mil dry film
H — High, 1.10 to 2.50 at 1 mil dry film
MH — Medium to high, 0.85 to 1.10 at 1 mil dry film
VH — Very high, 2.50 and over at 1 mil dry film

[b]Rating key

#1 Excellent
#2 Good
#3 Fair
#4 Poor

[c]Normally applied at heavier films.

[d]At dry film thickness of 6 mil dry or more.

conformance with ever advancing coating technology. The chart, with full instructions on its use and interpretations, appears herein.

B. COIL COATING FINISHES COMPARATIVE PROPERTIES AND PERFORMANCE CHART

1. Preface

This "Comparative Properties and Performance Chart" is a composite evaluation and rating reference covering the various finishes used by the coil coating industry.

This chart has been prepared for general information purposes only. It is not a standard or a specification in any sense of those words. It describes only average coatings of the generic type listed and is intended to serve only as a guide for selection of coating systems suitable for end product requirements. Among coatings in each generic category there are, of course, varieties that are both much better and much worse than the ratings given here. This chart has been prepared to serve only as a cross index to illustrate the general comparative characteristics of the coatings listed. If used as such, it should prove to be a valuable tool to the coil coater.

As a further aid in the proper use of this chart, the instructions that were given to the committee that rated the various coatings for the properties listed are attached. It is suggested that any person using this chart first read these instructions. For example, the instructions point out that generally the ratings here are for one-coat systems on properly treated aluminum. Obviously, changing to a two-coat system, or changing the substrate, could conceivably substantially change the ratings given here for the various coatings. We therefore strongly urge you to consider the purpose of this chart and to use it as intelligently as possible for information purposes.

In all cases, final recommendation of a finishing system should be obtained from the coil coater and/or the finishes supplier.

2. Addendum: Instructions for Ratings

The following guides should be used for rating the various coatings for the properties listed in Table 4-1.

1. Each type coating should be rated as a general class for each property. Example: Vinyl coatings provide flexible films, therefore they would rate high in forming properties, impact resistance, embossing, etc. Epoxy-type coatings chalk rapidly and therefore would not be recommended for exterior durability, particularly in colors and tints, etc.
2. The rating for each coating should be in terms of performance as a coil finish and not in terms of maintenance finishes, appliance finishes, or other uses involving the same generic type of coating.
3. A fair rating must be made for each type of coating for each property. To do this it will be necessary to compare the performance of one type coating to another. You will find a more valid appraisal can be made if you occasionally check back and review the rating you gave to another class of coating.
4. For rating purposes, assume all the various coatings are pigmented for the best exterior durability.
5. For rating purposes, assume all the various coatings are over the same substrate, namely, properly treated aluminum.
6. For rating purposes, assume all the various coatings are applied at equal film thickness (1.0 mil dry film). Exceptions to this would be the plastisols and vinyl laminates which may be applied as heavy films for embossing purposes.
7. The following rating key should be used to rate all properties:

   #1 is excellent
   #2 is good
   #3 is fair
   #4 is poor

Exceptions to the above are as follows:

Minimum Temperature: Indicate temperature. Example: 100°C, 150°C, etc.

Thermoplastic: Rate yes or no

Thermosetting: Rate yes or no

Unit Finishing Cost Range: Rate as cost at cents per square foot at 1.0 mil dry film thickness as follows:

L is low or 0.65 cents or less at 1 mil dry film
M is medium, or 0.65 to 0.85 cents at 1 mil dry film
H is high, or 1.10 to 2.50 cents at 1 mil dry film
MH is medium to high, or 0.85 to 1.10 cents at 1 mil dry film
VH is very high, or 2.50 cents and over at 1 mil dry film

NOTE: The information and ratings in Table 4-1 were developed by and represent the concensus of the Classifications and Definitions Committee of the National Coil Coaters Association. The chart has been approved by the Technical Representatives of member companies and the Board of Directors of the N.C.C.A. with the understanding that it is for informational use only and is NOT to be regarded as a standard or specification. The conditions of any given application of a coating finish, the intended end use of a coated coil, and other important considerations are not within the knowledge or control of the National Coil Coaters Association. No warranties or guarantees of results are made by the Association.

Table 4-1 clearly indicates the major advantages, as well as disadvantages of each generic coating. Since we now have a better understanding of the film properties of each type coating, we can better associate them with specific end uses or end products. Several of the coatings are widely used for major market areas such as building components, mobile home and truck sheathing, appliances, and other commodities. Other coatings have limited use and some, understandably, overlap into many markets or classes of commodities. Table 4-2 indicates the major end uses for the various coil coatings under present use.

Specific primers are often required for the two-coat systems indicated in Table 4-2. Primers, usually epoxy types, can also be used on an optional basis with all the coatings to upgrade general performance.

Table 4-2. End Uses for Coil Coatings

Type coating	Major end uses
Amine alkyd	Aluminum residential sidings, awnings, canopies, mobile home sheathing
Polyesters (oil free and oil modified)	Awnings, canopies, mobile home sheathing, coated strip for general use
Vinyl alkyd	Roof deck coatings (limited use, interior only)
Solution vinyl	Residential siding, rain-carrying equipment, awnings, canopies, under eave soffit systems, refrigerator and freezer inner liners, coated strip for general use
Organosol (two-coat system, requires primer)	Residential steel and aluminum siding, steel building panels
Thermoset acrylic	Residential aluminum siding, awnings, canopies, mobile home sheathing, rain-carrying equipment, coated strip for general use
Silicone alkyd (primer recommended)	Residential aluminum and steel siding, steel building panels, residential and commercial aluminum roof shingles
Silicone acrylic (primer recommended)	Residential aluminum and steel siding, steel building panels, residential and commercial aluminum roof shingles
Silicone polyester (primer recommended)	Residential aluminum and steel siding, steel building panels, residential and commercial aluminum roof shingles
Polyvinyl fluoride (primer recommended)	Residential aluminum and steel siding, other building components
Polyvinylidene fluoride (primer recommended)	Residential aluminum and steel siding, steel building panels, aluminum roof shingles
Polyvinyl chloride laminate films (requires bonding adhesive)	Interior automotive and aircraft trim, luggage, interior decorative products
Polyvinyl fluoride laminate films (requires bonding adhesive)	Residential aluminum and steel siding and steel building panels
Acrylic laminate films (requires bonding adhesive)	Residential aluminum and steel siding and steel siding and steel building panels

4-5. PRECOATED METAL STRIP — WHAT ARE THE MARKETS?

The greatest growth area for the use of precoated metal in the last ten years has been the building component field. Residential aluminum and steel siding has maintained the lead position in being the volume product made from precoated metal. In 1964 the metal residential siding industry utilized one billion square feet of coated stock. By 1969 this figure increased to three billion square feet. A continual growth is expected as the need for additional homes is ever increasing. Metal residential siding is now being used at a rapidly increasing rate for new home construction. Ten or fifteen years ago prefinished metal siding was mainly used for the improvement or renovation of older homes. Today it is recognized as a high quality building product suitable for construction of new homes.

As the use of metal residential siding increases, there is, of course, a similar increase in the need for other products. This includes the siding accessories and attendant products such as corner sections, facia, trim, window caps, and rain-carrying equipment. All of the attendant products are also formed from precoated metal.

A second large market for precoated metal is in the fabrication and manufacture of building panels or curtain-wall products. The panel sections are used in the construction of pre-engineered buildings. This market is also expanding rapidly, since there is great demand for this type of product in the commercial building field today. We have all seen this product used for new commerical buildings as supermarkets, automobile agencies, store fronts, industrial buildings, factories, warehouses, farm buildings, and aircraft hangers.

The use of prepainted, post-formed building panels has also taken a more aesthetic role in today's building field. This product is now often used for schools, community buildings, recreational buildings, and even for churches. Very often the panels are used with other building products as brick and stone to enhance the styling and to accentuate design features. Building panels in various designs and shapes are also used in multistory monumental office and apartment buildings. The architect today is very interested in this new concept in building products and design.

A third large and impressive market for precoated metal is the mobile home and recreational vehicle industry. The industry embraces the manufacture of mobile homes, travel trailers, campers, truck campers, and self-contained motor homes. This industry has enjoyed a steady 20 to 25% annual growth during the last decade. It is predicted this growth level will be maintained, or will possibly even increase, through 1975.

Just a few years ago the mobile home and travel trailer industry finished their units by spray painting. Each unit had to be finished separately, entailing special preparation such as masking, paint handling, and cleanup. Today the entire operation has been simplified by using wide precoated metal for the exterior sheathing. The benefits realized by using precoated stock were: (a) a more durable and better looking finish, (b) lower material and operational cost, (c) no need for in-plant finishing equipment, and (d) generally simplified operations resulting in higher production.

It is estimated that 97% of the mobile home and recreational vehicle industry uses precoated metal on their units. In-plant spray painting has virtually disappeared in this industry.

Another rapid growth area for precoated metal is the appliance industry. In 1963 the Amana Corporation pioneered the use of prepainted aluminum coil stock for the inner liners of domestic refrigerators and commercial deep freeze units. The prepainted inner liners replaced porcelain. This program was so successful that several other refrigerator manufacturers have adopted the same approach. As time goes on, it is conceivable that an increased number of appliance parts will be fabricated from prepainted coil stock. The appliance unit of the future, be it refrigerator, clothes washer, dish washer, or dryer, may be entirely fabricated and assembled from precoated, post-formed sections. When the appliance industry advances to this stage it is logical to assume that the same concept will be considered for air conditioning units, metal furniture, file cabinets, kitchen cabinets, lockers, and other utilitarian products.

A few of the major markets served by the coil coating industry have been enumerated. There are many other products that are formed from prefinished coil. For example, the automotive industry is potentially another large user of coil stock. At the present time several components of the modern automobile are made from precoated strip. This includes miscellaneous braces, brackets, voltage regulator covers, interior trim, instrument panel parts, and dashboards. Each year more prepainted parts will be added.

As you can see, the use of prepainted coil stock has advanced at a phenomenal rate. This can best be illustrated, and further emphasized, by reviewing the following end product list prepared by the National Coil Coaters Association. The products have been grouped into major categories representing large market areas. (See Table 4-3.)

4-6. INDUSTRY GROWTH

The coil coating industry has experienced swift and steady growth since 1959.

An industry survey in 1969 revealed that there are sixty-five companies actively engaged in coil coating in the United States. This figure includes

Table 4-3. Current or Suggested End Uses of Pre-coated Metal Strip (roll formed — brake formed — stamped-drawn)

Appliances (large)

1. Air conditioners
2. Clothes dryers
3. Dish washers
4. Furnaces
5. Gas or electric ranges
6. Radio and phonograph cabinets
7. Refrigerator and freezer liners
8. Refrigerator and freezer — doors and shells
9. Space heaters
10. Vending machines
11. Washing machines
12. Water coolers
13. Water heater jackets

Appliances (small)

1. Business machine housings
2. Can openers
3. Clock faces and housings
4. Dehumidifiers
5. Electric fan blades
6. Floor waxers
7. Hair dryers
8. Homogenizers
9. Humidifiers
10. Knife sharpeners
11. Miscellaneous parts for appliances (braces, brackets, etc.)
12. Sound recording equipment
13. Vacuum cleaners

Construction

1. Accessories for sidings, facia, trim, corners, etc.
2. Awnings and canopies
3. Baseboard heating covers
4. Bathroom cabinets
5. Building soffit systems
6. Bus stop shelters
7. Carports, boat shelters
8. Car wash booths
9. Ceiling tile
10. Commercial building marquees
11. Curtain wall and building sheet (supermarkets, aircraft hangers, factories, schools, etc.)
12. Decorative chimney
13. Decorative shutters
14. Doors
15. Doors and window frames
16. Ductwork
17. Electrical switch and outlet plates
18. Elevator paneling
19. Fencing

Table 4-3. — Continued

20. Fireplaces
21. Garage doors
22. Gutters and downspouts
23. Interior partitions and trim
24. Kitchen cabinets
25. Lighting reflectors and housings
26. Louvered vents
27. Patio covers and supports
28. Radiator fin stock
29. Refreshment booths (to house vending machines)
30. Residential siding
31. Roof decking
32. Roof flashing
33. Roof shingles and sheet
34. Screen frames
35. Shower stalls
36. Silo roofs
37. Stadium seats
38. Storage sheds, tool sheds
39. "T" bar hangers for tile
40. Telephone booth paneling
41. Toilet partitions
42. Walkway covers and supports
43. Wall tile

Farm and garden equipment

1. Animal shelters
2. Farm storage bins
3. Feed troughs
4. Garden equipment
5. Grain dryers
6. Mowers
7. Snowblowers
8. Spreaders
9. Tools

Furniture (residential and commercial)

1. Cabinets (storage, beverage, functional)
2. Card tables
3. Chairs
4. Clothes hampers
5. Coat racks
6. Desks
7. Display cases
8. Filing cabinets
9. Fireplace accessories
10. Institutional furniture
11. Ironing boards
12. Juvenile furniture
13. Lamps and shades
14. Lawn furniture
15. Library shelving
16. Lockers
17. Metal drawer dividers

Table 4-3. — Continued

18. Radiator covers
19. Shelving
20. Store fixtures
21. Tubular products, legs, stands, etc.
22. T.V. trays
23. Waste baskets

Packaging

1. Bulk containers
2. Cans and containers
3. Caps and closures
4. Drums
5. Edging for cartons
6. Film canisters
7. Semirigid container (T.V. dinner trays, etc.)
8. Strapping

Recreational equipment

1. Aluminum boats
2. Bar-B-Q grills
3. Basketball backboards
4. Camping equipment (ice boxes, camp stoves, etc.)
5. Exercising equipment
6. Fabricated play houses
7. Folding camp cots and chairs
8. Golf carts
9. Picnic jugs
10. Playground equipment
11. Portable swimming pool frames and sheathing
12. Prefabricated baseball dugouts

Transportation

1. Aircraft, bus, and train ceilings
2. Aircraft trim
3. Arm rests
4. Automotive trim
5. Baggage racks
6. Bicycle fenders
7. Commercial truck sheathing
8. Highway guard rails
9. Instrument panels
10. Interior door panels and trim
11. License plates
12. Miscellaneous parts, horn shells, voltage regulators, oil caps, braces, oil filters, canisters, etc.
13. Mobile home sheathing and interior components
14. Recreational vehicles
15. Snowmobiles
16. Station wagon flooring
17. Window frames

Table 4-3. — Continued

Miscellaneous	
1. Blackboards (metal)	10. Instrument panels
2. Bread boxes	11. Luggage
3. Camera shells and parts	12. Mail boxes
4. Casket handles	13. Measuring tapes (metal)
5. Dispensing machines, towels, etc.	14. Metal signs (interior and exterior)
6. Drapery fixtures and curtain rods	15. Picture frames
7. Games and toys	16. Tool and tackle boxes
8. House numbers	17. Window blinds, venetian blinds, pivot shades, and accessories
9. Instrument gauge faces, clocks, thermometers, etc.	

many of the major prime producers of aluminum and steel as well as numerous independent coaters. Several of these companies have more than one coil coating plant. In addition to this, many coaters are operating more than one coating line in a given plant. Considering multiple plants and multiple lines, it is estimated that there are 115 to 125 active coil coating units operating in this country. This would, of course, include a variety of lines from the narrow venetian blind type up to the massive 60 in. width unit.

Coil coating has also grown internationally and is ever increasing. Growth has been particularly rapid in Canada, Sweden, Germany, France, the United Kingdom, and Japan. Increased activity is expected in Australia, New Zealand, South America, and Mexico. As of January, 1970, the membership of the National Coil Coaters Association included a total of seventeen coil coaters from other countries.

The total market growth and demand for prepainted metal can best be illustrated by the following figures compiled by the N.C.C.A. Statistical Committee.

1. The year-to-year increase in tons of metal (aluminum and steel) precoated is as follows:

   1962 - 464,000 tons
   1963 - 637,000 tons

1964 - 799,000 tons
1965 - 1,059,000 tons
1965 to 1969 - 1,762,000 tons

The above figures indicate a volume increase of 72% between 1962 and 1964, a 40% increase from 1964 to 1965, and a 67% increase from 1965 to 1969.

2. The total gross sales of aluminum and steel coated coil stock surpassed $550 million for the year 1967.

3. The total projected gross sales of coated coil stock for the year 1970 is $1 billion.

4. It was estimated that the coil coating industry consumed over 7 million gallons of coatings in 1965. (Value, $35 million, based on a $5.00 /gal average cost.)

5. The estimated total market value of paint for the coil coating industry in 1970 is $55 million.

6. The total market value for coatings should reach $80 million by 1973 if the coil coating industry's year-to-year growth continues as it has since 1962.

4-7. A DECADE OF GROWTH — A BRIGHT FUTURE

The foregoing information has related the scope of the coil coating industry and provided some insight into its dynamic growth in the last decade. Certainly this young industry has advanced at an impressive rate.

We must acknowledge that this advancement has been due partially to the efforts of several industries closely affiliated with, or in service to, the coil coating industry.

The equipment concerns have engineered and produced improved roller coating units, ovens, strip slitters, laminating equipment, roll formers, and other equipment used in the total operation.

The prime metal producers have provided improved aluminum and steel coil stock. Advanced metal rolling techniques at the mills have made possible the supply of wide metal strip possessing uniform flatness and minimum camber so vital to the coil coating operation.

The metal cleaning and treatment companies have developed more efficient cleaning and treatment systems. Many proprietary chromate and/or phosphate metal treatments are now available for adequate preparation of the metal prior to coating. Technological advancements in the realm of metal treatments, or conversion coatings, have provided significant improvement in performance of the total finishing system. The most important upgrading in performance relates to film adhesion, film flexibility, corrosion resistance, and general weatherability.

The paint industry has provided more flexible and durable coatings as a result of constant, intensified research and development efforts. The new high quality, durable coatings, such as plastisols, silicone-modified products, and the fluorocarbons, have greatly assisted in developing important markets for precoated metal.

In addition to the industries we have just reviewed, we must not overlook the role of several basic raw material suppliers. They, too, have offered new and improved products that have materially aided the technological advancement of the coatings industry. I have reference here to the resin manufacturers, pigment suppliers, and concerns that supply solvents, plasticizers, stabilizers, and other components and intermediates used in coil coatings.

Last, but certainly not least, a credit-line must also go to the many coating firms. They have developed new markets for coil coated stock. They have also endeavored to provide a quality product, and in so doing, have fostered the adoption of prefinished coil stock by other industries.

All the industries just mentioned have made substantial improvements on their products. This alone, however, did not account for the rapid advancement of coil coating. All the advancements and improvements were meshed by a cooperative spirit among coaters, supplier groups, equipment concerns, and others involved. People — and concerns — worked together with a high level of rapport. This, above all, helped to move the coil coating industry ahead at a rapid rate. A healthy competitive condition also encouraged keen participation by those industries involved.

The coil coating industry today has many favorable attributes that could lead it to be one of the foremost industries in the world. In brief, it is dynamic, competitive, progressive, creative, interesting, and challenging. Additionally important is its potential to reach into many major commodity markets, assuring a high and lasting demand for the basic product - precoated coil stock.

4-8. COIL COATING GLOSSARY

Over a period of time all major industries usually adopt certain terms, expressions, or vernacular to describe specific phases of the industry involved. These terms or expressions may describe equipment, concepts, processes, conditions, characteristics of the product, performance of the product, and other matters or items related to the specific industry. This, in a sense, becomes the language of theindustry. As with other industries, the coil coating industry has acquired many terms and expressions that have been accepted and used on a universal basis. Several of these expressions have been used in this chapter. A clear and brief definition of specific terms can certainly provide a much better understanding of the total subject matter. The following glossary, taken from the National Coil Coaters Association Bulletin No. IV, 2 (2), is included as a reference tool for the reader.

Accumulator — A series of fixed and movable rolls which act as a reservoir of strip in a continuous coating line. Usually accumulators are found at both the beginning and the end of a line. Their purpose is to provide enough strip to avoid the necessity of shutting down the line when attaching a new coil at the entrance or removing a coil at the exit end of the line.

Applicator roll — The roll in a roller coater which applies the paint, conversion coating, or other liquids to a moving strip of metal.

Backing(er) coat — A thin coating applied to the back or unexposed side of coated strip. The backing(er) coat is applied for such reasons as appearance, durability, providing lubrication for roll forming operations, and protection of the top coat. This coating is controlled for either color, gloss, or applied film or combination thereof.

Baffle — A device used for deflecting a material in a desired direction.

A. 1. A metal piece placed in a roller coater pan to direct the flow of paint to the pickup roll.
   2. A piece of metal extending under the surface of the paint in the coating pan to direct air bubbles away from the pickup roll.

B. A metal panel used in an oven to direct the flow of air in a desired direction.

Blocking — Sticking of one layer or wrap of a coil of painted metal to the adjacent layer or wrap.

Body — The consistency or viscosity of a paint.

Bridle — A series of rolls that provide the tension and/or driving force to the strip in a continuous coating line.

Camber	The deviation of a side edge of a metal strip from a straight line. The measurement of the amount is taken on the concave side using a straight edge. Long lengths of unsupported strip having a large amount of camber tend to deviate from the horizontal, or slope to one side.
Center stretch (full center)	A phenomenon whereby the center of the strip is longer than the edges, so that the strip is not flat. (See oil can.)
Chatter	A transverse row of marks or lines of varying film thickness on a painted strip. The marks are usually due to vibration of the coating rolls, strip, or to an eccentric roller.
Clock spring	The slipping action of one wrap on an adjacent wrap that can occur in a coil when the strip is being recoiled or uncoiled.
Coater or roll coater	An apparatus that applies paint, conversion coating, or other liquids to a strip of metal. The coater consists of rolls for (a) supporting the strip through the apparatus, (b) the pickup, metering, and deposition of material onto a moving strip of metal.
Concentric roll	A roll with a perfectly circular cross section. A roll with all points on the circumference equidistant from the axis of the roll when it is revolving.
Conversion coating	A chemical treatment normally applied to the metal strip prior to final finishing. It is designed to react with and modify the metal to produce a surface suitable for paint or adhesive bonding.
Cratering	A surface coating defect related to surface tension. It is characterized by small pock marks or indentures surrounded by a ring of coating material projecting above the general plane of the coating. In severe instances, the area in the center of the crater may show the substrate.
Deposition or coating weight	The amount of conversion coating or other material deposited by a metal treating compound. It is measured in weight per surface area, such as milligrams per square foot of surface area. Can coating weights are also measured in a similar manner, such as milligrams per square inch.

Direct coating	Coating with coating roll revolving in the same direction as the travel of the strip, sometimes referred to as forward coating.
Durometer hardness tester	An instrument for the determination of hardness of elastomers or other compressible rubber-like materials, commonly used for hardness measurement of roll coverings and heavy films of plastic (e.g., plastisols). Test results are displayed as a percentage of scale deflection from 0 to 100 for any given scale employed (most commonly used: scale "A").
Exude (exudate)	The migration of a component or substance in a coating toward the surface during curing or storage.
Fast solvent	A solvent that evaporates quickly.
Fish eye	An elongated crater. (See cratering.)
Floating	A process of flooding in which the final color is not uniform and homogeneous, i.e., it may be streaked, spotty, or otherwise nonuniform. (See streaking.)
Flooding	Separation of one or more pigments in a coating during curing so that a nonuniform color or an unintended color is produced.
Flop	The condition where two painted panels appear to match in color when viewed at one angle, and not to match in color at all other angles, or where one panel changes color as the viewing angle changes.
Flow	The property of a paint which manifests itself in the degree of leveling or in an ability to flow or move under applied stress.
Gloss	The luster, shininess, or reflecting ability of a surface.
Gloss meter	An instrument for measuring gloss, usually at 60° or 20° from the vertical.
Hardness, pencil	A method to determine the hardness of a paint film. Each of a series of drawing pencils, calibrated for hardness, is sanded to a blunt point and then held at a 45° angle and pushed forward and downward against the panel to be tested. The hardness designation of the pencil which just fails to cut the film is the pencil hardness of the film.

Journal	The ends of a roller which slip into the bearings. The journals ride on the bearings as the roller is revolved.
Leveler or roller leveler	An apparatus containing a series of steel rolls which flatten or level a metal strip as it passes between the rolls.
Leveling	The phenomenon by which paint flows out to minimize surface irregularities and approaches a smooth surface.
Live center	A point at the very center of each end of a roll. This is usually a center of the journal in which the roll turns. The roll is driven from this point and considered concentric from the same point.
Metal marking	Black marks left on a painted strip when bare metal is drawn across its surface.
Metamerism	A color state in which two or more painted panels match in color under one type of light but do not match under a different type of light. This usually occurs when different pigment combinations are used to produce the same color.
Metering roll	A roll used to apply a uniform coating of paint, conversion coating, or other liquids to a transfer roll or an applicator roll.
Micrometer	An instrument used for mechanically measuring the dry film thickness of paints or the thickness of a metal strip.
Migration	The transfer of an ingredient of a paint film from within the film to another part of the film, the surface of the film, or to another film in contact with the film containing the migrating substance.
Mil	A unit of measuring thickness, 0.001 in. Generally applied to paint films.
Oil can	A localized out-of-flat condition often seen as buckles toward the center of an otherwise flat strip.
Paint	In the sense of roller coater application, the term paint is used to denote a liquid composition which is converted to a solid film after application as a thin layer. Usually the liquid will be pigmented.
Pan	An open container at a roller coater which holds

	the paint, conversion coating, or other liquids and within which the pickup roll revolves.
Pick-off	Tendency of a paint to be picked up in very small pieces from one side of a painted strip and held by the coating on an adjacent wrap of the strip. It can also refer to transfer of coating material to tension and bridle rolls.
Pickup roll	A roll which revolves within the pan and is partially submerged in the paint, conversion coating, or other liquids. The pickup roll picks up paint from the pan and applies it to the transfer or applicator roll.
Pits	Very small craters.
Polishing	An apparent increase in gloss of a paint film caused by the rubbing of the top coat and backing coats during the recoiling of a strip or by contact with roll forming rolls or other smooth moving objects.
Popping	A defect in a paint film usually caused by organic solvents trapped during the curing of the film (fine blistering).
Pressure marking or mottling or streaking	An uneven pattern often seen as glossy spots which are usually caused by pressures within a painted coil. These become visible during the uncoiling process.
Recoiler	The apparatus used to recoil the strip after it is painted.
Reverse coating	Coating with the coating roll revolving in a direction opposite to that of the strip.
Ribbing	Longitudinal streaks that do not flow out on a painted strip (also called roping).
Roll former	An apparatus that forms a continuous strip of painted metal into various shapes by a series of revolving metal wheels or rolls.
Roll grinder	A special lathe used in grinding down rubber applicator rolls.
Roping	See ribbing.
Shear	The viscous force acting parallel to the strip at the point where the applicator roll meets the strip

	on a roller coater. Shear also occurs at all roll-to-roll contact points. Can also refer to metal cutting equipment.
Sheen	A measure of gloss at low angles, usually 85° from the vertical.
Skipping	An irregular paint application usually occurring when improper contact is made between the applicator roll and the strip. (See starving out.)
Slitting	A process by which wide strip is slit or cut into narrower widths.
Slopping over	A term used when some of the finish used on the top side of a continuous strip finds its way to the bottom side.
Slow solvent	A solvent that does not evaporate quickly.
Spool	See uncoiler.
Starter strip or night strip	A length of metal threaded through a line after shutdown. This is used repeatedly with new coils attached to thread the new coil through the line.
Starving out	Irregular film thickness applied to the strip, usually the result of insufficient liquid level in the coating pan, adverse reaction of the coating to the shear actions, improper speed ratio between strip and coating rolls, or improper wetting of the rolls by the coating material.
Stitcher	A stapling device used to connect the end of one coil to the beginning of another coil.
Streaking	A type of floating that forms a longitudinal pattern visible as a variation in gloss or color.
Tandem line	A roller coat line with two coaters capable of applying and baking two coats of paint to each side (i.e., primer and topcoat) prior to recoiling.
Thermoplastic	Capable of becoming soft or plastic upon the application of heat. Pertains to a fused film which softens when heated, then regains its original properties on cooling.
Thermosetting	Having no tendency to soften upon heating. Pertains to a curing or converted film which retains its hardness when heated.
Trough	See pan.

Transfer roll	The roll between the pickup and applicator roll which transfers paint, conversion coating, or other liquids to the applicator roll.
Uncoiler	An apparatus at the beginning of the line used to payoff the strip. This is also used to control strip tension.
Vortex	Usually a six-sided color float pattern visible under magnification, caused by evaporation of solvent from the film.
Wash coat	Same as backing(er) coat except that it is not closely controlled for color, gloss, or applied film. Usually a functional coating.
Water cooler or quench	The mechanism used to apply water to painted strip directly after leaving the oven to cool the strip prior to recoiling. Means for drying before recoiling are provided.

REFERENCES

1. National Coil Coaters Association Technical Manual, Technical Bulletin No. IV, 3.
2. National Coil Coaters Association Technical Manual, Technical Bulletin No. IV, 2.

GENERAL REFERENCES

National Coil Coaters Association Technical Manual, Technical Bulletin No. IV, 4.

National Coil Coaters Association Technical Manual, Technical Bulletin No. V, 1.

H. K. Darby, "Pre-painted Coil Progress Report," Modern Metals, November, 1962.

H. K. Darby, "Coated Coil Business Booming," Modern Metals, June, 1962.

"Painting Before Fabrication, Who's Doing It? How Much? and Why?" Metal Products Manufacturing, November, 1962, pp. 3-6.

R. Harrison, "Coil Coatings: A Look to the Future," Metal Products Manufacturing, November, 1962, p. 8.

"Coil Coaters Expand Facilities to Keep Pace with Booming Sales," Modern Metals, November, 1965, pp. 68-74.

"Coil Coaters Count Up Solid Sales Gains ... Shoot for More," Modern Metals, June, 1965, pp. 27-36.

R. W. Carson, "Prepaint of Postpaint," Product Engineering, February 28, 1966. Reprint 781.

F. L. Church, "Improved Paints, Laminates Help Keep Coated Coil Boom Rolling," Modern Metals, June, 1970, pp. 37-58.

L. R. Rakowski, "Billion Dollar Recreational Vehicle Industry Geared for More Growth" (Coil Coated Stock), Modern Metals, August, 1970, pp. 74-91.

F. L. Church, "Coil Coaters in High Gear for Another Record Year," Modern Metals, November, 1969, pp. 25-33.

B. D. Wakefield, "Action Builds in Coil Coating," Iron Age, July 16, 1970.

Chapter 5

EXTERIOR HOUSE PAINT

G. G. Schurr

The Sherwin-Williams Company
Chicago, Illinois

5-1. INTRODUCTION

In this chapter exterior house paints are discussed from the standpoint of their aesthetic and protective functions, the types and techniques of formulation, the importance and determination of performance characteristics, and laboratory and field techniques for predicting durability.

Exterior house paints are defined for the purpose of this chapter as air dried coatings that are applied to exterior surfaces of structures for decorative and protective purposes. Traditionally, the term "house paint" has implied application to wood substrates. The influence of the substrate, emphasizing wood but also including masonry and metals, is discussed. Since caulks and putties also have important influences on the performance of house paints, they are also discussed.

Vannoy has given an excellent history of house paints in Volume III of Mattiello's Protective and Decorative Coatings under "Exterior Coatings" (1). Vannoy has covered house paints through the early 1940s. Since that time the major developments have been

1. The use of greater proportions of highly bodied oils instead of unbodied oil with small amounts of "Q" bodied oil. This allowed the use of higher pigment volume paints with less actual oil and gave substantial improvement in flow and appearance and resistance to cracking. The adoption of this type of formula was hastened by the necessity for conserving oil during World War II. Quite similar formulas are still in use today.

2. In the late 1940s it was recognized that zinc oxide and leaded zinc oxide pigments in oil primers for wood substrates contributed substantially to the susceptibility of such primers to moisture blistering. Since that time most oil primers have therefore been zinc pigment free even though it has been recognized that this has led to reduced mildew resistance.

3. Of course, the outstanding development has been the adoption of latex paints as exterior paints. This began in the 1950s on masonry substrates and gradually expanded to all substrates until today more than half of the exterior house paints are latex based. Early exterior latex paints found their popularity from the fact that they were water paints and consequently gave easy application and clean-up properties. However, today latex paints also give better performance.

Although there have been some attempts to merchandise highly durable but expensive house paints such as those based on silicone alkyds or epoxidized oil/polyester vehicles, this approach has not as yet made much headway. There have also been some attempts to popularize the use of application methods that are faster than brushing, such as spraying or roller coating, but these efforts also have failed to make inroads. There has been a trend toward merchandising one-coat exterior paints. However, there do not seem to have been any technological trends accompanying this merchandising other than to increase the concentration of hiding pigments and improving the rheology during application.

5-2. SUBSTRATES

Four fundamental properties of substrates are important with regard to paint holding properties. These are: dimensional stability, chemical stability, interaction with water, and surface energy. Unfortunately wood, which is the most common substrate for exterior house paints, ranks low for three out of four of these properties. In addition, house paints are exposed to an environment containing powerful ultraviolet radiation, a strong oxidizing agent (oxygen), a good catalyst (water), severe temperature changes, and various atmospheric pollutants. The paint formulator has a formidable job then, in attempting to devise paints that will last under such conditions.

A. WOOD

Browne (2) and Shur et al (3) ,have emphasized that paint-holding properties of wood seem to correspond to the proportion of summer wood to spring wood in a particular species or on a particular board. Since summer wood is denser than spring wood, density also has been used as a criterion for paint holding properties. However, Miniutti (4) has pointed out that at least part of the effect of summer wood is an indirect effect. Spring wood, being less dense and softer, is compressed more than summer wood during planing operations. Subsequent wetting of the painted wood causes more swelling of the spring wood and consequent pulling off of the paint from the adjacent summer wood area.

1. Dimensional Instability

It would appear that dimensional instability of wood with varying moisture content (5) is the characteristic that has a direct effect on paint-holding properties of wood. Browne's findings illustrate the good correlation between paint-holding properties of wood (6) and the dimensional stability of wood with moisture content (7). Table 5-1 shows such a correlation for four woods commonly used on the exterior of dwellings. Figure 5-1 shows a similar correlation for all of the woods on which Browne gave data. Thus, it appears that the better the dimensional stability of wood with changing moisture content, the better are the paint-holding properties of that wood.

TABLE 5-1. Paint-holding Properties vs Dimensional Change

Wood	Paintability classification[a]	Dimensional change 20%-0% H_2O[b] (%)		
		Radial	Tangential	Volume
Cedar	Group 1	1.6	3.3	4.5
Redwood	Group 1	1.7	2.9	4.5
Ponderosa pine	Group 3	2.6	4.2	6.4
Southern pine	Group 4	3.7	5.2	8.1

[a] By Forest Products Laboratory Publication R 1053.

[b] From Forest Products Laboratory Publication No. 736.

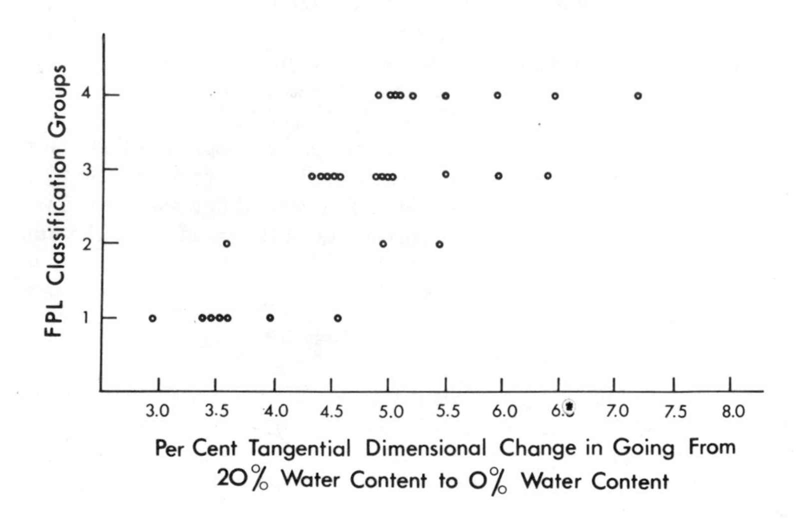

FIG. 5-1. Paint holding properties vs dimensional change.

2. Chemical Stability

The chemical stability of relatively dry wood is excellent, except on the surface. Unfortunately, however, wood is not stable when irradiated with ultraviolet energy. Kalnins (8) has shown that wood loses methoxyl groups and lignin when irradiated, with a corresponding increase in acidity and the evolution of formaldehyde and methanol. This degradation of the surface leaves a residue on the surface of wood which is so mechanically weak that paints have a difficult time adhering. Thus, weathering of wood, either prior to painting or through paint films with inadequate hiding (clears, thin films, or eroded films),can lead to a loss of adhesion by the coating and subsequent peeling (9) (see Fig. 5-2).

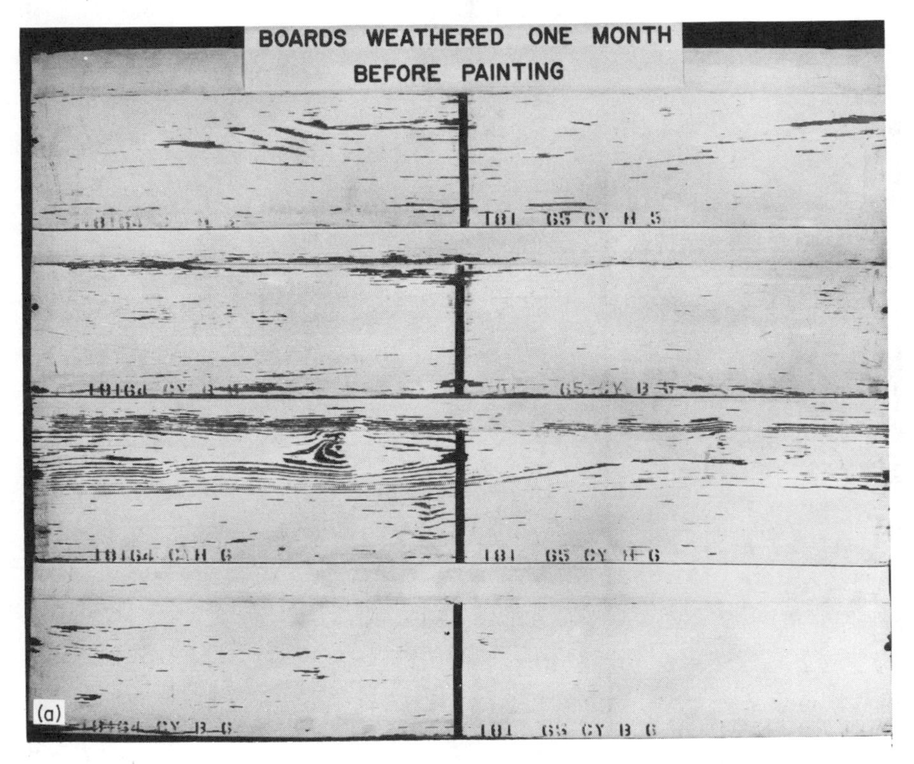

FIG. 5-2. (a-c) Effect on wood of weathering before painting.

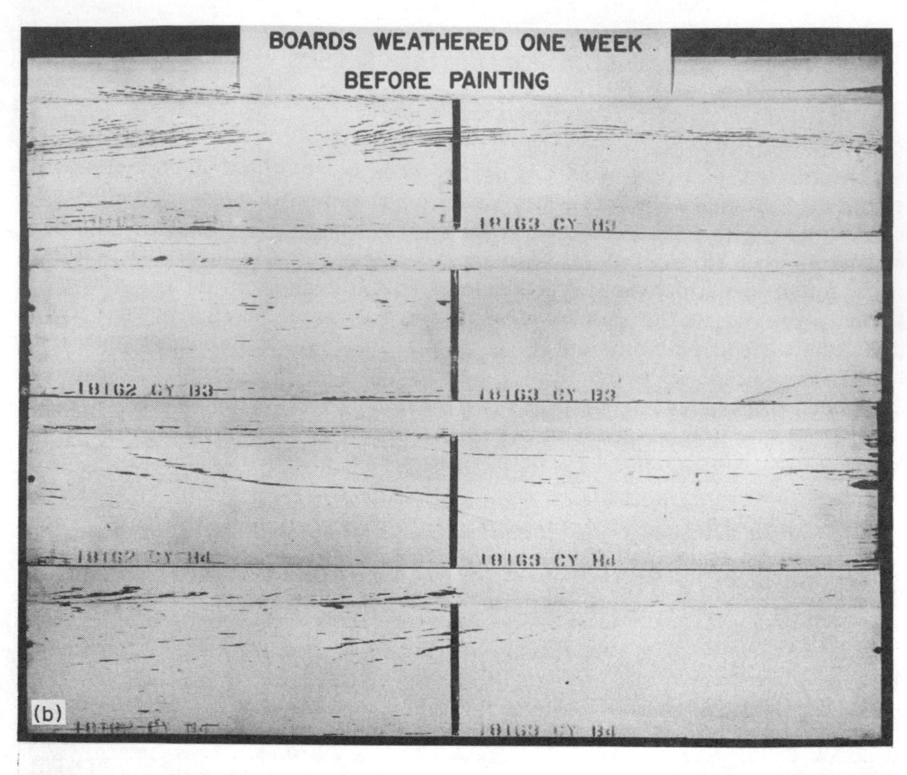
BOARDS WEATHERED ONE WEEK
BEFORE PAINTING
18163 CY H3
18162 CY B3
18163 CY B3
18163 CY H4
(b)

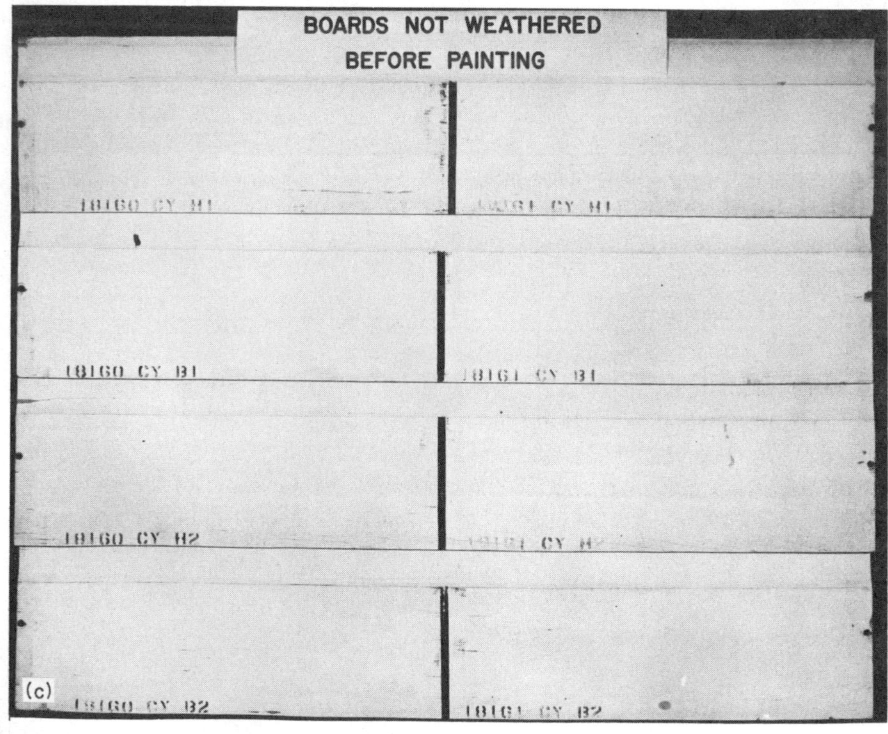
BOARDS NOT WEATHERED
BEFORE PAINTING
18160 CY H1
18160 CY B1
18160 CY H2
(c)
18160 CY B2

3. Surface Energy

The surface energy of wood is high enough so that adhesion to sound wood is not a problem with most coatings, provided the moisture content is reasonably constant. Zenker (10) has shown that high moisture content of wood when it is painted is not particularly deleterious to the performance of paint. This type of evidence has been available in the writer's laboratory. However, changing moisture content such as results from structural defects, particularly when end grain is exposed, is very deleterious, presumably because of the resulting dimensional changes.

Given these properties for wood, a paint system should be capable of a fair degree of elongation not only originally, but after many years of weathering; it should be relatively impermeable to and nonreactive with water; and it should shield the wood substrate from ultraviolet energy. It should possess the ability to elongate, not only at normal temperatures, but also at those temperatures that prevail outdoors during the winter season.

B. MASONRY

Traditionally in the paint industry, those surfaces that tend to be alkaline in nature but are dimensionally stable with exposure to moisture are lumped together under the heading of "masonry." This term then includes not only brick and stone surfaces, but also concrete, concrete and cinder blocks, cement asbestos shingles, and stucco.

Although masonry substrates are dimensionally stable, they are chemically reactive, particularly when fresh, and can be quite permeable and partially soluble in water. Three characteristics of masonry surfaces make them peculiar for proper painting practices. One of these is their alkalinity. Paint systems for masonry should be resistant to hydrolysis at pH's well above 12. Another characteristic of masonry is its propensity to degrade from weathering and to form a loosely bound chalk on the surface. This requires a paint coating that will penetrate such a layer and bind it to a sound surface underneath the chalk layer, or in severe cases, the removal of such a layer by wire brushing, sandblasting, etc. The third characteristic is that many masonry surfaces contain water-soluble materials. These materials can be leached out of the masonry substrates by water that penetrates paint coatings and deposited on the surface of the coating. The resulting condition often resembles fading and can be quite unsightly. This process is referred to as efflouresence.

Thus, satisfactory paints for masonry are easy to formulate from the standpoint of retention of flexibility (none is needed), but are difficult to formulate from the standpoint of alkali resistance, penetration and binding of chalky layers, and resistance to effloresence. Polymers of styrene and butadiene have been popular in masonry paints, both in solution and latex form. They are low in cost, have good to excellent alkali resistance, and readily control effloresence. They are, however, somewhat deficient in chalk penetrating and binding power. For this reason, aged masonry surfaces are often treated with surface conditioners based on alkyd resins prior to topcoating.

Additional information on types of masonry and its preparation for painting is given in Refs. 11 and 12.

C. METALS

Four metals commonly found in exterior house construction are iron, aluminum, galvanized iron, and copper. As paint substrates they can be characterized as relatively dimensionally stable, impermeable to water, and possessors of high surface energy. The problems in coating metals arise mostly because of their chemical instability (corrodibility) and the use of surface treatments to control corrosion. Special primers are often needed for satisfactory paint performance on metals. Unfortunately they are not used very often.

1. Iron

Most of the iron or steel found around a house is of the hot rolled type. Hot rolling results in the formation of mill scale, a dense oxide of iron, Fe_3O_4. Although mill scale is an excellent substrate for painting if it is intact, it is almost never found in the intact state. Usually mechanical damage and corrosion have removed a fair proportion of the mill scale with the formation of rust prior to painting. Ideally, the remaining mill scale and the rust should be removed prior to painting. Practically, however, this is seldom done. The remaining mill scale is difficult to remove even with sand blasting. Usually, all that is done in preparation for painting is wire brushing. This normally does not remove the intact mill scale at all and only partially removes the rust. It is desirable, therefore, to have primers meant for iron to be slow in drying and low in viscosity to maximize penetration of the rust so that a bond can be formed with the sound metal substrate. Primers based on mostly unbodied linseed oil satisfy these characteristics.

A tendency for iron or steel to corrode continues even after painting Mayne stated (13) that paint films do not act as barriers to O_2 or H_2O, but rather act as barriers to the passage of ions and thus prevent corrosion by preventing the flow of electrical current. Corrosion inhibitive pigments can serve to reinforce the ion barrier property of the film but can also serve to form a film on the metal (sometimes composed of the metal oxide) or to impart alkalinity.

There is voluminous literature on corrosion inhibition and it is covered further in the chapter on primers for steel beams and construction metal. Therefore, additional discussion here is limited to the effects of water sorption and desorption characteristics. Hay and Schurr (14) have pointed out that low rates of water desorption from paint films can detract markedly from their value as corrosion-preventive coatings. Such low rates of water diffusion out of a film can result when topcoats control diffusion better than primers (such as a low PVC, glossy enamel used over a relatively high PVC low gloss primer). Thus, topcoats on metal must be selected with this effect in mind.

2. Aluminum

Aluminum is a good substrate for paint. Besides being relatively dimensionally stable, impermeable to water, and possessing high surface energy, it is also relatively chemically stable. Although it ranks above iron in the electromotive series, its tendency to form a tightly bound oxide layer in most atmospheres makes it relatively resistant to corrosion. Provided the aluminum is reasonably free from contaminants, most house paint primers (or topcoats) will perform satisfactorily. However, Rigg and Skerrey (15) have reported that in marine atmospheres, red lead, Pb_3O_4, should be avoided in primers for aluminum. They advocate the use of zinc chromate pigments.

3. Galvanized Iron

Zinc metal treatments on iron are common. Both hot-dipped and electrolytically applied zinc are used. The former usually gives a spangled appearance unless rolled and annealed; the latter gives a flat matte appearance. Both are difficult surfaces on which to maintain adhesion of aged paint films. The chemical reactivity of zinc metal with the formation of soluble zinc salts or soaps causes loss of adhesion during exterior exposure. Sometimes this loss of adhesion can be very rapid (a few months) and sometimes it can be quite slow (even up to several repaints). In addition, the producers of galvanized iron often use inhibitive treatment of the zinc to prevent the formation of "white rust." These silicate and chromate treatments apparently lower the surface energy to a point where initial adhesion is difficult to attain. Unfortunately, these inhibitive treatments are difficult to detect (16) and to remove. Extensive weathering prior to painting appears to be the best way to remove such passivating treatments.

Special primers have been developed that perform well on galvanized iron. One of these is based on zinc dust/zinc oxide pigmentation. Since it is a two-package paint with the zinc dust packaged separately, it has not been used extensively in the house paint area. Another is based on the inclusion of Portland cement in the formulation. A possible mechanism by which the combinations function is that the zinc dust or the Portland cement react with acidic components that may form in the paint film before such components can react with the zinc metal and weaken the interfacial bond. More recently, certain latex-based paints have been found to perform well over galvanized iron, particularly if they are used with an all-latex system.

A particularly bad practice has existed in the painting of galvanized iron gutters and downspouts. These are often painted by the sheet metal craftsman with paints identified as "tinner's red." Generally these have been low-cost paints based on iron oxide in an alkyd vehicle and have shown poor adhesion to galvanized iron, with or without a treatment to prevent "white rust." Latex-based "tinner's reds" are now appearing. They are much to be preferred.

4. Copper

Copper is much like aluminum. It is an easy metal to paint because it forms a protective oxide. There has been very little attention paid to painting copper in the literature, probably because of its limited use for exterior architectural applications and the fact that painting of copper has not been a problem.

5-3. FORMULATION

The formulation of exterior house paints is based on several general requirements which are: capability of being applied by brush, capability of drying or curing under atmospheric conditions, resistance to the destructive elements of weathering (moisture, temperature changes, ultraviolet light, and pollutants), capability to attain and maintain adhesion to imperfectly cleaned surfaces, and capability to impart a pleasing appearance to the surface. All of this must be done at a reasonable raw material cost in a formula that is not critical for manufacturing control and that is stable for extended periods ($\geq$ 3 years) in the package under uncontrolled storage conditions.

These requirements present quite a challenge to the house paint formulator. For many years he relied almost entirely on oil vehicles to give him easy brushing and atmospheric curing. In the 1930s and 1940s alkyds were gradually introduced, particularly in the dark colors, with a marked improvement in properties such as color and gloss retention. Then in the 1950s latex-based paints began to appear and twenty years later they hold more than half of the market. During this same period a few other more durable vehicles had some very minor usage such as silicone-modified alkyds and polyesters reacted with and plasticized with epoxidized oils. In all cases mildew resistance has been a continuing problem that has not been solved completely. This is particularly true in the warm, damp climates of the Gulf of Mexico coast.

A. TYPES

1. Oil

The original white house paints as we know them today were based on white lead-in-oil. These did a reasonably good job of protecting wood and metal substrates but left much to be desired from the aesthetic standpoint (although some usage still persists today). Lead-in-oil paints were easy to apply by brushing, in fact, with a high-shear viscosity of about 1.8 P, they brushed so easily that thin films were applied (17). Hiding was so poor that three coats were needed for good initial appearance. Then the appearance rapidly declined because of excessive dirt collection and discoloration by the formation of lead sulfide upon reaction with hydrogen sulfide in the atmosphere. In addition, lead-in-oil films checked and alligatored very badly. When they did start to clean up through degradation and sloughing off of the top layers, the appearance became quite spotty because of uneven cleanup.

Gradually other prime pigments and inert fillers were introduced into the basic lead-in-oil formula. The prime pigments included zinc oxide, leaded zinc oxide, lithopone, zinc sulfide, and finally titanium dioxide. Now lead pigments are being phased out completely because of federal, state, and local legislation calling for increasingly stringent requirements from November, 1971 onward.

a. Vehicle

From the beginning linseed oil has been the primary vehicle in house paints. Its widespread availability and relatively good drying properties (iodine value, 170 to 190) have kept it popular in house paints throughout the years (18). Even today it is used in the bulk of drying oil house paints still on the market. Several other oils have been used in house paint, but to a much lesser extent than linseed oil and often in conjunction with linseed oil. The more important oils have been: (a) tung oil, fast drying because of its conjugated unsaturation. It is used very little now because of uncertainty in supply and cost; (b) perilla oil, higher in unsaturation than linseed (iodine value, 185 to 205) but in short supply; (c) soybean oil, too low in unsaturation (iodine value, ~ 135) to be used by itself. At the present time it is higher in cost than linseed oil; (d) oiticica oil, similar to tung oil in unsaturation and supply; (e) fish oils, high in saturated acids and poor in odor; (f) safflower, between soya and linseed in unsaturation (iodine value ~ 150) but higher in cost; and (g) dehydrated castor oil, doubly conjugated and good drying, but high in cost.

Today's oil-based house paints generally use combinations of unbodied and bodied oil. The usage of major amounts of bodied oil in house paints dates back to 1943 when wartime shortages necessitated conservation of drying oils by government edict. This change provided better paint properties (albeit unknowingly) by increasing the critical pigment volume concentration (CPVC) (19). The pigment content could be increased and the binder content decreased while maintaining or improving critical properties of the paint film. Bodied oils, with their increased polarity over unbodied oils, tend to be better pigment dispersants, thereby increasing the ability of pigment particles to attain close packing and consequently increasing the CPVC. Since the net result of using bodied oils was a better house paint in such important properties as flow, drying, and reduction in dirt collection, checking, and cracking tendencies (19), this wartime change was never abandoned.

Typical formulations for oil-based white house paints are shown in the U.S. Government paint specifications (Formulations 5-1 to 5-3).

b. Pigments

(1) Titanium Dioxide. Titanium dioxide is the white pigment used in house paints (and essentially all other types of paint) because of its relatively high refractive index (~ 2.70), availability, and chemical inertness. TiO_2 can be anatase or rutile in its crystalline form. Compared to rutile, anatase TiO_2 has a lower refractive index (~ 2.55) and lower ultraviolet light absorption. It has more of a tendency to form peroxides (20) or reduced oxides (21) which cause chalking. Therefore, the use of anatase TiO_2 is confined almost completely to paints for building exteriors where chalking is desirable to assist in self-cleaning. Even rutile TiO_2 has some tendency to form peroxides or reduced oxides. Therefore, pigment suppliers have produced rutile pigments with various kinds and amounts of surface treatments. Alumina, silica, and zinc oxide were used in combination to make the most chalk-resistant sulfate process (du Pont's R 610, Titanium Pigment's RANC, Glidden's R-88, New Jersey Zinc's R-760, American Cyanamid's OR 640) grades of TiO_2. Larger quantities of silica have been used on chloride process TiO_2 - as much as 14% - to give even more chalk-resistant pigments (duPont's R 960 and R 966, and Titanium Pigment's CL-NC, now discontinued) with a premium cost. In oil house paints the advantages of such premium pigments are so minor as to be difficult to detect. Therefore, standard chalk-resistant grades are used normally.

The lower refractive index of anatase TiO_2 means that anatase pigments have 20 to 30% less hiding power than rutile pigments. Since they are only about 10% lower in cost, they are used only when their special properties (tendency to chalk and blueness) are desired. In house paints they are used for their chalking properties. Since they are less economical than rutile pigments for obtaining hiding power, the anatase pigments that can be used

FORMULATION 5-1. Regular White Paint:[a]
Federal Specification TT-P-102, Class A[b]

Materials	lb/100 gal	gal
Leaded zinc oxide (58.7% basic lead sulfate)	537.2	10.67
Anatase titanium dioxide	145.5	4.48
Talc	256.4	10.79
Raw linseed oil	359.6	46.31
Z-2 Bodied linseed oil	95.4	11.86
Mineral spirits	96.0	14.75
Pb naphthenate (24%)	5.6	0.58
Mn naphthenate (6%)	1.7	0.21
Co naphthenate (6%)	2.8	0.35
	1500.2	100.00

[a]PVC is 30.8; wt/gal is 15.0 lb; consistency is 82 KU; pigment is 62.6%.

[b]"Comments: Formula provided. Being a free-chalking paint, it is unsuitable for tinting because of color fading. Paints of this type are frequently used as primer as well as top coats, although better results are obtained with primers that contain no zinc oxide. Anatase titanium dioxide is used for chalking and cleanup. Proprietary formulas usually employ a mixture of anatase and rutile titanium dioxide. The large proportion of raw to bodied oil promotes easy brushing."

at the lowest concentration to get chalking are to be preferred. In addition, the TiO_2 suppliers have offered "water dispersible" grades of anatase TiO_2 at somewhat lower prices. These grades are usually used in house paints. The proportion of anatase to rutile TiO_2 varies somewhat depending upon individual preferences with most paint producers using from 50 to 100% of the TiO_2 as anatase. Usually the higher percentages of anatase are used with higher levels of zinc oxide or leaded zinc oxide.

(2) Zinc Oxide. Essentially all oil house paint finish coats carry substantial amounts of zinc oxide or leaded zinc oxide (from 2 lb/gal on up calculated as zinc oxide). Since zinc oxide is not nearly as economical as TiO_2 as a white hiding power pigment (22), it is used for other properties. These are

FORMULATION 5-2. White Base for Light Tints:[a]
Federal Specification TT-P-102, Class B[b]

Materials	lb/100 gal	gal
Leaded zinc oxide (58.7% basic lead sulfate)	536.1	10.66
Rutile titanium dioxide	107.8	3.08
Talc	293.8	12.35
Raw linseed oil	347.2	44.72
Z Bodied linseed oil	106.3	13.29
Mineral spirits	95.8	14.71
Pb naphthenate (24%)	5.5	0.57
Mn naphthenate (6%)	2.2	0.27
Co naphthenate (6%)	2.8	0.35
	1497.0	100.00

[a] PVC is 31.0%; wt/gal is 15.0 lb; consistency is 85 KU; pigment is 62.5%.

[b] "Comments: The main difference between this formula and the preceding one is that the titanium dioxide has been changed from anatase to rutile type to reduce chalking and improve color permanence. Although the amount of titanium dioxide is reduced, there is but small sacrifice in hiding power because of the superior hiding of the rutile type. Moreover, hiding power is substantially improved by most of the color pigments that are used for tinting. By increasing titanium oxide for better hiding, this type formula produces white trim paint that does not cause whitening of lower areas of brick and stone."

its fungistatic action and its ultraviolet light absorbing properties (23). Although zinc oxide slows down the erosion rate of oil house paints considerably, it also helps to control mildew because of the unusual detergent action of water-soluble zinc salts formed with some of the oxidation products in weathering paint films (24).

However, zinc oxide in oil house paints also contributes some undesirable properties. Most of these have to do with the water sensitivity of oil paint films containing zinc oxide. This water sensitivity causes excessive swelling of the film in the presence of water (25, 118, 119) which probably contributes to a tendency toward blistering (25). Leaching of these water solubles (or $ZnSO_4$ formed by the action of atmospheric SO_2 on the ZnO) and deposition on the surface so that it subsequently lies at the interface between a repaint film and the original film, probably causes undereave intercoat or "tissue paper" peeling (26).

FORMULATION 5-3. Fume-resistant White Paint:[a] Federal Specification TT-P-103[b]

Materials	lb/100 gal	gal
Anatase titanium dioxide	82.6	2.54
Rutile titanium calcium, 30% titanium dioxide	253.7	9.36
Lead-free zinc oxide	316.2	6.70
Talc	173.6	7.31
Raw linseed oil	302.4	38.95
Z Bodied linseed oil	128.8	16.02
Mineral spirits	103.6	15.91
Ca naphthenate (4%)	21.3	2.73
Mn naphthenate (6%)	3.9	0.48
	1386.1	100.00

[a]PVC is 32.0; wt/gal is 13.86 lb; consistency is 82 KU; pigment is 59.6%.

[b]"Comments: This formula is designed for industrial areas that have sulfur fumes in the air. Since lead compounds turn dark when exposed to sulfur fumes, the formula contains no white lead and no lead driers."

Although ZnO acts as a fungistat for exposed areas of a structure, it can accelerate mildew growth in protected areas. Again water sensitivity, which tends to keep the film wet for longer periods of time in protected areas, is the probable cause (27). Zimc oxide also appears to aggravate the appearance of nailhead rusting on structures by "fixing" the rust stain.

Nevertheless, in spite of all of the unfavorable effects from zinc oxide, it is almost essential to use zinc oxide in oil house paints. Without ZnO they would be too subject to mildew in exposed areas and would erode too rapidly even without any anatase TiO_2 pigment.

(3) Lead Salts. Basic white lead pigments (carbonate, sulfate, and silicate) have also been commonly used in oil house paints, presumably for their chemical reactivity. However, most paint manufacturers have also produced fume-resistant grades of white oil house paint which are essentially lead-free. When these lead-free paints are loaded with anatase TiO_2 rather than rutile TiO_2 to promote cleanup, their durability is not much below that of lead-

containing paints. At any rate, lead pigments and driers are disappearing from paints with the enactment of governmental regulations.

(4) Other White Pigments. Other white pigments have enjoyed some usage in oil house paints over the years, but this was limited and has essentially disappeared. These included antimony oxide, lithopone ($ZnS \cdot BaSO_4$), and zinc sulfide.

(5) Colors. The selection of color pigments and formulation into oil-based house paints has been treated by Fuller (28). Exterior durability of organic pigments has been discussed by Vesce (29). However, very few of the organic pigments are used in oil house paints, partly because of cost and partly because of insufficient durability. The phthalocyanines are an exception.

BLUES: Copper phthalocyanines are blue when partially chlorinated and green when fully chlorinated. Both shades are quite resistant to ultraviolet radiation. The phthalocyanines are relatively high in color strength so that it is not necessary to use very much for tints. This is helpful from the standpoint of cost, but harmful from the standpoint of color retention. The harm comes from the fact that extender pigments mask the color of the phthalocyanine pigments in a chalk layer. Gray oil house paints made with carbon black or lampblack suffer from this same shortcoming. In industrial coatings, where less chalking occurs, the phthalocyanine pigments give tints with excellent color retention.

There is one class of inorganic blue pigments that has been used to some extent in oil-based house paints, i.e., iron blue, Chinese blue, etc. which are sodium or ammonium ferriferrocyanide. These are durable pigments but they can suffer from reduction in the package or reaction with alkali – even with mildly alkaline pigments like calcium carbonate. The only inorganic green pigment still of interest is chromic oxide. This is a very durable pigment with high absorption in the ultraviolet light region so that paints made from it are especially durable. It is an olive green tone, however, that is not pleasing to everyone. Chrome green —a mixture of lead chromate and iron blue – cannot be used any longer because of lead restrictions.

IRON OXIDES: Synthetic iron oxide pigments provide a range of low chroma colors from yellow to brown to red. They are all very durable, stable, high ultraviolet absorbing pigments, giving good performing oil house paints. Their drawback is their lack of high chroma or brightness. However, this is not often desired in house paints. Yellow iron oxides are hydrated ferric oxide, while the reds and browns are not hydrated. Venetian red will give somewhat brighter reds with its content of 60% calcium sulfate.

REDS: Cadmium selenide pigments are rich in color in the orange-red to maroon range. They are durable although expensive. They probably will be banned in the U.S. as hazardous substances by the Food and Drug

Administration. Organic reds are either not durable, except masstone (toluidine red), or in very expensive forms (quinacridone reds).

YELLOWS: Yellows, aside from those dull yellows produced by ferrite pigments, are either expensive or toxic. Lead chromate is now a banned ingredient in paint. It has been the major yellow pigment for bright yellows even though it does not resist attack by atmospheric SO_2. Silica surface treatment (duPont's Krolor process) greatly reduces such attack. Cadmium sulfide could be a color alternative to lead chromate in spite of its high cost, but it also may be banned as a hazardous substance. Hansa yellows and other newer organic yellows show good color retention when used full strength, but not in light tints. All are expensive.

(6) Extenders. In a modern paint plant there are probably stocks of more kinds of extender pigments than any other single class of raw materials. There can be easily be 100 to 200 different extenders actually in use in a paint company that makes a complete spectrum of organic coatings. The necessity for each of these extenders will be stoutly defended for some specific formulations. Oil house paints, then, are no exception and many different extenders are recommended for different effects on performance. This is particularly true for flat or low sheen house paints where proportion of extender pigment is high and the precise level of pigmentation is more critical.

(a) CPVC Concepts. Much of the proliferation of extender pigments is unnecessary. This proliferation has resulted from a failure to recognize the principle of "saturation value" promulgated by Armstrong and Madson in 1947 (30) and elaborated by Vannoy (31) as "equal free binder" and by Asbeck and Van Loo (32) as "critical pigment volume concentration (CPVC)" in 1949. A cursory survey of six years of the Journal of Paint Technology (1966-1971) revealed at least six studies in which attention should have been paid to this principle. In four of the six papers this principle of evaluating pigments at equal PVC/CPVC ratios was ignored (33-36). In two of the papers it was handled correctly (37,38). Work done in the area of oil house paints probably suffers from this same lack of attention to PVC/CPVC ratios. A survey in the early 1960s of the extender pigments used in oil house paints, both self-cleaning and trim and tinting whites made by ten different major paint manufacturers revealed the use of silica, magnesium silicate, aluminum silicate, calcium carbonate, calcium sulfate, and a 70:30 combination of calcium sulfate and TiO_2. In spite of this variety of extender pigments, the performances were quite uniform except for chalk rate. The chalk rate depended primarily on the balance between anatase and rutile TiO_2 rather than on PVC or type of extender.

These were all gloss paints. As Asbeck and Van Loo have pointed out (32), paint properties tend to change slowly with PVC when the PVC/CPVC ratio is considerably less than unity. This is illustrated in Fig. 5-3.

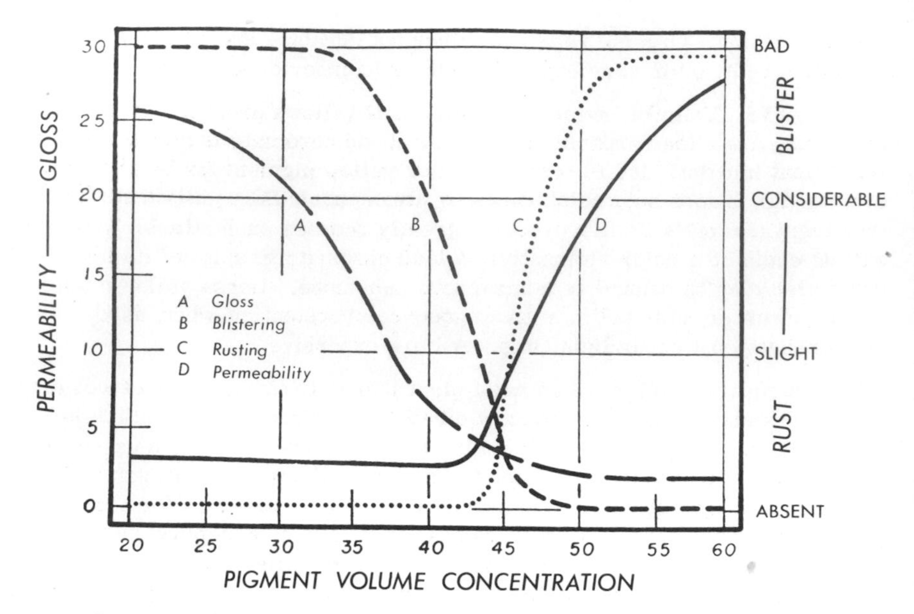

FIG. 5-3. Effect of CPVC on paint characteristics.

With flat house paints the extender is again not critical if the PVC/CPVC ratio is maintained correctly (~ 1.1) and if the SO_2 content of the atmosphere is relatively low (<0.05 ppm).

Sulfur dioxide has been found to attack most extender pigments in oil paints except silica in oil paints. This effect has been shown at both du Pont's exposure fences in Delaware and Titanium Pigment's exposure fences in New Jersey. It also has been demonstrated in a gas-controlled Weather-Ometer in the author's laboratory. Calcium carbonate is particularly susceptible to attack by SO_2. Calcium sulfate and barytes ($BaSO_4$) presumably would not be attacked.

(b) Rutile Yellowing. Rutile yellowing, an unexplained tendency for chalk faces of rutile TiO_2 to turn a fairly bright yellow color in the presence of ultraviolet light and moisture, seems to be accentuated by calcium carbonate. The reasons for this are not known. However, excellent correlation between the level of calcium carbonate and rutile yellowing has been demonstrated on exterior exposures in the Chicago area by the author's laboratory.

(c) Function. Traditionally, extender pigments have been used in oil house paints to control gloss (flat paints only), suspension, and viscosity.

Additives can be used to accomplish the last two purposes quite adequately. If flatting is their only function, the question is why should extenders be used at all in gloss oil house paints. Some of the answers that have been given in the past have been to reduce permeability, to increase filling properties, and to reinforce the film to prevent checking and cracking. In the author's opinion, none of these really justify the use of extenders in gloss oil house paints. Permeability does not decrease very much with increasing PVC below the CPVC, as demonstrated in Fig. 5-3. Checking and cracking are avoided by relaxation of a film, not by tensile strength. Filling of surface irregularities is not a property expected from gloss house paints.

There is some tendency to discontinue the use of extender pigments in house paints. This is discussed in the section on alkyd house paints (Sec. 5-3, A, 2).

c. Driers

Oil house paints tend to dry slowly. Most producers recommend a two-day drying period before recoating oil paints in good weather conditions and four days under adverse conditions. Drying by oxidative polymerization has been catalyzed using a combination of lead, cobalt, and manganese driers and the presence of basic lead pigments (see formulations 5-1 to 5-3). Food and Drug Administration regulations published in the March 11, 1972 Federal Register, par 191.9, in effect ban the use of lead pigments after December 31, 1972 and lead driers after December 31, 1973. Although lead pigment-free oil house paints have shown good performance (31) obtaining adequate dry and retention of drying after prolonged storage in the package poses a problem. Lead is a rather mild-acting drier that acts as a polymerization catalyst as well as an oxidation catalyst. Cobalt and manganese are considerably more active oxidation catalysts and cause top drying and wrinkling if used in too high a concentration). Therefore, it is not possible to employ sufficient cobalt and manganese to allow for drier absorption by finely divided organic pigments without the danger of wrinkling with fresh paints before the drier absorption has taken place.

Many paint producers have marketed, for many years, oil-type fume- and mildew-resistant white paints that are free of added lead. These usually attain their dry by the use of organic salts of manganese along with fairly large amounts of calcium. Terachlorophenol has been used in addition to the metallic driers to assist both drying and mildew resistance. Even so, these paints are slow in drying, particularly to a hard film. Fortunately, they are sold and used mostly in warm climates, i.e. in the southern part of the United States where drying conditions usually are good. Whether they could be marketed successfully on a national basis is still a question.

In the meantime, the paint industry is working feverishly to find suitable substitute driers. At the moment combinations of cobalt and calcium appear to be the most promising. Unfortunately, the industry's trend away from oil paints during the past decade has not inspired much work on driers. The Philadelphia Society for Paint Technology in two publications (39) reported that in addition to the type of metal, the most important variables were the type of vehicle, temperature, and storage in the package. Wheeler (113) has demonstrated that heterocyclic, polycyclic polyamines (particularly 1,10-phenanthroline) can be effective driers in oil paints when used in conjunction with cobalt or manganese driers.

d. Solvents

Turpentine was the accepted thinner for paint up until the 1930's. At that time mineral spirits, a blend of aliphatic, naphthenic, and sometimes aromatic hydrocarbons, was adopted for cost and odor reasons. Originally it was thought that the terpenes in turpentine had some beneficial effect on the drying of oil paints, but this could not be demonstrated. Little attention was paid to the aromatic content of mineral spirits until Los Angeles put Rule 66 into effect in 1966 (40). Activity in the U.S. spread to other localities. Because of San Francisco Bay area Regulation 3 and Philadelphia's Rule 5 and the federal requirement that states and other municipalities establish their own rules to regulate air pollution, most paint companies are reformulating their oil house paints to use odorless (~0% aromatics) or low odor (2-8% aromatics) mineral spirits. Usually this causes no technical problems, but it does increase the cost.

e. Additives

Normally, very few additives such as anti-oxidants or fungicides are used in oil house paints. However, anti-settling agents such as aluminum stearate have been used in the past. These were shown to be responsible for seeding in the presence of zinc oxide and have fallen into disfavor. They have been replaced by materials that impart a high degree of thixotropy such as amine-treated bentonite clays (Bentones from National Lead) and highly bodied oils (such as the Thixatrols from Baker Castor Oil). The function of these additives is to control the rheological profile for good package stability, brush pickup, application properties, and flow. In general, they cause substantial increases in low shear viscosities, such as during pigment settling in the package, without greatly affecting the high shear viscosity such as during brushing (32). Patton (41) has a good discussion on this area. More is said in the following section on alkyd paints.

In the late 1930s a major paint company introduced the first alkyd house paint. Unfortunately, they experienced numerous complaints of failure by intercoat or "tissue paper" peeling. It is assumed that this alkyd paint contained zinc oxide for mildew resistance and perhaps for color retention. As has been pointed out (26), after the fact, zinc oxide could be expected to give such problems if used in an alkyd house paint. Unfortunately, there were no other effective fungicides for use in alkyds at that time, nor for many years afterward. A lack of knowledge of the mechanism of intercoat peeling and consequent fear that the alkyd itself was the cause, plus the lack of an effective fungicide, kept alkyd house paints off the market for many years. They began to appear in the 1950s, but only in dark colors where mildew was not obvious. In the 1960s white alkyd house paints were marketed by a few paint companies. Generally they used phenyl mercury oleate or other aryl mercury derivatives as a fungicide. These paints either had insufficient mildew resistance in severe mildew areas or were formulated to chalk too rapidly for good durability. There were some effective organic fungicides introduced in the 1960s that could be used successfully in alkyd house paints (particularly R. T. Vanderbilt's Vancide PA and Dow's Dowicil S-13). They had to be used in such high concentrations (~ 1% or more) that the cost of such alkyd paints was considerably higher than the traditional Pb/Zn/Ti oil house paints. For this reason most paint companies continued to market oil house paints for their whites and light colors.

It is expected that in the 1970s alkyd house paints will replace oil house paints because of the ban of lead as a hazardous substance. Adequate drying of alkyds can be attained with cobalt and calcium soaps. Adequate drying can also be obtained by the use of 1,10-phenanthroline metal chelates as pointed out by Stearns (114) and elaborated by Myers et al (115). Although alkyd paints with adequate mildew resistance will be expensive to produce, there will be a number of advantages to be gained besides the elimination of lead. Alkyd formulas should give faster drying for less dirt and insect collection during drying; better gloss and tint retention because of less ultraviolet light absorption by the alkyd than the oil (42); slower rate of erosion for longer durability; less nail head corrosion staining because of the absence of zinc and its "fixing" action; less tendency for intercoat peeling if applied over zinc-free paints or over old paints that have been washed and rinsed properly (26), and less tendency to mildew in protected areas of a structure because of less tendency to retain water in the film.

a. Rheology (See Vol. 2, Parts I and II)

When alkyd vehicles were developed by the paint industry in the late 1920s there was a general feeling by paint formulators that they yielded paints that were considerably more "sticky" in brushing characteristics than oils. The feeling was based on fact — i.e., the paints that were made from alkyd vehicles were more difficult to brush — but the effects can be changed by understanding the rheology involved and formulating so as to get more favorable rheological profiles for the alkyds.

Probably the major reason that alkyds gave sticky brushing paints was the fact that they were more nearly Newtonian (their viscosity was less dependent on the shear rate at which the viscosity was measured) than oil paints. Oil house paints were formulated at a PVC of about 35%. Their CPVC was about 45%. The tendency was to formulate alkyd paints at the same PVC of 35%, but their CPVC was about 55% because pigment wetting was more thorough and pigment packing was facilitated. Paint vehicles like oils and alkyds are usually Newtonian (viscosity independent of shear rate). The addition of pigments to these vehicles imparts shear sensitivity, in the form of shear dependence of viscosity, and thixotropy. As the PVC approaches the CPVC of the liquid paint, thixotropy increases, at least up to a point. In addition, the lower volume solids of alkyd paints would tend to make them more Newtonian.

In the 1930s and 1940s the viscosity of house paints was measured with flow cups or paddles. These viscometers operate at low shear rates (from 100 to 1000 sec^{-1}) compared with brushing shear rates of 5000 to 20,000 sec^{-1}. If alkyd paints are adjusted by holding out thinner to a desired viscometer reading, they will not necessarily brush easily; alkyd paints are less thixotropic so that they are likely to have too high a high shear viscosity for easy brushing. This is illustrated in Fig. 5-4,

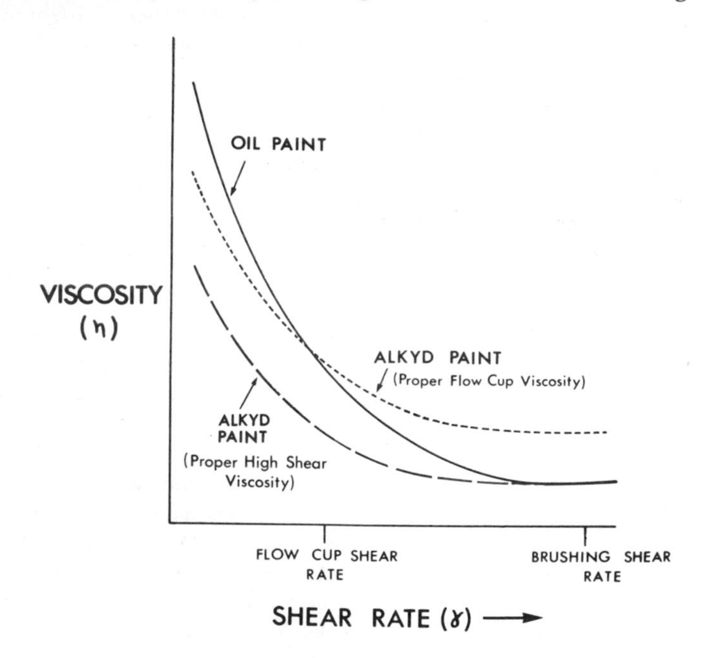

FIG. 5-4. An illustration of the rheological problems with alkyd paints intended for brushing.

where the dotted line represents the rheological profile of a relatively high solids alkyd paint which possesses the proper low shear viscosity and the dashed line represents a similar profile for a lower solids alkyd paint with proper brushing viscosity but "too low" flow cup viscosity.

Certain additives came on the market that can increase the low shear viscosity of a paint without substantially changing the high shear viscosity. Some of these were highly bodied oils or alkyds, bodied leaded oils, bodied modified (amide or stearamide) alkyds, and amine-modified clays. Unfortunately, some of these obtained their bodying action by flocculating the pigment. This was deleterious because hiding power decreased and because flow decreased since the rate of viscosity recovery after shearing (brushing) was too rapid. Others, like the amine-treated bentonites, gave poor flow also, again because of too rapid recovery of viscosity after shearing.

Still others, like amide-modified alkyds, gave excellent flow, but tended to sag. It has been found to be possible, however, by judiciously combining bodying agents in alkyd formulations, to arrive at a rheological profile for alkyd paints very similar to that for oil paints with proper rates of viscosity recovery after shearing for good flow, without sagging, and with good hiding.

More is said about the measurement of viscosities, controlling flow, and the measurement of flow in the section on latex paints (Sec. 5-3,A,3).

b. Formulations

Alkyd house paints, aside from the rheological aspects, are very easy to formulate. Selection of the polyol and acid species is not critical except that the percentage of fatty acid (called the oil length of the alkyd) has been considered critical for ease of brushing and flexibility. A discussion on brushing characteristics was presented under "Rheology." Using those techniques it is possible to get good brushing paints with any alkyds within the medium to long oil range (55 to 75% oil). Initially, longer oil alkyd paints are more flexible, but their films lose some flexibility on exposure. Claims have been made by raw material suppliers that one polyol is outstanding for durability. Other suppliers claim that isophthalates are more durable than orthophthalates, etc. Since it is difficult, if not impossible, to hold other critical variables constant while varying one component like the polyol, such claims really should be limited to the specific alkyds under study. Alkyd resins have been discussed by Martens (43), Patton(44), and Blegen (45).

All three authors mention the need for long oil alkyds (about 70% oil) for good flexibility. It may be true intitially that long oil alkyds give better flexibility, but it is not necessarily true upon aging. As indicated by the figures that Van Loo (42) presented (see Table 5-2), oils plus their necessary driers tend to absorb considerably more ultraviolet light than do alkyds plus their necessary driers. Since ultraviolet light has enough energy to break carbon-carbon bonds, it should be expected that the oil portion of an alkyd

TABLE 5-2. Absorption of Ultraviolet Energy
Samples Dissolved in Benzene Without Removal of Solvents

No.	Vehicle	Calculated energy absorbed (%)
1	Raw linseed oil	5.2
2	Alkali-refined linseed oil	9.9
3	Alkali-refined linseed oil plus driers	11.8
5	Heat-bodied linseed oil	15.9
8	Blown linseed oil	38.1
19	Phthalate alkyd	8.6
20	Phthalate alkyd plus driers	9.0

is more subject to attack by ultraviolet light. Furthermore, alkyds containing less oil should be more stable in the presence of ultraviolet light. This has been shown to be true by stress/strain studies (46) and by actual panel exposures. In the latter, low PVC (about 20%) extender-free paints showed less cracking for a medium oil length alkyd (about 56%) than for a long oil length alkyd (about 66%). Furthermore, the low PVC extender-free paint showed less cracking than did a paint with the same alkyd but at a higher PVC (about 27%). This latter point demonstrates that extenders are not necessary nor even desirable to prevent cracking and checking. It also demonstrates the need for careful formulating with alkyds.

Extender-free alkyd paints were found to have one other advantage. This was in color retention. The trim paints were found to give substantially better color retention than similar paints containing extenders. This is due to the gain in hiding power of extenders in air over that in resin. In resin, extenders have essentially no hiding power since their refractive index is about the same as that of the resin. However, in air, they have considerable hiding because of the differences of ~ 0.5 in refractive indices. The consequence is the masking of the absorption of light by the color pigments due to the refraction of light by the extender pigments.

3. Latex Paints

At the present time, at least 60% of the exterior house paint market is filled by latex paints. Synthetic latexes such as rubber replacements were a World War II development, and the polyvinyl acetate latexes were then used in exterior paints in Germany. Shortly after the war styrene/butadiene latexes were adapted for usage in interior flat wall paints. In the 1950s similar paints, but with lower PVCs, were introduced as exterior masonry paints, but not for wood substrates. They were found to be too brittle after relatively short periods of weathering for the latter use. In the late 1950s latex house paints for wood and masonry substrates began to appear on the market. Generally, they were based on acrylic latexes such as Rohm and Haas' AC-33. These paints were so low in brushing viscosity (~0.5 P versus 2 to 3 P for solvent-based paints) that they were spread too far and recovered their low shear viscosity so rapidly (< 10 sec) that they had very poor flow. Nevertheless, they were well accepted by the public because of the ease of application and clean-up with water.

The early latex house paints did not contain any oil or alkyd modification and, therefore, did not penetrate chalk layers on the repaint surface. This meant that their adhesion to weathered paint surfaces was so poor as to be unsatisfactory. The normal recommendation was for the use of a standard, oil-type house paint primer to bind the chalk and then two coats of the latex house paint for optimum durability. Paint companies tried various means of circumventing the use of an oil primer for repainting. Some introduced latex paints modified with oil or alkyd at about 25% of the vehicle solids. These were deficient in mildew resistance when used as topcoats so some companies introduced such modified latex paints as primers only. Others marketed an oil or alkyd additive that was stirred into that part of the topcoat that was used as the first coat by the customer. Others simply advocated washing the chalk off before repainting.

Common practice now is to market latex paints modified with solvent-type vehicles (usually long oil alkyds) but fortified with 10 to 20% of the pigment as ZnO for mildew resistance (47). The use of additional fungicides, such as a phenyl mercurial or other organic compounds has been quite common. The application and flow rheology has been improved (usually through the latex) to such an extent that one coat of the latex paint is usually sufficient for adequate appearance. Thus, latex paints are rivals of solvent based paints for flow and leveling, but there is still room for considerable improvement.

Although such alkyd-modified latex paints "dry" well, even without metal driers present, as judged by some type of tack test, Stearns et al. (116) have pointed out that using a performance test such as scrub resistance demonstrates the need for driers. Furthermore, he has shown that regular metal drier soaps tend to lose their drier activity in modified latex paints during storage in the package. He advocates the use of 1,10-phenanthroline with cobalt for retention of drying activity after aging in the package.

a. Rheology

At high shear latex paints tend to have very low viscosities of approximately 0.5 P. As such they have such little resistance to brushing that the average amateur painter tends to spread them too far, e.g., about 600 ft^2/gal instead of the recommended 400 ft^2/gal. This is unfortunate for two reasons. First, although the wet film hiding may be adequate, the dry film hiding often is not. Second, the film thickness is so thin (about 1.0 mil) that durability suffers, particularly in the valleys of the brush marks.

(1) Application. The high shear or residual or base viscosity of latex paints arises from the presence of solid particles (latex and pigment) in the continuum (water). The volume solids of latex paints is relatively low, ~40%. This does not give very high viscosity. However, successful attempts have been made by the latex suppliers to increase the base viscosity by undisclosed principles.

(2) Flow and Leveling. With these latexes it is possible to formulate practical paints with high shear viscosities of 1 to 2 P. Peculiarly, viscosities above 1.5 P usually lead to paints that are hard brushing, although solvent-based paints at 1.5 to 2.0 P are "sloppy" in brushing characteristics. Bullett (48) has recognized this difference and suggested that the elastic forces in the film causing normal stresses, in addition to the usual viscous forces, may affect the spreading qualities of paint under a brush.

(3) Measurement. The low shear viscosity of latex paints comes mostly through the action of the thickener on the pigment and latex particles. The cellulosic thickeners generally used in latex paints are relatively large molecules. It is theorized that they stretch across the distance between particles and form a loosely bound, cross-linked structure (48). Shearing action readily breaks down this structure, but it reforms rapidly after the shearing has stopped, so rapidly, in fact, that little flow can take place (<10 sec). Increasing the thickness of the wet film that is applied by increasing the high shear viscosity does help flow somewhat, since flow is proportional to the third power of film thickness, as pointed out by Smith et al. (49), derived in detail by Rhodes (50), and elaborated by Patton (41). Other at least partially successful approaches toward improved flow have been the use of extra amounts of coalescents, such as high boiling ether alcohols, and the use of thickeners other than the cellulosics.

Instrumentation for measuring the high shear viscosity of paints varies from quite simple and inexpensive instruments like the Brushometer (18,500 sec^{-1}), to quite complex instruments like the Haake Rotovisko (variable up to 13,700 sec^{-1}) or the Weissenberg Rheogoniometer (variable up to 40,000 sec^{-1}). The ICI Cone and Plate Instrument is quite convenient since it has built-in temperature control, uses only a few drops of paint, and is very easy to clean. The Weissenberg Rheogoniometer has two unique features: (a) it can superimpose a reciprocating motion at low shear rates on a rotational motion at high shear and go instantaneously from the latter to the former thus simulating brushing, and (b) it can measure normal forces at various shear rates. It is more expensive than other common rheology instruments. Pierce has described most of these instruments and their application to the paint industry (51).

Instrumentation for measuring viscosities that are meaningful for predicting flow is not as well defined. Beeferman et al. (52) have described a technique using the Haake Rotovisko with a stress relaxation technique. This has been criticized by Dodge (53) for insufficient high shear rate and the method of calculating viscosity. Patton (54) has suggested a somewhat similar technique but using a Wells-Brookfield Cone/Plate viscometer. Again, Dodge (53) has criticized this method for these same reasons and for the delay in transferring the sample from the device in which it is subjected to high shear to the Brookfield.

Dodge (53) has suggested a method that does not use viscosity. Instead he lays down a film under high shear conditions but with uniform "brush marks" by using a threaded bar. This film is allowed to dry and the surface roughness determined with a Talysurf. The surface roughness can be used as a numerical expression of flow. Dodge's method has certain similarities to that of the New York Society (55) in that both attempt to measure actual flow, but has the advantage of uniform shear rates.

b. Formulations

In contrast to the simple formulation of alkyd paints, the formulation of latex paints is complex. The formulator has many additional factors to consider such as film formation from discrete particles of vehicle rather than from a continuum, preventing premature coalescence of the latex particles in the package because of freezing or mechanical agitation, dispersing pigments in a highly polar substance (water), thickening the very low viscosity continuum (water), preventing attack by microorganisms in the package, and preventing the formation of stable foams during manufacture and/or application. The formulator selects different ingredients for latex paints to minimize each of these factors, in addition to trying to achieve the normal durability, flexibility, color retention, and mildew resistance expected of all house paints. It is no wonder that so-called "bed sheet" formulas evolve. Typical formulations for an acrylic latex house paint from the Rohm and Haas Company and for a similar vinyl acetate house paint from Grace Organic Chemicals follow (Formulations 5-4 and 5-5).

(1) Latex. There is a formidable array of latexes offered to the paint industry for use in exterior latex paints. Generally they are based on a combination of monomers designed to give polymers with the correct glass temperature for coalescence at ambient temperatures with suitable film hardness to avoid tackiness and dirt collection in the film. Hardness is gained through the use of such monomers as styrene, α-methyl styrene, methylmethacrylate, vinyl acetate, vinyl chloride, or acrylonitrile. These are copolymerized with "softer" monomers in proportion to give suitable glass temperatures.

Formulation 5-4. Alkyd-modified White Topcoat with Zinc Oxide: Rohm and Haas Formulation W-88-7A

Materials	lb	gal
Hydroxyethyl cellulose (2.5% solution)	85.0	10.30
Water	62.5	7.50
Tamol 850 (30%)	10.5	1.05
Triton CF-10	2.5	0.28
Potassium tripolyphosphate	1.5	0.07
Antifoamer	1.0	0.13
Ethylene glycol	25.0	2.69
Preservative (57% mercury)	0.1	0.01
Nonchalking rutile titanium dioxide	237.5	7.22
Zinc oxide	50.0	1.07
Talc	187.7	8.11
Grind the above materials in a high speed mill (Cowles, 3800 to 4500 ft/min for 20 min) and let down, at a slower speed, as follows:		
Rhoplex AC-388 (50%)	390.8	44.39
Long oil alkyd[a]	30.8	3.69
Antifoamer	1.0	0.13
Ammonium hydroxide (28%)	2.0	0.27
Tributyl phosphate	9.8	1.21
Propylene glycol	35.0	4.07
Water	65.1	7.81
	1197.8	100.00

Pigment volume content	40.0%
Solids: volume	41.0%
weight	58.6%
Alkyd: volume solids/binder volume solids	15.0%

Formulation 5-4. —Continued

Materials	lb	gal
Approximate pH	9.4 to 9.7	
Approximate viscosity (KU)		
Initial	70 to 74	
Equilibrated, unsheared	81 to 85	
Equilibrated, sheared	70 to 74	
Approximate raw material cost	$1.74/gal	

[a]Drier treated with 0.5% by weight of 6% cobalt, 0.5% of 6% manganese, and 1.4% of 24% lead, prior to incorporation of alkyd.

The more common latexes today can be roughly classified as follows:

Acrylics
Polyvinyl acetates
- Dibutylmaleate copolymers
- Ethylene copolymers
- Acrylate copolymers

The resins on the market in these categories are durable enough with the possible exception of some vinyl acetate/ethylene copolymers. The choice then must be made on the basis of cost and rheological properties. There does seem to be a correlation between flow and latex particle size, with flow improving with larger size latexes. Unfortunately, hiding power efficiency of titanium dioxide tends to decrease with increasing latex particle size.

The acrylic latexes tend to be the highest cost on a volume solids basis. They do have two distinct advantages over the vinyl acetate latexes, however. One is better resistance to efflorescence on masonry substrates. Efflorescence results from the penetration of the film and substrate by water, leaching out water-soluble salts from the substrate and deposition of these salts on the surface of the film. The better alkali resistance of acrylic latex paint appears to be the major factor in their better resistance to efflorescence. As the alkali resistance of vinyl acetate latexes improves, their efflorescence resistance also improves. The other advantage is more subtle. The acrylic latexes appear to be less critical in respect to anatase TiO_2 content. With vinyl acetate latexes the amount of anatase TiO_2 that gives the proper degree of clean-up in the Great Lakes area appears to chalk too rapidly in the Gulf of Mexico coast area and vice versa. This is a draw-

Formulation 5-5. Alkyd-modified White Topcoat with Zinc Oxide: Grace Formulation No. 1372 L

	lb	gal
Dispersion		
1. Water	100.0	12.00
2. Potassium tripolyphosphate	1.0	0.05
3. Surfynol TG	5.0	0.60
4. Victawet 35-B	6.0	0.64
5. Igepal CO-630	5.0	0.57
6. Foamicide 581-B	2.0	0.29
7. Ethylene glycol	20.0	2.16
8. Carbitol acetate	8.0	0.97
9. Zinc oxide XX-503	100.0	2.14
10. Titanox RANC	200.0	5.72
11. Ti-Pure FF	25.0	0.77
12. Mica 325 Mesh	25.0	1.06
13. Asbestine 325	50.0	2.14
Reduction		
14. 3% Cellosize QP-15,000 solution	100.0	12.00
15. Water	70.0	8.40
16. Everflex BG (52%)	345.0	38.76
17. Alkyd resin mix (Items 18-21)	115.0	13.76
	1177.0	102.03
Alkyd resin mix		
18. ADM 1271	550.0	65.87
19. Igepal CO-630	28.0	3.17
20. PMO-10	10.0	1.22
21. Advacar (24% lead)	8.0	0.83
22. Advacar (6% cobalt)	3.0	0.37

FORMULATION 5-5. —Continued

	lb	gal
23. Advacar (6% manganese)	3.0	0.37
24. Water	398.0	47.78
	1000.0	119.61

Physical constants

Pigment volume concentration	39.1% (BG to pigment)
Total solids	54.6%
Total volume solids	37.3%
Viscosity	85 ± 5 KU
Raw material cost (approximate)	$1.84/gal

Procedure

1. Combine items 1 through 8 in a Cowles Dissolver, then add pigments in order and grind.
2. Combine items under "Alkyd resin mix" (items 17 through 23) in a Cowles Dissolver and mix until homogeneous.
3. Add items under "Reduction" with slow agitation to the dispersion.

Raw material sources

3. Airco Chemical Company
4. Stauffer Chemical Company
5. Antara Chemicals Company
6. Colloids, Inc.
8. Union Carbide Chemical Company
9. New Jersey Zinc Company
10. Titanium Pigment Corporation
11. E. I. duPont DeNemours and Co., Inc.
12. English Mica Company
13. International Talc Company
14. Union Carbide Chemical Company
16. Dewey and Almy Chemical Division
18. Archer-Daniels-Midland Company
19. Antara Chemicals Company
20. Nuodex Product Company
21. Carlisle Chemical Works, Inc.
22. Carlisle Chemical Works, Inc.
23. Carlisle Chemical Works, Inc.

back for companies marketing a national product. This may not be as true for the vinyl acrylic latexes.

(2) Pigmentation. (a) Hiding. The selection of white hiding pigment for latex paint is not critical. Almost any standard chalk-resistant titanium dioxide pigment, either sulfate process or chloride process, is satisfactory. Although a slight advantage for the most chalk-resistant premium grades can be detected on exposures, this does not seem to be a sufficient advantage to justify the added cost. From the standpoint of hiding power efficiency, the selection of the dispersant/surfactant system, the process of dispersion, and the rheology affecting application and leveling are all more important than the selection of a particular titanium dioxide pigment for particle size distribution.

Even though the total solids and the vehicle solids are normally considerably lower for latex paints than solvent paints, the number of pounds of TiO_2 per gallon of paint is usually higher. This additional TiO_2 in latex paints is an attempt to make up for poorer application (too easy and therefore spread too far) and flow properties in latex paints. High quality latex house paints generally use 2.00 to 2.50 lb of TiO_2/gal. This results in a hiding pigment (TiO_2)PVC of about 24%. Alkyd paints with approximately the same number of pounds of TiO_2 per gallon have a hiding pigment volume concentration (HPVC) of about 17%, and oil paints have an HPVC of only about 10%. Since mono-dispersed, optimally spaced TiO_2 has the highest efficiency of hiding power at HPVCs of 10% or less, and maximum absolute hiding power at an HPVC of 30%, it is evident that hiding power can be increased more in solvent paints than in latex paints by adding more TiO_2 (56). Thus, we find a maximum of about 3.50 lb/gal of TiO_2 in premium latex paints (HPVC ~ 30%) but 3.50 lb or more in premium solvent paints (HPVC ~ 21% in Federal Specification TT-P-103, proposed).

(b) Extenders. Some of the properties of extender pigments are discussed in the section on oil (Sec. 5-3, A, 1). Generally the same remarks apply to latex house paints also. However, when evaluating extender pigments in latex house paints, failure to consider changes in CPVC can have much more drastic effects than in oil or alkyd paints. This is mostly because latex house paints are generally formulated to be flat and, therefore, have PVC/CPVC ratios closer to 1.0 and closer to the critical area where properties change drastically with PVC/CPVC ratio (see Fig. 5-3). In latex house paints it is even more important to know the CPVC and the change in CPVC when making formula changes and to take this change into account when setting the PVC.

Unfortunately, it is difficult to determine the CPVC of latex paints. The cell method proposed by Asbeck and Van Loo (32) will not work since the latex particles in the liquid paint act as pigment particles. Becker and Howell (57) investigated various properties of PVC ladders of latex paints to find one property that would have a suitably sharp inflection point to be used to determine the CPVC. They rejected gloss and permeability since gloss tends to be low and permeability high for all latex paints. Other properties that they investigated were tensile strength, elongation, resistance to scrubbing, removal of stains, and exterior or Weather-Ometer durability. They found that tensile strength was the property with the sharpest inflection point and advocated its use to determine CPVC. Berardi (58) also investigated PVC ladders of latex paints for the same purpose. He tried tensile strength, elongation, scrub resistance, and permeability. He advocated the use of tensile strength and elongation as the preferred properties to measure in determining the CPVC.

Stieg (59) suggested the use of spatula rub-up oil absorptions on individual pigments as a simple means of determining the approximate CPVC of solvent paints. The Philadelphia Paint and Varnish Club (60) carried on this work using binary pigment mixtures in an attempt to refine Stieg's work. Liberti and Pierrehumbert (61) discarded oil absorption as unsuitable for latex paints. They substituted water sorption, determined more like a Gardner-Coleman oil absorption. They found agreement within ± 3% of CPVCs determined by water sorption and those determined by tensile strength, elongation, or scrub resistance.

More recently, Cole (62) has shown that the measurement of the specific gravity of a PVC ladder of latex paints can be used to determine the CPVC. This method is perhaps the most direct method according to the definition of CPVC offered by Asbeck and Van Loo (32), "It is that point in a pigment-vehicle system at which just sufficient binder is present to fill completely the voids left between the pigment particles incorporated in the film after volatilization of the thinner." The specific gravity method plots the ratio of the calculated specific gravity of a film to the measured specific gravity against the PVC. That PVC at which this ratio shows a sharp upward inflection (voids in the film make the measured value less than the calculated value) is taken as the CPVC. Although this is a direct, accurate method, it is not used much because of the tedious, delicate techniques that are necessary.

Still another method that depends on the change in a property with a change in PVC has been discussed by Shaller (63). This method depends on the fact that hiding power from light scattering decreases in efficiency as the PVC increases up to the point where voids start to be present in the paint film. The voids add to light scattering and hiding power and, therefore, make the hiding efficiency increase. This PVC is the CPVC. This is also a tedious but accurate method. Salt spray resistance of a PVC ladder of latex paints applied to cold rolled steel panels is a much less tedious method and perhaps is accurate enough. Since latex paints in which the PVC is greater than the CPVC corrode very rapidly (a few hours in the salt spray cabinet), it can be accomplished quite quickly without too many hours of actual work.

(c) Colors. The remarks on color pigments under the section on oil apply equally well to latex paints. There is one added hazard with latex paints, however. This comes about in oil- or alkyd-modified latex paints. They have a liquid (water) external phase and another liquid, but internal phase (the oil or alkyd). The pigments are normally dispersed in the water phase. However, organic pigments tend to like the organic (oil or alkyd) phase better. This can lead to migration of the organic pigments into the organic phase and subsequent loss of tinting strength of the organic pigments. The result can be a color drift both during the tinting operation because of the agitation and in the package. Minimizing the agitation during tinting, increasing the polarity of the organic phase, or decreasing the polarity of the water phase by the proper choice of coalescent can be used to minimize this problem.

(3) Chalk Binding. The necessity for using a liquid film former in latex paints for attaining adhesion on chalky substrates has already been introduced. However, the selection of a particular liquid resin and the determination of the proper quantity of such liquid resin deserves some discussion. As with most formulating of paints, the use of resin modifiers in latex paints entails compromise. That is, the attainment of chalk adhesion requires a sacrifice of other properties, especially resistance to mildew growth of the dried film and color retention of the dried film. Mildew resistance is probably the most important of these sacrifices in performance.

Originally unbodied oils such as linseed or tung were used as modifiers, usually at a rate of one quart per gallon of paint. However, this was found to cause discoloration (yellowing) in protected areas and allowed excessive mildew growth, so not much paint was marketed with this type of modification. Very long oil alkyds (85 to 90%) were tried next. These had such low viscosities that they could be used at 100% NVM. The 100% solids was felt to be an advantage since the normal reducing solvent, mineral spirits, was thought to be so nonpolar as to be difficult to keep emulsified in the latex paint. These very long oil alkyds were found to be quite efficient in giving chalk adhesion and to detract from mildew resistance much less than oils. Many commercial latex house paints use them today. However, they do detract from color retention and mildew resistance. The prohibition of mercurial compounds as fungicides will put more pressure on improving the inherent mildew resistance of such additives.

Generally a tape adhesion test along the lines of Federal Test Method 630 . has been used to test the efficacy of various modifiers for latex paints. It is the author's opinion that this test is too severe, particularly if the paint is applied to the chalky substrate by drawdown rather than brushing. The tendency has been to formulate paints to pass this type of test rather than to do only well enough in this test to eliminate failures in the field. Actually latex paints based on small particle size latexes (~0.4μ) do reasonably well on most chalky substrates without any modifier, and some have been used this way. Therefore, almost any improvement over the performance of unmodified latex paints in tape test results is sufficient to get good field performance. Since the side effects of modifiers are deleterious to performance (mildew resistance, color stability during production and storage, and color retention during exposure), it is logical to use the minimum amount of modifier.

Water-dispersible or water-soluble esters of polyallyl alcohols have been advocated as latex paint modifiers because of better mildew resistance. Considerable testing in the author's laboratory has met with variable success. It does appear that with some esters in some formulations better mildew resistance can be attained. Mention should also be made that the possibility of designing latexes that in themselves have better chalk-binding properties has been pursued by some latex suppliers.

(4) Thickener. Since latex paints attain most of their low shear viscosity through the use of thickeners, they are discussed as a separate subject rather than being lumped together with additives. As mentioned in the section on rheology (Sec. 5-3, A, 2, a), the low shear rheology of paints determines, to a large extent, the flow properties. This makes thickener selection of critical importance.

The first latex paints designed for use as interior wallpaints used protein thickeners such as casein. They were quite good for flow, but suffered from susceptibility to both bacterial and fungal attack. Therefore, most latex house paints have used cellulosic thickeners such as hydroxy ethyl cellulose or methyl cellulose. Generally the higher molecular weight cellulosics (higher viscosity grade) have been preferred. Polysaccharides and water-soluble acrylic polymers have been used to a far lesser extent.

The mechanism by which thickeners function has not been completely determined. Some hold that thickeners function by bridging between latex or pigment particles. Others feel that they function more by intertangling with themselves. Lalk (64) has discussed this subject and the interaction of thickeners with ionic and nonionic surfactants. He favors the intertangling theory since the thickeners are effective in water without latex or pigment particles. However, it is also true that the fewer the particles in a latex paint the better the flow — as, for instance, with clears or paints made from large-particle latexes.

Thickeners do affect properties of latex paints other than those dependent on rheology. They can affect color acceptance and uniformity, water resistance, and the state of pigment flocculation in the package and in the dry film. There is a need to understand the mechanisms by which thickeners function both rheologically and as protective colloids. It is quite possible that significant advantages could accrue by designing new thickeners that would give improved flow, water resistance, and hiding.

(5) Additives. (a) Pigment-dispersing Agent. Usually anionic surfactants can and do act as pigment dispersants also, even though they are usually referred to as stabilizers or simply surfactants. Anionic surfactants are usually the sodium or potassium salt of a compound containing phosphate, sulfonate, or carbonyl groups, or of a polycarboxylic acid. The selection of the proper dispersant is difficult since there are no known methods of systemizing the search. Usually such a selection is simply a matter of trial and error. The judgment as to the efficacy of a dispersant includes not only the consideration of actual pigment dispersion, but also the stability of the dispersion and the ability to accept tinting colors.

Once, the dispersant has been selected, then the proper amount to be used must be determined. Too little dispersant will give poor dispersion stability; too much is apt to give foaming problems. One method proposed by Rohm and Haas can be useful. It is designated as the concentration-aggregation value. Briefly, it consists of making a solution of anionic dispersant with equal parts of nonionic surfactant in water. This is used to titrate a stiff pigment/water paste until it is just fluid or levels when it is disturbed. More of the surfactant solution is added slowly until a few drops of the fluid paste appear stable when mixed with a solution of an ionic thickener. From 30 to 35% of this final concentration of dispersant is considered optimum for paints. Rohm and Haas gives concentration-aggregation values for a number of pigments (65).

Reactive pigments can cause special problems of dispersion and dispersion stability. Zinc oxide is commonly used in latex paints, even though it is troublesome from this standpoint. The use of a phosphate dispersant, such as potassium tripolyphosphate or tetra potassium pyrophosphate, along with the dispersant has been found useful. There was a zinc oxide available from a supplier that had been treated with a phosphate (New Jersey Zinc's XX-505) that showed much better dispersion stability in latex paints. It is no longer available. But the use of a phosphate surfactant seems to accomplis the same job.

(b) Surfactants. The term "surfactant" is usually applied to the nonionic surface active agent that is used in a latex paint. However, surfactants are also used in the manufacture of the latexes themselves in order to solubilize the monomers and to prevent the coalescence of the latex particles. Unfortunately, latex suppliers do not usually reveal the composition of the surfactants that they use, so the paint formulator more or less ignores their presence. They do have considerable bearing on the tint acceptance of the final paint, however.

In paint formulation the surfactant is used to stabilize the dispersion in the package and to ensure that the surfactant demand of the pigment is satisfied. If this demand is not satisfied, the latex surfactant is apt to migrate to the pigment surface, and the system may lose its mechanical stability.

Stewart (66) has advocated the use of Atlas's HLB, or hydrophile-lipophile balance, approach to balance the dispersant/surfactant system of the paint with that of tinting materials to ensure good color compatibility. He feels that a single surfactant will rarely provide the proper balance, but that a combination of anionics and nonionics can be found with a proper balance. In this way, each pigment particle can be coated so that its surface is just like that of all other pigments, thus minimizing color float or rub-up.

(c) Antifoams. Unfortunately there do not seem to be any suitable combinations of dispersants and surfactants that do not result in foam problems in latex paints. Therefore, the use of antifoamers, or defoamers, is necessary. As with surfactants, there does not seem to be any systematic way to select antifoam agents. Consultation of suppliers' literature is helpful but some trial and error experimentation is necessary. The mechanism by which defoamers function is complex. Kitchener and Cooper (111) have covered the broad field of defoamers, while Hudson (112) has discussed aqueous paint systems specifically. The use of defoamers can easily lead to cratering, so care must be exercised that the minimum amount of antifoam is used.

(d) Coalescing Agents and Freeze-thaw Stabilizers. Latexes designed for use in house paints generally are formulated to have glass transition temperatures in the 5 to 10°C range. This is desirable in order to get films that are not too soft for exterior paints, but it is undesirable from the standpoint of film formation, particularly at low temperatures. Thus, it usually is necessary to use solvents that act as temporary plasticizers during film formation to assist in coalescing. These are called coalescing agents. They are strong solvents that are slowly volatile. Ether alcohols, tributyl phosphate, and pine oil have been used as coalescing agents.

Unfortunately, coalescing agents are also effective in the package. Thus, they promote instability through coalescence if the package freezes, particularly if it freezes slowly, allowing the latex particles to come in contact with one another. Thus, freeze-thaw additives are necessary. Generally, glycols are used. They are poor solvents for the latex but are readily water-soluble and lower the freezing point of water, as well as retard the removal of water from the latex particle. Ethylene glycol is commonly used at a concentration of 20 to 30 lb/100 gal of paint.

(e) Preservatives. There is an additional hazard with latex paints that does not occur normally with solvent paints, that is, attack by microorganisms in the package. The presence of water, the normally alkaline condition, and the presence of thickeners as food all are conducive to the growth of microorganisms. The effect with proteinaceous thickeners is an offensive, putrid odor; with cellulosic thickeners there is a loss of viscosity. The latter effect comes from the action of enzymes or chemicals produced by the microorganisms to help them assimilate their food. Enzymes are very difficult to destroy or neutralize and, therefore, it is difficult to recover a batch of paint that has become infested.

The best protection against microorganism attack is prevention through cleanliness. Most raw materials are received in a paint plant free from infestation. However, microorganisms like the bacteria *Pseudomonas* are very common and contamination will occur if conditions are right. Therefore, it is imperative to avoid accumulations of latex or thickener solutions. Drums or buckets used to handle thickener solutions should be scrupulously cleaned. Pipelines should be flushed out if they are going to remain idle for any length of time. Tank covers, as well as tanks, should be cleaned. Steam cleaning is advocated because it sterilizes as it cleans. Goll et al. (67) have described the danger of contamination of raw materials which contain water even in trace amounts if it is present as free water.

The most commonly used preservatives have been the phenyl mercurial compounds. Besides being effective at very low concentrations, they apparently are capable of neutralizing the action of enzymes on cellulosic thickeners. The EPA order, which was published in the Federal Register on March 29, 1972, bans the use of alkyl mercurial compounds and withdraws the registration of all other mercury compounds. This will eventually prevent the use of phenyl mercurial compounds in paint. Many nonmercurial preservatives are being offered to the trade, but their efficacy has not been tested in practical situations for long enough periods to make any firm judgments. A rather thorough laboratory study has been reported by the Chicago Society for Paint Technology (68). One preservative, Dow's Dowicil 100, a quaternary ammonium compound that releases formaldehyde slowly, has been on the market for some time. It does seem to be effective in sterilizing paints, but it is not effective in neutralizing enzymes already present.

(f) Fungicides. Like solvent paints, the dried films of latex house paints are subject to disfiguring mildew growth. Unmodified latex films on proper unprimed substrates are quite resistant to mildew growth. On the other hand, alkyd-modified latex paints over solvent-type primed or previously painted wood can be quite subject to attack by mildew, even though not to the extent of unprotected oil or alkyd films. Since these latter conditions are those under which latex paints normally are used, an effective mildew-control agent is very desirable.

Zinc oxide, which is usually called a fungistat, is probably the most commonly used agent in latex as well as in oil paints. Generally, somewhat lower levels are used in latex paints, however – usually 0.5 to 1.0 lb/gal. Phenyl mercurials have also been used extensively in spite of their tendency to give sulfide discoloration in protected areas of a structure. Zinc oxide often is not sufficiently active to prevent early mildew growth. As previously explained, government regulations probably will prevent the use of the phenyl mercurials. This leaves a need for an effective, long-lasting organic mildewcide in latex paints. There are many candidates on the market. Many of these are discussed in the section on mildew resistance (Sec. 5-3, D).

There is a problem of hydrolytic stability with fungicides in latex paints that prevents the use of certain fungicides that are effective in solvent paints. One such product is trans 1, 2 bis (n-propylsulfonyl) phthalimide. It is quite effective in solvent paints, but hydrolyzes very rapidly in alkaline latex systems. Others, such as 2, 3, 5, 6-tetrachloro-4-(methyl sulfony) pyridine, hydrolyze more slowly in alkaline latex systems and may be stable enough in nearly neutral or acid systems to be useful.

4. Others

House paints based on other than solvent (oil and alkyd) or latex have received some attention, but have not made much headway in the market place. In the early days of latex house paints, there was considerable effort on the part of oil suppliers to introduce emulsified oil and alkyd-based house paints. These suffered from the effect of water leachable antioxidants in certain wood substrates and from an inability to avoid "flashing" during application. There have also been attempts to promote more durable vehicles such as silicone-modified alkyds or epoxidized oil/polyester combinations. The silicone alkyds were high priced if they contained sufficient silicone to really show an advantage in durability (25% silicone). The epoxidized oil/polyesters suffered because their fast reactivity required two packages.

B. CLEARS

The natural beauty of wood is desirable to many people; they would prefer protective finishes that do not mask this beauty. Transparent stains are one way of accomplishing this. They are covered in a separate chapter. Clear finishes are another way to show off wood. Unfortunately, clear finishes on wood have shown very poor durability on the exterior. This poor performance has led to a disenchantment of consumers so that the market for clears for exterior wood is quite small at the moment.

Most investigators in the field of clear finishes on wood have concentrated on finding resins that are durable when not protected by pigments. For instance, Ashton (69) studied various clear finishes and found tung/phenolic varnishes along with a cold-mixed varnish and a castor-cured urethane to perform best, but even they last only two years. The Golden Gate Society (70) investigated 31 different clear varnishes based on catalyzed epoxies, polyurethanes, and polyesters. They found all to be unsatisfactory. The Philadelphia Society (71) attempted to protect various clear varnishes (tung oil, oil-modified urethane, and alkyd) with 11 different organic ultraviolet absorbers. They found that specific absorbers helped speeific varnishes but could not correlate performance with ultraviolet absorption of the clear coating. Rothstein (72) found a substituted benzotriazole to be the most effective organic ultraviolet light absorber in protecting the resins in clear coatings.

However, according to work in the author's laboratory and work published more recently, the real problem with clear finishes on wood is destruction of the wood/coating interface by ultraviolet light. Miniutti (73) showed that although silicone varnishes did not themselves degrade, they failed on wood by the destruction of the top layer of the wood substrate. He details the physical destruction of various parts of the wood cell structure by ultraviolet light. He shows the formation of diagonal microchecks through pits that are concentrated in the area of the junction between spring and summer growth, which accounts for the onset of failure in this area. He advocates the use of ultraviolet absorbers in the coating to protect the wood surface. Schneider (74) refers to an earlier unpublished research report (1965) in which it was shown that with low ultraviolet-absorbing clear finishes the wood/coating boundary separated, but with more highly ultraviolet absorbing finishes, the finish itself deteriorated.

Unpublished work in the author's laboratory has shown that organic ultraviolet absorbers alone in a clear finish cannot protect the wood substrate completely, especially for extended periods of exterior exposure. Presumably ultraviolet light destroys some of the organic structure. The inclusion of just sufficient inorganic ultraviolet absorbers to ensure complete screening of ultraviolet light from the wood substrate has been found to give "clear" finishes with excellent durability and good appearance in the dry film. Since these inorganic absorbers are usually iron oxide pigments, the wet finish is colored and looks like a paint.

C. PRIMERS

A paint primer has two basic functions. The most fundamental and important function is to attain and maintain adhesion to the substrate. The second primary function is to seal or uniform the substrate. Unfortunately, primers are also expected to contribute to the hiding of a paint system. Of course, there are numerous other properties of a primer that influence the performance of a house paint system and, therefore, must be controlled by formulation. Some of these properties are moisture sorption/desorption characteristics, mechanical properties, ability to serve as a food for fungi, and freedom from water-soluble material.

1. Adhesion

Generally, there is no problem in attaining initial adhesion with house paint primers. Zisman over the years (summarized in Ref. 75) has promoted

Young's equation, $\gamma_{SV} - \gamma_{SL} = \gamma_{LV} \cos \theta$, and the importance of surface tensions of the substrate and the coating for obtaining spreading and adhesion. Gans (76), Gray (77), and others have argued with Zisman on the applicability of this concept for adhesion prediction. However, for house paint primers such arguments are probably superfluous. Other influences such as non-uniformity of the substrate because of weathering and changes in the mechanical properties of the coating as a result of weathering are so great as to overshadow such laboratory-type measurements.

a. Weathering of Substrates

Weathering of substrates, particularly wood (78), generally leads to the formation of weak, loosely bound material at the surface of the substrate. One illustration of this is shown in Fig. 5-2. In this case 12 matched, unexposed Southern pine pattern 108 siding boards were furnished in sealed containers by the Southern Pine Association. One set of two heart-side panels and two bark-side panels was painted with two different oil house paint systems. A similar set of four panels was exposed on a vertical test fence facing south for one week in the fall in Chicago -and then painted with the same two systems. A third set of similar panels was exposed for one month before painting. All panels were then exposed on a vertical test fence facing south for four years in Chicago.

It can be readily seen that exposure for as little as one week under relatively mild conditions has a definitely deleterious effect on the maintenance of adhesion over extended periods of time. Exposure for one month has a very deleterious effect on the maintenance of adhesion. Similar experiments with other types of forest product substrates has demonstrated similar but usually less severe effects of weathering. Douglas fir shows an effect about like that of southern pine. Cedar, hardboard, and paper overlay plywood take longer (6 months versus one month) to degrade to the same extent.

b. Penetration

There is a widespread belief, particularly among painters, that the way to overcome the deleterious effect of weathering on paint adhesion is to add extra oil to the primer so as to penetrate the weak surface layer and bond to sound substrate material. Experimental evidence has shown that this does help, but only marginally. A far better way to correct the situation is to sand the wood surface, even lightly, before priming. Or, if this is not feasible, considerably better performance over weathered woods can be obtained by using an all latex system. Presumably, the reason for improvement with such a nonpenetrating system is that the latex system shrinks less during exposure and, therefore, puts less stress on the adhesive bond.

c. Mechanical Properties.

The shrinking of topcoats during exposure is the primary "change in mechanical properties" referred to previously. Any coating that continues to polymerize or oxidize during exposure (such as oils or alkyds) tends to shrink. If the coating does not have a sufficient ability to relax, such shrinkage can put a tremendous stress on the primer and the adhesive bonds. Newman (79) pointed this out as far back as 1939. Long and co-workers demonstrated one method of minimizing shrinkage of alkyd vehicles by a "pre-oxidizing" technique (117-119). Pierce (80) re-emphasized this point in 1966 when he made the point that primers and topcoats should be "matched" in torsion modulus values. Latex films, with their pre-polymerized, non-oxidizing vehicles tend to put less stress on adhesive bonds. Latex films, however, do not have the same torsion modulus as solvent-based films and should not be followed by solvent-based topcoats when they have been used either as primers or as repaint materials.

Zisman (75) has pointed out that where incomplete wetting of a substrate occurs, it is better to have a rough substrate to avoid the possibility of having a weak plane that could peel off because of an "unzippering" effect when stress is applied. Perhaps an intuitive feeling by formulators has led to the generally accepted concept that primers should be relatively low in gloss rather than glossy and smooth. But the avoidance of Zisman's "unzippering" effect does not require completely flat finishes. In fact, flat primers should be avoided because they tend to have very low flexibility (42) and, therefore, have little ability to relax. A PVC/CPVC ratio of 0.6 to 0.7 has been found advantageous with solvent-type primers for general use and even lower for use on woods that tend to check such as Douglas fir plywood. Ratios of 0.6 to 0.7 result in a 60° gloss of 15 to 25. This has been found to be satisfactory.

2. Hiding

House paint primers are expected to maintain adhesion for the life of the structure. For this reason it is unfortunate that they are also expected to contribute to the hiding of a paint system. Primers do have the advantage of being protected by topcoats during their service life. But this is only partial protection where the primer contributes part of the hiding of the system, since some of the ultraviolet light does penetrate the topcoat and attacks the primer. It would tend to improve the long-term adhesion of primers if they were formulated to be a contrasting color from the topcoat. This would encourage the application of sufficient topcoat to give better protection of the primer from ultraviolet light. It would also indicate the need for repainting before the topcoat had eroded to the point of giving almost no protection to the primer. However, the demands of the market place have not allowed such an approach to primer formulating as yet.

3. Sorption/Desorption

In the section on metal substrates (Sec. 5-2, C), mention is made of the work by Hay et al. (14) on the effects of the moisture sorption/desorption on corrosion resistance. This same characteristic plays a role in mildew resistance. Thus, if the rate of desorption of water is low, the coating system tends to stay wet for longer periods of time. Since moisture is necessary for mildew growth, the susceptibility to mildew will be greater with such slowly desorbing systems. This is one of the major reasons why oil primers are poorer than alkyd primers, which in turn are poorer than latex primers for mildew resistance. Normally, it is thought that this progression is because of the ability of the mildew to use these particular film formers as food. Although the nutritional value undoubtedly plays a part, the moisture sorption/desorption characteristic is also important.

This means that primers should not be formulated so as to be able to hold water better than subsequent topcoats. In turn, this means that primers should not have substantially higher PVC/CPVC ratios than topcoats nor should they be more hydrophilic than topcoats.

Since we are talking about mildew resistance of primed systems, it would be well to mention that, other factors being equal, it is best to divide any fungicide that is used in a house paint system between the topcoat and the primer rather than to put it all in the topcoat.

4. Blistering

Blistering and subsequent peeling of house paints has been an interim type of problem. When lead and oil systems were used as the house paint system, blistering was a very minor problem. With the introduction of zinc oxide as first a white pigment and then as a reactive pigment, this problem increased. However, blistering has decreased as a problem since the early 1950s when it was realized that water-soluble zinc compounds were the major hazard in paint films. Since that time there has been little usage of zinc oxide in primers, and blistering has substantially decreased. Improved use of vapor barriers in building construction has also helped to alleviate the problem. Perhaps it is sufficient to say here that water-soluble materials and ingredients that can form water-soluble materials should be kept out of primers.

D. MILDEWCIDES

Several comments have already been made on the mildew resistance of solvent-type and latex-type house paints and the influence of primers. This section will cover some comments on the influence of chalk on mildew resistance, the problem of permanency of mildewcides, the influence of particular type of wood substrates, and the influence of protected areas of exposure.

1. General

a. Chalk

White house paints have been formulated with a percentage of anatase TiO_2 for many years in order to get self-cleaning properties. This chalking, even in trace amounts, is influential in obtaining mildew resistance of a paint.

The use of materials, such as barium metaborate, that severely reduce the amount of chalking is difficult for this reason. In addition, some fungicides, such as the arsenicals and some of the sulfur/nitrogen-containing organics, get part of their efficiency as fungicides from the fact that they increase the degree of chalking. For this reason, careful chalk ratings by weight loss should always accompany mildew ratings when assessing the efficacy of fungicides.

B. Permanency

Hoffman, as early as 1960 (81), suggested that the reason phenyl mercury compounds did not give lasting protection against mildew was that they were inactivated by sulfides, either in the form of pigments, or as a gas in the atmosphere. In subsequent work (82-84) he found that all phenyl mercury compounds tended to decompose, even with sulfides, and disappeared from the films by volatilization or water leaching. He further found (85, 86) that other organic fungicides also disappeared from paint films quite rapidly particularly under humid tropical conditions. He summed up his work in 1971 (87).

Thus, Hoffman has delineated one of the major problems in the protection of house paints from mildew - the inability to find fungicides that remain in a paint, either because of activity in the wet paint in the package, or because of volatility, water solubility, or reactivity in the dry paint film. There are literally hundreds of organic and inorganic compounds on the shelf that are very effective against fungi that attack paint (primarily *Aureobasidium pullulans*), but very few that are effective in the dry paint film on the exterior for extended periods of time.

c. Substrate

It is generally recognized that cedar is quite mold-resistant by itself and that pine is not. Fig. 5-5 shows that paint films applied over these substrates also show varying degrees of mildew resistance, depending upon the substrate (and the type of coating). It is thought that not only the type of inhibitive chemicals found in the wood, but also the moisture-holding properties of the wood play a role here. Pine substrates have been found to be so variable in mildew resistance from panel to panel, and even within one panel, that they are not suitable for test substrates for mildew resistance of paints. Redwood has been selected as a compromise between severity and uniformity as a test substrate.

Masonry substrates are generally considered to be less severe than wood substrates for mildew resistance of paints applied over them. The CDIC Paint Society (88) has pointed out that fungi prefer an acidic environment. It has been thought by many that the fact that masonry surfaces tend to be alkaline accounts for their better mildew resistance. It is proposed that two other factors play an important role. First, the latex paints that are gen-

FIG. 5-5. Exposed for one year in Lousiana.

erally used on masonry substrates now are usually not used over a primer, whereas on wood substrates they usually are used over a primer. Since primers tend to both furnish nutrient for mildew and act as a moisture reservoir to keep the system wet, the lack of a primer on masonry substrates tends to give better mildew resistance on such substrates. Second, the masonry substrate itself tends to act as a moisture sink and, therefore, to keep paints applied on such substrates dry. This again would favor the paints applied over masonry substrates for mildew resistance.

2. Specific Mildewcides

Over the years, literally hundreds of compounds have been evaluated as paint fungicides in the author's laboratories. In all cases this was by exterior exposure in multiple locations, and in most cases the fungicides were tested in both solvent and latex systems. Table 5-3 gives a rough indication of the results on the more important compounds. It should be

TABLE 5-3. Efficacy of Various Mildewcides[a]

Compounds	Activity	Comments
Organic		
Butadiene sulfone	C	
Butadiene polysulfone	C	
Biocide 203	C	
Chemocide	C	
Dowicides (phenolics)	B	
Cosan P, Fungitrol 11, Advacide TMP n-trichloromethyl thiophthalimide	A	Questionable stability to hydrolysis
Cosan S 3,5 dimethyl tetrahydro 1,3,5, 2H thiadiazine-2-thione	C	
Fungitrol Alpha 2,4 dichloro 6-(0-chloranilino)-3-triazine	B	
Diaphene TBS-80 mixture of di and tribrom salicylanilide	B	
Flurophene 3,5-dibromo-3'-trifluoromethyl salicylanilide	B	
Salicylanilide, Shirlan	C	
Dow SA-1013 (S-13) 2,3,5,6-tetrachloro-4-(methylsulfonyl) pyridine	A	
Irgasan FP 5,6-dichloro benzoxazolinone-2	C	
Irgasan BS-200 3,5,3',4' tetrachloro salicylanilide	C	
Onyxide 172 ethylbenzyl dimethyl alkyl ammonium cyclohexyl-sulfamate	C	
Karathane dinitro-1-methyl heptyl phenyl crotonate	C	
Mergal 10-S (Riedel-deHaen)	C	

TABLE 5-3. — Continued

Compounds	Activity	Comments
Metasol TK-100 (Merck) 2-(4-thiazolyl) benzimidazole	B	
p-Toluene sulfonamide	C	
Sulfur (fine particle size)	C	
Bayer fungicide A-1 (Preventol)	C	
Biocide C-333 n-(3-chlorophenyl) itaconimide	C	
Tuex tetra methyl thiuram disulfide	B	
Vancide PA trans 1,2 bis (n-propylsulfonyl) ethylene	A	Poor hydrolysis resistance
Vancide 89 n-trichloromethyl thiotetrahydro phthalimide	B	
Captan 50-W 50% active material, similar to Vancide 89	B	
Industrial Biocide DS-2787 2,4,5,6, tetrachloro-isophthalonitrile	A	
Proxel CRL (ICI America) aqueous amine solution of 1,2 benzisothiazolin-3 3-one	C	
Skane M-8 2-n-octyl-4-isothiazolin-3-one	A	
Amical 77 p-chlorphenyl diiodomethyl sulfone	B	
Mercury compounds		
Phenyl mercury acetate	B	
Phenyl mercury propionate	B	
Phenyl mercury saccharinate	C	
Phenyl mercury borate	B	
Phenyl mercury lauryl mercaptide	B	
Phenyl mercury-2-ethylhexyl maleate	B	
Chlor methoxy propyl mercury acetate	B	
Pyridyl mercuric acetate	B	
Phenyl mercury diphenyl acetate	C	

TABLE 5-3. — Continued

Compounds	Activity	Comments
Ethyl mercuric thiosalicylic acid	C	
Phenyl mercury oleate	B	
Phenyl mercury dodecenyl succinate	B	

Most mercurials give appreciable protection which is roughly proportional to the content of metallic mercury. New pollution standards and impending legislation will eliminate mercurials entirely within a short time. Sulfide staining is a problem in exterior paints.

Zinc

Compounds	Activity	Comments
Zinc oxide	A	
Zinc borate	B	
Zinc pentachlor phenate[b]	B	
Zinc ethylene bis dithiocarbamate (Dithane), (Dianol)	B	
(Dithane M-22 related Mn compound) (Maneb)	C	
Zinc methane arsonate	C	
Zinc pyridine thione	B	
Zinc dimethyl dithiocarbamate	C	

Zinc oxide is an effective fungistat which has proved quite useful in many formulations, but may cause instability in latex paints. Many zinc compounds show at least some degree of activity.

Organo arsenic

Compounds	Activity	Comments
10,10'-Oxy-bis-phenoxarsine (Dow ET 546)	A	
Arsenoso benzene	B	
8-Hydroxyquinoline arsanilate	B	
Arsanilic acid	C	
Carbarsone	C	
Phenyl arsonic acid	C	
3-Nitro-4-chloro phenyl arsonic acid	A	

Dow ET-546 gave good protection to latex paints. Compounds rated "B" also gave a useful degree of protection. Due to toxicity, arsenicals are of limited interest as mildewcides.

TABLE 5-3. — Continued

Compounds	Activity	Comments
Organo tin		
Tributyl tin oxide	B	Poor package stability
Tripropyl tin oxide	B	Toxic, odorous
Carcide T-10 tributyl tin neodecanoate	C	
Tin compounds showed fair activity in some cases. Tributyl tin oxide had poor package stability. Tripropyl tin oxide showed fairly good activity but is quite toxic, difficult to handle, and has a disagreeable odor.		
Copper		
Cuprous oxide	A	Colored, chalks
Copper-8-quinolinolate	A	Colored, chalks
Copper aceto arsenite (Paris green)	B	Colored, chalks
Copper methane arsonate	B	Colored, chalks
Copper antimonate	C	Colored, chalks
These are usually quite effective but limited in use to a few specialty products due to color, toxicity, and side effects, such as chalking.		
Borate		
Barium metaborate	A	
Zinc borate	B	
Glycol borates	C	
Potassium fluoborate	C	
Barium metaborate has proven useful in some applications, but has limited use in other areas. Glycol borates showed no activity.		

[a]Activity is rated A, B, C on the following basis: A is good to fair; B is fair to poor; C is poor to none. The ratings are only approximate values and do not take into account the various side effects encountered in certain paint systems.

[b]Tends to discolor in latex paints.

noted that there are 11 organic materials containing both sulfur and nitrogen. As a group these materials perform the best as mildewcides, averaging "B" activity.

5-4. TESTING

Progress in improved composition and performance of house paints has been understandably slow. This has been due to the large monetary risks involved if an incompletely proven house paint is introduced. Not only can there be large inventories built up at the retail outlet level, but also it is possible that large numbers of buildings can be painted before weaknesses in a new product become apparent. If laboratory or accelerated test methods could be developed that would adequately prove a house paint's performance, it would be possible to make considerably faster progress. Although much effort has been expended toward developing laboratory test methods, completely adequate ones do not exist in many areas.

A. LABORATORY TESTING

1. Package Properties

Testing that is done on liquid house paints is generally for the purpose of quality control during the manufacturing process rather than for the purpose of predicting performance. Most of these are listed in ASTM Designation D 2833-69. "Standard Index of Methods for Testing Architectural Paints and Coatings." As an example ASTM Designation D 1475 gives a method for determining the density of a paint using a weight per gallon cup. Generally, as the pigment loading of a paint increases, the density increases. Since pigments can be fillers or prime pigments, the density has no consistent relationship with performance, except within one paint type. From the standpoint of quality control, however, density is very useful in assuring that the specified ingredients actually were used.

"Fineness of Dispersion of Pigment-vehicle Systems," ASTM Designation D 1210, describes a gauge that allows the determination of the film thickness at which pigment particles begin to protrude through the wet film. Since the gauge uses an inclined trough that goes from 0 to 100 μ with 10-μ graduations, it is far too coarse an instrument to really measure dispersion (TiO_2, for example, has a particle size between 0.2 and 0.3 μ). But this method is useful to assure that approximately the same dispersion is obtained from batch to batch. It is also true that no particular micron "grind" can be said to be correct for house paints. A 2 Hegman (75 μ) might be satisfactory for a lead/titanium/zinc/oil house paint, whereas a 6 Hegman (25 μ) might be better for a titanium/alkyd house paint. Both paints could give satisfactory performance.

There are, of course, many other properties of liquid paints that can be tested which must be satisfactory from a practical standpoint such as settling, foaming, freeze-thaw stability, resistance to bacteria, and the presence of coarse particles. These are usually self-evident and will not be covered in detail here.

2. Application

This section could also be headed "Rheology" since it is the rheology of a liquid paint that determines its application and flow properties. Since application and flow are so critical and are determined by formulation, this subject has already been covered, particularly under the sections on alkyd and latex (Secs. 5-3, A, 2 and 5-3, A, 3).

3. Appearance

House paints, like most other paints, have an aesthetic as well as protective function. Since "aesthetic" properties tend to be judged subjectively and, therefore, somewhat differently by different individuals, it is important to have nonsubjective laboratory test methods. Color, gloss, uniformity, and hiding are the most important appearance properties. Nonsubjective laboratory test methods have been developed for each of these except uniformity. Color measurement is a science unto itself and is covered in a separate chapter.

Gloss is generally measured as the specular reflectance of light from a planar paint film. The common angles of incidence and reflectance that are used in the paint industry are 60° or 85°. A 60° gloss generally indicates the "shininess" of a surface, while 85° gloss indicates the low-angle sheen. Instruments for measuring gloss at other angles are available, but are seldom used. Goniophotometers are also available for special purposes. Flat paints generally have an 85° sheen of less than 10, semigloss paints have a 60° gloss of 20 to 70, and gloss paints are usually 80 or above.

Uniformity is important from an aesthetic viewpoint. As mentioned, there are no objective laboratory methods for measuring uniformity. The lack of uniformity generally results from penetration of the binder into the substrate—either unpainted wood, masonry, or rusty metal—or into the old, chalky paint film, or even into lapped areas of the new paint. Stieg (89) has pointed out that all types of paints show the least tendency for binder penetration into substrates when the PVC equals the CPVC. Of course, binders themselves show varying tendencies to penetrate, depending upon their surface tension (wetability), viscosity, and rate of viscosity increase (drying). Latex paints in general show very little if any penetration into many substrates, but apparently do penetrate into themselves as evidenced by nonuniformity at lapped areas.

Hiding is the essence of what a paint film does from an aesthetic viewpoint. Hiding power seems like an obvious property with simple side-by-side comparisons being adequate for determining relative hiding powers. Indeed, ASTM D 344 describes just such a method, where the test paint and a control paint are brushed out on a black and white background at prescribed spreading rates with relative hiding subsequently judged visually. Such a method is fine for quality control purposes but more quantitative, accurate methods are needed for development purposes, particularly in the evaluation of titanium dioxide pigments and of dispersion methods.

The human eye can detect only a 5 to 10% difference in hiding power. It could be (and has been) argued that if the eye cannot see a difference then the difference is not of practical significance. But if it is considered that 5% of the TiO_2 used by the paint industry in a year is well over $100,000,000 in value then it is evident that a 5% difference in hiding efficiency is important. For this reason, methods like ASTM D 1738 and further refinements have been developed.

D 1738 specifies application of uniform films (film thickness measured by weighing) and instrumental measurement of reflectances to the fourth significant figure. It also specifies a correction for light absorption so that the light-scattering efficiency of TiO_2 can be determined. Mitton et al. (90) further refined this method by using Carrara black glass substrates and weighing the dry film to determine film thickness. He also simplified the calculations by presenting charts from which SX could be read directly. Bruehlman and Ross (91) developed a still more accurate method based on transmission rather than reflectance and using Mylar tape as a substrate with weighing as the method for determining film thickness. All of these refinements have lead to a possible accuracy of ± 1%, which is about the limit of accuracy of the Kubelka-Munk theory on which all are based.

4. Predicting Performance

It is in the area of predicting performance from the standpoint of retention of appearance and durability that the most trouble is encountered with laboratory tests. Unfortunately, it also is the area where laboratory tests are most needed, since service-type tests are so long and drawn out. Probably it is the very fact that service tests are long and drawn out that make them difficult to reproduce in the laboratory. Diffusion in a dried paint film can be important to performance, but diffusion is a difficult thing to accelerate without distortion. In addition, the chemical reactivity leading to oxidative polymerization, and cross-linking or chain scission is difficult to accelerate without changing its type. These and similar factors, plus the complexity of conditions to which house paints are subjected, make the whole picture of predicting performance very complex and difficult.

In fact, the area of predicting performance of exterior house paints is so complex that it is difficult to organize for discussion. One type of organization would be simply to discuss the different kinds of apparatus that have been developed and are being used to simulate part or all of the elements present in the field. These would include ultraviolet cabinets (ASTM designation D 822-68), blister boxes (ASTM designation D 2366-68), tropical chambers (92), sunlight-concentrating instruments (93), refrigerated exterior panels (94), and temperature change devices (ASTM designation D 1211-68). The problem with this type of approach is that generally the correlation with exterior performance has not been good. This is particularly true with light- and water-exposure apparatus (95-101).

Another type of organization would be to select a specific characteristic that affects performance and to measure that characteristic. Examples of such characteristics are ultraviolet absorption by polymer molecular structures (42), resistance to chemicals that are apt to be in the atmosphere such as ozone (102) or sulfur dioxide (103), the rate of water sorption and desorption by paint films (14), and the magnitude of a paint film's elastic modulus (104). In these cases there generally can be good correlation, but with only a narrow range of formulations. Ultraviolet absorption, for instance, works quite well for clear coatings, but is complicated when the coatings are pigmented. Resistance to sulfur dioxide is pertinent only when pigments that are susceptible to SO_2, like chrome green, are present. Resistance to ozone results can be justified only when a priori knowledge of other influencing factors exists. The rate of water sorption and desorption is pertinent for corrosion and mildew resistance, but has only a side effect on the resistance to ultraviolet light.

Still another type of organization is by mechanisms of failure and a discussion of test methods that have been designed around such mechanisms. Examples of such mechanisms are the loss of intercoat adhesion in protected areas on a structure (26), the cracking of a topcoat over a soft primer (79), cracking because of cross-linking or chain scission (46), the loss of corrosion or mildew resistance because of a tendency to absorb water (14), erosion because of vehicle degradation on the surface from ultraviolet light (105, 106), and the loss of adhesion because of moisture from inside a structure or structural defects (9). Since the real goal of accelerated testing is to determine the propensity for failure, and since the mechanistic approach has the advantage of testing a priori principles, this is the organization that is followed here.

a. Cracking

There is no difficulty in designing paints that perform satisfactorily on houses in the initial stages. Failures come with age. Thus, it appears that if we could keep a paint film from changing with age we could control failures. A primary change that takes place in paint films is a reduction in the ability to relax and, therefore, a reduction in the ability to conform to dimensional changes in substrates, particularly wood. The chemical changes that lead to this reduction in relaxation can be cross-linking, chain scission, or a loss of plasticizer. The effect of these changes is summed up by stress/strain properties. Thus, if we can measure changes in stress/strain properties we can estimate the degree of chemical changes that have taken place in the paint film (46). Although flexibility, or strain, is the critical property in stress/strain from the standpoint of performance, tensile strength or stress, is the easiest to measure accurately. Therefore, tensile strength is the property we have chosen to follow to indicate chemical changes.

Most paint films tend to gain in tensile strength during their early life. But at some point in time, chain scissions, which reduce tensile strength, begin to predominate, resulting in an inflection point in a plot of tensile strength versus time of exposure. The sharpness of this inflection point, its location in respect to time of exposure, and the slope of the curve beyond the inflection point all serve to indicate the degree of chemical activity within the film. They, therefore, can be used to predict durability. Although many workers (95-101) have found difficulty in correlating Weather-Ometer exposures with exterior exposures when the results are judged by surface characteristics, the correlation has been good when judged by a bulk property like tensile strength (46). For instance, in one case in the author's laboratory, a group of about 20 experimental latexes in an exterior house paint formulation were ranked correctly by stress/strain measurements during 2000 hours of exposure in a Twin Arc Weather-Ometer as judged by five years of multiple exposures on wood in various exterior locations. In fact, one of the experimental latexes was chosen from the 20 on the basis of the stress/strain results and short-term (one-year) exposures and cost considerations for use in a major line of paints, with good results.

b. Mildew and Corrosion

Mildew and corrosion resistance of paint films are related in that both require the presence of water in order to occur. In fact, both seem to require about the same level of relative humidity (60 to 65% at room temperature) in order to occur at all. Accelerated laboratory tests for corrosion resistance (such as salt spray cabinets) and mildew resistance (such as petri dish tests or tropical chambers) are designed to maintain constant high levels of moisture. Exterior conditions, on the other hand, vary considerably in RH levels with alternate wet and dry periods. Paint films vary in their rates of water sorption and particularly in water desorption, depending upon the hydrophobic/hydrophilic character of the film itself and of the substrate (old paint, primer, wood, etc.). Thus, laboratory tests that maintain constant levels of moisture negate one of the characteristics of paint systems that is important for exterior performance, and, therefore, are often misleading. Measurement of water sorption/desorption rates gravimetrically under carefully controlled conditions can be used, therefore, to gain insight into corrosion and mildew resistance (14).

For instance, it is well known that oil primers tend to give much poorer mildew resistance under latex topcoats than do alkyd primers. Generally it is thought that this is due to the fact that they are a better source of nutrition for the fungi. However, another factor is moisture sorption/desorption characteristics. Laboratory measurements show that the oil/latex system sorbs water faster and to a greater extent than does the alkyd/latex system. But even more important, they show that the oil/latex system desorbs water considerably slower than does the alkyd/latex system. In support of this mechanism it has also been observed that if oil primers are weathered for about six months to a year before being topcoated with a latex, the system has mildew resistance about equivalent to an alkyd. The oil primer continues to be susceptible to mildew after such exposure if it is not topcoated, demonstrating that it is still good nutritive material for fungi. However, sorption/desorption measurements show that the weathered oil/latex system tends to dry out faster after being wet than does an unweathered oil/latex system. Thus, a weathered oil/latex system becomes similar to an alkyd/latex system in sorption/desorption properties and in mildew resistance.

c. Intercoat Peeling

Intercoat or tissue paper peeling is a failure that occurs primarily on protected areas of a structure such as under eaves, on carport ceilings, etc. The primary difference in these areas in comparison to the exposed areas of a structure is the fact that they are not washed by rain. This indicates that perhaps water solubles formed in or on paint films during weathering might be important. In addition, the fact that intercoat peeling occurs only in the northern part of the United States indicates that freezing weather is important. Thus, it is theorized that water solubles are leached out of a paint film in protected areas by dew and deposited on the surface. These water solubles remain at the interface when structures are repainted. Then, if dew or condensation wet the repainted surface and the water permeates the film to the interface, the water solubles could cause a loss of adhesion. Furthermore, if that water evaporates at normal temperatures, adhesion is regained through the surface tension of the water; but if the water sublimates during sub-freezing temperatures, adhesion cannot be regained and intercoat peeling results. A cabinet where dew formation, freezing, and sublimation takes place has been constructed. Testing the results of pre-weathered films in such a cabinet demonstrates that indeed the theory is correct (26). Subsequent use of the cabinet and related testing has shown that atmospheric SO_2 plays a large role in the formation of water solubles. Those ingredients in a paint film capable of forming soluble sulfates, such as zinc oxide and extenders based on calcium, magnesium, or other metals, are the primary contributors to the formation of water solubles and should be eliminated to avoid the problem of intercoat peeling.

B. TEST FENCES

Some of the earliest paint test fences were established in the very early 1900s at the North Dakota Agricultural College by the paint industry as a result of North Dakota's "pure paint" law, requiring any ingredient in a paint other than white lead or oil to be shown on the label as an adulterant. Interestingly, this led to the establishment of a paint chemistry course at the school which still exists today as well as leading to the entry of many North Dakotoans into the technical side of the paint industry. Unfortunately most paint test fences today do not differ much from those early test fences.

1. Climate

It has been recognized for many years that different climates affect a paint in different ways. Van Loo (107), for instance, pointed out in 1949 that Florida exposures tend to accentuate checking and cracking and erosion: Kansas exposures accentuate checking and cracking also, but show little erosion; while Chicago exposures accentuate erosion, but show little checking or cracking. He did not explain why. Obviously the amounts of ultraviolet, moisture, and temperature changes vary between the different locations as he pointed out. A possible explanation for increased erosion in Chicago is the fact that wet conditions are followed more often by freezing temperatures, causing minute fissures in the film surfaces because of the expansion of water as it freezes. At any rate, it is necessary to use multiple exposures in different climates to adequately assess a paint film. Details on how to conduct normal exposure tests for paints on wood are given in ASTM designation D 1006-65 and in Van Loo's paper (107).

2. Micro-climate

Several references have been made in this chapter to the fact that the performance of paint is different in protected areas of a structure than on exposed areas (see Sec. 5-4,A,4 Intercoat Peeling and Mildew Resistance, Protected Areas). This is because the micro-climate is considerably different in protected areas. Rain does not reach these areas, but dew is more apt to form there, particularly on inverted horizontal areas like undereaves. This should be taken into consideration in exposure fence testing as discussed by the author (26). It should also be taken into account that an exposure fence is a small structure with little mass, which consequently cools off rapidly at night. This means that there is more of a tendency for dew to form on a test fence than on a house. Since the presence of moisture tends to accelerate degradation from ultraviolet light (94), this accounts for the fact that test fences tend to give accelerated failures compared with a house. These two factors — the absence of protected areas on test fences and the greater deposition of dew on test fences — tend to explain the differences that Ross (108) has found between test fences and houses.

It should also be pointed out that the kind of testing described in D 1006 is in one way milder than actual conditions on a house. This is because the end grain of the panel is well sealed with paint as the panel is prepared for exposure. On a house this is not true — particularly where siding is butted together and on window and door frames. Water tends to enter the siding or framing joints and easily travels into the end grain wood. This causes a considerable degree of dimensional change and often leads to the first peeling failures on a house. Such peeling does not usually occur on a simple test panel. For this reason, fences containing windows with standard framing are proposed as a test method. Such testing has been done by Walters and Peterson (109).

3. Substrates

There are, of course, many different types of substrates that are used on houses. Metal and masonry substrates are less variable and less important from a painting standpoint than are forest product based substrates. It is with forest product substrates that the greatest problem arises in choosing panel types for paint testing. For instance, house paints are apt to be used over cedar, redwood, southern pine wood siding, low-, medium-, or high-density hardboard siding (which may be factory primed), Douglas fir and southern pine plywood, and various kinds of pine trim. The problem is how to cover all of these substrates with a reasonable number of tests. The answer is usually a compromise for test fence exposures backed up by a limited number of test houses, and eventually by test marketing on a limited scale.

In the author's laboratory, a fairly standard set of substrates has evolved, based on the type of paint characteristic that is under investigation. Selected edge grain cedar beveled siding has been found to give the most uniform substrate. Therefore, this has been chosen for general testing of such characteristics as chalking, fading, gloss retention, etc. Although the pines are the most severe substrate for mildew resistance tests, they have been found to be too variable from board to board, and even within one board, to be a good test substrate. Cedar, on the other hand, is not severe enough to be a good substrate for mildew resistance testing. Redwood has been found to be a good compromise with intermediate reproducibility and severity for mildew resistance testing. Douglas fir plywood has been selected as a substrate for evaluating adhesion and resistance to wood checking and for undereave exposures.

These three substrates, then, comprise the generally used test substrates. Others are used in special instances.

4. Controls

Since wood and wood products are such nonuniform substrates for paint testing, it is mandatory that control paints be used on each panel. Generally, this is done by selecting two paint systems—one that is expected to fail and one that is expected to perform well—and to use these on small areas of each panel. The alternative is to use multiple panels. The author prefers the use of control paint systems since the multiple panel approach really requires more panels per test system than can be included at reasonable cost (ten or more).

5. Test Fence "Houses"

Figure 5-6 illustrates a kind of test fence that the author's laboratory has found useful for evaluating paint systems that are near the final development

FIG. 5-6. Test fence "houses".

stage. It shows two vertical fences, each of which has a protective overhang on the north side. The far fence has 16 small windows built into the vertical section of the fence under the overhang. The south side (not visible) of the near fence has been completely sided with beveled siding. "Run-of-the mill" redwood siding on the vertical portions of the fence, Douglas fir plywood on the undereave portions, and pine window frames were used. The construction of the fence resembles that of a house wall. This kind of exposure fence really represents 16 test fence "houses" per 48-foot fence that are essentially identical. All of the kinds of exposures that are found on a house are incorporated on the fence. The one exception is that of moisture arising from inside the house, but this can be tested on blister boxes such as described in ASTM D-2366.

The particular fences in the illustration have been repainted. In the case of the north side of the fence in the foreground, the 16 original test paints were repainted with themselves on the left half and with a standard chalking white oil paint on the right half. In the case of both the north and south exposures of the test fence "houses," two standard paints and one newly developed paint were applied to each test section. This then allows comparison of these three paints over 16 different original paint systems, or the equivalent of 16 different test houses. Obviously, this is an expensive method of test fence exposures, but it is to be preferred over test houses because it eliminates the differences between houses and allows the differences between paint systems to be evaluated.

5-5. CAULKS

This section on caulks is included here only because of their importance to the performance of paint, particularly on wood substrates (110). It does not cover the formulation of such products.

Most houses are repainted because of failures around wood joints such as around window and door frames, corners, and where siding butts together. This is because wood is so much more permeable to water along the grain than across the grain. Thus, if water can get in contact with the end grain on a piece of wood, it absorbs very quickly, causing swelling; it also desorbs quickly when drying conditions occur, causing shrinking. Repeated swelling and shrinking weakens the paint adhesion in these areas and leads to cracking and eventual peeling.

Of course, an attempt is made to prevent such failures by the use of caulks. The problem is that home owners and contractors do not recognize the importance of long-term durability of such caulks, so they use the lowest-cost caulks instead of the best caulks. Oil-based caulks are by far the cheapest and, unfortunately, the least durable. They tend to harden and lose their flexibility, which causes cracking and loss of adhesion within about one year. Butyl rubber caulks are the next lowest-cost, but they fail in three or four years. Latex caulks, at about twice the cost of oil caulks, are expected to last ten years. Polysulfide caulks are still more costly and more durable, but they are difficult to apply.

Certainly the use of a good caulk on a structure is well worthwhile in terms of paint durability.

REFERENCES

1. J. J. Mattiello, Protective and Decorative Coatings, Vol. III, Wiley, New York, 1943.
2. F. L. Browne, Behavior of House Paints on Different Woods, U.S. Department of Agriculture, Forest Products Laboratory, Dec., 1934; "Why Some Wood Surfaces Hold Paint Longer Than Others," Leaflet 62, U.S. Dept. of Agriculture, Sept., 1930.
3. E. G. Shur, H. G. Guentsler, and P. Rosa, "Effect of Type of Grain on Performance of Exterior Coatings on Plywood," Forest Prod. J., 16 (5), (1966).
4. V. P. Miniutti, "Properties of Soft Woods That Affect the Performance of Exterior Paints," Offic. Digest Fed. Soc. Paint Technol., 35(460), 451 (1963).
5. F. E. Dickinson, "Dimensional Changes in Wood, Plywood, Reconstituted Wood, Metals with Changing Temperature and Moisture Contents," Am. Paint J., 48(8), 54 (1964).
6. F. L. Browne, "Wood Properties That Affect Paint Performance," Forest Prod. Lab. Report, No. 1053, Revised (1951).
7. J. D. MacLean, "Effect of Moisture Changes on the Shrinking, Swelling, Specific Gravity, Air or Void Space, Weight, and Similar Properties on Wood," Forest Prod. Lab. Report, No. R 1448 (1944).
8. M. A. Kalnins, "Surface Characteristics of Wood as They Affect Durability of Finishes," Forest Prod. Res. Paper, FPL 57, Part II (1966).
9. G. G. Schurr, "Proper Coatings For Wood Exteriors," Am. Painting Contractor, 46(12), 18 (1969).
10. R. Zenker, "Effects of Wood Moisture, Drying Process, and Age on Durability of Exterior Paint Finishes," Plaste Kautschuk, 12, 560 (1965).

11. Anon., "Some Observations on the Painting of Masonry," Paint India, 31(8), 17 (1967).

12. NACE Technical Committee Report,"Surface Preparation of Concrete for Coating," Mater. Protection, 5(1), 84 (1966).

13. J. E. O. Mayne, "The Mechanism of Protection by Organic Coatings," Metal Finishing J., 12, 143 (1966).

14. T. K. Hay and G. G. Schurr, "Moisture Diffusion Phenomena in Practical Paint Systems," J. Paint Technol., (Scheduled for May, 1971.)

15. J. G. Rigg and E. W. Skerrey, "Priming Paints for Light Alloys," Paint Manufacture, 24(3), 75 (1954).

16. Anon., Ind. Finishing, 39, 87 (1962).

17. W. K. Asbeck and M. Van Loo, "Residual Viscosity of Paint Systems at Infinite Shear Velocity," Ind. Eng. Chem., 46, 1291 (1954).

18. J. J. Mattiello, et al., Protective and Decorative Coatings, Vol. I, Wiley, New York, 1941.

19. M. Van Loo, "Twenty-five Years of Paint Testing," Ind. Eng. Chem., 41, 267 (1949).

20. A. E. Jacobson, "Chalking of Titanium Dioxide Pigments," Paint Varnish Prod., 55(2), 29 (1965).

21. R. D. Murley, "Some Aspects of Physical Chemistry of Titanium Pigment Surfaces," J. Oil Colour Chemists' Assoc., 45(1), 16 (1962).

22. W. H. Madson, "White Hiding Pigments," Am. Paint J., 26 (June 10, 1963).

23. H. E. Brown, "Zinc Oxide Rediscovered," The New Jersey Zinc Company, New York, 1957.

24. S. Werthan, "Exterior House Paints," Offic. Digest Fed. Soc. Paint Technol., 21(293), 311 (1949).

25. N. A. Brunt, Verfroniek, 33, 93 (1960).

26. G. G. Schurr and M. Van Loo, "Undereave Peeling of House Paints," J. Paint Technol., 39(506), 128 (1967).

27. T. K. Hay and G. G. Schurr, "Moisture Diffusion Phenomena in Practical Paint Systems," J. Paint Technol., 43(556), 63 (1971).

28. W. R. Fuller, Understanding Paint, American Paint Journal Company, St. Louis, Chap. VII (1965).

29. V. C. Vesce, "Exposure Studies of Organic Pigments in Paint Systems," Offic. Digest Fed. Soc. Paint Technol., 31(415), (2), 3 (1959).

30. W. G. Armstrong and W. H. Madson, "The Effect of Pigment Variation on the Properties of Flat and Semigloss Finishes," Offic. Digest Fed. Soc. Paint Technol., 19(269), 321 (1947).

31. W. G. Vannoy, "Current House Paints," Offic. Digest Fed. Soc. Paint Technol., 21(292), 235 (1949).

32. W. K. Asbeck and M. Van Loo, "Critical Pigment Volume Relationships," Ind. Eng. Chem., 41, 1470 (1949).

33. D. E. Brody, "Performance of Flatting Pigments in High-speed Dispersion," J. Paint Technol., 40(525), 439 (1968).

34. L. E. Brooks, P. Sennett, and H. H. Morris, "Kaolin, Five Years on the Test Fence," J. Paint Technol., 40(520), 240 (1968).

35. O. J. Mileti, "Pigment Settling Rates: Effect of Vehicle Types," J. Paint Technol., 42(450), 55 (1970).

36. D. F. Johnson, "Calcium Carbonate in Flat Wall Paint," J. Paint Technol., 43(552), 109 (1971).

37. H. S. Ritter, W. A. Blose, and T. A. Demski, "Influence of Fine Particle Extenders on Hiding Pigment Performance," J. Paint Technol., 38(500), 508 (1966).

38. T. P. Dobkowski, "Calcined Aluminum Silicate Pigments in Latex Paints," J. Paint Technol., 41(535), 448 (1969).

39. J. R. Garland, "Driers, What Do We Really Know About Them ?" Part I, J. Paint Technol., 40, 528 (1968); Part II, ibid., 41(538), 623 (1969); Part III, D. J., Engler, ibid., 43(553), 62 (1971).

40. E. C. Larson and H. E. Sipple, "Los Angeles Rule 66 and Exempt Solvents," J. Paint Technol., 39(508), 258 (1967).

41. T. C. Patton, Paint Flow and Pigment Dispersion, Wiley (Interscience), New York, 1964.

42. M. Van Loo, "Physical Chemistry of Paint Coatings," Offic. Digest Fed. Soc. Paint Technol., 28(383), 1126 (1956).

43. C. R. Martens, Alkyd Resins, Reinhold, New York, 1961.

44. T. C. Patton, Alkyd Resin Technology, Wiley, New York, 1962.

45. J. R. Blegen, "Alkyd Resins," Federation Series on Coatings Technology, Unit V (1967).

46. G. G. Schurr, T. K. Hay, and M. Van Loo, "Possibility of Predicting Exterior Durability by Stress/Strain Measurements," J. Paint Technol., 38(501), 591 (1966).

47. R. C. Trueblood, "Three-Year Study of Mildew Growth on Coated Surfaces," J. Paint Technol., 43(554), 77 (1971).

48. T. R. Bullett, "Rheology in Painting," Rheol. Acta, 4(4), 258 (1965).

49. N. D. P. Smith, S. E. Orchard, and A. J. Rhind-Tutt, "The Physics of Brushmarks," J. Oil Colour Chemists' Assoc., 44, 618 (1961).

50. J. F. Rhodes, Ph.D. Thesis, Ohio State University, 1968.

51. P. E. Pierce, "Rheology of Coatings," J. Paint Technol., 41, No. 533, 383 (1969).

52. H. L. Beeferman and D. A. Bergren, "Practical Applications of Rheology in the Paint Industry," J. Paint Technol., 38(492), 9 (1966).

53. J. S. Dodge, "Quantitative Measures of Leveling," J. Paint Technol., 44(564), 72 (1972).

54. T. C. Patton, "Viscosity Measurement Using a Cone/Plate Spring Relaxation Technique," J. Paint Technol., 38(502), 656 (1966).

55. F. Liberti, "Practical Methods for Evaluating Leveling of Latex Paints," J. Paint Technol., 43(553), 72 (1971).

56. W. H. Madson, Federation Series on Coatings Technology, Unit VII, "White Hiding and Extender Pigments," Federation of Societies for Paint Technology, Philadelphia, 1967.

57. J. C. Becker and D. D. Howell, "Critical Pigment Volume Concentration of Emulsion Binders," Offic. Digest Fed. Soc. Paint Technol., 28(380), 775 (1956).

58. P. Berardi, "Parameters Affecting CPVC of Resins in Aqueous Dispersions," Paint Technol., 27(7), 24 (1963).

59. F. B. Stieg, "Color and CPVC," Offic. Digest Fed. Soc. Paint Technol., 31(379), 695 (1956).

60. R. G. Alexander, "Predicting Oil Absorption and CPVC of Multicomponent Pigment Systems," Offic. Digest Fed. Soc. Paint Technol., 31(418), 1490 (1959).

61. F. P. Liberti and R. C. Pierrehumbert, "CPVC of Vinyl Emulsion Paints by Pigment Water Sorption Method," Offic. Digest Fed. Soc. Paint Technol., 31(409), 252 (1959).

62. R. J. Cole, "Determination of Critical Pigment Volume in Dry Surface Coatings," J. Oil Colour Chemists' Assoc., 45, 776 (1962).

63. E. J. Shaller, "Critical Pigment Volume Concentration of Emulsion Paints," J. Paint Technol., 40(525), 433 (1968).

64. R. H. Lalk, "The Functions and Interactions of Additives in Latex Paints," Offic. Digest, Fed. Soc. Paint Technol., 37(489), Part 2, 49 (1965).

65. Rohm and Haas Company, "Rhoplex, Acrylic Emulsions for Outdoor Paints," Progress Report 12, p. 19 (June, 1969).

66. W. J. Stewart, "Paint Driers and Additives," Federation Series on Coatings Technology, Unit Eleven (1969).

67. D. H. Snyder, E. M. Watson, and M. Goll, "Microbiological Involvement in Paint Raw Materials," ACS, Div. of Organic Coatings and Plastic Chem., Preprint of Papers Presented at the Atlantic City Meeting, p. 6 (September, 1965).

68. R. A. Bergfeld, "Efficacy of Non-Mercurial Preservatives in Aqueous Coatings," J. Paint Technol., 43(563), 80 (1971).

69. H. E. Ashton, "Clear Finishes for Exterior Wood," J. Paint Technol., 39(507), 212 (1967).

70. N. Estrada, "Exterior Durability of Catalyzed Clear Coatings on Redwood," J. Paint Technol., 39(514), 655 (1967).

71. J. R. Kiefer, "Ultraviolet Light Absorbers in Clear Coatings for Wood," J. Paint Technol., 39(515), 736 (1967).

72. E. C. Rothstein, "Compatibility and Reactivity of Ultraviolet Absorbers in Clear Wood Coatings," Paint Varnish Prod., 58(2), 39 (1968).

73. V. P. Miniutti, "Microscale Effects of Ultraviolet Irradiation and Weathering on Redwood Surfaces," J. Paint Technol., 41(531), 275 (1969).

74. M. H. Schneider, "Coating Penetration into Wood Substrate Studied with Electron Microscopy Using Replica Techniques," J. Paint Technol., 42 (547), 457 (1970).

75. W. A. Zisman, "Surface Energetics of Wetting, Spreading, and Adhesion," J. Paint Technol., 44(564), 42 (1972).

76. D. M. Gans, "Positive Final Spreading Coefficients," J. Paint Technol., 42(550), 653 (1970).

77. V. R. Gray, "Surface Aspects of Wetting and Adhesion," Chem. Ind., 23, 969 (1965).

78. M. Van Loo, "Exterior Finishes for Wood," unpublished. Presented to Engineering Institute, University of Wisconsin, Madison, Wisconsin (Nov. 30, 1961).

79. L. W. Newman, "Some Studies in the Shrinkage of Paint Films," Unpublished Thesis, North Dakota State University (June, 1939).

80. P. E. Pierce and R. M. Holsworth, "Torsion Modulus," J. Paint Technol., 38(496), 263 (1966).

81. E. Hoffman and O. Georgoussis, "Phenylmercury Compounds as Fungicides," J. Oil Colour Chemists' Assoc., 43(11), 779 (1960).

82. E. Hoffman and B. Burtsztyn, "Phenylmercury Compounds as Fungicides, Part II," J. Oil Colour Chemists' Assoc., 46(6), 460 (1963).

83. E. Hoffman and B. Burtsztyn, "Phenylmercury Compounds as Fungicides, Part III," J. Oil Colour Chemists' Assoc., 47(11), 871 (1964).

84. E. Hoffman, A. Saracz, and J. R. Barned, "Phenylmercury Compounds as Fungicides, Part IV," J. Oil Colour Chemists' Assoc., 49(8), 631 (1966).

85. E. Hoffman, A. Saracz, and J. R. Barned, "Formulation of Fungus Resistant Paints; Addition of Pentachlorphenol," J. Oil Colour Chemists' Assoc., 49(7), 551 (1966).

86. E. Hoffman, A. Saracz, and J. R. Barned, "Formulation of Fungus Resistant Paints: Addition of Paratoluene Sulfonamide," J. Oil Colour Chemists' Assoc., 50(6), 516 (1967).

87. E. Hoffman, "Inhibition of Mold Growth by Fungus Resistant Coatings Under Different Environmental Conditions," J. Paint Technol., 43(558), 54 (1971).

88. R. C. Trueblood, "Some Further Observations on Mildew Growth," J. Paint Technol., 40(527), 582 (1968).

89. F. B. Stieg, "Particle Size as a Formulating Parameter," J. Paint Technol., 39(515), 703 (1967).

90. P. Mitton and A. E. Jacobson, "New Graph for Computing Scattering Coefficient and Hiding Power," Offic. Digest Fed. Soc. Paint Technol., 35(464), 871 (1963).

91. R. J. Bruehlman and W. D. Ross, "Hiding Power from Transmission Measurements: Theory and Practice," J. Paint Technol., 41(538), 584 (1969).

92. E. D. Aman, P. A. Wolf, and F. J. Bobalek, "Fungistats for Paint Films: Laboratory Evaluation Procedures," J. Paint Technol., 43(562), 76 (1971).

93. C. R. Caryl and A. E. Rheineck, "Outdoor and Accelerated Weathering of Paints," Offic. Digest Fed. Soc. Paint Technol., 34(452), 1017 (1962).

94. Ralph J. Wirshing, "Some Causes of Paint Film Failure," Ind. Eng. Chem., 33, 234 (1941).

95. J. W. Tamblyn and G. M. Armstrong, "Modification of Atlas Twin-Arc Weather-Ometer," Anal. Chem., 25(3), 460 (1953).

96. R. C. Hirt, R. G. Schmidt, N. Z. Searle, and A. P. Sullivan, "Spectrums of Ultraviolet Light Sources," J. Opt. Soc. Am., 50, 706 (1960).

97. W. F. Singleton, "Accelerated Weathering Machines," Am. Paint J., 41, 92 (1957) May 20.

98. L. J. Nowacki, "An Evaluation of Various Weather-Ometers for Determining the Service Life of Organic Coatings," Offic. Digest Fed. Soc. Paint Technol., 34(454), 1191 (1962).

99. K. G. Martin, "Influence of Radiant Energy Source on Bitumen Oxidation," J. Appl. Chem., 14(11), 514 (1964).

100. R. Epple, 'The Value of Durability Investigations of Paints by Accelerated Weathering Tests," F.A.T.I.P.E.C., 8, 401 (1966).

101. E. A. Kanevskay, "Investigation of Lacquer Coating Aging Processes Under Artificial Light Sources and Natural Conditions," F.A.T.I.P.E.C. 11 Congr., Review, 41, 314 (1968).

102. K. Gutfreund, "Studies of the Deterioration of Paint Films by Measurement of Their Mechanical Properties," NPVLA, Scientific Section Circ., 793 (March, 1965).

103. G. L. Holbrow, "Atmospheric Pollution; Its Measurement and Some Effects on Paint," J. Colour Chemists' Association, 45(10), 701 (1962).

104. E. M. Corcoran, "Determining Stresses in Organic Coatings Using Plate Beam Deflection," J. Paint Technol., 41(538), 635 (1969).

105. C. J. Berg, W. R. Jarosz, and G. F. Salathe, "Performance of Polymers in Pigmented Systems," J. Paint Technol., 39(510), 436 (1967).

106. F. B. Stieg, "Weathering and Titanium Dioxide," J. Paint Technol., 43(553), 82 (1971).

107. M. Van Loo, "Twenty-five Years of Paint Testing," Ind. Eng. Chem., 41, 267 (1949).

108. L. A. Weinert and R. T. Ross, "Modified Barium Metaborate — A Versatile Pigment: 1. Limitations of Test Fences in the Evaluation of Exterior House Paints," Am. Paint J., 51(7), 46 (1966).

109. C. S. Walters and K. R. Peterson, "Exposure Test of Painted, Pressure Treated Millwork," Forest Prod. J., 14(2), 87 (1964).

110. G. G. Schurr and C. R. Martens, "Caulk Your Way to a Better Paint Job," Popular Mechanics, 120 (1971) Sept.

111. J. A. Kitchener and C. F. Cooper, "Current Concepts in the Theory of Foaming," Quart. Rev. Chem. Soc., 13, 71 (1959).

112. P. S. Hudson, "Foam Control Agents in Aqueous Paint Systems," Paint Manuf., 27 (1969) June.

113. G. K. Wheeler, "Effect of Nitrogen-Containing Compounds on Drying of Paints," Ind. Eng. Chem., 39, 1115 (1947).

114. M. E. Stearns, "1,10-Phenanthroline Promotes the Drying of Paints," Offic. Digest, Fed. Soc. Paint Technol., 26(356), 817 (1954).

115. R. R. Myers, W. H. Canty, and C. K. Wheeler, "Drier Catalyst Activity in Organic Coatings," Ind. Eng. Chem., 52, 67 (1960).

116. M. E. Stearns and W. H. Canty, "1,10-Phenanthroline Promotes the Drying of Paints: II", Offic. Digest Fed. Soc. Paint Technol., 30(396), 58 (1958).

117. J. S. Long, "Keynote Address," Offic. Digest Fed. Soc. Paint Technol., 32(428), 1119 (1960).

118. J. S. Long, M. Schwartz, Y. S. Chiang, R. E. Schuler, and R. Sturgeon, "Surface Forces and Dimensions," Offic. Digest Fed. Soc. Paint Technol., 35(456), 11 (1963).

119. J. S. Long, S. Thames, and O. Smith, "Surface Forces and Dimensions," J. Paint Technol., 39(507), 169 (1967).

Chapter 6

FLAT WALL PAINTS

Fred B. Stieg

Titanium Pigment Division of
N L Industries, Inc.
South Amboy, New Jersey

6-1. INTRODUCTION

The most commonly accepted meaning of the term flat wall paint is "a lusterless, or relatively lusterless, pigmented coating designed for application to interior walls and ceilings for protective, decorative, or sanitary purposes" (1).

Despite this seemingly straightforward definition, all of the following descriptive terms have at one time or another appeared on the labels of flat wall paints:

Acrylic
Alkyd
All-oil
Calicoater
Ceiling white
Deep tone
Egg shell
Flat enamel
Fume proof
Latex
Mill white
Nonyellowing
Odorless
Oleo-resinous
One-coat
Primer-flat
Rubberized
Self-sealing
Spraying
Stippling
Synthetic
Texture
Vinyl
Vinyl-acrylic
Washable

Some of these terms refer to the composition of the flat wall paint, others to performance, and still others to the special applications for which these products have been designed. Some are identical in meaning and several might be applied to a single formulation without contradiction as is explained, but a sufficient number of distinctly different classifications remain that require the application of basically different formulating techniques.

6-2. VOLUME RELATIONSHIPS

All paint formulation might be said to revolve around the relationship of pigment to vehicle binder. This is not a chemical relationship, but a physical one based upon the relative volumes of pigment and binder present in the dry paint film. If the binder dominates, the films produced tend to be glossy; if the pigment dominates, the films tend to be lusterless, or "flat."

This relationship is generally expressed, particularly in the trade sales paint industry which produces flat wall paints, as the pigment volume concentration (PVC). Numerically, the PVC is the volume percentage of pigment present in the dry paint film and all flat wall paints may be generally classified as being formulated above 50 PVC. This is not precisely true, because some few "flat enamels" and "eggshell flats" have been formulated below that range, as have an equally small number of flat latex paints.

The reason for the high pigmentation required of flat wall paints is, of course, related to the fact that, by definition, they must develop a lusterless surface.

The surface tension of paint binders tends to produce a smooth, level surface which, because of the absence of surface irregularities and a refractive index higher than air, will produce light reflection like a still pool of water. The presence of pigment within the binder makes little difference until so much is present that the air interface of the paint film begins to conform to the texture of the population of solid particles lying below the surface. As soon as this happens, the gloss of the paint film is reduced by diffuse reflectance from the surface irregularities— just as the mirror image in a pool of water is destroyed by a breeze ruffling its surface.

Considerable surface texture is required to produce a truly "lusterless" film, and if the level of pigmentation were gradually increased, it would be found that its appearance would only gradually change from a condition of high gloss at low PVC's of 10-20%, and pass through an intermediate stage of between 35 and 40% PVC, commonly referred to as semigloss, before actually becoming lusterless. As a matter of fact, it has been a common practice for painters to physically mix gloss and flat wall paints of similar vehicle types to produce their own semigloss paints. Because of the possibility of vehicle incompatibility, however, the practice is not recommended.

The exact point in the PVC range at which a paint film passes from a semigloss to a flat wall paint may be quite different for different pigmentations. This is because the ability of the solid pigment particles to roughen, or flatten, the surface of the paint film is a function of their particle size and shape, as well as their volume. It is also a function of how the many different sizes and shapes which may exist in the same pigmentation pack together. This latter is a very significant factor, for the volume that a given amount of pigment will actually occupy is made up, not only of the total volume of the particles themselves, but also of the volume of the spaces between particles. This can be visualized if we consider a cord of wood cut in fireplace-size logs which are laid edge-to-edge flat on the ground. Four or five inches of snow could completely cover it. But if the same weight of wood in the form of the lopped-off branches were piled over the same area, the brush heap formed would require four to five feet of snow to cover it because of the looseness of its structure.

Pigment particles do not, of course, differ as much in shape as the difference between a fireplace log and a leafy branch, but they do differ substantially in size.

6-3. EXTENDER PIGMENTS

At this point it becomes necessary to point out, for the benefit of the neophyte, that the pigmentation of flat wall paints, as of most other paint products, consists of two distinctly different classes. One, made up of the white and colored "prime" pigments, produces the color and opacity of the paint film. They are relatively expensive pigments and are consequently used only in sufficient quantities to provide the required opacity and color. Since this does not add up to a large enough volume to produce a lusterless (or even semigloss) film, other pigments known as "extenders," or "fillers," are required to provide the additional volume. These are relatively inexpensive.

Another reason, in addition to cost, that prime pigments are never used alone to produce flat wall paints is that they are produced in fine particle sizes to obtain maximum opacity and color development and because they must be usable in high gloss enamels as well as flats. This fine particle size does not produce enough surface irregularity, even at high PVCs, to eliminate gloss.

It is the extender portion of the pigmentation of flat wall paints that is produced in a particle size range large enough to efficiently reduce gloss. As a matter of fact, many of the coarser varieties - those with fibrous, cellular, or acicular shapes - are known as "flatting extenders." Table 6-1 lists a selection of common extender pigments with their approximate particle sizes.

This listing is only a minute selection used to illustrate the ranges of particle size available in each variety. For a more complete listing, it is suggested that the reader consult the Raw Materials Index, Pigment Section, published by the National Paint and Coatings Association.

6-4. PIGMENT PACKING

The effect of particle size on the packing of pigment particles in a paint film can be illustrated by a simple analogy. If we fill a barrel with basket-

TABLE 6-1. Common Extender Pigments

Trade name	Type	Diameter (μ)
Atomite	Calcium carbonate	2.5
Oolitic C	Calcium carbonate	5.75
Camel-KOTE	Calcium carbonate	11.0
Surfex MM	Calcium carbonate	0.2
Vicron 45-3	Calcium carbonate	17.2
Zeolex 80	Silica (amorphous)	0.03
Gold Bond "R"	Silica (amorphous)	8.5
Santocel C	Silica (amorphous)	5.0
Celite 281	Silica (diatomaceous)	6.0
Dicalite 395	Silica (diatomaceous)	2.8
Asbestine 3x	Talc (fibrous)	1.5
Nytal 200	Talc (fibrous)	10.0
ASP 100	Clay	0.55
Al-Sil-Ate W	Clay	1.5
Lorite	Comb. diat. silica and $CaCO_3$	10.0

balls, it is "full" when we have perhaps put in five to seven. There is a lot of unoccupied space, however, and if we now decide to add baseballs, the barrel will not again be "full" until we have added 15 to 25 of the smaller spheres. There is still space remaining, which we could successively fill with perhaps a few hundred marbles and, finally, with a few thousand BB's. Now the barrel would be too full to lift — but would it really be full? Not at all! We could still pour in 20 gallons of water before it reached the brim and overflowed.

This packing effect of progressively smaller particles has been extensively studied by civil engineers who have been interested in the process as a means for formulating concrete compositions of maximum density.

Furnas (2) has considered the case of mixtures of particles of different sizes where the voids formed by the larger particles are assumed to be occupied by smaller ones. He has plotted curves from which the minimum voids can be read for a specified number of monosized components, and specified ratios of the smallest-to-the-largest particle size in the mixture. An example of two such curves, for components having initial void volumes of 40%, is reproduced as Fig. 6-1.

It will be noted that residual void volume, that is, the volume of the system not occupied by pigment, decreases as the ratio of particle sizes increases. Since vehicle binder is required to fill in those void spaces not occupied by smaller pigment particles, the more uniform the particle size of the pigment, the more binder will be required to produce a continuous film.

This is one of the most important discoveries to have been made with respect to the formulation of flat wallpaint — the proper selection of extender particle size can completely alter paint film characteristics at any given PVC.

6-5. THE CRITICAL PVC

The first clue to this important discovery was provided by Asbeck and Van Loo (3), who observed a common transition point of film properties such as gloss, blistering, vapor permeability, and rusting at a pigmentation level which they referred to as the "critical" PVC (CPVC). Their initial findings are illustrated by Fig. 6-2 from their original publication.

Asbeck and Van Loo found that the CPVC was not the same for different pigmentations. They proposed that it represented a volume relationship of pigment and binder which provided just enough binder to satisfy the pigment surface and fill in the void spaces between particles. At or below the CPVC, the continuous film produced resisted the passage of water vapor, making the film vulnerable to blistering, but excluding the water which would pro-

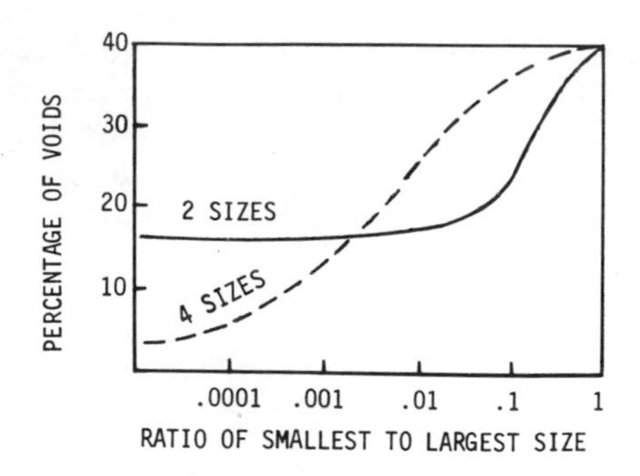

FIG. 6-1. Furnas computations.

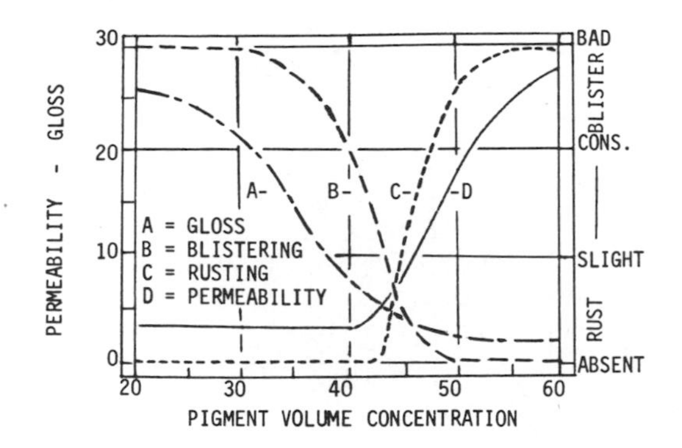

FIG. 6-2. Transition points for film properties.

duce rusting of a ferrous metal substrate. Above the CPVC, insufficient binder was present to completely fill the void spaces between pigment particles, and the resulting air pockets provided a passageway for water vapor, relieving the pressure-producing blisters, but admitting moisture to the substrate.

Stieg (4) proposed that the CPVC could be determined experimentally by measuring the oil absorption of dry pigment blends and calculating the volume percentage of pigment present at the end point.

$$\text{CPVC} = \frac{\text{vol 100 lb pigment}}{\text{vol 100 lb pigment} + \dfrac{\text{O.A.}}{7.75}}$$

By determining the oil absorption of a number of combinations of the same two pigments, CPVC curves were produced which visually demonstrated the changes in CPVC with composition. Examples of these CPVC curves are given in Fig. 6-3 to 6-5.

These curves demonstrated, for paint pigmentations, the same general principle illustrated by Furnas for concrete compositions. The larger the particle size of the extender, the higher the peak of the CPVC curve representing a condition of maximum packing (and therefore of minimum voids).

The CPVC curve may be considered as a boundary line which separates those pigmentation levels which produce flat wall paints with good washability and enamel hold-out, from those which are porous and develop "dry hiding" — a phenomenon which is discussed at a later point.

The CPVC is also the PVC at which a given combination of prime pigment and extender will produce the lowest angular sheen attainable with that par-

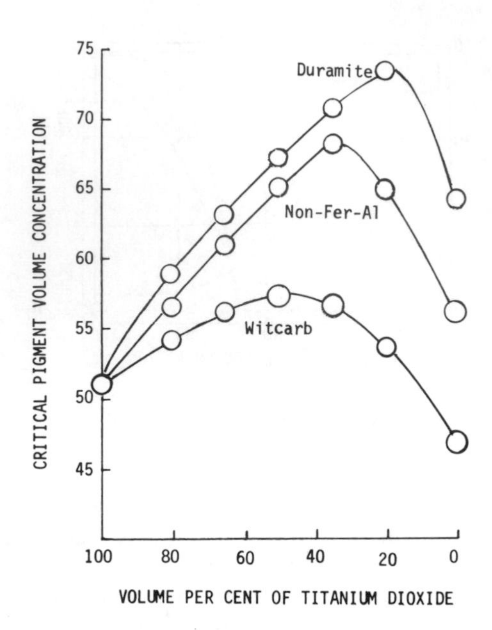

FIG. 6-3. Critical PVC curves.

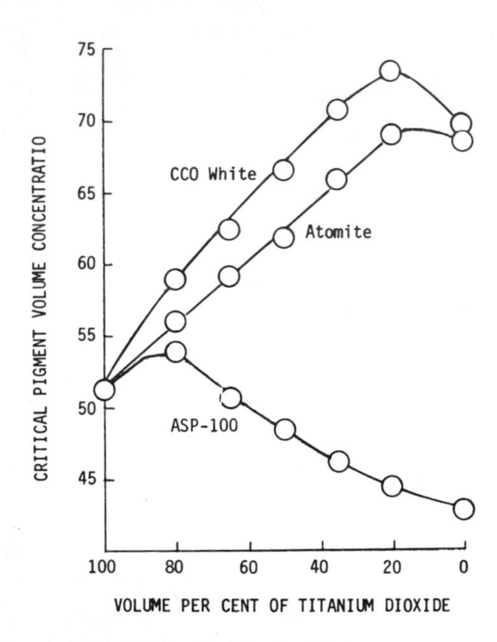

FIG. 6-4. Critical PVC curves.

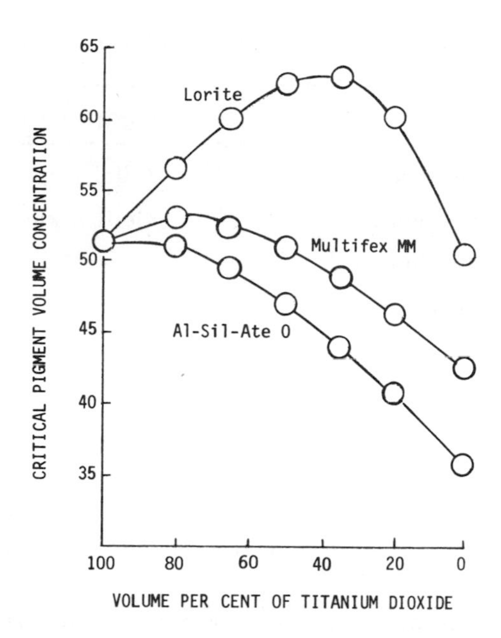

FIG. 6-5. Critical PVC curves.

ticular combination. This has not been appreciated by some formulators in the past, who have assumed that lower sheen could always be obtained by raising the PVC. This is only true if the starting point is below the CPVC. Angular sheen will then decrease as the PVC approaches the CPVC, and increase again when it is exceeded.

The reason for higher sheen, or pigment sheen, above the CPVC is due to the effect of refractive index on reflectance. So long as the pigment particles are covered by a layer of binder, the reflectance of the paint film is determined by the difference in refractive index between the binder and air. When the CPVC is exceeded, however, pigment particles are pushed up into the air interface because of the deficiency of binder. If the pigment has a higher refractive index than the binder, which will be true of any white prime pigment, the reflectance of the film will be increased and gloss will also be increased. For gloss is only directional reflectance when the angle of incidence and the angle of reflectance are the same.

There is also a secondary effect based upon the additional crowding of the pigment particles as binder is removed. The closer they are packed together, the smaller the surface irregularities produced by individual particles.

Table 6-2 classifies a few of the flat wall paint descriptions listed at the beginning of this chapter in terms of their probable relationship to the CPVC.

TABLE 6-2. Flat Wall Paints and the CPVC

At or below the CPVC	Above the CPVC
Deep tone	Calcicoater
Flat enamel	Ceiling white
Mill white	One-coat
Primer flat	Stippling
Self-sealing	Texture
Washable	

Obviously, this list cannot include any of those descriptive terms which apply to composition, as the CPVC relationship is independent of vehicle type.

The presence of "deep tone" flat wall paints in the listing "At or below the CPVC" is related to some other transitions in film properties which take place at the CPVC, but were overlooked by Asbeck and Van Loo. This was undoubtedly due to the fact that their work was oriented toward exterior formulations, rather than flat wall paints.

6-6. COLOR AND SHEEN UNIFORMITY

One of the major problems encountered in formulating a satisfactory "deep tone" flat wall paint is that of obtaining uniformity of appearance over adjacent sealed and unsealed wall surfaces. The primary cause of such variation in porosity is the presence of repaired cracks in plasterboard construction. However, these areas could easily be sealed with a good wall primer, so that the true cause of the difficulty is an attempt to do a one-coat job over nonuniform surfaces.

This variation in substrate porosity can affect the appearance of deep tone flat wall paint in one of two ways: it may appear to be either darker or lighter in shade over a sealed area due to a variation in luster; or it can be uniform in luster, but appear to be lighter in color over an unsealed area. Rather than search at random for such areas on an actual wall surface, the paint formulator will generally use either a section of wallboard panel which has been partially primed with a shellac stripe or a commercially prepared "penetration chart" (5).

Similar color and sheen differences occur in pastel tints and light colors as well as deep tones, but are generally not noticeable because the difference in brightness of color produced is too small a percentage of the normal brightness of the color to be detected by the human eye.

6-7. VEHICLE PENETRATION

Since the cause of the defect is substrate porosity, the use of a nonpenetrating vehicle is mandatory — but the formulator will discover that "nonpenetration" is a relative term. The same vehicle will not penetrate in some formulations, but will penetrate in others.

Stieg (6) has demonstrated that color and sheen uniformity go through a transition at the CPVC — and so does vehicle penetration. By running PVC "ladders" with pigmentations which differed in their total extender content and in the type of extender used, it was shown that both vehicle penetration, as measured by the distance it would wick outwards from the point of contact through a piece of filter paper, and color uniformity, as measured by the brightness differential over sealed and unsealed substrates as a percentage of the sealed brightness, correlated with CPVC.

Table 6-3 lists some comparative values for two pigmentations which differed only in the oil absorption of the calcium carbonate used. The extender comprised 27.9% of the total pigmentation. The vehicle used was a "nonpenetrating" gelled alkyd designed for deep tone flat wall paints.

TABLE 6-3. Penetration and Color Uniformity

	High oil extender	Low oil extender
CPVC (from O.A.)[a]	55	65
Penetration, 70 PVC	5/32 in.	3/32 in.
Penetration, 60 PVC	2/32 in.	0
Penetration, 50 PVC	0	0
Color variation, 70 PVC	5.3%	2.4%
Color variation, 60 PVC	3.8%	0.8%
Color variation, 50 PVC	0.0%	0.0%

[a]O.A. refers to oil absorption.

Generally speaking, a color variation of more than 1.0% is quite visible.

It was observed, however, that even the most uniform flat wall paint, under normal drying conditions, would exhibit excessive color variation if caused to dry in a closed container so that the air space above the film quickly became saturated with solvent. The same effect could be produced by drying under conditions of high humidity or low temperature.

6-8. DARCY'S LAW

Darcy's law (7) for the fluid conductivity of a porous material offers an explanation for this behaviour:

$$\frac{V}{t} = \frac{d^2 \times k \times p}{\eta L}$$

where V is the volume of vehicle penetrating per unit area
t is the time
d is the mean square pore diameter
p is the substrate capillary pressure
η is the vehicle binder viscosity
L is the mean continuous pore path
k is a constant

For a specific paint system, the factor t is regulated by the solvent evaporation rate, or the time required for the residual vehicle's viscosity to become so high that flow ceases. To produce color uniformity over surfaces of varying porosity, the volume V must be restricted to that which does not raise the PVC of the drying film enough to visibly increase its brightness.

Any circumstance which retards the drying of the paint film (increases t) —such as high humidity, solvent saturation of the atmosphere, or low temperature—will cause the vehicle to penetrate the substrate, since V must increase to keep the formula in balance.

Solvent selection can be important not only because of its effect on drying time, but also because of its effect on vehicle viscosity. Therefore while a slow solvent, which happens to be a poor solvent for the vehicle binder, may not adversely affect the color uniformity of a given deep tone flat wall paint because η remains high, a slow solvent with a higher solvency for the binder could be highly detrimental.

This is the basic reason why odorless flat wall paint vehicles should not be used with normal hydrocarbon solvents. The odorless solvents generally have lower solvency than their "normal" counterparts.

The important terms in this equation for the formulator to remember, however are d (the mean square pore diameter) and L (the mean continuous pore path). If a paint is formulated above the CPVC, it will, by definition, contain an insufficient volume of vehicle solids to fill the void spaces between pigment particles. These voids can then only be filled, or partially filled, with air — making the pore path through the film discontinuous, and the factor L very small. This is why vehicle penetration occurs above the CPVC, causing poor color uniformity.

The mean square pore diameter, d, is a function of both the amount of binder present and the size of the particles making up the pigmentation. If the PVC is far below the critical, the excess of binder separates the individual pigment particles, making d large and causing the vehicle to penetrate until enough has drained away to constrict the pores sufficiently to cause the flow to cease. This penetration also causes a variation in the PVC of the paint film laid down over a porous surface, as compared with a sealed substrate, but since both film areas are below the CPVC, only sheen or gloss differences result. This condition, however, is just as detrimental to uniformity of appearance, so that only the PVC's immediately above and below the CPVC are suitable for deep tone flat wall paints.

Particle size has an obvious effect upon the mean diameter of the pore path between particles. A pigmentation need not be made up entirely of small particles to produce a small value for d, however. The smaller particle size component of a mixed pigmentation will always occupy the voids of the larger particle size component, so the smaller particle size component really determines the residual void size.

Titanium pigments perform this function in most flat wall paint formulations, but if hiding considerations do not require enough titanium dioxide to produce a condition of maximum packing — or if cost considerations do not allow it — some fine particle size extender must be added to insure that all of the large voids of the flatting extender system are occupied.

The smaller the mean square pore diameter d, the lower the viscosity, η, of the binder which will produce acceptable nonpenetration. Conversely, the lower the viscosity of the available vehicle binder, the more care must be taken to supply enough fine-particle-size pigment to provide for close packing.

6-9. FILM POROSITY

The importance of the CPVC is not limited to deep tone flat wall paints, of course, but applies to any flat wall paint which is required to be nonporous, such as a number of those appearing on the first page of this chapter under the following descriptive terms:

Eggshell	Primer-flat
Flat enamel	Self-sealing
Mill white	Washable

A knowledge of the CPVC is also necessary if the porosity of the flat wall paint film is to be controlled.

Many problems of flat wall paint formulation will, for example, involve matching the porosity characteristics of a known formulation with new raw materials. For this purpose it is most convenient to calculate the porosity index (P. I.) of the original so that it may be duplicated without laborious trial and error testing.

The porosity index is derived from the following formula (8):

$$P.I. = 1 - \frac{CPVC(1 - PVC)}{PVC(1 - CPVC)}$$

The CPVC of the original pigmentation is first determined by making a dry blend of its pigmentation and determining its oil absorption by the spatula rub-out method (9), utilizing the driest possible endpoint. The P. I. is then calculated using the CPVC equivalent to this endpoint, as previously described, and the PVC of the formulation.

The CPVC of the proposed substitute pigmentation is then similarly determined and the PVC necessary to obtain the same value for P. I. calculated. If the new PVC is significantly higher or lower than the original, it may then be advisable to consider a change in the proposed substitute.

This comes about because a significantly higher PVC will mean less flexible binder in the dry paint film, and therefore a more brittle film. On the other hand, it will also mean a film with better sanding characteristics, which may be desirable if the flat is to be used as an enamel undercoater.

A significantly lower PVC will mean more binder in the dry film and higher cost than necessary, since all paint binders are more expensive than extenders. It may also mean poorer brushability.

In either case, the original PVC level may be approached— if there is more than one extender involved— by determining the CPVC of dry blends of each of the proposed extenders with the prime pigment, using the same volume percentage of TiO_2 as in the original pigmentation, and calculating the combination which will provide the desired CPVC. This is only possible if the CPVC's observed with the individual TiO_2-extender blends bracket the desired value. If they do not, a different selection must be made.

6-10. FORMULATING FOR EQUIVALENT POROSITY

As an example of the process, consider the reformulation of an odorless flat wall paint such as that shown in Table 6-4.

The PVC of this formulation is 67.0 and the total solids by volume is 40% (26.9 gal pigment and 13.1 gal vehicle solids). It is desired to replace the existing extender combination with a blend of Camel-Carb calcium carbonate and Lorite.

A dry blend of the pigmentation ingredients is first prepared using the same weight ratios to provide a total of 5 g for the oil absorption test:

$\frac{190}{665} \times 5 = 1.43$ g rutile TiO_2 = 0.340 cc

$\frac{365}{665} \times 5 = 2.74$ g Lesamite = 1.012 cc

$\frac{60}{665} \times 5 = 0.45$ g ASP-100 = 0.174 cc

$\frac{50}{665} \times 5 = 0.38$ g Celite #28 = 0.165 cc

5.00 g blend 1.691 cc

TABLE 6-4. Odorless Alkyd Flat Wall Paint

	Nonvolatile comp. (lb)	Total weight (lb)	Gal
Rutile TiO_2		190.0	5.4
Lesamite		365.0	16.2
ASP 100		60.0	2.7
Celite 281		50.0	2.6
Odorless alkyd	117.0	334.0	47.4
Odorless M.S.		160.0	25.2
24% Lead drier		2.9	0.3
6% Cobalt drier		0.8	0.1
Anti-skin		0.1	0.1
Totals	117.0	1162.8	100.0

The oil absorption test requires 0.96 cc of linseed oil to produce a dry paste from the 5 g of blended pigment. The CPVC is therefore:

$$\frac{1.69}{1.69 + 0.96} = 0.638 = 63.8\%$$

Similar dry blends are next made with the same rutile TiO_2 (if this is to be used) and Lorite, and rutile TiO_2 and Camel-Carb in the ratio of 5.4/21.5 by volume (pigment-to-extender ratio in original pigmentation), the oil absorptions determined, and the CPVC's calculated as before.

The CPVC of 20% rutile TiO_2 and 80% Lorite by volume is 59.9%, that of a similar combination of rutile TiO_2 and Camel-Carb is 71.3%. The required blend of Lorite and Camel-Carb to produce a CPVC of 63.8% can now be calculated. (The following example uses the "rectangular rule.")

$$\begin{array}{lcr} 71.3 & \nwarrow\quad\nearrow & 63.8 \\ \underline{63.8} & \times & \underline{59.9} \\ 7.5 & \swarrow\quad\searrow & 3.9 \end{array}$$

Volume of Lorite to be used is:

$$\frac{7.5}{7.5 + 3.9} \times 21.5 = 14.15 \text{ gal}$$

Volume of Camel-Carb to be used is:

$$21.5 - 14.15 = 7.35 \text{ gal}$$

The new formulation in Table 6-5 will have the same P.I. as the original because both the CPVCs of the pigmentations and the PVCs of the two formulas are the same.

This substitution presupposes that the characteristics of the original formulation are satisfactory insofar as total pigment content and porosity are concerned. This might not necessarily have been the case. The formulation, for example, is very slightly porous:

$$\text{P.I.} = 1 - \frac{0.638 \times 0.330}{0.670 \times 0.362} = 0.132$$

If it were desired to eliminate this porosity, the relative amounts of Lorite and Camel-Carb could be altered to raise the CPVC of the pigmentation to equal the PVC of the formulation (see Table 6-6):

$$\begin{array}{lcr} 71.3 & \nwarrow\quad\nearrow & 67.0 \\ \underline{67.0} & \times & \underline{59.9} \\ 4.3 & \swarrow\quad\searrow & 7.1 \end{array}$$

Volume of Lorite to be used is

TABLE 6-5. Odorless Alkyd Flat Wall Paint

	Nonvolatile comp. (lb)	Total weight (lb)	Gal
Rutile TiO_2		190.0	5.4
Lorite		300.0	14.1
Camel-Carb		165.0	7.4
Odorless alkyd	117.0	334.0	47.4
Odorless M.S.		160.0	25.2
24% Lead drier		2.9	0.3
6% Cobalt drier		0.8	0.1
Anti-skin		0.1	0.1
Totals	117.0	1152.8	100.0

TABLE 6-6. Odorless Flat Wall Paint

	Nonvolatile comp. (lb)	Total weight (lb)	Gal
Rutile TiO_2		190.0	5.4
Lorite		170.0	8.0
Camel-Carb		300.0	13.5
Odorless alkyd	117.0	334.0	47.4
Odorless M.S.		160.0	25.2
24% Lead drier		2.9	0.3
6% Cobalt drier		0.8	0.1
Anti-skin		0.1	0.1
Totals	117.0	1157.8	100.0

$$21.5 \times \frac{4.3}{4.3 + 7.1} = 8.1 \text{ gal} = 170 \text{ lb}$$

Volume of Camel-Carb is

$$21.5 - 8.1 = 13.4 \text{ gal} = 300 \text{ lb}$$

It is also apparent that the original porosity of the film could be retained while raising the level of pigmentation. The combination of extenders just calculated for zero porosity could also be used, for example, at a higher PVC level:

$$\text{P.I.} = 0.132 = 1 - \frac{0.670\ (1\text{-PVC})}{\text{PVC} \times 0.330}$$

$$\text{PVC} = 70.2\%$$

6.11. HIDING POWER

It has also been assumed that the same amount of titanium dioxide will provide the same amount of hiding power with different extender combinations. This, too, might not necessarily be true.

The hiding efficiency of titanium dioxide is a function of its dilution in the dry film (10) and of the porosity of the film. Only those extenders possessing particle diameters of the same order of magnitude as titanium dioxide (0.2–0.3 μ) can serve as diluents, and it must therefore be recognized that the removal of a fine-particle-size clay (such as ASP-100) and its replacement with coarser extenders, will result in the loss of some white hiding power and tinctorial strength.

Similarly, a reduction in porosity, such as in the second example, will result in a further loss of hiding power because of the replacement of pigment/air interfaces with pigment/vehicle interfaces.

Dry hiding power, the additional increment of white hiding and tinctorial strength which is contributed by film porosity, is the least expensive form available and the paint formulator should always carefully appraise the true need for a nonporous film before committing his employer to unnecessarily high costs.

As Dr. J. S. Long was once heard to observe, "Nobody walks on a ceiling but flies!" The ceiling white, the calcicoater, and the one coat flat are all examples of products which either do not require a high degree of film integrity, or which require too much hiding power to obtain it economically from a nonporous film.

Porosity, if not excessive, may also be a desirable feature in flat wall paints which will be required to cover spackled cracks in plaster and joints in dry wall construction. Such "joint fillers" usually contain water and are all-too-frequently painted over before all of the water has had a chance to evaporate. When this occurs, the water will be sealed in by a nonporous flat wallpaint, producing unsightly sheen and color differences. A slightly porous film however, will permit this residual water to evaporate without harm to the finish.

Since this problem is greatest in tinted formulations, it is common practice to test for color uniformity over spackled joints by cutting a shallow 2-in.-wide groove in a plasterboard panel, filling it with water-base joint filler, and applying a coat of the flat wall paint to be tested as soon as the joint filler has flatted off.

Dry hiding in flat wall paints is obtained only by pigmenting above the CPVC. There are, however, a considerable number of choices as to how this may be done. Taking the nonporous formula of Table 6-6 as a starting point, all of the changes listed in Table 6-7 could be made without changing the total solids by volume. For the same change in CPVC, example F will produce more dry hiding power than example E because of the dilution effect.

TABLE 6-7. Development of High-Dry Hiding

			For same PVC increase
A	+ fine extender	PVC goes up	Highest increase in P.I.
	- vehicle solids	CPVC comes down	
B	+ Lorite	PVC goes up	Next highest increase in P.I.
	- vehicle solids	CPVC comes down	
C	+ Lorite + Camel-Carb	PVC goes up	Intermediate increase in P.I.
	- vehicle solids	CPVC stays same	
D	+ Camel-Carb	PVC goes up	Lowest increase in P.I.
	- vehicle solids	CPVC goes up	
E	+ Lorite	PVC stays same	—
	- Camel-Carb	CPVC comes down	
F	+ fine extender	PVC stays same	—
	- (Lorite + Camel-Carb)	CPVC comes down	
		TiO_2 dilution improved	

The amount of fine extender which may be introduced will generally be limited to 5 to 10% on the volume of the pigmentation because of the large viscosity increases that result from its high oil absorption.

Obviously, it is of considerable importance how the porosity index (P.I.) is raised as to the amount of dry hiding that may be expected for the sacrifice of an equal amount of film integrity.

Plots of hiding power versus P.I. for various pigmentations always have been found to produce straight lines. Different pigmentations produce different slopes to the line, however, and it is possible to select the type of pigmentation which offers the greatest increment of dry hiding power per unit of porosity by comparing these slopes. Fig. 6-6 illustrates the typical difference for similar pigmentations containing large and small particle size extenders.

6-12. RELATIVE HIDING POWER

If test formulations are sufficiently porous to begin with, the slope of the hiding-porosity line may be experimentally determined by simply adding vehicle to lower the PVC — and the P.I. When determining the hiding power of the let-down formulation, it is important to remember to correct for the

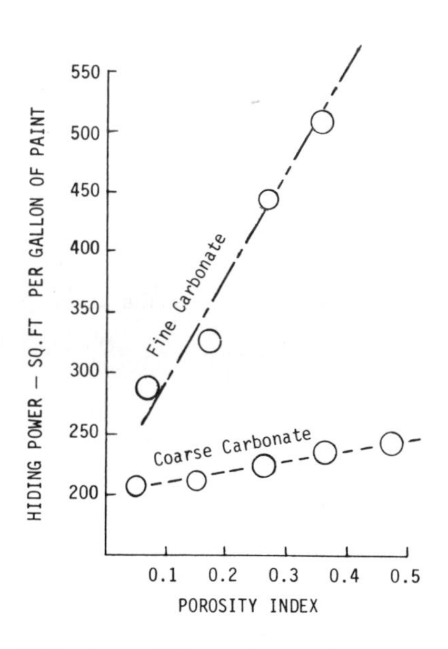

FIG. 6-6. Effect of extender size.

change in total solids and volume. This may be avoided, however, if the relative hiding power is determined by adding an increment of colorant and obtaining the scattering coefficient S derived from the Kubelka-Munk relationship (11):

$$\frac{K}{S} = \frac{(1-R_\infty)^2}{2R_\infty}$$

where K is the absorption coefficient
S is the scattering coefficient
R_∞ is the reflectance of tinted original

For the purpose of determining the slope of the hiding power-porosity curve it may be assumed that the absorption coefficient K does not change with dilution. This may not be quite true if the added vehicle is dark in color, but the absorption of the colorant (black is recommended) is so much greater that the vehicle contribution may be ignored.

It is only necessary to tint the original formulation before the addition of vehicle and to then make drawdowns or brushouts of both the original and the letdown. Reflectance readings are then made on the dry films and the ratio of the scattering coefficients calculated for the two PVC levels:

$$\frac{S_1}{S_2} = \frac{(1-R_2)^2 \times R_1}{(1-R_1)^2 \times R_2}$$

where S_1 is the scattering coefficient of original
S_2 is the scattering coefficient of letdown
R_1 is the reflectance of tinted original
R_2 is the reflectance of tinted letdown

If $P.I._1$ and $P.I._2$ are then calculated from the CPVC of the test pigmentation and the PVCs of the original and the letdown, a plot may be made of porosity versus scattering coefficient by assigning the value of unity to the original and the calculated S_1/S_2 ratio to the letdown.

It should be noted that the PVC of the letdown should never be lower than the CPVC of the pigmentation. If it were, we would no longer be measuring dry hiding, and the slope of the hiding power/porosity curve obtained would be incorrect.

6-13. HIDING POWER AND THE EFFECTIVE PVC

In discussing dry hiding power we have perhaps paid insufficient attention to the problem of obtaining sufficient hiding power in nonporous flat wall paints. This is more of a problem than it will appear to the neophyte, who may assume that it is only necessary to add more titanium dioxide if more hiding is required. This does not happen to be true.

As previously described, the finer pigment particles in a flat wall paint film are not allowed to distribute themselves uniformly throughout the dry film but are packed into the void spaces between the larger extender particles. There is a limit to this packing, however. Since the pigment particles must be wet with vehicle, they can never be packed any more closely together than the geometric arrangement existing at their own CPVC.

The lower the CPVC of a given prime pigment, the tighter its particles can be packed together in a flat wall paint film and the less efficient they will become in producing hiding power (12).

It has been common knowledge for many years that the hiding power which a pound of titanium dioxide is capable of producing is a function of the PVC at which it is dispersed. It has been less widely understood that it is no longer the PVC of the paint formulation which is involved once any significant volume of extender has been introduced, but the effective PVC at which the titanium dioxide is distributed.

There are two ways that the formulator may control this distribution. Since the CPVC of the prime pigment is a limiting factor, he can lower the limit by selecting a high-oil-absorption titanium dioxide. The effect of oil absorption is given by the formula (12):

$$\text{Hiding per pound} = C\left(1 - \sqrt[3]{\frac{\text{CPVC}}{0.74}}\right)$$

By arbitrarily setting the hiding power of a low-oil-absorption rutile TiO_2 at 100, it is possible to demonstrate the relative effect of oil absorption by calculating the data of Table 6.8.

TABLE 6-8. Hiding Power as a Function of Oil Absorption

Oil absorption	CPVC	Relative hiding power
20	52.3	100
25	45.6	130
30	41.2	155
35	37.5	174

In practice, the hiding power advantage provided by oil absorption in a nonporous film will be not quite as great as suggested by Table 6-8 because higher oil absorption is obtained in a titanium pigment by adding inorganic surface treatments (as much as 20% in some grades). This necessarily reduces the amount of titanium dioxide remaining. The overall effect, however, is still highly beneficial insofar as high-PVC hiding power is concerned, since 80% of 174 is still 139.

The second way that the formulator may control the distribution of the prime pigment is by adding a diluent — such as a fine-particle size extender.

Assuming, for example, that 200 lb of the low oil absorption (20) pigment of Table 6-8 have been used in a flat wall paint using only large-particle-size extender, the pigment will produce no more than the indicated relative hiding power. If, however, an equal volume of coarse extender were replaced with 50 lb of fine-particle-size calcium carbonate, it can be assumed that the titanium dioxide has now been diluted with an additional 2.22 gal of diluent (the volume of 50 lb of calcium carbonate):

$$\text{Effective PVC} = \frac{200 \times 0.0286}{\dfrac{200 \times 0.0286}{0.513} + 2.22} = 42.75\%$$

At this PVC, rutile titanium dioxide will have a relative hiding power of 147 as compared to 100 at a CPVC of 51.3 — an increase of 47%.

For the calculation to apply, however, the 50-lb replacement of fine-particle-size extender for the original coarse extender must not have lowered the CPVC of the combined pigmentation below the PVC of the formula. If this should happen, the hiding power increase would be even greater, but the film would become porous.

6-14. COMPROMISE IN FORMULATION

It will have become apparent by now that the formulation of a flat wall paint involves a great many compromises insofar as different properties are concerned. Any basic formulation may be modified by the formulator to accentuate those properties felt to be most desirable by his sales department — but generally not without the sacrifice of some other. More hiding power can be produced, for example, without raising cost, or cost may be lowered without affecting hiding power, but only at the sacrifice of washability, enamel hold-out, and color uniformity in tints.

6-15. SUMMARY

The optimum balance of all paint properties required is the responsibility of the paint engineer. The principles laid down in this chapter are his tools.

6-16. GLOSSARY OF SYMBOLS

c — an empirical constant characteristic of each pigment; for rutile TiO_2, c = 370; for anatase, c = 298

CPVC — critical PVC. Per cent pigment by volume in a pigment/binder mixture which provides just enough binder to satisfy the pigment surface and to fill in the voids between particles (%)

d — mean square pore diameter (cm)

k — a dimensionless constant

K — Kubelka-Munk absorption coefficient (cm^{-1})

L — mean continuous pore path (cm)

O.A. — oil absorption by spatula rub-out methods (lb oil/100 lb pigment)

P — substrate capillary pressure (dyn)

P.I. — porosity index. Volume of air in dry paint film as per cent of total binder demand of pigment (%)

PVC — pigment volume concentration. Per cent pigment by volume in total nonvolatile (%)

R — green reflectance. Per cent of incident light reflected ($45^\circ : 90^\circ$) (%)

R_∞ — reflectance of infinitely thick film (%)

S — Kubelka-Munk scattering coefficient (cm^{-1})

t — time (sec)

η — vehicle binder viscosity (poise)

V — volume of vehicle binder penetrating unit area of substrate (cc)

REFERENCES

1. F. B. Stieg, Paint Logic, April--May (1953), 25, No. 4-5, 20 (1953).
2. C. C. Furnas, Ind. Eng. Chem., 23, 1052 (1931).
3. W. K. Asbeck and M. Van Loo, Ind. Eng. Chem., 41, 1470 (1949).
4. F. B. Stieg, Official Digest Fed. Soc. Paint Technol., 28, No. 379, 695 (1956).
5. Leneta Co., P.O. Box 576, Ho-Ho-Kus, N.J. 07423.
6. F. B. Stieg and D. F. Burns, Official Digest Fed. Soc. Paint Technol., 26, 81 (1954).

7. F. B. Stieg, J. Paint Technol., 39, No. 515, 703 (1967).
8. F. B. Stieg and R. I. Ensminger, Official Digest Fed. Soc. Paint Technol., 33, No. 438, 792 (1961).
9. F. B. Stieg, Am. Paint J., 43, No. 2, 106 (1958).
10. F. B. Stieg, Official Digest Fed. Soc. Paint Technol., 31, No. 408, 52 (1959).
11. P. Kubelka and F. Mulk, Z. Tech. Phys., 12, 593 (1931).
12. F. B. Stieg, Official Digest Fed. Soc. Paint Technol., 34, No. 453, 1065 (1962).

Chapter 7

INTERIOR FINISHES

Oliver Volk and Murray Abriss

E. I. du Pont de Nemours Co.
Marshall Laboratory

7-1. INTRODUCTION

The diversity of end uses as well as the different types of vehicles available for formulation provide a good basis for partitioning this chapter. However, there are a great number of common requirements for interior finishes. A section regarding these might help to set the stage for specific discourse on the individual products.

Interior products are not required to have exceptional durability characteristics related to exposure to sunlight, rain, temperature extremes, wind, or others. However, they are normally expected to perform satisfactorily as a decorative finish for the life of the surface to which they are applied. For interior surfaces, the span of service may encompass 50 years or more. For example, the woodwork trim on doors, windows, and other surfaces is a relatively permanent part of the dwelling. The first coat of paint applied to such a surface must adhere and provide a stable base for all subsequent applications of decorative coatings. When you take into account variation in substrate types of wood, plaster, metals and others, combined with types of topcoat and service and maintenance conditions, the requirements for an interior finish are formidable.

The coating must impart protection, decoration, and ease of maintenance as well as eventual recoatability to the surface on which it is applied. Since interior surfaces encompass a wide variety of materials of differing textures, porosity, shape, and size, and since consumers differ in their appreciation of gloss or sheen, a discourse on interior finishes needs to be divided into short items dealing with specific types of finishes for the various end uses.

This chapter has therefore been segmented to include information on primers and undercoats for walls and woodwork, and also interior flats, semigloss, and gloss finishes. Since the technology of solution-type and latex-based finishes is very different, the chapter has also been divided to deal independently with formulating techniques for each of these finish categories.

7-2. SOLUTION-TYPE FINISHES

This category of paints uses vehicles in which the binder is in solution in a solvent. The purpose of the binder is to bind the pigment particles into a coherent film and provide adhesion of that film to the surface on which it is applied. The purpose of the solvent is to provide fluidity to permit easy brush or roller application properties. In order for the paint to become a serviceable film, the solvent must evaporate and the binder must harden to a tack-free state. The mechanisms of oxidation, drying, etc., have been covered in other chapters and are not repeated here.

Interior solution finishes have undergone a transition over the years depending on (a) the state of the art, (b) the development of the science of modern polymers, and (c) the availability and cost of raw materials. Until the early 1930s, the vehicles used were based on bodied and heat-treated oils usually combined with rosin, ester gum, or other natural gums. With the development of alkyd resin technology, starting about 1927, formulators were able to prepare a variety of architectural enamels for interior and exterior uses. However, the bulk of enamel production for architectural use was made with bodied oil and/or varnish as the binder until after 1945. After World War II, paint manufacturers quickly switched to alkyd resin based formulas and they remain the most utilized type of solution type formulas to this day.

As previously mentioned, the role of the volatile solvent in a solution-based finish is to solubilize the resin binder to provide fluidity and workability to the paint during application. On evaporation, the solvent becomes a part of the atmosphere and is concentrated primarily in the area in which the paint is being applied. Aromatic solvents and normal mineral spirits are penetrating, odorous, and unpleasant, and a great deal of ventilation is required when paints containing them are being applied. With the tremendous increase in "do-it-yourself" consumer painting and with the advent of low-odor, latex-based paints, the petroleum industry developed an odorless mineral spirit for use with consumer paints. The solvent is composed of essentially straight and branched chain aliphatic hydrocarbons, C_{10} to C_{12}. It is a relatively poor solvent compared with regular mineral spirits and use of the odorless mineral spirits in consumer paints required reformulation of the resin binders to provide compatible solutions. While adequate ventilation is still required for safety, the odor during application is low enough to permit prolonged exposure without undue discomfort.

The following paragraphs describe formulating principles for primer, undercoats, flat wall finishes, and gloss and semigloss enamels. Many variations are possible for the formulas presented, and each formulator will adjust ingredients to provide the balance of film properties or cost that he desires. It should be borne in mind that architectural paints are sold principally through consumer outlets and are often subject to long periods of storage before they are sold and used. Consequently, formulations developed for this end use must have good stability in the package and still perform as intended when finally used. Also, each formulator must know the specific requirements for the product he develops and must adjust the brushing properties, color, drying speed, odor, hiding, gloss, flow, leveling, and many other properties to provide the best balance of properties for his customer.

In most cases, an adjustment in the formula ingredients to affect an improvement in one or more of these properties will also have a deleterious effect on one or more other properties of the paint. When the raw material cost and processing are added, a formula adjustment rarely results in an improvement in all properties in addition to an equal or lower cost.

A. WALL PRIMER AND SEALER

These products are specifically made for application to interior wall surfaces comprising plaster, plaster board, masonite and composition board, brick, concrete, and others. The purpose of the coating is to provide uniform porosity and low penetration of the successive topcoats applied thereon. Since it is generally topcoated, the film is not required to be completely opaque, but the composition must be compatible with common tinting materials on the market so it can be adjusted to match the color of the topcoat being applied.

For most architectural uses, an alkyd resin-based vehicle is very satisfactory and preferred for property balance as well as cost. However, for those applications where extreme moisture or alkaline conditions exist, special chemically resistant vehicles are required to prevent reaction and loss of adhesion of the primer.

The product should be easy to apply and should require no longer than overnight drying to set to a hard finish capable of being recoated. It should provide a relatively smooth surface so as not to introduce texture into the surface of the paint system.

A typical primer-sealer formula for use on interior plaster and plaster board is shown in Formulation 7-1.

FORMULATION 7-1. Alkyd Primer-sealer[a]

Ingredients	lb/gal
Rutile titanium dioxide [1]	1.40
Zinc oxide [2]	0.20
Med. particle size calcium carbonate [3]	4.40
Soya lecithin [4]	0.10
Long oil soya oil alkyd resin, 45% solids	3.80
Lead tallate drier, 24% metal	0.08
Cobalt naphthenate drier, 6% metal	0.02
Odorless mineral spirits	1.60
Gal weight	11.60

[a]See Table 7-2 for references.

This combination of ingredients should be dispersed by sand mill or high-speed dispersion equipment to a fineness of 1.00 mil and will give a Stormer viscosity in the range of 70 KU. Volume solids is 44% and pigment volume concentration is 55%. Proper selection and processing of the ingredients is important. TiO_2 should be a high oil absorption type normally recommended and used in flat wall paint formulation. This type generally has a high degree of surface treatment to give good hiding in high pigment volume concentration (PVC) formulations. The zinc oxide should be small particle size. It functions to react with acidic components of the vehicle to form soaps which control flow and sag resistance at a desirable balance, and also prevents hard settling of the other pigments. Calcium carbonate should have a relatively small to medium particle size to produce a smooth surface for the primer film and permit application of succeeding coats with little or no texture from the primer. Gloss of the paint film will be in the range of a high-sheen flat wall paint with enough sheen to make it easy to spot low-gloss areas where the porosity of the substrate was greatest. Despite the lower gloss of the sealer film at these points, the surface is normally sealed, unless the low-gloss is the result of a "holiday," or exceptionally thin film of paint.

Choice of resin is the most critical of the primer ingredients. Since it must "seal" the surface, it must either penetrate and fill the small voids in the surface to prevent penetration of topcoats, or it must bridge them and thereby prevent further penetration. Most present day primer-sealers are of the latter type. The resin used in Formulation 7-1 is a high-viscosity (Z to Z-3) long oil soya, linseed, or tall-oil pentaerythritol alkyd resin at a solids of 45% and is reacted until it is fairly close to the gel point. Such resins build viscosity rapidly upon loss of solvent and quickly become insoluble as they dry, thereby preventing penetration. It is important to maintain use of "odorless" mineral spirits in formulations which utilize resins formulated and cut in odorless mineral spirits. Some loss of sealing occurs when thinners with higher solvency are used.

The above formula may be modified in a number of ways to achieve a different property/cost balance. An increase in PVC will bring the formula closer to the critical pigment volume concentration (CPVC) and additional hiding will be obtained by virtue of the pigment air interfaces developed. This would allow reduction in TiO_2 level and lower cost but at a sacrifice in sealing. The drier balance should also be adjusted to conform to the drying characteristics of the actual resin being used. It might be necessary to include a small amount of an inhibitor to prevent skinning and gelation of partially filled containers. Proprietary additives on the market can be used to replace soya lecithin for flow control. They should be investigated if a higher degree of sag resistance, thixotropy, and viscosity are desired. Use of different types of lead driers can also produce effects in pigment wetting and flow and should be investigated to provide optimum product performance.

Primer-sealers of this type should be tested for sealing ability by applying them over a substrate which is known to be very nonuniform in porosity. Application of a semigloss tint over the dried coat of primer-sealer and observation of any gloss and color variations will indicate the ability of the sealer to provide a uniformly sealed substrate.

B. ENAMEL UNDERCOATS

An enamel undercoat must fill a variety of needs in the task of obtaining a good paint finish. The undercoat seals and fills small surface defects, such as wood grain, in the substrate. It provides color and hiding to enrich the appearance of the system and it provides a good adhesive link between the substrate and topcoat. A good undercoat will provide fullness and depth and is a stabilizing intermediate coat which minimizes such surface defects as alligatoring, crawling, and wrinkling caused when multiple coats of enamel finishes are used. Since it is an important part of the finishing system, the formulation of an undercoat should be regarded with equal importance to that of formulation of finish coats.

An undercoat should be formulated to have good brushing properties in order to permit easy application with wide brushes on large surfaces and yet be workable with small tools on wood moldings and trim work. It should have good flowing properties so as not to interfere with the smoothness of the finish system, yet it should have enough resistance to sagging so it will not run in the normally thicker coats, which occur in moldings or in corners. It should be dry hard enough overnight to permit sanding and topcoating the following day. It should provide enough hiding so that only a single coat of topcoat will normally be needed to provide complete hiding and color uniformity. It should be compatible with available tinting colors to permit custom shading where desired. It is also desirable that it have good color retention so as not to interfere with the color retention of the topcoat and the system.

A typical undercoat formulation is shown in Formulation 7-2.

Here too, proper selection of the amount and types of the raw materials can have a big effect on product performance. The rutile titanium dioxide selected should be one of the types recommended for enamels. At the pigment volume of this formulation, the use of enamel grade titanium dioxide will not sacrifice any significant amount of hiding but will contribute to flow and leveling. The diatomaceous silica is the medium particle size type and is used to provide good sanding and good "tooth" for adhesion characteristics. Medium particle size calcium carbonate is used as the inert pigment since it is easily dispersible and contributes good sanding properties at the PVC of this formula. Fine particle size zinc oxide is

FORMULATION 7-2. Alkyd Undercoat[a]

Ingredients[b]	lb/gal
Rutile titanium dioxide [20]	2.30
Med. particle size, diatomaceous silica [5]	0.50
Med. particle size, calcium carbonate [3]	3.40
Zinc oxide [2]	0.20
Long oil soya alkyd resin, 45% solids	2.55
Medium oil soya alkyd resin, 55% solids	1.45
Soya lecithin [4]	0.05
Lead naphthenate drier, 24% metal	0.08
Cobalt naphthenate drier, 6% metal	0.04
Calcium naphthenate drier, 4% metal	0.08
Odorless mineral spirits	1.25
Gal weight	11.90

[a]See Table 9-2 for references.

[b]The combination of above ingredients should be dispersed by sand mill or high-speed dispersion equipment to a fineness of 1.00 mil and should result in a product having the following constants:
Stormer viscosity is 85 KU, volume solids is 48%, pigment volume concentration is 53%.

used for many reasons, but its main functions are to minimize penetration, thereby improving holdout properties, to promote hardness of the films, and to aid in control of flow and leveling properties. It also improves suspension of the pigment and prevents excessive yellowing of the film when exposed to light. Use of zinc oxide also improves brush application properties, probably as a function of the zinc soaps formed by reaction with acidic components of the resin vehicle. Care should be exercised to use the proper amount and grade of zinc oxide to prevent excessive hardening and embrittlement of the film and also to prevent excessive reaction and viscosity increase with high acid vehicles.

Selection of the proper resin is important in order to reach the desired property balance. Use of the same type of alkyd resin in Formulation 7- would provide a product with good sealing, but it would be deficient in filling and sanding properties due to low solids and slow dry. On the other hand, use of harder drying, higher solids resins would result in poor sealing of the formulation over porous surfaces.

A blend of two such resins has therefore been used to provide optimum performance of the undercoat for sealing, filling, dry, and sanding. This technique of blending resins to provide optimization of various properties has been found to be a very useful tool in consumer paint formulating work. Many times the property balance achievable by a blend of two vehicles is not obtained when a single cook resin with the same composition is used. It is also of benefit in a manufacturing operation since a separate resin for each product need not be stored and a complete line of consumer paints can be made with only two or three standard alkyd resins.

The resin blend used in Formulation 7-2 is a 60/40 blend by weight on the solids of the alkyd resin used in the primer-sealer Formulation 7-1, combined with a medium oil soya pentaerythritol alkyd resin, both cut with odorless mineral spirits. The first resin provides sealing and holdout, while the second provides volume solids and hardness for filling and sanding. The soya oil resin may be replaced with resins made using linseed or tall oils, but consideration should be given to the possible yellowing effect of the film. The undercoat as formulated provides an enamel-like film and can be used as a topcoat if desired. Yellowing of the film with linseed resins would then be a disadvantage, as well as when used as an undercoat for white or light topcoats, which do not have complete hiding.

Soya lecithin is used as a dispersing aid, to inhibit pigment settling, and to control flow and leveling. The drier combination is satisfactory to produce the desired drying characteristics but should be modified to conform to the drying properties of the specific resins used. If needed, the lead naphthenate can be incorporated in the pigment dispersion to provide improved dispersibility. Drying inhibitors such as thymol and butyraldoxime can be added if resistance to skinning is required.

Undercoat formulations should be tested for a number of properties to ensure proper performance. Sanding should be tested at 12, 16, and 24 hrs. to determine if the finish "powders" satisfactorily or if it is "gummy" and causes the sandpaper to fill with soft cheesy paint. If too much powdering results, the film may be too hard and may become brittle and lose adhesion. If it is too soft, it may need a rebalancing of the driers, a harder resin, or other formula adjustment. Sealing should be checked as described in Section 7-3 A on primer-sealers. Adhesion should be checked by knife and crosscutting then applying a strip of adhesive plastic tape over the cut area and pulling it off with a sudden yank. Adhesion of topcoats to the sanded and unsanded surface of the undercoat may be tested in the same manner.

C. FLAT WALL FINISHES

In the past 30 years the formulation of flat wall solution-type finishes has undergone many changes. Some of the types of vehicles used include heat-bodied oil/varnish blends, processed-bodied oils, heat-bodied and blown oils combined with alkyd resins, alkyd resins, and gelled alkyd resins. Since 1950, however, the vehicles most used for interior flat wall finishes have been the long oil alkyds cut with odorless mineral spirits. These paints have offered a balance of performance characteristics and cost which has gained universal popularity.

Flat wall paints are used to periodically coat the interior wall surfaces of homes, office buildings, schools, hospitals, and others, and many coats of paint accumulate on a wall in the course of the years. It is very important, therefore, that these paints be formulated to have a high degree of permanent intercoat adhesion to prevent massive failure of the paint surface over the large interior surfaces. The main causes of adhesion failures on interior wall surfaces are embrittlement of the film and reactivity with damp, highly alkaline surfaces, such as new plaster. The solution to the problem of embrittlement can be obtained with proper formulation, but the solution to the problem of alkali reactivity can be achieved either by (a) using special alkali-resistant primers, (b) allowing the plaster or masonry surface to age and dry out, or (c) using latex-based finishes that do not react. All of these are used as solutions to the problem, but since the majority of solution wall paint used is over properly aged surfaces, most manufacturers use alkyd resins for wall paint formulations.

Since most wall paint is used over existing surfaces which must be patched with spackle or plaster, it is desirable that the applied paint dry to a uniform appearance despite the wide differences in surface porosity. This is accomplished through proper choice of alkyd resin binder and the type and amount of extender pigments, as well as through formulation below the critical PVC for the pigment/resin combination being used. This prevents a large increase in hiding power when a small portion of the binder is absorbed into a porous area and maintains color and gloss uniformity of the flat finish. Pigmentation of flat wall paints is covered rather extensively in Chapter 6 on flat wall paints. Therefore, this chapter deals only cursorily with the pigmentation used in the suggested formulations.

The surfaces on which flat wall paints are applied are generally non-uniform in texture, smoothness, or "planeness," and the lack of gloss reflection from the flat finish can do a lot to hide or mask such surface imperfections. Most flat wall paints are formulated with an 85° gloss of less than 10 and perform satisfactorily in this regard. Formulation at a higher sheen to achieve certain aesthetics or cleanability generally results in a loss of uniform appearance over rough surfaces. Flat finishes are subject to marring and burnishing when the surface is scraped with a hard object or rubbed repeatedly with a brush or cloth during a cleaning operation. The result is a change in gloss and color of the affected area. Proper choice of pigments can help to minimize the above effects and provides the formulator with a means of optimizing the property balance of his paint. A suggested flat wall paint formulation is shown in Formulation 7-3.

The film will have good hiding and color holdout, low gloss, good scrubbability, and nominal raw material cost. For production of a line of colors, the formula can be modified to lower the TiO_2 content considerably, since hiding lost due to loss in refraction of the TiO_2 removed can normally be offset by the absorption of the colored pigments used to produce the tints. Care should be observed in selection of tinting pigments to provide nontoxic formulations with color stability and permanence. Most tints found in a normal "line" of flat wall paints can be made using the following abbreviated list of pigments (see Table 7-2 for references):

Carbon black [8] for gray, brown, and grayed colors
Ferrite yellow [9] for low brightness yellows, oranges, and greens
Dalamar yellow high-strength mono-azo yellow [10] for bright yellows, oranges, and greens
Phthalocyanine blue [11] for blues
Phthalocyanine green [12] for greens
Synthetic iron oxide [13] for low-brightness reds, pinks, brown and others
Naphthanil red azo pigment dye [14] for bright pinks

The usual practice followed in manufacture of tints is to prepare separate single pigment dispersions of the colored pigments using sand mill, ball or pebble mill, or other good dispersion equipment. These dispersions are then added to the base wall paint batch in small amounts to produce the tints. The batch must be thoroughly mixed and tested to ensure complete incorporation and uniformity of the color.

Flat wall paint should be tested to assure the formulator that it has the desired level of scrub resistance and cleanability. Federal Specification TT-P-141-B, Methods 614-1 and 614.2 can be used to check scrub resistance after four days' drying time, and stain removability after five days' drying time.

Self-sealing and color holdout can be tested by application of a coat of the flat over a surface having nonuniform porosity. Test boards such as this can be made by applying 3-in. stripes of paints with widely different sealing and porosity characteristics onto 1/4 in. composition board. When thoroughly dry, the board is coated with the flat wall paint (tinted to a medium green color) and allowed to dry. Uniformity of color and gloss over the various subcoatings indicates the level of color holdout of the paint. This

FORMULATION 7-3. Alkyd Flat Wall Paint[a]

Ingredients[b]	lb/gal
Rutile titanium dioxide [1]	1.90
Med. particle size calcium carbonate [3]	2.10
Med. particle size calcium carbonate [6]	2.10
Zinc oxide [2]	0.20
Small particle size diatomaceous silica [7]	0.20
Med. particle size diatomaceous silica [5]	0.60
Long oil soya alkyd resin, 45% solids	3.20
Lead tallate drier, 24% metal	0.07
Cobalt naphthenate drier, 6% metal	0.03
Odorless mineral spirits	1.80
Gal weight	12.20

[a]See Table 7-2 for references.

[b]The combination of ingredients should be processed in high-speed dispersion equipment to a maximum fineness of 2.5 mils and will provide the following constants: Stormer viscosity is 90 KU; PVC is 65%; volume solids is 46%.

type of testing simulates the porosity differences encountered on walls which have been repaired with plaster or spackle.

Formulation 7-3, as well as Formulations 7-5 and 7-6, can be modified by TiO_2/extender adjustment to provide light and deep bases for custom color systems. Tinting strength tests must be run to assure uniform color production from batch to batch and between various gloss qualities when tinted. Also, the drier level must be balanced and inhibitor added to prevent skinning in partially filled cans.

Formulation 7-3 can also be modified through choice of extender pigments of low flatting efficiency to provide finishes with "eggshell" or high angular sheen (85° gloss, 15 to 30). Care must be used to assure uniform gloss in the lapped areas on application of these finishes.

The pigmentation shown in Formulation 7-3 has been chosen because of its good color holdout properties. Many different pigmentations can be used in formulation of good alkyd flat wall paints and all have a slightly different balance of properties. Some of the types of pigmentations used for flat wall paints on the market can be seen in Table 7-1. This has been constructed from label analysis of flat wall paints produced by some of the large national paint manufacturers and the composition of Formulation 7-3 is shown for comparison.

Enamel-type flat wall finishes with excellent scrub resistance and cleanability, good flow and hiding, and very smooth texture have also been made. These are generally low-pigment volume formulations, where the low gloss is achieved by use of high flatting extender pigments. A formula of this type is shown in Formulation 7-4.

FORMULATION 7-4. Flat Wall Enamel

Ingredients[b]	lb/gal
Rutile titanium dioxide [20]	3.50
Fine particle size diatomaceous silica [7]	1.60
Zinc oxide [2]	0.30
Long oil soya alkyd resin, 45% solids	2.40
Med. oil soya alkyd resin, 55% solids	1.20
Soya lecithin [4]	0.05
Lead naphthenate drier, 24% metal	0.07
Cobalt naphthenate drier, 6% metal	0.02
Butyraldoxime inhibitor [15]	0.01
Odorless mineral spirits	1.95
Gal weight	11.10

[a]See Table 7-2 for references.

[b]Pigment dispersion is accomplished on a sand mill and fineness should be 1.0 mil maximum. The formulation has the following constants:
Stormer viscosity is 80 KU; volume solids is 40%; PVC is 50%.

TABLE 7-1. Flat Wall Paint Compositions

Pigmentation	Paint A	Paint B	Paint C	Paint D	Paint E	Paint F	Formulation 7-3
Titanium dioxide	14.8	—	4.8	10.1	9.6	11.5	15.6
Titanium calcium	—	34.5	29.4	—	37.3	28.9	—
Calcium carbonate	32.4	34.5	12.4	44.9	2.1	7.2	34.4
Zinc oxide	—	—	—	1.6	—	—	1.6
Silica	—	—	—	8.0	4.3	—	6.6
Silicates	—	—	—	—	—	18.8	—
Silica and silicates	11.3	—	10.0	—	—	—	—
% pigment	58.5	69.0	56.6	64.6	53.3	66.4	58.2
Vehicle							
Alkyd resin	10.0	7.8	10.9	9.9	13.6	8.1	11.8
Drier	—	—	—	—	2.3	0.2	—
Mineral spirits	31.5	23.2	32.5	25.5	30.8	25.3	30.0
% vehicle	41.5	31.0	43.4	35.4	46.7	33.6	41.8

The applied film has the application properties, flow, and smoothness associated with enamel-type coatings, but holdout is marginal and requires a farily well sealed substrate for it to provide uniform color and gloss.

D. SEMIGLOSS FINISHES

Formulation of semigloss finishes requires consideration of their use on trim, doors, moldings, and other surfaces, as well as considering that they are used on large wall areas where high sheen, good wearability, and cleanability are desired. Semigloss finishes are often used for walls and ceilings of kitchens and bathrooms because of their easy maintenance. One of the most critical problems in formulation of semigloss finishes is nonuniformity of gloss of lapped areas on large surfaces. This can be overcome by use of relatively high boiling solvents to keep the film open and fluid for a long period to permit lapping during this time, and also by maintaining a 60° gloss above 30. Semigloss formulations in the eggshell range below 30 are prone to development of gloss and color differences in the lap areas.

Semigloss products are normally formulated for application over previously painted or primed surfaces. Since they do not necessarily have to be self-sealing, the formulator can use whatever latitude in choice of ingredients is necessary to provide the smoothness and uniformity desired for these finishes.

A typical formula for an alkyd semigloss enamel is shown in Formulation 7-5.

Care should be exercised to provide uniform dispersion and mixing of each batch to assure gloss reproducibility from batch to batch. Small adjustments in gloss may be made by adding the 55% solids alkyd resin to raise the gloss or by adding a dispersion of the diatomaceous silica to lower the gloss. Large adjustments should not be attempted since the hiding power and other properties might be undesirably affected. Compensating batches with adjusted dispersions should be used for large gloss adjustments.

The suggested formula may be modified in a number of different ways to increase or decrease hiding power, gloss, solids, and cost. Tints can be made as suggested in Section 7-2C flat wall finishes. The revised formula should be thoroughly tested to assure that the desired balance of properties and stability has been achieved. Laboratory tests as well as field tests involving practical applications of consumer formulations are desirable to insure against some unforeseen failure under a given set of conditions and substrate.

E. GLOSS ENAMELS

Gloss enamels can be used for all interior surfaces provided the surface has been properly prepared. They are not designed to be self-sealing,

FORMULATION 7-5. Alkyd Semigloss Enamel[a]

Ingredients[b]	lb/gal
Rutile titanium dioxide [20]	3.50
Zinc oxide [2]	0.30
Med. particle size diatomaceous silica [5]	0.40
Fine particle size calcium carbonate [16]	1.60
Long oil soya alkyd resin, 45% solids	2.40
Med. oil soya alkyd resin, 55% solids	2.00
Soya lecithin	0.10
Lead naphthenate drier, 24% metal	0.07
Cobalt naphthenate drier, 6% metal	0.03
Odorless mineral spirits	1.20
Gal weight	11.60

[a] Refer to Table 7-2 for references.

[b] The ingredients should be dispersed to yield a fineness of 1.5 mil maximum. The following constants will result: 60° gloss is 45; stormer viscosity is 90 KU; volume solids is 45; PVC is 45.

or filling. In fact, their very composition tends to conflict with any concept of sealing. Since they are always used as topcoats on wood, metal, and other surfaces, their principal function in the home is that of providing a decorative surface. In industrial applications they also function to provide surfaces which retard dirt and dust pickup and which are easily cleanable. The film must have a uniform high gloss and a lot of build or depth of appearance. To accomplish this it must be formulated with a low pigment volume with little or no extender pigment, and with higher solids/lower viscosity resins than are used for sealers and flats.

Two of the most important requirements of a gloss enamel film are (a) that it provide and maintain a high uniform gloss, and (b) that it retain its original color. Since a great proportion of the gloss enamels sold are used as "white" on cabinets, doors, and trim, it is important that the formula be resistant to yellowing. This is accomplished by paying close attention to the following variables in the formulation:

1. Use of a soya alkyd or one with equivalent yellowing resistance.
2. Inclusion of zinc oxide.
3. Minimum drier content.
4. High titanium dioxide content.

Maintenance of high gloss requires formulation along the following lines:

1. Low PVC.
2. Good dispersion.
3. Low anti-flow and drier content.
4. Compatibility of resin with volatile ingredients.

A suggested formulation for an all-purpose gloss enamel is shown in Formulation 7-6.

FORMULATION 7-6. Utility Gloss Enamel[a]

Ingredients[b]	lb/gal
Rutile titanium dioxide [20]	2.80
Zinc oxide [2]	0.20
Med. particle size calcium carbonate [3]	0.50
Medium oil soya alkyd resin, 55% solids	4.30
Long oil soya alkyd resin, 70% solids	1.40
Saturated short chain oil acids [17]	0.03
Lead naphthenate drier, 24% metal	0.06
Cobalt naphthenate drier, 6% metal	0.02
Calcium naphthenate drier, 4% metal	0.03
Thymol inhibitor, 10% solution [18]	0.01
Blue tinting pigment [19]	Trace
Odorless mineral spirits	1.05
Gal weight	10.40

[a] See Table 7-2 for references.

[b] Constants for the formula will be as follows:
Visc. #15 Parlin cup is 20 sec; PVC is 23%; vol. solids is 48%; 60° gloss is 85 minimum.

The composition of Formulation 7-6 is dispersed in a sand mill, pebble mill, or other equipment capable of achieving a fineness of 0.5 mil. The oil acids are included in the grinding portion of the formula to produce zinc soaps for flow and sag control. Care must be taken to insure uniform moisture content of pigments and other ingredients, since water can cause a decrease in flow and gloss of the film. The calcium drier in the formula provides drier stability and prevents formation of seed. An inhibitor, such as thymol, prevents skin drying and helps provide a sharp gloss in addition to its function as a skinning inhibitor for package stability.

Blue or purple, red, and black tinting pigments, such as Monastral and ultramarine blues, are normally used in white products to increase their apparent whiteness. They change the hue of the formulation from yellow-white to blue-white, a color most people perceive as being "whiter." It also brings the color in line with other white appliances and porcelain sinks, used in the home. The blue offsets, to a large extent, the normal yellowing effect of the oil/alkyd vehicle on aging. Tints and colors can be made using the colored pigments suggested in Section 7-2C on flat wall finishes.

Formulation 7-6 will dry to a high gloss in 6 to 8 hours and will have good flow and color retention. It can be tinted to provide a range of colors and can be used for most painting chores around the house including furniture, bicycles, wagons, woodwork, walls, and appliances. A higher solids and higher hiding modification of it is more suited for professional painting where speed and foolproof application is required (see Formulation 7-7).

Formulation 7-7 can be dispersed like Formulation 7-6. It will have a higher hiding power and gloss and easier application properties (particularly over large surfaces), a softer, slower drying film, and better color retention than Formulation 7-6. It can be modified to replace titanium dioxide with calcium carbonate extender to provide a high quality product for maintenance work on schools, offices, plants, etc.

In the preceding formulations an attempt has been made to give a numerical reference in parentheses for specific ingredients. Table 7-2 shows typical material grades, identification, and source of supply for the materials so marked. This list is not meant to be discriminatory in any way. Comparable qualities can be obtained from many other suppliers. The list is supplied only as a guide to one of the many suppliers of these materials to the paint industry. In the formulations there are also alkyd resins, driers, and odorless mineral spirits used which have no indicated reference but which are available universally from many different suppliers. The choice of supply source is left to the discretion of the formulator. Adjustments in ingredient amounts should be made when specific resins are chosen to insure substitution on an equivalent weight solids or volume solids basis to obtain satisfactory performance in the suggested formulation.

FORMULATION 7-7. Architectural Gloss Enamel[a]

Ingredients	lb/gal
Rutile titanium dioxide [20]	3.30
Zinc oxide [2]	0.20
Medium oil soya alkyd resin, 55% solids	3.50
Long oil soya alkyd resin, 70% solids	1.70
Saturated short chain oil acids [17]	0.02
Lead naphthenate drier, 24% metal	0.03
Cobalt naphthenate drier, 6% metal	0.02
Manganese naphthenate drier, 6% metal	0.01
Calcium naphthenate drier, 4% metal	0.02
Blue tinting pigment [19]	Trace
Odorless mineral spirits	1.40
Gal weight	10.20

[a]See Table 7-2 for references.

7-3. INTERIOR LATEX FINISHES

Interior latex paints include primer-sealers, flat wall paints, semigloss, and gloss products. The flat wall paints and primer-sealers were the first to be marketed, followed more recently by the semigloss and gloss products. Today, better than 95% of all interior wall paints used by the home owner are latex-based. Within the past several years the semigloss latex enamels have also begun to displace the conventional alkyd semigloss products used in the home, with full gloss latex products just beginning to scratch the surface of the interior finishes market.

The popularity of water-based latex paints, particularly with the home owner who prefers to do his own interior painting, is attributable to the ease of application, fast drying characteristics, ease of cleanup with soap and water, and lack of solvent odors. In addition to these advantages, the paints may be applied to damp surfaces without adverse effects and may be

TABLE 7-2. Raw Material Suppliers

Reference	Ingredients	Source of supply
[1]	Ti Pure R-931 titanium dioxide	E. I. du Pont de Nemours & Co.
[2]	Black Label #20 zinc oxide pigment	St. Joseph Lead Co.
[3]	Gamaco calcium carbonate pigment	Georgia Marble Co.
[4]	N. lecithin	Cargill Corporation
[5]	Celite 375 diatomaceous earth	Johns-Manville Products Corp.
[6]	Kalmac calcium carbonate pigment	Georgia Marble Co.
[7]	Celite 266 diatomaceous earth	Johns-Manville Products Corp.
[8]	Soft Raven Beads #11 carbon black pigment	Columbian Carbon Co.
[9]	Marpico Yellow 1150 iron oxide pigment	Columbian Carbon Co.
[10]	Dalamar Yellow high-strength mono-azo yellow pigment YT-717-D	E. I. du Pont de Nemours & Co.
[11]	"Ramapo" Blue FR resinated phthalocyanine blue pigment	E. I. du Pont de Nemours & Co.
[12]	Monastral Green GT-751-D phthalocyanine green pigment	E. I. du Pont de Nemours & Co.
[13]	Kroma Red RO-3097 pure iron oxide pigment	C. K. Williams & Co.
[14]	Naphthanil Red RT-531-D azo pigment dye	E. I. du Pont de Nemours & Co.
[15]	Exkin #1 anti-skinning agent	Tenneco Chem. Inc.-Nuodex Prod. Div.
[16]	Gama-sperse 140 calcium carbonate pigments	Georgia Marble Co.
[17]	"Neo Fat" #8 fatty acids	Armour Industrial Supply Co.
[18]	Thymol NF standard anti-skinning agent	Givaudan Corp.
[19]	Ultramarine Blue 6130 pigment	Holland-Suco Color Co.
[20]	Ti-Pure R-900 titanium dioxide pigment	E. I. du Pont de Nemours & Co.

recoated the same day or, under good drying conditions, within several hours. Water, as the volatile component in the paint, offers ready availability, it is inexpensive, nontoxic, odorless, and nonflammable.

The early water-based interior paints were alkyd resin emulsions stabilized with large amounts of casein and other stabilizers. The alkyds used were generally long oil vehicles and the paints generally had poor emulsion stability. They were subject to bacterial attack with resultant putrefaction, poor water resistance, and usually did not exhibit any good properties associated with alkyds. Further, since the oil-modified alkyds used in these paints were only partially polymerized during manufacture, the films still had to convert by the processes of oxidation and polymerization, as though the films had been laid down from an organic solvent or solution paint. Modern latex paints were introduced shortly after World War II as an offshoot of the synthetic rubber programs developed during that period. They have now completely displaced the alkyd emulsion paints from the interior architectural finishes market. The first of these modern latex paints was based on butadiene-styrene latex and were commonly called rubber-based paints. They easily achieved popularity in spite of some early defects. This popularity stimulated the development of improved latex coatings. Today, there are a multitude of latexes available for use in interior and exterior architectural products.

The film-forming component of a latex paint is a high molecular weight polymer produced in an emulsion polymerization process (1, 2). The resin exists in the latex as small, discrete, spherical particles approximately 0.05 to 0.25 μ in diameter, and is stabilized against coagulation or settling by surrounding each particle with a layer of emulsifying agent. In resin solutions the solids content is limited by the solubility and viscosity of the polymer in the solvents used. These properties are directly related to the molecular weight of the polymer. In general, the higher the molecular weight, the higher the viscosity and the lower the solubility, the limiting factor being practical considerations of reasonable viscosity and solids. Synthetic latexes do not suffer these limitations since the latex particle is suspended and not dissolved in the aqueous phase. Consequently, the viscosity is primarily a function of the external water phase, with some contribution by the resin particles, and high solids and high molecular weight are readily achieved. Inherent in high molecular weight is good durability and mechanical properties.

The three types of latex polymer most commonly used in the manufacture of latex paints are: (a) styrene-butadiene types, (b) polyvinyl acetate types, and (c) the acrylics. More recently, copolymer blends of styrene and acrylate have been introduced, combining the most durable features of each monomer into a single polymer having superior properties.

The styrene-butadiene latexes were the first commercially available types used for interior architectural finishes. They are the lowest-cost polymers and have the largest share of the wall paint market. Normal ratio of styrene to butadiene is 67:33, which is approximately the inverse ratio of GR-S rubber. The polymer is chemically unsaturated and will subsequently cure further, after forming a strong continuous film, by oxidative processes. This oxidation also results in yellowing of the film, and, in an exterior exposure, in eventual embrittlement and failure. The alkali resistance of this polymer is the greatest of all the emulsion polymers. For this reason it is used very often in severe chemical exposures, i.e., in strongly alkaline or acid media. The latex can be blended with a large variety of alkyds, oils, hydrocarbon mixtures, etc., after suitable prior emulsification, to obtain films for a variety of end uses. Similarly, by alteration of the styrene-butadiene ratio, and incorporation of a small amount of acid monomer, a different balance of properties for other end uses can be developed. Styrene-butadiene latexes tend to have a somewhat characteristic odor, probably due to low molecular weight polybutadiene formed during the polymerization.

Polyvinyl acetate (PVA) latexes have been in existence since the late 1930s, but their earliest uses were as adhesives rather than paint. These latexes are intermediate in cost between the styrene-butadiene and acrylics, and can be utilized in exterior coatings as well as in interior paints. Polyvinyl acetate homopolymer forms a tough, horny film, but because it is a hard polymer, it requires plasticization. This is particularly true if it is to be utilized in an exterior exposure. Plasticization may be via copolymerization with a suitable monomer such as dibutyl maleate, or post-plasticized with a chemical plasticizer such as dibutyl phthalate or some polymeric plasticizer. For obvious reasons the plasticizer used for exterior exposures must be a permanent type that will neither leach out nor volatilize from the film. For interior usage, this is not a stringent requirement, but loss of the plasticizer results in embrittlement of the film and reduction in elongation properties.

PVA latexes are manufactured in a variety of ways and the use of certain protective colloids or stabilizers will affect their subsequent performance, particularly over surfaces containing free calcium salts or ions. Of the three major types of latex polymer, polyvinyl acetate has the greatest water sensitivity and the least resistance to alkali. The latter property is more important in exterior exposures where the alkali level may not remain high for any extended time, than in an interior application. The combination of alkalinity and weathering will degrade polyvinyl acetate films more rapidly than it will acrylic films.

Polyvinyl acetate emulsions tend to be acidic, whereas the styrene-butadienes and acrylics are alkaline. These acidic latexes and paints are consequently more corrosive than the alkaline materials, and unless precautions are taken to prevent attack on unprotected metallic storage tanks and cans, the resultant cations liberated can cause coagulation or discoloration of the latex or paint. Color retention of PVA latexes and paints is very good; their resistance to grease and oils is also very high.

Acrylic latexes are the highest in cost of the three types discussed, but they have an excellent combination of properties for the development of high quality paints. Their films are tough but flexible, have excellent color retention and durability, are oil, grease, moisture, and alkali-resistant, dry rapidly, and develop good water resistance shortly after application.

The acrylic latexes are generally copolymers of methyl methacrylate and any one of a number of acrylic esters. The choice and level of comonomer used is dictated by the end use requirement, and a wide range of copolymer compositions is possible (3). Other combinations of monomers, such as styrene and acrylic esters or vinyl acetate and acrylic esters, have recently come into the market. These modified acrylic or vinyl acetate copolymers combine several of the best properties of each monomer into a single polymer. Styrene will impart greater water resistance to an all-acrylic system than the methyl methacrylate normally used. Similarly, use of butyl and hexyl acrylate will upgrade the water resistance of a polyvinylacetate homopolymer as well as its flexibility, low-temperature coalescence, and gloss. Physical blends of the different polymers do not result in a latex having the combined properties of each, but most often results in the latex having those properties of the polymer present in greater proportion. Copolymerization, on the other hand, yields a polymeric film that is completely different from a simple mixture of two latexes.

Before getting into the details of specific formulations, it is important to understand the mechanism of film formation of a latex film. In an emulsion polymerization process, the polymer that is formed exists as discrete spherical particles having diameters of the order of 0.05 to 0.25 μ. By modifications of the process, it is possible to not only alter the particle size distribution, but to also create particles as small as 0.02 μ and as large as 1.0 μ. This latter dimension is usually outside of the normal size range for use in latex paints.

During the process of film formation (4), as the water evaporates it generates a driving force to bring the latex particles into close contact. As the particles begin to contact each other, a network of capillaries is formed, and as the last of the water evaporates through these capillaries, the large force exerted on the particles deforms the spheres so that the particles fuse together and fill in the voids left by the water. A continuous film results. The forces generated by the water evaporating through the capillaries has been calculated to exert from 110 psi, when particles are 1.0 μ in diameter, to 11,000 psi, when particles are 0.01 μ. In the calculation of these forces, all things being constant, the particle size of the latex is the dominant variable with the smaller particle size latexes coalescing to tighter, more coherent films.

The modulus of elasticity of the polymer is another important characteristic to be considered. This plays an important part in the coalescence of the film as well. If the polymer is too rigid to deform under the driving force of the water evaporating from the capillaries, then the particles will not fuse into each other and the resulting film may be powdery or lack adequate cohesion to perform satisfactorily. Associated with this property of the polymer particles is the glass transition temperature (T_g) or, more meaningful from a practical standpoint, the minimum film-formation temperature of the latex. At temperatures above this minimum, the polymer particles tend to be rubber-like and require relatively little force to deform them. Below this temperature, the particles become harder and more rigid, requiring large forces to cause them to coalesce. Therefore, temperatures become an important factor in film formation. For this reason all latexes used in trade sales paints have relatively low minimum film-forming temperatures of 2° to $10^\circ C$. Obviously it is preferable to apply latex paints above these temperatures to insure optimum coalescence and drying.

Relative to the effect of temperature on coalescence are other environmental factors. Application of a latex paint to a surface whose temperature is below the minimum film-forming temperature of that of the latex paint will result in a coating having very poor durability. Similarly, poor coalescence often results when the latex paint applied to a surface, although above the minimum filming temperature, is exposed to strong drafts. In this instance the chilling effect of the evaporating water from the film will drop the surface temperature to a point below which good coalescence will not occur. Conversely, if the painted surface is exposed to a hot, dry, air movement it may not coalesce satisfactorily due to the rapid evaporation of water. This too rapid loss of water does not allow adequate time for the particles to coalesce into a tight paint film.

Temperature is only one of the external factors affecting the time required for film formation by evaporation of the water. The relative humidity at the time of application will accelerate or retard the rate of drying, depending on whether it is low or high, as will the film thickness applied and the porosity of the substrate. Porous surfaces tend to wick the water out of the film very rapidly and consequently require higher application temperatures. The use of small quantities of coalescing agents is often used as an aid in film formation (5, 6). These may be plasticizers, as the phthalate esters, or volatile solvents, as ethers or esters, that act to soften the polymer particles during the last stages of drying. In effect, these agents act to lower the minimum filming temperature of the particles.

Despite the fact that a latex film may coalesce satisfactorily under the most ideal conditions, it nevertheless is far more permeable to water vapor than a film deposited from a solvent system. This ability to "breathe" is the reason latex paints seldom blister when applied over damp surfaces, allowing the substrate to dry with no harm to the coating. This is one of the properties that is making latex paints so popular, particularly for exterior painting.

From the preceding discussion it becomes obvious that the formulation of finished latex paints will be somewhat different in many important respects, from the formulation of an oil or alkyd solution paint. In fact, the only points of similarity are the pigment volumes utilized to achieve certain film properties, and these are only approximately the same. Formulating a latex paint involves specific problems requiring a reasonable knowledge of colloid chemistry.

The pigment in a latex paint is not incorporated into the resin but is dispersed in water, utilizing the most efficient dispersants and/or wetting agents for the particular pigment or pigments. The subject of wetting agents is very extensive (8, 9), and one which is beyond the scope of this chapter. Suffice it to say that the wetting agent, which may be nonionic or anionic, is used to reduce the surface tension of the water for more rapid and complete wetting of the pigment. The dispersant's function is to absorb on the pigment surface and prevent reagglomeration or flocculation of the pigment after it has been dispersed. The choice of which wetting agent or dispersant to use is one that can only be determined by trial and error, since very often the material giving the optimum dispersion may interact with one of the many other components used in the latex paint and result in an unstable paint. Consequently, the final choice may be a compromise between maximum dispersion and some related physical, film, or rheological property of the paint.

The list of available surfactants to choose from is very large (10), but in actual practice is quite limited in number as determined from experimental work and practical experience. Some of the better dispersants are polymeric carboxylates in the form of their sodium or ammonium salts, and polymeric phosphates. The latter are particularly efficient for titanium dioxide pigments. Very often a combination of both is used to achieve optimum dispersion with a mixture of pigments. Based on the types of latex available in the market today, the use of cationic dispersants in latex paints is not extensive because of adverse interactions, and in future discussions only the anionic or non-ionics will be covered. In any event, after the final choice of dispersants, surfactants, defoamers, etc., is made, it is wise to establish some sort of limited program to determine if the final product has the necessary stability for commercialization. Although these intermediates are only used in small quantities relative to the total composition, their effects are comparatively large.

The preferred grade of titanium dioxide used in interior products is the rutile type. However, manufacturers of this pigment have introduced a large number of modifications of the basic pigment in order to obtain varying results in specific kinds of formulations. Thus, there is a form of rutile titanium dioxide preferred for gloss or semigloss paints and several other forms preferred for primers or flats. In general, these variations are based on particle size and specific surface treatments on the pigments.

Depending on the quality of latex paint being formulated, the choice of the correct grade of rutile pigment is important.

Rutile titanium dioxide has the highest hiding power of the available white pigments, and as a result is the primary white pigment used for interior products. Lithopone, zinc sulfide, zinc oxide, or the basic carbonates, sulfates, or silicates of lead are never used except in exterior products. Extended titanium dioxide pigments, that is calcium sulfate or barium sulfate coprecipitated rutiles, are very seldom used today because of poor money value.

The choice of extender pigment is only limited by the type and quality of product to be formulated. Calcium carbonates, clays, micas, diatomaceous earths, silicas, silicates, and practically any other type of extender may be used in the development of a latex paint. The treatments imparted to the pigments may be as varied as the kinds of extender pigments available. They may be dry ground or water ground, natural or synthetic, and many are calcined, resulting in a considerable change in physical properties. The extender pigments will affect flow and leveling, gloss, cleanability, scrub resistance and appearance. The type of extender chosen can also affect package stability; its dispersion and the manner in which it packs with the other pigments will also affect hiding properties of the dry film as well as color development and rheological properties in the wet film.

The colored pigments used in emulsion paints must have a high order of alkali resistance because not only are they often used in highly alkaline systems, but they are also often applied over highly alkaline surfaces like fresh plaster or cement. For pigments used in polyvinyl acetate systems, acid resistance may be the requirement. The general classes of pigment excluded from latex paint are the chromes (chrome yellows and molybdate oranges, excepting chromium oxide) and the iron blue pigments. Certain organic red pigments are unstable in aqueous systems, and with some carbon blacks it is difficult to obtain stable dispersions in water. Generally, all the earth pigments such as iron oxide yellows, blacks, browns, and reds, are satisfactory for use in latex paints, as are the toluidine reds, Hansa yellow pigment, Dalamar high-strength mono-azo pigment, phthalocyanine green, chromium oxide green, and many others.

Along with the latex, surfactants, and pigments, there are several other important additives that must be incorporated into the formulation to obtain a stable and satisfactory product. Thickeners in the form of water-soluble protective colloids are necessary to control the application and flow properties of the paint. Since the viscosity of the latexes is normally very low and the addition of a dispersed pigment phase does not generally give an increase in the viscosity of the combined phases, the addition of an additive to adjust the viscosity characteristics of the paint is required. The viscosity should be a balance between good application properties, flow and sagging, and good pigment-settling resistance in the package. The thickeners used today for this purpose are essentially all synthetic, with perhaps some small useage of natural thickeners, such as casein, still remaining. The exact mechanisms involved in the thickening process are not precisely understood and are complex, probably involving more than one simple mechanism (7). Casein thickeners tend to lose viscosity on aging and are prone to putrefy in the paint. Consequently, in spite of the fact that this class of thickener imparts excellent flow and leveling properties to wall paints or trim paints, its use as a paint thickener has decreased over the years. Thickeners most commonly used are cellulosics, notably hydroxyethyl cellulose and methylcellulose, polyacrylates, polyacrylamide, polyvinyl alcohols, and many others. In each of these categories there are ranges of molecular weights available to give wide variations in thickening efficiencies as a function of the solids of the thickener.

The quantity of thickener generally used is relatively small, usually of the order of 1% or less, but its contributions to the overall final properties of the paint are much greater proportionately. In addition to the abovementioned characteristics imparted to the paint, the thickener can also affect brush drag and ease of spreading during application, holdout properties, uniformity of appearance over porous or nonuniform substrates, gloss, and enhanced emulsion and freeze-thaw stability. It becomes obvious, then, that the proper choice of thickener is important, and more than a superficial study should be made in adopting a candidate for use in the final formulation.

Latex paints contain large amounts of water and as such provide a favorable environment for fungi and bacteria to grow. To prevent spoilage in the can, which may manifest itself by loss of viscosity, gellation, or objectionable odors, a preservative is incorporated into the paint. Bacterial contamination could be introduced by one of the raw materials or intermediates or picked up during manufacture, and the preservative should be incorporated into the manufacturing cycle at the earliest practical point.

The most effective of the in-can preservatives, based on cost-performance, are the mercurials in the form of phenyl mercury salts or other modified organic mercurials. The quantity used for in-can preservation is usually from 0.01% mercury metal, based on total paint, to 0.05%. Depending on whether mercury is used and its percentage in the paint, precautionary warning notices may have to be printed on the product labels warning against application of the paint on furniture or other interior surfaces which may be chewed by children. The percentages are specified in existing federal statutes and their associated regulations and cover other toxic metals as well. Normally, for in-can preservation, the low levels of mercury used will only require the minimum type of warning on the label; but if higher percentages are required such as for special mildew conditions, then the more extensive warnings must be added to the labels and these will be spelled out by the regulations.

There are many other nonmercurials available as preservatives, but any of these should be given a thorough microbiological evaluation in the specific formulation that may be used. Also, good housekeeping practices and cleanliness in the producing plants is mandatory in minimizing bacterial contamination. Packaging the paint in a sterile condition also reduces the demand on the preservative to maintain this condition over long periods of time.

Defoaming and antifoaming agents are additional adjuncts in latex paints. They are required to suppress or eliminate any foam that tends to develop during manufacture or in the container-filling operation. Very often, the defoamer, which acts to knock down any foam that has developed in the paint processsing, has a short-lived action, and then an antifoamer is added to prevent or minimize foaming during application by brush or roller. Again, the type and quantity of defoamer and/or antifoamer required in a formulation will vary depending on the pigment volume and type of latex and surfactants used. A trial-and-error method is used to determine if the additive has any merit in the specific formulation. Caution should be exercised in the use of these materials; if used in excess they may cause pitting or crawling in the films.

Several other components of latex paints that may or may not be required are glycols, coalescents, and pH adjusters. Glycols — ethylene and propylene—serve as freeze-thaw stabilizers (11) and to increase the wet-edge and open time of the wet films. Because latex films tend to set up very rapdily after application, lapping into a partially dried film can result in objectionable film distortion; the longer the film remains fluid, the longer the time available to lap into previously painted areas and for the film to flow out and level. An excess of the glycols in some formulations will lead to excessive flow and sagging on vertical surfaces, particularly under conditions of high temperature and humidity. For a given formulation, the optimum amounts are determined experimentally, but in general, much less glycol is required to give a freeze-thaw-resistant paint than to yield a product with a long lap time.

Coalescents are additives incorporated into a latex paint in order to optimize the coalescence of the latex particles. They can be of the temporary variety, which evaporate from the film after it has dried, or they can be of the permanent type, becoming a part of the film. In the former instance, water-miscible solvents, or even water-immiscible organic solvents, can be used. They function by absorbing onto the polymer surface and softening it so that less driving force is required to coalesce the particles. After the film has coalesced, the solvent evaporates without affecting film continuity, although flexibility is adversely affected. Among these are the ethers of ethylene glycol or mixed ester-ethers of ethylene glycol. The permanent coalescents or plasticizers can be alkyd resins or chemical types such as dibutyl phthalate or the phosphate esters. In time, the low molecular weight plasticizers also volatilize to some degree, leaving the film somewhat embrittled. Coalescents and plasticizers are, for the most part, used with polyvinyl acetate latex rather than with acrylics or styrene-butadiene because in the latter two the plasticizer is part of the polymer and generally no further aids for coalescence are required. Polyvinyl acetate can also be internally plasticized so that no additional coalescents are required. In any event, if it is desired to coalesce a hard polymer at room temperature, the use of temporary coalescents is often employed.

Butadiene-styrene and acrylic latexes achieve their maximum stability in alkaline systems. Consequently, the systems are usually adjusted to a pH of 9.0 to 9.5. Aqueous ammonia is most commonly used for this, but other amines, as the ethanol-amines, can also be used. For optimum color retention and water resistance of the films, ammonia is the preferred alkali. Alkali-metal hydroxides, being nonvolatile, remain in the films with resultant water sensitivity. Polyvinyl acetates, on the other hand, are usually acidic, in the range of 4.5 to 5.5. The pH can be adjusted to 8.0 to 8.5 with ammonia but it is difficult to maintain it in this region since the polymer hydrolyzes slowly, releasing acidic vinyl acetate. In time, the system will drift back into the acid range.

A. PRIMER-SEALERS

Latex paints are particularly well suited for use on porous surfaces such as masonry or plaster. The discrete latex particles, although very small, are still large enough so that they will not penetrate into the porous surface in the same manner as a polymer molecule in solution, and consequently are excellent sealers. One requisite of a primer-sealer is to provide maximum holdout of topcoats for gloss, color, and hiding uniformity over nonuniform surfaces.

Large particle size (approximately 0.3 to 1.0 μ) polyvinylacetate latexes have been used for making pigmented or nonpigmented primer-sealers. These latexes are relatively low in cost, can be made easily in large particle sizes, and have high surface tension — all of which make for a good sealer. Good primer-sealers can be made with acrylics or butadiene styrene resins, but the overall advantage for this end use lies with the polyvinyl acetate because of cost/holdout balance.

Freshly plastered surfaces, or new masonry, are high in moisture and highly alkaline. An alkyd primer-sealer on these surfaces would quickly disintegrate due to saponification by the alkali. Such surfaces must cure for 60 to 90 days before they can safely be sealed with an alkyd. Latex sealers would not be affected by the alkali or the water and could be applied to these same surfaces after 2 to 3 weeks of aging. However, under the following conditions the latex sealer may not be as efficient as the alkyd: (a) over very powdery surfaces resulting from improper troweling or insufficient gauging, (b) over "rotten" plaster, which is plaster that has dried out too fast or has dried under conditions of high humidity for too long, resulting in an overly soft surface; under such conditions the gypsum does not crystallize properly, and (c) efflorescence of soluble salts. If time is a factor, then two coats of latex sealer will often suffice to cover these surface defects, but if the surface is too powdery, adhesion will be adversely affected.

Latex sealers may be clear or pigmented. The pigmented primer-sealers can vary from 20% pigment volume to as high as 50% pigment volume. Within limits, the lower the pigment content, the better the holdout of the topcoats, so the choice of system will depend on the properties desired and the economics. A clear sealer based on an acrylic latex is shown in Formulation 7-8.

A primer-sealer having excellent sealing properties and based on a polyvinyl acetate homopolymer having a particle size distribution of 1 to 3 μ is shown in Formulation 7-9.

The pigment dispersion for Formulation 7-9 is made in a Cowles Dissolver or similar high-speed impeller equipment. The ingredients are added in the order shown and mixed 15 to 20 minutes after the last of the pigment has been added. The remaining ingredients in the let-down are added with low speed agitation. The viscosity is adjusted with the 2% methylcellulose solution to the desired viscosity.

In both sealer formulations given, a volatile plasticizer is used to improve film coalescence. With the acrylic latex, hexylene glycol is used, and with the polyvinyl acetate homopolymer, Carbitol is the preferred material.

B. LATEX WALL PAINTS

In the formulation of a latex wall paint there is very wide latitude in the choice of raw materials to use, and the resulting properties of the paint can be just as broad. Cost is the factor that most effectively predetermines the quality of the finished product and the raw materials that will be used. With this as a basis, the formulating parameters fall into more restrictive ranges, with greater limitations on the choice and quantity of raw materials to be used. Obviously, where the raw material cost of a product is to be relatively high, the amount of hiding, as determined by the titanium dioxide levels and the volume solids of the paint, and the film properties that can be achieved, will be at a higher level than when lower raw material costs are specified.

The vehicle of choice in the formulation of a flat wall paint is generally dictated by cost. The least expensive is butadiene-styrene, both on a cost per pound basis as well as on a volume-solids basis. Poor color retention in whites and light tints is a drawback. The polyvinyl acetates are slightly

FORMULATION 7-8. Clear Sealer Based on Acrylic Latex

Ingredients	lb	gal
Acrylic latex, 44.5% solids (1)[a]	409.0	46.21
Ethylene glycol	15.0	1.62
Water	148.0	17.77
Hexylene glycol	30.0	3.93
Ammonium hydroxide, 28%	1.2	0.16
2.5% hydroxyethyl cellulose (2)	250.0	29.28
Defoamer (3)	2.0	0.26
Preservative, 30% solids (4)	0.8	0.10
	856.0	100.03

Physical constants

Wt/gal	8.6
Total solids	22.0%
Viscosity, krebs units	82 KU

Raw material sources

1. Rhoplex AC-22 acrylic latex, Rohm & Haas Co.
2. Hydroxyethylcellulose QP-4400, Union Carbide Co.
3. Colloid 600, Colloids, Inc.
4. Phenyl mercury acetate, Troy Chemical Co.

[a]Number refers to raw material sources.

higher in cost but have excellent color retention. Polyvinyl acetates also have high polymer densities, making their bulking volumes somewhat lower than butadiene-styrene, thereby further increasing their costs. The most expensive latexes are the acrylics, but they have excellent color retention, durability, and achieve a higher level of water resistance more rapidly after film formation than the other latexes. Under normal usage as flat wall paint vehicles, all the latexes have good alkali resistance and are permeable to water vapor. However, specific latexes may vary in their ability to resist coagulation by polyvalent cations. This property is import-

FORMULATION 7-9. Pigmented Primer-sealer Based on Polyvinyl Acetate Homopolymer

Pigment dispersion	
Water	143.8
2% methylcellulose solution (1)[a]	210.0
Nopalcol 6LW, surfactant (2)	6.5
Mica, 325 mesh (3)	89.4
Rutile titanium dioxide (4)	125.0
Let-down	
Polyvinyl acetate latex, 55% solids (5)	376.0
Mono-ethyl-ether diethylene glycol (6)	50.0
{Mono-ethyl-ether diethylene glycol (6)	22.2
Pentachlorphenol (premix) (7)}	3.1
	1026.0
Physical constants	
Wt/gal	10.2 lb
Total solids	41.8%
Pigment volume	26%
pH	5.0-6.0
Viscosity, krebs units	85 KU

Raw material sources

1. Methocel, 4000 cps, methylcellulose, Dow Chemical Co.
2. Nopco Chemical Co.
3. #325 wet ground mica, Diamond Mica Co.
4. Ti-Pure R-901, titanium dioxide, E. I. du Pont de Nemours
5. Elvacet 81-900, homopolymer, E. I. du Pont de Nemours
6. Carbitol solvent, Union Carbide Co.
7. Dowicide 7 preservative, Dow Chemical Co.

[a]Number refers to raw material sources.

ant when applying the wall paint to bare plaster or over the joint cements used in dry wall construction. Additional considerations in the choice of latex are the particle size and size distribution, and the compatibility of the stabilizers in the latex and the surfactants used for the pigment dispersions. Often, particularly when the paint manufacturer purchases the latex from an outside source, these considerations become more important in that the final paint formulation is, of necessity, tailored to the latex. The critical pigment volume concentration (CPVC) and its relationship to the actual pigment volume concentration (PVC) of the formulation is directly affected, along with the corollary film properties. Adverse interactions can cause the dispersed pigment to flocculate to varying degrees; since hiding power is directly related to dispersion, among other factors, flocculation will reduce the hiding power of the paint.

The pigments normally used in latex paints are nonreactive and have essentially no solubility in water. The water-soluble salt content in the pigments should be kept very low in order to minimize the introduction of any adverse ions. Rutile titanium dioxide is the white pigment used for hiding, but there are many grades of rutile available, depending on the end use requirement. Thus, there are special grades for gloss products, exterior paints, interior paints, and in interior latex paints, special grades for high and low PVC paints. The difference in these grades of titanium dioxide is in the surface treatment given the pigment by the manufacturer. Some of these treatments can affect the stability of the latex. With these factors in mind, the type of titanium dioxide pigment is chosen to give maximum hiding, color, and stability.

Flat wall paints are low-gloss products, the low luster being achieved by the choice of appropriate extender pigments. Almost any class or type of extender pigment may be used in latex paints, and combinations of pigments are used to obtain a balance between application and film properties and stability. Particle size and size distribution, particle shape, surface treatments, or any other treatments that could have altered the surface characteristics of the pigment, are pigment properties that must be considered in choosing the extender. Steig, in his chapter on flat wall paints, discusses the relationships of the CPVC in solution paints, and, in a general way, these relationships are valid for latex paints. However, a point of departure is that since a latex is a polymer dispersion, then the polymer spheres also pack into the voids along with the other fine pigments, instead of filling the voids as a continuum of solution polymer. For this reason it is very difficult to determine the CPVC of a latex paint with the same degree of accuracy that it can be determined for the solution paint; in general, the CPVC of a latex paint will be lower than that of a solvent paint with the same pigmentation, and as the particle size of the latex increases, the CPVC decreases.

The general categories of extender pigments that are used in latex paints include clays, calcium carbonates, silicates, diatomaceous earths, silicas, barytes, talcs, etc. These pigments come in a multitude of particle sizes, size distributions, shapes, surface treatments, and in hydrous or calcined forms. They can be blended in an almost infinite variety to achieve different property balances. However, it is beyond the scope of this chapter to go into these. Many investigators have worked in this area (12-17) and have developed considerable data on various pigmentations. However, for any given set of parameters, the pigmentation finally decided upon will very likely have been determined by experimentation and evaluation. Ceiling paints, for instance, are not required to be very washable but should have good uniformity of appearance, easy application, and good one-coat coverage. Wall paints, on the other hand, should be highly washable and stains should be easily removed; they should have good application properties, good lapping and touchup properties, good holdout over substrates of varying porosity, and good color retention. The different requirements for each of these products would dictate different PVC/CPVC ratios, and often significantly different pigmentations.

Earlier in this chapter the subject of surfactants was mentioned with reference to the dispersion of white and color pigments in paints, and that surface-active agents are used to aid in pigment wetting by lowering the surface and interfacial tensions, and that dispersants are used to prevent flocculation after the pigments are dispersed. The choice of surface-active agents for pigment dispersions is normally a tedious trial-and-error process because of the large number of available surfactants and the tremendous number of possible pigment blends that can be used. However, this process has been somewhat simplified by the HLB (Hydrophilic-lipophilic-balance) system in which surfactants are given a number relative to their dispersibility in water and in oil. The HLB values are assigned only to non-ionic surfactants, since it is difficult to assign values to anionic emulsifiers. However, the anionics have very high HLB values. A low HLB value indicates that the surfactant is oil-loving; high values indicate its affinity for water. Investigators have determined that for many pigments there is a narrow range of HLB values in which optimum dispersion takes place (18). With changes in vehicle, thickener, or other components of the paint, it may be necessary to change the chemical type of surfactant used, but the HLB value will remain essentially constant.

To determine the optimum amount of dispersant required for a particular pigment or blend of pigments, a tensiometric titration method has been used. The method essentially tells when the pigment surface is saturated with surfactant and when the surface tension of the aqueous mixture has reached a minimum value. A mixture of the pigments in water is titrated with the dispersant and at given increments the surface tension of the mixture is measured. After the pigment surfaces have been saturated, further additions of surfactant will have little effect on further reducing the surface tension and a minimum value will have been reached. Usually at this point the viscosity of the pigment mixture will have also reached a minimum value and is taken as further evidence of the saturation of the pigment surfaces. In practice, a slight amount in excess of this value is used to assure complete saturation. However, large excesses are to be avoided since this may cause the system to flocculate.

With the preceding discussion of general principles as a guide, Formulations 7-10 through 7-13 are given as examples of the range of compositions that can be developed. In no way are these formulations intended to be restrictive in their pigmentations, pigment volumes, types of pigment, vehicles, thickeners, or any of the other ingredients used in making latex paints. The pigment dispersion bases are formulated to be made with high-speed mills or impellers; a change to another type of disperser may require revisions to accommodate the formula to it for optimum dispersion.

In the let-down stage of Formulation 7-10, the intermediates should be added in the order shown. Agitation should be carefully controlled to avoid incorporating foam into the dispersion. Lined tin cans are suggested for packaging paints based on polyvinyl acetate emulsions.

The paint in Formulation 7-11 is prepared in the same manner as the paint in Formulation 7-10 and the same precautions should be observed to inhibit any foam development. The copolymer latex exhibits good low temperature coalescence and eliminates the need for costly coalescing solvents in the formulation. Although the pigment volume concentration is high, the paint has good application properties, exhibiting very little brush drag and good leveling. Scrub resistance is satisfactory but not outstanding.

A good quality butadiene-styrene latex wall paint is illustrated in Formulation 7-12. This paint is thickened with methylcellulose, which exhibits inverse solubility characteristics, i.e., it is less soluble in hot water than in cold water, but is more resistant to bacterial decomposition than the hydroxyethyl celluloses.

Formulation 7-12 gives excellent scrubbability, flow, and leveling. Application properties and coverage are also very good.

In recent years the "dripless" or "gelled" or "thixotropic" latex wall paints have become increasingly important. Their unique application properties have considerable appeal to the "do-it-yourself" market since "dripping and running" are minimized. These products most often also have excellent settling resistance in the container and generally do not require any stirring prior to use. A thixotropic latex wall paint based on an acrylic latex is illustrated in Formulation 7-13. This type formulation will usually have the highest raw material costs, all things being equal.

Veegum, a refined clay, is used to develop the thixotropy in Formulation 7-13. It is necessary to prepare the Veegum dispersion separately in a high-speed disperser using water at 120° to 130°F (49° to 54°C). A small

FORMULATION 7-10. High Performance Interior Latex Paint — Polyvinyl Acetate[a]

Pigment grind	lb/100 gal
Water	113.0
Tamol 731, dispersant, 25% solids (1)[b]	10.0
Potassium tripolyphosphate	1.0
Propylene glycol	25.0
Troysan PMA 30, preservative (2)	0.5
Hercules defoamer 357	1.0
Ti Pure R-901, titanium dioxide pigment (3)	274.0
Satintone #1, calcined clay (4)	75.0
Snowflake whiting, calcium carbonate (5)	50.0
Amorphous silica, 1160 (6)	75.0
Methocel 90 HG-DG 4000, 2% soln. (7)	75.0
Let-down	
2-Amino-2-methyl-1-propanol (8)	2.0
Methocel 90 HG-DG 4000, 2% soln.	199.0
Hercules defoamer 357	1.0
Water	7.0
Ethylene-vinyl acetate copolymer latex (9)	276.5
	1185.0
Physical constants	
Pounds/gallon	11.8
Pigment volume	50.0
Nonvolatiles by weight	53.9
pH	8.4
Viscosity, krebs units	85-90 KU

FORMULATION 7-10. — Continued

Raw material sources	
1. Tamol 731, dispersant	Rohm & Haas Company
2. Troysan PMA 30, phenyl mercury acetate, 30% soln.	Troy Chemical Corp.
3. Ti Pure R-901, titanium dioxide	E. I. du Pont de Nemours
4. Satintone #1, calcined clay	Engelhard Minerals & Chem
5. Snowflake Whiting, calcium carbonate pigment	Whittaker, Clark & Daniels
6. Silica 1160	Illinois Minerals Co.
7. Methocel, 90 HG methylcellulose	Dow Chemical Co.
8. 2-Amino-2-methyl-1-propanol	Commercial Solvents Corp.
9. Ethylene-vinyl acetate latex, Elvace 1942	E. I. du Pont de Nemours

[a]Courtesy E. I. du Pont de Nemours & Company, Electrochemicals Dept.

[b]Number refers to raw material sources.

amount of Super Ad-It, 0.01% of total weight, should be added to the Veegum dispersion before dispersing for 15 minutes. For reproducible results, the Veegum dispersion must be prepared as described, otherwise variations in thixotropy of the finished dripless paint will likely occur. This rheological property, which is characteristic of dispersions or colloidal systems, is dependent on other factors too, such as the volume fraction of the dispersed phase and the average particle sizes of the latex and pigments, and their respective size distributions. The type of thickener and its molecular weight also contribute to the rheology of the system. As a consequence, formulations of this type must be very carefully balanced if the desired final results are to achieved.

In the application of wall paints, there are certain characteristic differences between alkyd or solution paints and latex paints. Over dirty or greasy surfaces the water paints may exhibit more crawling and poorer adhesion than the solution paints. Latex paints cannot be applied over old calcimine or starch paints unless these are completely washed off; over wall paper it is important to make sure that the edges are tightly adhering to prevent the water from opening up the seams. Additionally, the adhesive used on the paper should not be easily redissolved by the water, otherwise the paper will detach from the wall. This latter condition tends to occur more often with ceiling papers when the weight of the paint pulls the wet paper from the ceiling.

FORMULATION 7-11. Low-cost Polyvinyl Acetate Interior Paint[a]

Pigment grind		lb/100 gal
Water		115.0
Tamol 731, dispersant		10.0
Potassium tripolyphosphate		1.0
Propylene glycol		25.0
"Colloid" 581-B, defoamer (1)[b]		2.0
Nuodex PMA-18, preservative (2)		0.5
Ti Pure R-911, titanium dioxide		200.0
Satintone #1, calcined clay		75.0
ASP 400, clay		75.0
Snowflake Whiting, calcium carbonate		160.0
Celite 281, diatomaceous silica		25.0
Methocel 65 HG-4000, 2% soln.		50.0
Let-down		
2-Amino-2-methyl-1-propanol		2.0
Methocel 65 HG 4000, 2% soln.		230.0
Water		67.0
Ethylene-vinyl acetate copolymer latex		166.5
		1204.0
Physical constants		
lb/gal	12.0	
Pigment volume	66.3	
Nonvolatiles by weight	53.2	
Viscosity, krebs units	80 KU	
pH	8.5-9.0	
Sources of supply		
1. Colloid 581-B, defoamer	Colloids, Inc.	
2. Nuodex PMA-18, phenyl mercury acetate, 18% mercury	Tenneco Chemicals, Inc.	

[a]Courtesy E. I. du Pont de Nemours & Co., Electro Chemical Dept.

[b]Number refers to raw material sources.

FORMULATION 7-12. High Quality Butadiene-styrene Latex Paint

Pigment grind	lb/100 gal
Methocel 65 HG 4000, 2% soln (1)[a]	295.0
Water	71.8
Ethylene glycol	20.0
Tamol 731, dispersant	6.5
Triton X-100, wetting agent (2)	3.2
Balab 748, defoamer (3)	1.0
Dowicide A, bactericide (4)	2.0
Titanox RA-46, titanium dioxide pigment (5)	244.0
"Duramite," calcium carbonate pigment (6)	113.0
Hydrite D, aluminum silicate pigment (7)	108.0
Let-down	
Butadiene-styrene latex, 48% solids (8)	297.5
Balab 748, defoamer	1.0
	1163.0
Physical constants	
lb/gal	11.6
Pigment volume concentration	48%
Nonvolatile (weight)	53.4
Nonvolatile (volume)	34.6
pH	8.5-9.5
Viscosity, krebs units	85-90 KU

Source of supply

1. Methocel 4000 cps, methylcellulose	Dow Chemical Company
2. Triton X-100, wetting agent	Rohm & Haas Company
3. Balab 748, defoamer	Witco Chemical Company
4. Dowicide A, bactericide	Dow Chemical Company
5. Titanox RA-46, titanium dioxide pigment	Titanium pigment Corporation
6. "Duramite"	Thompson, Weinman Company
7. Hydrite D, extender pigment	Georgia Kaolin Company
8. Gen-Flo 355, latex	General Tire & Rubber Company

[a]Number refers to raw material sources.

FORMULATION 7-13. Dripless Acrylic Latex Wall Paint[a]

Pigment grind	lb/100 gal
Water	86
Tamol pigment dispersant # 731	15
Triton wetting agent CF-10 (1)[b]	3
"Colloid" 600, antifoamer (2)	1
Cellosize hydroxyethyl cellulose QP4400, 2.5% soln. (3)	50
Hexylene glycol (4)	30
Ethylene glycol (5)	35
Super Ad-It, preservative (6)	1
Ti-Pure, titanium dioxide pigment R-901	180
1160 Silica (7)	52
Icecap K, calcined clay (8)	130
Zeolex 80, sodium silica aluminate pigment (9)	27
Let-down	
Rhoplex acrylic latex AC-22 (10)	255
"Colloid" 600, antifoamer	4
Veegum "T" clay, 4% dispersion (11)	125
2.5% Hydroxyethyl cellulose	95
Water	26
	1115
Adjust pH to 9.5 with ammonium hydroxide (28%)	
Physical constants	
lb/gal	11.1
Pigment volume	57%
Nonvolatile	46.3%
Viscosity, krebs units	92 KU

FORMULATION 7-13. — Continued

Source of supply	
1. Triton wetting agent CF-10	Rohm & Haas Company
2. Colloid 600	Colloids, Inc.
3. Cellosize hydroxyethyl cellulose QP4400	Union Carbide Company
4. Hexylene glycol	Union Carbide Company
5. Ethylene glycol	Union Carbide Company
6. Super Ad-It preservative	Tenneco Chemicals
7. 1160 Silica	Illinois Minerals Company
8. Icecap K calcined clay pigment	Burgess Pigment Company
9. Zeolex 80, silicate pigment	J. M. Huber Corporation
10. Rhoplex AC-22, acrylic latex	Rohm & Haas Company
11. Veegum "T" clay	R. T. Vanderbilt Company

[a] Courtesy Rohm & Haas Company, Philadelphia, Pa.

[b] Number refers to raw material sources.

Latex paints do not penetrate powdery or chalky surfaces as well as the solution paints. This often results in a loss of adhesion. Very often, over new plaster, efflorescence will occur with a latex paint. Because of its greater permeability, and under proper conditions of temperature and humidity, water coming from the plaster through the film will carry dissolved salts along, which are left as white crystalline deposits of sodium sulfate on the film surface. To prevent this unsightly appearance from occurring, either all sodium ions should be avoided in the formulation, or else a sufficient amount of potassium ion is introduced to give a 3:1 ratio of potassium to sodium. Potassium sulfate does not form hydrated crystalline salts, and the 3:1 mixture of potassium and sodium sulfates forms a double salt that does not effloresce.

Occasionally, the presence of soluble salts of calcium, or aluminum, or zinc (the latter two being introduced as alkali neutralizers) will cause what appears to be similar to the "alkali-burn" that occurs with oleoresinous systems over highly alkaline substrates. However, it is not alkali affecting the pigments, but rather the polyvalent cation causing flocculation in the wet film. The choice of appropriate surfactants and protective colloids in the latex and pigment dispersions will usually overcome this. Application of the latex paint at low temperatures will sometimes give the same effects, but more often results in the paint never fully developing its designed film properties, including adhesion to the substrate. The differences in brush-drag between alkyds and latex paints is very obvious, the latter being considerably easier to apply. However, this ease of brushing tends to allow the user to spread the paint too thin and often results in inadequate film thickness being applied. This is an important consideration in exterior paints where the film thickness and durability are closely related.

C. INTERIOR SEMIGLOSS AND GLOSS LATEX PAINTS

Earlier, a discussion of the film-forming process brought out that one of the unique characteristics of latex paints is that the film is not deposited as a continuum, but rather is formed from discrete particles. It is this mechanism that is basically the cause of the difficulty in producing high-gloss finishes. Electron microscope replicas of the surface of a coalesced latex film indicate that the film surface is not smooth on a microscopic scale; rather, there is a very bumpy surface, the crests of which correspond with the location of latex particles prior to drying (14, 15). It is this microscopic irregularity that results in poor gloss of the film.

One of the most effective methods of improving gloss in a latex paint is to markedly reduce the size of the latex and pigment particles. Smaller latex particles lead to improved coalescence because of the greater interfacial compressive forces, and the surface irregularities become smaller. However, small particle sizes lead to stability problems. Consequently, a compromise is necessary to achieve commercial products. In any event, the high gloss products that can be developed in alkyd systems have not been achieved in latex paints.

In addition to the use of small particle size latexes and pigments, an effective way to obtain gloss in a latex paint is to maintain a low PVC, but this will effect the level of hiding that can be obtained, and again, a compromise between these properties must then be established. There is also often a problem of surface tack with the thermoplastic latexes. Although it might seem logical to increase the hardness of the polymer (or increase its glass transition temperature) to over come this, the fact that the particles must coalesce at room temperature places a limiting value on this. A further technique in reducing the tack is to increase the pigment content, but this in turn reduces the gloss again. In essence then, the development of a semigloss or gloss latex paint is based on a series of compromises. Because of this, in addition to the limitations spelled out above, the volume of high-gloss latex paints sold for interior use is not very large. The semigloss latex paints on the other hand, being able to achieve a better balance of properties, are taking over a larger share of the interior market, displacing the alkyd trim paints.

An important application property of the gloss and semigloss latex paints affecting their formulations is the need for long wet-edge times. In order to lap into previously painted areas without leaving the surface distorted with brush or roller marks, the paints must stay wet longer than flat latex wall paints. They must remain wet enough so that there will be adequate reflow in the lapped areas. This is accomplished by using relatively high levels of propylene glycol in the formulation, but this must be balanced against the requirement of good flow and leveling versus sag resistance under varying conditions of temperature and humidity. The propylene glycol also functions to increase the plasticity of the system, so that a slight increase in the hardness of the polymer can be tolerated for reduction in surface tack.

In general, semigloss latex paints, based on acrylic or styrene-acrylic latexes, exhibit better color retention, retain better flexibility than the alkyds, and are much more alkali resistant. They can be used in bathrooms and kitchens, exhibiting good gloss uniformity and good adhesion. Their drawbacks are generally poorer hiding than the alkyds and the use of large amounts of water-soluble solvents making the film water sensitive in its early stages of dry. Early recoat may also be a problem. Occasionally, over bare wood, it may be necessary to apply a first coat of alkyd primer in order to achieve adequate holdout of the subsequent coats as well as to minimize the grain-raising and fiber-lifting in the wood that the water causes.

Representative formulations of semigloss and gloss latex paints are shown in Formulations 7-14 and 7-15. The pigment dispersions may be made either by a high-speed disperser, such as a Cowles mill, or sand grinder. The latter piece of equipment tends to yield slightly higher gloss.

Compositionally, the semigloss and gloss enamels are very similar, except for the incorporation of a small amount of extender pigment in the semigloss paint for adjustment of the gloss level. The grade of titanium dioxide used is one that yields the maximum gloss in latex paints in preference to other grades that may give more hiding but lower gloss. High gloss grades of titanium dioxide from several manufacturers give comparable performance.

The thickener used must not only provide good thickening efficiency, but also must not unduly impair flow and leveling, and should impart sufficient brush drag to achieve adequate film build. Flocculation effects resulting from interactions between the latex and thickener, and surfactants and thickener, can often be noted by the loss of film clarity and/or a drop in the gloss. Replacement of any of these intermediates in a formulation by substitute products should always be thoroughly investigated for their effects on these as well as other properties.

FORMULATION 7-14. White Semigloss Latex Paint

Pigment grind	lb/100 gal
Water	80.0
Tamol 731, pigment dispersant, 25% solids (1)[a]	10.0
Nopco defoamer NDW (2)	1.8
Propylene glycol	20.0
Ti-Pure, high-gloss titanium dioxide R-900 (3)	250.0
Aluminum silicate (4)	20.2
Let-down	
Water	42.2
Propylene glycol	50.0
Acrylic latex, 46.5% solids (5)	486.0
Nopco defoamer, NDW	1.0
Preservative, 10% mercury (6)	1.0
Hexylene glycol	25.0
Triton wetting agent GR-7 (7)	1.8
Hydroxyethylcellulose, 2.5% solution (8)	75.0
	1064.0
Physical constants	
Gallon weight	10.6 lb
Weight solids	47.1%
Volume solids	32.3%
PVC	25.5%
Gloss (60°)	45 ± 5
Viscosity, krebs units	85 ± 5 KU

Source of raw materials

1. Rohm & Haas Co.
2. Nopco Chemical Co.
3. E. I. DuPont de Nemours & Co.

FORMULATION 7-14. — Continued

4. ASP 170, Minerals & Chemicals Co.
5. Rhoplex AC-490, Rohm & Haas Co.
6. Super Ad-It, Tenneco Chemical Inc.
7. Rohm & Haas Co.
8. Cellosize WP-4400, Union Carbide Co.

[a]Number refers to source of raw materials.

FORMULATION 7-15. White High-gloss Acrylic Enamel

Pigment grind	lb/100 gal
Water	90.0
Potassium tripolyphosphate	2.0
Tamol 731, pigment dispersant	6.4
Triton X-100, wetting agent	1.2
Preservative, 10% mercury	0.4
Propylene glycol	20.0
Ti-Pure titanium dioxide pigment R-900	275.0
Napco defoamer NDW	1.5
Let-down	
Propylene glycol	50.0
Acrylic latex, 46.5% solids	544.0
Nopco defoamer NDW	1.0
Hexylene glycol	25.0
Hydroxyethylcellulose, 2.5% solids	57.5
	1074.0

(Adjust pH to 9.0 to 9.5 with 28% ammonium hydroxide solution.)

FORMULATION 7-15.—Continued

Physical constants	
Gallon weight	10.7 lb
Weight solids	50.0%
Volume solids	34.9%
PVC	23%
Gloss (60°)	75 ± 5
Viscosity, krebs units	85 ± 5 KU
Source of raw materials	
See White Semigloss Enamel	

Hexylene glycol is used in these particular formulations as a volatile coalescent for the latex in order that the paint may be used at the low application temperatures often encountered. Other coalescents that are effective with the specified latex are Butyl Cellosolve and Butyl Carbitol, but since the paints are designed for interior end uses, odor and toxicity considerations generally preclude their use. Along with the propylene glycol, the coalescents generally function to increase the wet-edge time. Their usage is balanced to insure good sag resistance at varying temperatures and humidities against good lapping properties and the time required for the film to reach its ultimate hardness.

The preceding formulations only represent a very small number of possible formulations that can be developed based on the general principles outlined previously and with the wide range of raw materials available to the formulator. Although specific commercial raw materials were used in the examples given, it must be realized that products of other manufacturers may perform equally well and that the formulations represented should be used only as guides to future development work.

REFERENCES

1. F. A. Bovey, I. M. Kolthoff, A. I. Medalia, and E. J. Meehan, Emulsion Polymerization, Wiley (Interscience), New York, 1900.

2. A. H. Loranger, T. T. Serafini, W. Von Fischer, and E. G. Bobalek, Official Digest Fed. Soc. Paint Technol., 31, No. 411, 482 (1959).

3. E. H. Riddle, Monomeric Acrylic Esters, Reinhold, New York, 1954.
4. W. A. Henson, D. A. Taber, and E. B. Bradford, Ind. Eng. Chem., 45, 735 (1953). G. L. Brown, J. Polymer Sci., 22, 423 (1956).
5. R. B. Green, Ind. Eng. Chem., 45, 726 (1953).
6. R. E. Dillon, E. B. Bradford, and R. D. Andrews, Ind. Eng. Chem., 45, 728 (1953).
7. W. N. MaClay, Official Digest Fed. Soc. Paint Technol., 39, No. 506, 156 (1967).
8. P. Becher, Emulsions—Theory and Practice, Chap. 6, ACS Monograph No. 135, Reinhold, New York, 1957.
9. Resin Review, Vol. XIII, No. 1, Rohm & Haas Co., Phila., Pa., 1963.
10. J. W. McCutcheon, Detergents and Emulsifiers, Annual Publication.
11. F. A. Digioia and R. E. Nelson, Ind. Eng. Chem., 45, 745 (1953).
12. W. K. Asbeck and M. Van Loo, Ind. Eng. Chem., 41, 1470 (1949).
13. F. B. Stieg, Official Digest Fed. Soc. Paint Technol., 28, No. 379, 695 (1956).
14. F. B. Stieg and R. I. Ensminger, Official Digest Fed. Soc. Paint Technol., 33, No. 438, 792 (1961).
15. J. C. Becker and D. D. Howell, Official Digest Fed. Soc. Paint Technol., 28, No. 380, 775 (1956).
16. F. B. Stieg, Official Digest Fed. Soc. Paint Technol., 34, No. 453, 1065 (1962).
17. R. Pierre Humbert and C. Boyce, Can. Paint & Varnish Mag., Sept., 1965.
18. R. H. Pascal and F. L. Reig, Official Digest Fed. Soc. Paint Technol., 36, No. 475, 839 (1964). G. L. Weidner, Ibid., 37, No. 490, 1351 (1965).

GENERAL REFERENCES

H. Payne, Organic Coatings Technology, Vol. 11, Wiley, New York, 1961.

"Film Formation in Colloidal Systems," Symposium, Div. Organic Coatings and Plastics Chemistry, American Chemical Society, 25, No. 1, 313 (1965).

Official Digest Fed. Soc. Paint Technol., Symposium on Physical Chemistry of Interfaces as Related to Coatings, 34, No. 448, 465-543 (1962).

Chapter 8

MARINE PAINTS

David M. James

International Red Hand Marine Coatings
Gateshead, Co.
Durham, England

8-1. CHARACTERISTICS OF THE MARINE ENVIRONMENT

The peculiar requirements for any paint system are determined by the environment in which it is to be used, so we must start from there.

Marine paints are used above and below the surface of the sea. The sea is wet, salty, full of life, and around harbors it often stinks; in these four observations lie the essentials of the problem.

A. WETNESS

Certainly marine paints must be capable of surviving in water. However, it is necessary to distinguish between fresh and seawater. Paints are commonly tested for inland use by exposure in a humidity chest or immersion in tap water, both of which are tests against nearly pure water. Seawater is a dilute solution of salt in which the activity of the water is lower than that of fresh water, a fact which is expressed by its osmotic pressure of approximately 25 atmospheres against pure water. Since the primary attack of water on paint films is osmotic in character, seawater is less aggressive than fresh water toward the film itself; it is, in compensation, much more aggressive toward metals, which are the most commonly painted substrates.

It follows that "water resistance" without qualification is too vaguely defined a quality for marine paints. If tests are to be meaningful they must be conducted in seawater itself, or at least in an artificial seawater containing all the major constituents and at the correct pH.

Waters of other compositions are also met with, such as sea spray, which may become concentrated on a surface by evaporation and diluted by rain. River water of all salinities, from that of the open sea to fresh water, and all degrees of pollution is also encountered. This may complicate the problem, but does not reduce the primary importance of ocean water.

B. SALTINESS

The salinity, defined as total salts in grams per kilogram, is normally about 35. The surface salinities vary with time and place; in Northern summers they lie between 30 and 41 over the greater part of the oceans; but in the Baltic sea, for example, large areas are as low as 7, and in the inner gulfs the figure may be as low as 1.

However the total concentration may vary, the proportion of constituents making up the salt varies very little. At a chlorinity of 19 the composition is given by Sverdrup, Johnson, and Fleming (1) as follows:

	g/kg
Cl^-	18.980
Br^-	0.065
SO_4^{2-}	2.649
HCO_3^-	0.140
F^-	0.001
H_3BO_3	0.026
Mg^{2+}	1.272
Ca^{2+}	0.400
Sr^{2+}	0.013
K^+	0.380
Na^+	10.556
	34.482

There are also traces of a large number of other elements.

When the seawater is in equilibrium with atmospheric CO_2, the pH is 8.1 to 8.3. The salts exert some buffering action, but the pH varies between 7.5 and 8.4 or even higher, the former at great depths or in diluted or polluted waters, the latter when photosynthesis has reduced the concentration of CO_2, which is the main factor determining the pH.

The presence of salts reduces the electrical resistance, which is about 30 Ω cm at 10°. This low resistance permits the transport of corrosion currents. The chloride ion is a stimulator of corrosion. For both reasons the corrosion of metals in the marine environment is much more rapid than in most inland environments.

C. MARINE LIFE

Structures immersed in the sea are rapidly covered with slimes consisting of bacteria and diatoms. If they are coastal structures such as piers, or if, like boats and ships, they are present for a greater or lesser proportion of their time in coastal waters, they become fouled with marine growths within a few days or weeks. In some cases these growths are barnacles, capable of disrupting protective coatings and allowing corrosion to proceed. Barnacles greatly increase the skin friction or "drag" of boats and ships with a consequent loss of speed and increase of fuel consumption.

This problem of marine fouling is considered in detail later in this chapter; it constitutes the only really unique feature of marine painting.

While fouling is common to all surfaces immersed in the sea, there is a particular problem with wood, that is, attack by marine borers. These are not susceptible to poisoning by antifouling paints, and therefore are not considered in the section on antifouling paints. We therefore devote a little space to them here.

The marine borers are animals which enter wood, multiply, and riddle it with holes. Their work may be seen on pieces of wood washed up on the shore, and they constitute a serious menace to wooden piers and boats. There are three common types described by Russell and Yonge (2):

1. The Shipworm (Teredo). This is a bivalve mollusk with shell valves at the head of a long, naked body and siphons at the rear. The siphons draw in water, the mouth at the head takes in food (plankton from the water and also wood), and the shells are used for boring. It settles from free-swimming larvae. The entrance in the wood is pinhead size, but once in its burrow the shipworm stays there and grows, and may burrow out a hole 1 cm in diameter.

2. The Gribble (Limnoria). This is a crustacean, like a woodlouse, and has very different habits from the Teredo. Its burrows are of uniform size, so that as the gribble grows it emerges from one burrow and starts a new one. It is always the fully grown animals which start new burrows. The young are hatched in the burrows and can start boring immediately. The burrows are quite shallow, and unlike the Teredo's damage, can be seen without cutting the wood. Another difference is that the gribble does not digest wood.

3. Martesia. This animal, like Teredo, is a bivalve mollusk. It is like a small mussel; its burrow is only the length of the animal. However, it is capable of boring through very hard material. Martesia is a tropical species.

The only way to defeat marine borers is to cover the wood with an intact coating. Hard paints, such as epoxy or polyurethane, are best. They may be reinforced with glass fiber. Gaps in the coating allow the borers to get into the wood. A deposit of fouling organisms such as barnacles or mussels has very little deterrent effect on them.

D. POLLUTION

The most dramatic cases of pollution at sea are oil spillages. Most ships acquire bands of oil on the hull which are very difficult to remove and are an infernal nuisance when repainting, but the damage to the underlying paint is seldom important, except in the case of some soft antifoulings.

Pollution of rivers and harbors by sewage and industrial waste is much more significant because it may generate sulfide, which can inactivate copper-based antifoulings by depositing insoluble black copper sulfide on them; and because it promotes the activities of sulfate-reducing bacteria, which stimulate the corrosion of steel. The severity of pollution is shown in the case of the Thames, which has been found to have less than 20% of the saturated concentration of dissolved oxygen for 10 miles above and 30 miles below London Bridge — a situation which worsened in the half century up to 1959, and then improved somewhat due to improved sewage treatment (3).

So much for the natural environment. When painting we must also think of the conditions determined by man.

The outstanding characteristic of marine painting is its roughness. Everything is rough: at best, the surfaces are much more uneven than those of many industrial manufactures, and at worst they are irregularly undulating masses of old paint, corrosion product, and fouling; the work is done outdoors, against time, and the worker who does it is usually unskilled. Under these circumstances it is amusing that the film thicknesses should be specified in microns (μ). In fact, paint films must be much thicker than those that will suffice for oven-cured jobs on smooth phosphated steel, such as on motor cars.

The picture painted above is the worst, but it accounts for a great deal of ship painting, especially maintenance painting in drydock. In the building yards conditions are better; work is sometimes done under cover. In yacht painting the conditions may be very good. The surfaces are smooth, and these workers are usually skilled and well educated.

8-2. THE SOURCES AND EFFECTS OF HULL ROUGHNESS

Ships are painted above water mainly to improve their appearance and for purposes of hygiene. They are painted below water for entirely different reasons. It is not a matter of preventing corrosion from holing the bottom, because, in fact, this would rarely occur even with bare steel. The main reason for painting below the water level, making it worthwhile to spend a good deal of money on bottom painting, is that corrosion and the growth of marine fouling roughen the surface of the hull, causing loss of speed and increased fuel consumption.

A. THE COMPONENTS OF SHIP PROPULSION RESISTANCE

This section is broken down into two parts: "skin-frictional" and "residual" resistance. The residual resistance arises from the energy dissipated by pushing the water aside, and is determined by the shape of the hull and its speed. The skin-frictional resistance arises from the dissipation of energy by flow of the water past the hull, and is determined by the smoothness of the hull as well as its wetted area and speed. It is therefore only the skin friction that can be affected by the presence of paint.

The distribution of the total resistance between skin friction and residual depends upon the speed/length quotient (speed/length$^{1/2}$). If this is expressed in knots and feet, as is usual, then up to 1.0 (e.g., 10 knots, 100 ft) the skin friction is 70 to 80% of the total. As the quotient rises to 1.4 (e.g., 20 knots, 200 ft) the proportion of skin-frictional resistance falls almost linearly to 30%, and then falls little further as the quotient continues to rise. A large tanker might have a quotient of 0.6 (20 knots, 1000 ft) and the skin friction would be 70% of the total resistance. It is therefore of major importance in the economy of running the ship.

B. SKIN-FRICTIONAL RESISTANCE

The dissipation of energy by skin friction occurs in the boundary layer, that is, the layer of water at whose outer boundary the velocity is barely greater than that of the surrounding sea. This layer is estimated to be of the order of one meter, and it is larger the faster and longer the ship.

It is important to understand that the energy is dissipated in the water, not by "slip" between the solid surface and the liquid. Failure to understand this underlies the many, and always unsuccessful, attempts to reduce skin friction by treating hulls with water-repellent coatings. It is worth a little space to explain why "slip" does not occur.

Fowkes (4) calculates that of the total surface-free energy of water, 0.073 J/m^2, the fraction due to dispersion forces alone is 0.022 J/m^2. The energy of adhesion of water to any solid surface will therefore be at least 30% of the energy of cohesion of the water itself. As the solid surface begins to move through the water, the layer of water molecules adsorbed on it will move with it, and the next layer will move at a very slightly lower speed. The fractional drop in speed per layer of molecules is indeed very slight, being of the order of the molecular mean free path divided by the thickness of the boundary layer, i.e., 10^{-9}. If the speed is now increased until it imposes a stress on the water that is an appreciable fraction of the tensile strength of the water, we may postulate that "slip" occurs between the solid surface and the first layer of water molecules. To be significant, this must be an appreciable fraction of the whole speed gradient, say 1%. The relative speed of the first layer of molecules past the surface is then 10^7 as great as that of the layers of water molecules past each other. The absorption of energy is correspondingly enormous because the intermolecular attractive forces are of the same order. But a process requiring an enormous absorption of energy will not occur if better alternatives are available. The proposition of "slip" is therefore self-contradictory. In fact, it has never been observed experimentally. At speeds much higher than ships' hulls can attain, on propellers for example, the tensile strength of the water is exceeded, and cavitation occurs.

At normal ship speeds, the boundary layer is turbulent for most of its thickness, but there is a laminar sublayer. Most of the energy is dissipated in the turbulent flow. The roughness of the surface affects the laminar sublayer. Since no ship surface is ideally smooth, what sort of roughness is significant? Lackenby (5) discusses this in terms of "admissible roughness," that is, the degree of roughness below which the skin friction does not decrease as the surface becomes smoother. The admissible roughness varies with the type of ship, especially its speed, being at the extremes 8 μ for a fast passenger ship and 20 μ for a tug. The surface of an evenly applied paint film will usually be smoother than the lower of these, but any roughness of the substrate, uneven paint application, or deposition of marine growth can easily exceed the limit.

The characterization of roughness in terms of length derives from the use of sand grains of different diameters to produce surfaces of graded roughness. However, the shape as well as the size of the roughness is important. Sasajima and Yoshida (6) found that roughness of low slope ("wavy") caused less frictional resistance than sharp roughnesses and that most painted surfaces were of the wavy type.

1. The Economic Importance of Skin Friction

A great deal has been published on the effects of hull roughness, the classical controlled ship experiments being the "Lucy Ashton" and the "Lubumbashi."

The "Lucy Ashton" experiments (7) were carried out by mounting jet engines on the deck of a small steamer and running the measured mile on a Scottish loch with various conditions of the hull. After extensive corrections for the physical conditions, the effects of hull roughness on resistance were calculated. Comparing the resistance with that predicted from a model for a smooth hull, they found that at 11 knots the smoothest hull (faired seams, aluminium paint) had 6% extra resistance; the use of a rougher paint (red oxide) produced 9% above the smooth hull; and sharp seams with the red oxide paint, 14%. Considering the small scale of the paint roughnesses (of the order of 70 μ) and the wavy nature of the surfaces (amplitude/wavelength: 1/50 maximum) the difference between the paints is surprising. However, it was small compared with the effects of fouling. When the ship had been out 24 days in the summer, the skin friction was 12% above the smooth hull prediction; after 30 days, skin friction was up 24%; after 58 days, it was up 48%; and the fouling was light. In the winter, after 40 days the skin friction was only up 5%, and the ship was only fouled with slime.

The "Lubumbashi" experiments (8) were concerned with the power consumption and speed of a vessel in service. Increases in frictional resistance of about 10% were found due to fouling between drydockings, in addition to which there was a steady increase due to corrosion, which was not recovered on repainting, and accounted for a power increase of 17% after five years. At this point the surface was very rough (corrosion cavities 50 mm square and 2.5 mm deep, blisters 2.5 to 5 mm high, and scale). The hull was sandblasted and repainted and the skin friction reverted to that of the ship when new.

A study of a group of tankers (9) showed differences between apparently identical ships to be due to hull roughness caused by corrosion. At 14.5 knots the shaft horsepower, 5450 on sea trials, rose to 7500 in service, which was associated with a measured roughness height of 650 μ. This would account for seven to eight lost days per annum, at constant fuel consumption, due to reduced speed.

The Netherlands Research Centre T.N.O. reported on the cost of hull painting compared with that of fuel consumption (10) and concluded that the extra costs due to fouling more than compensated for the difference in price between three common grades of antifouling. They estimated that better paints could reduce fuel costs by 10% and save 18,000,000 guilders annually for a fleet of 5,000,000 gross register tons.

Comparing two cargo ships, one painted at drydocking with a low quality and the other with a good quality bottom paint scheme, the fuel saving due to lower corrosion and fouling over one year was 12,000 guilders for an extra painting cost of 4200 guilders.

In 1960 the preparation of steel for shipbuilding by centrifugal shotblasting was just becoming established; the improvement in service performance due to blasting was commented on by Logan (11). It is interesting to read that "tankers which of necessity spend little time in port do not experience serious fouling, and it is considered reasonable to state that, in general, loss due to this cause is negligible in the normal course of events." By the middle 1960s the improvement in anticorrosive protection had brought out the actual losses due to algal fouling. This has been a major concern of tanker owners ever since.

2. How to Reduce Skin-Frictional Resistance

The answer to the question of how to reduce skin-frictional resistance is very simple: build and maintain a smooth hull. Implementation is not so easy.

Other answers have been given, but none have proved both sound and practical. Among the unsound ideas is that of inducing "slip" between the water and the hull, whether by painting the hull with a water-repellent coating, or by injecting air into the boundary layer, a confused emulation of the Hovercraft. One sound idea is to treat the water with a water-soluble high polymer; this has the effect of damping out eddies (12). The quantity of polymer required is so great, however, that it makes the process uneconomical for commercial exploitation. Another idea is to cover the hull with a double skin of rubber containing water. This again damps out the eddies (13). The problem here is not so much that of cost but of the mechanical damage which the hulls of ships treated in this way inevitably suffer.

Prevention of corrosion and fouling is the main requirement of the bottom paint scheme. If it does this, small differences in surface texture are unlikely to be significant in commercial shipping. (Yacht racing is a different matter.) Nevertheless, it is undesirable to apply a very rough paint scheme if a smoother one will do the job. Couch (14) found that hot plastic antifouling increased the skin-frictional resistance coefficient of a model by 40% compared with bare metal, whereas a smooth paint (zinc chromate primer) barely increased it at all.

8-3. THE CORROSION OF UNDERWATER HULLS AND THE PRINCIPLES OF ANTICORROSIVE TREATMENTS

The metal of chief concern regarding the corrosion of underwater hulls is steel. Other metals are used in the construction of yachts and are considered in a later section.

A. THE CAUSES OF UNDERWATER CORROSION

All corrosion of metals in aqueous solutions is electrochemical in character. Steel corrodes at anodic areas and releases iron (II) into the solution; the electrons move along the metal and are discharged at cathodic areas. At the low hydrogen ion concentration of seawater, the cathodic reaction is not the liberation of hydrogen but the reduction of dissolved oxygen and water to hydroxyl ions.

The cathodic areas may be part of the steel itself, which differ from the anodic areas by virtue of greater access of oxygen, or chemical composition, or state of strain. They may be remote, as for example in a bronze propeller, stainless-steel fittings, or the earth itself (in the case of welding sets earthed to the ship, which have caused stray-current corrosion), but they must be in electronic contact with the steel. The low specific resistance of seawater allows this action by remote cathodes and the flow of corrosion current can be large if the necessary anodic and cathodic areas are present.

Ffield (15) pointed out that in still seawater the corrosion rate of unpainted steel is quite low, 50 to 175 μ per year, and therefore corrosion is only dangerous if it is accelerated. The nature of the steel, including alloying metals (except Cr at over 3%), is unimportant in seawater. As the speed of the water relative to the surface increases from 0.6 to 6 m/sec, the rate of corrosion approximately doubles, but even this is not dangerous. The danger arises from the presence of cathodes whose area is large relative to the anodic areas. The conditions are then right for a high current density at the anodes and a rapid loss of metal, i.e., pitting. In most cases the corrosion is under cathodic control, i.e., the area and access of oxygen to the cathodes limits the corrosion current which can flow.

While pitting is undoubtedly the most dangerous form of corrosion, because it calls for docking and replacement of plates, the over-all light corrosion is also very important because of its roughening effect, as mentioned before.

Bimetallic contact is most commonly found in the form of the bronze propeller, which is more noble than the steel of the hull. Being much smaller in area, however, it cannot cause serious hull corrosion, and indeed, if it is insulated from the hull it suffers self-corrosion. It should therefore be kept in electric contact and cathodically protected by the hull. Another form of bimetallic contact is at welds. Uusitalo (16) described corrosion due to the development of potentials up to 40 mV between weld and plate metals in contact in seawater, with accelerated corrosion of the plates. This is not a common occurrence.

The cathodic reaction may occur in the absence of oxygen, the electrons being accepted by sulfate-reducing bacteria in the course of reducing sulfates to sulfides; the sulfides may also stimulate the anodic reaction according to Iverson (17). A case of severe bacterial corrosion of a ship lying in mud has been described by Patterson (18). In the open stream, however, pollution of rivers by sewage, which is the usual reason for the growth of these bacteria, generally reduces the rate of corrosion. This is because the concentration of dissolved oxygen is low, and the bacteria cannot efficiently act as a substitute (19). A case of severe pitting due to sewage pollution was described, however, by Corcoran and Kittredge (20). This occurred under the paint film in a band just below the waterline and was ascribed to the biological production of anodic and cathodic areas in close proximity.

Another feature of seawater which stimulates corrosion, in addition to its low electrical resistance, is the presence of chloride ion. This ion inhibits the repair of the primary oxide film and thereby stimulates the anodic reaction (21).

A picture of the electrochemical behavior of a ship at sea was drawn by Barnard (22) in 1948. He surveyed a naval ship over a period of nine months by fixing reference electrodes far from the hull and search electrodes near the hull; he then measured the potential difference between them. By this technique, fairly well separated anodic and cathodic areas could be distinguished. Initially, before drydocking, the hull was extensively corroded. The potential differences were therefore small and the "zincs" (fitted at the stern to give cathodic protection) were inactive. After drydocking, without painting, marked cathodic areas appeared, chiefly at intact painted steel and on the bronze propeller. However, the waterline, which had been weakly cathodic, had reversed its potential. After 15 days, the potential differences had become much smaller and the hull had become "passive" due to polarization of the cathodes. After the second drydocking, when new "zincs" were fitted, the picture was much the same, except that the "zincs" were anodic. After three days, however, the waterline became strongly cathodic, and only the "zincs" were strongly anodic. This behavior was ascribed to different rates of building up a resistant film on cathodic areas. The author concluded that there are three main cathodic areas which can stimulate corrosion: painted steel (the paint being conductive by waterlogging), the waterline, and the bronze propeller. His recommendations were that the quality and application of underwater paints should be improved and that more and more effective cathodic protection was needed than the "zincs" could supply, cathodic protection being the most important way of reducing the corrosion.

B. CATHODIC PROTECTION

Since the corrosion is due to the flow of an electric current, it may be stopped by applying a potential sufficient to produce an equal and opposite current. In practice, such a nice balancing of a variable process would be impossible. Therefore, a higher potential is applied, the steel is rendered cathodic, and the anodic reaction is concentrated on specially fitted anodes, where it can be rendered harmless.

If the anodes are made of a baser metal than steel, they generate the current and are consumed by the anodic reaction. This was the principle of the "zincs," but these were ineffective. It was not until the 1950s that the need for using high-purity zinc was appreciated as necessary to prevent the anodic reaction from being stifled by a resistant film. Magnesium and aluminum alloys are also used. This sytem is called cathodic protection by sacrificial anodes.

If the anodes are made of a noble material that is resistant to chlorine, the current can be applied from a generator. The current is not limited, as in the sacrificial anode system, by the number and size of anodes. Therefore the anodes can be disposed in the minimum number necessary to distribute the current, and the current can be varied according to the demands of the hull, keeping the hull potential to the minimum needed for protection. This system is called cathodic protection by impressed current.

The purpose of cathodic protection is to choose certain types of electrode reactions that are determined by the electrode potentials. To achieve a certain potential on a certain metal in a certain medium requires a certain current density. The current density determines the rate of the electrode reaction. This distinction between the potential, which determines the type of electrode reaction, and the current density, which determines its rate, is important.

Unpainted steel, freely self-corroding in seawater, assumes an over-all potential which is the average of the potentials of the anodic and cathodic areas. This is -0.43 V to the normal hydrogen electrode or, to use a more practical measure, -0.69 V to Ag/AgCl/seawater. In order to protect the steel, its potential must be raised at least to -0.80 V (Ag/AgCl/seawater). At this potential the cathodic reaction is the reduction of O_2 and H_2O to form OH^-. At -0.95 V the reduction of H_3O^+ to H_2 is possible, and will occur when oxygen availability is limited, the faster the higher the potential; again OH^- is produced.

The current consumption by bare steel in seawater is limited by oxygen availability, so it increases with speed. A "chalk" of calcium and magnesium salts is deposited, which also limits the current. Nevertheless, at 15 knots the current density to achieve a protected or protective potential may be 0.15 A/m^2 (23). A ship's underwater surface might be 10,000 m^2, requiring a current of 1500 A. There is, therefore, a powerful reason for applying an insulating coating. In practice a painted ship's bottom will take about 0.0023 to 0.014 A/m^2 (24), which would reduce the consumption for the abovementioned hull to 23 to 140 A.

Cathodic protection imposes a special stress on the paint because wherever the cathodic reaction occurs, alkali is liberated. This applies also to the cathodic areas of an unprotected hull. Bottom anticorrosives must therefore be alkali resistant, as is discussed later. When cathodic protection is applied, however, the current density at the cathodes is higher than in self-corrosion; therefore the rate of production of alkali is higher, which, under conditions of water flow, means a higher local concentration. Different paints have different degrees of alkali resistance, and it is found in practice that they will resist different current densities. It is difficult to measure current density on a ship's hull, or even on a painted specimen, where the area of exposed steel may be very irregular and uncertain, and the proportion flowing through the steel and that through the paint itself is uncertain. Fortunately, the potential of the protected hull is a very fair guide to the current density at the areas of bare steel because these take most of the current; and provided the seawater is well stirred and aerated, the current density at a protected steel surface is higher the more negative the potential. Therefore, in practice, paints may be rated according to their resistance to a maximum applied potential in seawater.

The resistance of the seawater path increases with distance, but only up to a maximum of about 3 m. This means that the potential of the hull near an anode will be higher than that of the hull as a whole. It will also be greater the higher the potential of the anode and the nearer to it. For this reason the hull around the anodes is usually coated with thick layers of insulating material. In the case of impressed-current anodes, which run at higher potentials than the sacrificial types, even normally resistant paints are inadequate when near them, and special mastics or plastics may be used.

C. PROTECTION BY PAINT

The mechanism of protection by paint has been investigated by Mayne (25), who sums up his findings as follows: "In order to inhibit corrosion it is necessary to stop the flow of current. This can be achieved by suppressing either the cathodic reaction or the anodic reaction or by inserting into the path of the current a high electrolytic resistance which will impede the movement of ions and thereby reduce the corrosion current to a vanishingly small value.

"In order to suppress the cathodic reaction it is necessary to prevent water and oxygen from meeting at the metal surface. It seems now acknowledged that paint films of normal thickness are so permeable to water and oxygen that they cannot inhibit corrosion by preventing either from reaching the surface of the metal i.e. they cannot inhibit the cathodic reaction.

"On the other hand there are two ways in which a pigment may modify the anodic reaction. If the electrode potential of iron is made sufficiently negative, positively charged ions will be unable to leave the metallic lattice (cathodic protection); alternatively, the iron may become covered by a film, which is impervious to ions (anodic passivation)."

This statement requires modification only in that the oxygen permeability of paint films has been found by Guruviah (26) to be only about one-tenth of the magnitude required for free corrosion of steel, and it could therefore exercise a rate-controlling effect. There is, however, little evidence in his paper that it did so, and it may still be accepted that the primary anticorrosive mechanisms are electrolytic resistance and interference with the anodic reaction. These principles are exemplified in the section on formulation.

Gay (27) investigated the blistering of paints in seawater. He found that the steel under the blisters was bright and the liquid in the blisters was more alkaline and lower in chloride than the seawater. He postulated that the blisters formed on cathodic areas, hence the brightness of the steel and the alkalinity. The paint film acquired a negative charge, which allowed positive ions to pass, but held back negative ions, hence the lower chloride content. This negative charge was also responsible for the passage of water to the cathode by endosmosis, the film being repelled from the cathode. The alkali in the blisters could attack the film and weaken its adhesion to the metal, as a result of which the corrosion could creep across the metal. The conclusion that the film should be as impermeable as possible to ions and water, and as resistant as possible to alkali, is clear, and again is exemplified in the following section.

8-4. THE FORMULATION OF ANTICORROSIVE PAINTS FOR SHIP BOTTOMS

When iron replaced wood for shipbuilding, the problems of corrosion and fouling were revealed sharply. Copper-bottoming, which had coped fairly successfully with both problems, was tried with disastrous effects, due to the anodic polarization and consequent accelerated corrosion of the iron. An excellent account of the arguments about what to do can be found in a book by Young (28), dated 1867. "Paints are of little use," he said, "for though above water, if the paint should be rubbed off, it can be laid on again, under water this cannot be done without docking or grounding the ship, consequently the iron must be left exposed to this destructive action until an opportunity can be afforded for so doing." His lack of faith in paint was justified, because, in addition to red lead in oil, he mentions "clay, fat, sawdust, hair, glue, oil, logwood etc. mixed" and "sugar, muriate of zinc, wax, soap, calcareous stones, phosphate of soda, sulfate of zinc and copper, and the syrup of potatoes or sugar with powdered marble, quartz or felspar," as compounds which had been used for painting ship bottoms. His remedy was to sheath the steel in zinc, which, from the anticorrosive point of view, was sound. He also maintained, however, that it would keep off fouling by exfoliation, i.e., shelling off layers as it underwent sacrificial corrosion; this was unsound.

The anticorrosive painting of ship bottoms was an unsatisfactory affair right up to World War II. Red lead in oil was the usual primer, and while this is excellent in many situations, it is unsuitable for ship bottoms because the vehicle is liable to saponification by cathodic alkali. Nevertheless, it was used and gave quite good results provided (a) that it was dried a long time, and (b) that it was overcoated with materials containing acidic resins, which were permeable to the alkali and allowed it to diffuse away. The alkali problem was not appreciated theoretically, as is shown, for example, by Cushman and Gardner in 1910 (29), who stated: "The first, or under coating, for ship plates should contain in the vehicle a good hard drying varnish to act as an excluder of water." Ragg (30), in his standard work of 1925, had little to add to this.

However, although a dried film of linseed oil is bonded partly by ester groups, prolonged oxidation introduces a proportion of unsaponifiable groups, which can to some extent act as a protection to the ester group; and provided the sodium ions can diffuse away, saponification need not be disastrous. If, however, a linseed oil film is covered with a film relatively impermeable to ions, the alkali can creep under it from breaks and saponify tbe linseed oil film. Coal-tar pitch coatings over red lead in linseed oil were especially disastrous.

Before World War I the red lead system worked fairly well. After the war, however, shipbuilding was speeded up, and the slow-drying red lead was not used so much. The economic crises dictated low prices, and the cheap rosin/oil "anticorrosives" (a term used to denote undercoats for antifoulings, as distinct from "bare plate primers") were now applied directly on the steel. The results were horrible. When the British Government initiated work on ship bottom compositions during the war, the first raft trials showed most commercial primers to last a few weeks (31).

Shortly before World War II, an aluminum-bituminous primer was developed commercially (this is considered later) and gave an excellent performance. During the war, research was done in England by the Marine Corrosion Sub-Committee of the British Iron and Steel Research Association, under government stimulus, which produced a good oleo-resinous primer (32). In America, again with government encouragement, workers from Bakelite Corp. produced the Wash Primer, and in the later 1940s the Vinylite schemes were developed. Nevertheless, bottom paints used in the Royal Canadian Navy up to 1950 had only four months' average protective life (33). Epoxies, coal-tar epoxies, and chlorinated rubber, together with the widespread use of cathodic protection, were introduced in the 1950s; at the end of that decade centrifugal shotblasting became the standard method of preparing ship plate, which greatly raised the level of performance of all types of coating.

The various types of bottom anticorrosives are now considered.

A. BITUMINOUS ALUMINUM

One of the most successful ship bottom primers has been sold under the name "Silver Primocon." This originated from a coal-tar pitch/bituminous composition which was used for many years before and after 1939 as an anticorrosive. As an undercoat for antifoulings whose function is to bind together the old paint and residues of fouling it is good, but as a bare plate primer it is not. Some time in the early part of this century, a composition was made using aluminum flake in this varnish, which gave some good results on bare steel, but proved unreliable and therefore got a bad reputation. When systematic raft trials started in 1929 this matter was investigated, and by 1939 it was established that the amount and grade of aluminum powder was all-important — there must be enough of it, and it must be fine. This became the International Paint standard bare plate primer and acquired several imitators. When airless spray application arrived in the 1960s it was made thixotropic and capable of application at adequate thickness (125 μ minimum) in two coats.

The success of this type of coating depends, above all, on its very low water permeability. Deterding, Singleton, and Wilson (34) found that it had a permeability constant to water one-twentieth that of a bituminous coating with a normal granular pigmentation, due to the leafing of the aluminum flake. Secondly, the medium is highly resistant to alkali. This is valuable not only because alkali is produced during corrosion, but also because it enables the paint to resist cathodic protection up to an applied voltage of about -0.90 to Ag/AgCl/seawater. Above this the aluminum is attacked by the alkali and the medium cannot adequately protect it. A number of other properties, such as ready displacement of water, and the good adhesion of antifoulings, arise from details in the formulation which make it rather complex.

While this will protect well-prepared steel almost indefinitely on a raft test, it is limited in practice because of its rather low mechanical strength, or "crumbly" nature. The problem is not so much that of mechanical damage, which no organic coating can resist, but of the building up of successive layers of antifouling and "anticorrosive" at each drydocking; eventually they pull off the system in pieces, leaving a rough surface. Sussex (35) of the Royal Australian Navy estimated that an aluminum/bituminous system would last about six years with cathodic protection, after which it must be blasted off.

B. OLEORESINOUS PRIMERS

An account of the M.C.S.C. wartime work, which led to the formulation of A/C No. 173, is given by Fancutt and Hudson (32). Media and pigments were examined in turn. Basic lead pigments were found to be desirable, basic lead sulfate being better than red lead, which caused brittleness. Several oil varnishes were good, and a 2/1:linseed stand oil/rosin modified phenolic varnish was chosen. This, pigmented with basic lead sulfate, white lead, Burntisland Red (an iron oxide with a basic reaction), and barytes, constituted No. 173. The evaluation was by raft immersion of steel plates, pretreated by pickling and weathering and also by simple weathering, then wirebrushed and coated with two coats of primer and one of antifouling. The qualities sought were toughness, elasticity, resistance to seawater and abrasion, acceptance of antifouling, easy application, quick dry, tolerance of weather conditions during application, and storage stability.

Later work (36) led to the formulation of A/C No. 185, using aluminum flake in place of the white lead in No. 173, and eventually to No. 655, using a tung oil/linseed stand oil/rosin-modified phenolic: 1/1/1 in place of the linseed oil varnish in No. 185. This became the British Admiralty standard bare plate primer and was used in conjunction with a wash primer pretreatment. Due to its water impermeability, this primer was not suitable as an "anticorrosive" for repainting. It tended to detach, and No. 173 was used for that purpose.

This type of primer had certain limitations. It tended to blister, a feature recorded on all the raft trials, and to overcome this required many coats. Due to its inability to through-dry in thick films, it was not suitable for adaptation to high-build application. And because of its limited alkali resistance, being based upon vegetable oils, it was unsuitable for use under cathodic protection.

The M.C.S.C. trials produced very good results with chlorinated rubber, but this was not recommended due to "practical difficulties." Later, however, chlorinated rubber came into wide use.

C. CHLORINATED RUBBER

Chlorinated rubber is highly resistant to water and alkali, may be formulated with unsaponifiable plasticizers, and dries by solvent evaporation. It is, however, sensitive to the surface on which it is applied, and unlike the two foregoing types, is not suitable for application on rusty steel, however well brushed it may be. It therefore had to await the arrival of centrifugal blasting before it could be used widely in ship painting. There are often national preferences in paints, and for some reason Holland and Norway led the way in appreciating the value of chlorinated rubber on ships. Its advantages — in addition to those listed, its relatively low price, hardness, pale color, and ability to be formulated into high-build primers — became more widely known in the late 1960s.

The principles of formulation have been described by Birkenhead, Bowerman, and Karten (37). Aluminum is the best pigment; basic lead sulfate and metallic lead also have value. With the aid of thixotropic agents, high-build primers can be formulated which will airless spray to give 75 to 125 μ per coat.

It is possible to use resins other than the chlorinated hydrocarbons ("Cereclor" and "Aroclor") in order to increase the solids contents of chlorinated rubber paints, which tend to be rather low, as is usual with resins drying by solvent evaporation only. If oxidizing resins are used, the intervals between coats must be controlled in order to avoid solvent attack by one coat on another.

As well as being a strictly bottom coat, chlorinated rubber has the great advantage of being suitable for boottop (intermittently immersed) paints, because it can be formulated to give bright colors and is resistant to cathodic protection, which is especially aggressive on the boottop, due to the availability of oxygen. The maximum potential to which chlorinated rubber coatings are resistant is -1.0 V (Ag/AgCl/sea water) for an aluminum and -1.2 V for a red lead pigmentation.

Chlorinated rubber has many virtues, and for an organic coating its only limitations are that it requires a clean surface and must be applied to a protective film of at least 130 μ or else it blisters.

D. POLYVINYL CHLORIDE

Polyvinyl chloride is in the same class as chlorinated rubber, but is generally of a high molecular weight and is most frequently used as vinyl chloride/vinyl acetate copolymers (e.g., Union Carbide's "Vinylite" resins), which are soluble in mixtures of aromatic hydrocarbon and ketone solvents.

The vinyl systems, starting with wash primer, continuing with red lead primer, and finishing with cuprous oxide antifouling, were adopted by the U.S. and Canadian navies in the 1950s. This is another case of national preferences, because they never attained popularity elsewhere. There were many difficulties with loss of adhesion which were overcome by better surface preparation, coating techniques, and avoidance of bad weather (33). The cost of these precautions, and of the coatings themselves, may explain their lack of success in commercial ship painting.

They are resistant to cathodic protection, but the weak link in the scheme is the wash primer. Brown (38) found that the use of an aluminum vinyl primer over the wash primer greatly improved the resistance to cathodic protection, raising the limiting potential to -1.2 V (Ag/AgCl/seawater).

In spite of their limited popularity, vinyl systems behave well when well applied. De Wolf and van Londen (39) found them to be good in raft trials, though prone to lack of adhesion. In ship trials they were much better than M.C.S.C. 655 or the U.S. Navy's oleoresinous formula No. 14.

Another vinyl resin which has been employed for ship bottom coating is "Rhenoflex," a post-chlorinated polyvinyl chloride. Experiments have also been reported by Hanson and Tucker (40) on polyvinyl chloride plastisols applied at the steel mill. This is an interesting idea, but does not seem to have progressed, probably because of the loss of adhesion reported from the trials.

E. COAL-TAR EPOXIES

In the early 1950s, when epoxy resins were evaluated for ship's tanks and found suitable, they were tested as bottom paints. The results were mixed, and they did not become popular. Toward the end of that decade, there appeared a coating composed of a blend of amine-cured epoxy resin with a special pitch made by digesting powdered coal with coal tar, under the name "Tarset." In the 1960s this was followed by a mass of imitations and variations, and coal-tar epoxy became one of the most popular bottom coatings for tankers, bulk carriers, and other large fast ships. Like chlorinated rubber, it was dependent upon the widespread use of grit-blasting for its popular success.

The undiluted epoxies were capable, at best, of giving good results, but they were expensive both in material and application cost, because only about 80 μ could be applied in one coat. The coal-tar epoxies were much cheaper because only 40 to 50% of the film consisted of cured epoxy resin and the tar was very cheap. They could be applied in smooth films by airless spray at as much as 250 μ in one coat, and because they cured by reaction throughout the film, there was no problem of skin drying. They had excellent resistance to seawater, and the films were tough and hard. They could resist cathodic protection up to -1.5 V (Ag/AgCl/seawater). Their evangelists claimed that coal-tar epoxies were perfect bottom coatings; but they are just material things in the material world.

Relatively minor drawbacks of coal-tar epoxies are the difficulty of attaining proper cure at temperatures below 15°C and the limited range of antifoulings that will adhere directly to the surface. In practice, the cost is not low, because a minimum thickness of 250 μ is required, double that of the bituminous aluminum type, and if this thickness is not attained, the films blister. The reason for this probably lies in the nature of the coal-tar component, which in order to be compatible with the epoxy resin has to be rather rich in compounds of low molecular weight, which detracts from its protective properties. The main trouble with the coal-tar epoxy has been its sensitivity to surface preparation. Much of the early work was done on old ships which were blasted in drydock. After one or two years' service, a large number of them failed by blistering. Since the coatings did not suffer this way in laboratory tests, the problem was difficult to solve, but it seemed likely that it was due to residual impurities not removed from the pits of the old steel. This was confirmed by workers at Shell Research, who found that chloride ion was tenaciously held in the pits in spite of prolonged blasting (41). It was confirmed in another way: performance of coal-tar epoxies on new ships has generally been good.

Other reinforced coal-tar materials have been introduced, principally with the idea of obtaining a quicker dry under adverse conditions, but they have not at the time of writing been long enough on the market to be properly assessed.

F. THE PRESENT POSITION

There is still a large number of ships sailing with rough and rusty bottoms, but the proportion of world shipping in this condition is declining. Cathodic protection and the universal preparation of surfaces by shotblasting have improved the condition of the bottoms of newer ships to the point where corrosion is no longer the main source of hull roughness, and attention has been concentrated on fouling. The remaining weakness of the anticorrosive paints is susceptibility to mechanical damage. The stresses due to a ship's rubbing against the sea bottom or underwater structures are far greater than any organic coating can withstand, and for this reason zinc silicates, which combine the ductility of a metal with the hardness of silica, are of great interest for the future.

8-5. MARINE FOULING AND ITS PREVENTION

"Marine Fouling and Its Prevention" is the title of one of the very few, and probably the best of the textbooks on this subject (42).

A. CLASSIFICATION OF FOULING

The classification and description of marine fouling, if properly done, would far exceed the space available here. Table 8-1 provides a classification according to susceptibility to poisoning and has the merit of presenting the problem in the order of difficulty for the paint chemist.

B. THE HABITS OF FOULING ORGANISMS

The habits of fouling organisms is an enormous subject, but a few of the vital points must be noted.

TABLE 8-1. Susceptibility of Marine Fouling Organisms to Poisoning

Resistance to poisoning	Organism and description	Classification
Very high	Bacteria (microscopic, generally colourless).	Bacteriaceae
	Diatoms of certain encrusting types (algae of microscopic size, forming slime fouling).	Diatomaceae
	Ulothrix (a minute, filamentous green seaweed which, with related types, forms green, weedy "slimes").	Chlorophyceae (green algae)
High	Tubularia (a "hydroid" growing in massive clusters projecting up to 6 in. or more).	Hydrozoa
	Ectocarpus (a brown seaweed of feathery form up to several inches long in normal growth, but forming felt-like "slimes" or "brown felts" when stunted by copper poisoning).	Phaeophyceae (brown algae)
	Enteromorpha (a green seaweed, growing as streamer-like ribbons or tubes up to several inches long; a very characteristic fouling of water-line areas).	Chlorophyceae
	Barnacles of several species, including the "Goose Barnacle" as well as some sessile species	Crustacea Cirripedia
	Hydroides (a tubeworm building a calcareous tube, circular in section).	Annelida Serpulidae
	Polysiphonia (a red seaweed of dark, brown-red color, branching and of coarser appearance than the olive-brown Ectocarpus).	Rhodophyceae (red algae)
	Polyzoa of some species (forming thin, calcareous mats).	Polyzoa (Bryozoa)
Moderate	Barnacles of some species.	Crustacea Cirripedia

TABLE 8-1. — Continued

Resistance to poisoning	Organism and description	Classification
	Hydroids of some species (e.g., Gonothyrea, a smallish frondose type with creeping basal stems).	Hydrozoa
	Polyzoa of some species (e.g., Bugula, a tufted, frondose type).	Polyzoa (Bryozoa)
	Ceramium (a red seaweed appearing as bright or dark red plumose fronds).	Rhodophyceae (red algae)
	Ulva (a green seaweed with broad, thin, leaf-like growths).	Chlorophyceae (green algae)
	Cladophora (a green seaweed appearing as dark green feathery tufts).	Chlorophyceae (green algae)
	Barnacles of some species.	Crustacea Cirripedia
	Hydroids of some species (e.g., Plumularia).	Hydrozoa
	Polyzoa of some species (e.g., Membranipora, forming thin, calcareous mats).	Polyzoa (Bryozoa)
Low	Ascidians or Tunicates (sea-squirts of several species).	Tunicata
	Mytilus edulis, the common mussel.	Mollusca Lamellibranchia
	Oysters of a few species.	Mollusca Lamellibranchia
	Laminaria (a brown seaweed with broad fronds growing to a large size).	Phaeophyceae (brown algae)
	Sponges	Porifera

All the important fouling organisms, with one exception, are denizens of the seashore. They, and the larvae or spores which colonize surfaces, are therefore to be found only along the coasts. The exception is the Goose Barnacle, which grows on floating objects and may be found in the open ocean.

The larvae and spores settle most happily on surfaces which are at rest relative to the water. It is commonly stated that a relative velocity of 3 to 4 knots is the maximum which will allow settlement, and therefore that settlement is only of importance in harbors. This is true of barnacles, but there is reason to doubt its truth for algae, though no definite evidence is available.

The larvae and spores of the organisms are extremely small (from about 5 μ for seaweeds to 1 mm for some barnacles). This fact has an important bearing on the mechanism of poisoning them.

They are phototropic, generally in the sense that the animals avoid the light and settle on the flat bottom of the ship, whereas the plants, which need light for photosynthesis, settle on the sides and tend to layer according to color — greens at the top, browns lower down, and reds lower still.

The fouling seaweeds are attached but are not rooted in the sense of land plants — they do not feed from the base; each section of the frond is capable of living and growing if cut off from the rest.

Once fouling organisms have attached and grown they remain attached, even if subsequently killed, and are extremely hard to remove. In time, bacteria will remove dead seaweeds, but the calcareous shells of barnacles and tubeworms remain attached.

The growth rates of the organisms vary according to their nature and the ambient temperature, but a few weeks suffice to produce fouling of great enough size to disrupt the laminar sublayer and reduce the speed of the ship.

The settlement of fouling is seasonal in the temperate latitudes — the higher the temperature, the longer the season. In tropical latitudes, the settlement of organisms may be almost continuous.

Most fouling organisms are sensitive to salinity, and a rapid change may kill them. The green weed Enteromorpha is remarkable for its tolerance of varying salinities.

C. THE SEVERITY OF FOULING

In 1867, at a time when effective antifoulings had not yet been discovered, Young (28) quoted some horrible cases of fouling: "One iron ship of nearly

800 tons register, which had been eight months in a warm latitude, had thirty cartloads of barnacles (an estimated weight of twenty-eight tons) removed from her bottom...." In another example Young wrote, "The 'Pekin,' which left England in February, 1847, was docked at Bombay in October. I can compare her to nothing else than a half-tide rock. The barnacles were nine inches long, the second strata being complete, with a feathering coral formation sprouting from cluster to cluster. The stench from the animal matter was so great that non-one could remain on board, and the paint was tarnished (no doubt white lead paint on the topsides, blackened by hydrogen sulfide — D. M. J.). The 'Pekin,' although a fast-sailing ship, had her speed reduced by fouling to six and a half knots per hour." Again Young wrote, "Between Bombay and China no ship should be longer than four months without examination. The fouling commences immediately after undocking."

There has been some progress in 100 years, and cases of such severity are seldom found. Moreover, periods between docking are seldom less than one year, and frequently as much as two years. Nevertheless, fouling is still the main obstacle to increasing the time between dockings. The size of the "admissible roughness" shows that even short growths of weed or shell can seriously affect ship performance.

An example of barnacle fouling, severe by modern standards, is shown in Fig. 8-1.

The severity of fouling differs greatly in different places. A cooperative test of antifouling paints carried out by the O. E. C. D. (43) showed that paints which lasted about 20 months in Poole Harbor, England, lasted 30 to 40 months at Den Helder, Holland; about six months at Abidjan, Ivory Coast; two or three months at Genoa, Italy and Miami, Fla. Not only are the seasons of fouling settlement longer, but the fouling itself is more vigorous and resistant to poisons in the latter places.

It is commonly thought that fouling is severe only when ships are delayed in port. Certainly delays increase the problem, but the growth of weed on tankers, which spend very little time stationary, has been shown to continue merrily while the ships are under way.

D. METHODS OF PREVENTING FOULING

To date, the only successful method for preventing fouling has been to poison the larvae at the time of settlement by coating the hull with a paint containing a poison which can slowly dissolve in the seawater.

This statement took a long time to establish. The "exfoliation" theory, that copper sheathing, and later antifouling paints, acted by shedding layers

FIG. 8-1. Barnacle fouling on an antifouling which had run beyond its proper time.

with the fouling attached, was popular in the nineteenth century, and was still popular enough for Ragg (30) to devote time to arguing against it in 1925. However, the correct mechanism was described by Holzapfel (44) as long ago as 1904, in words which cannot be improved upon:

"My experience of compositions extends to personal observation over a period of nearly thirty years, and I have thoroughly satisfied myself that the principal cause of the efficacy of compositions is due to the fact that the mercurial and copper compounds they contain form, by contact with sea water, a very thin layer or film of chloride solution of the respective metals, and that these chloride solutions are destructive to the organisms which try to attach themselves to a ship's bottom. These organisms consist chiefly of spores, larvae, and others. The chloride solutions of copper and mercury have the effect of coagulating albumen, and it is probably this which causes the destruction of the organisms at the moment they try to attach themselves to the bottom of a vessel coated with an effective composition."

Other methods have been tried, one being a variant of the poisoning theme which was marketed at the end of the 1950s under the names "Toxion" (45) and "Anfo" (46). Based on the premise that fouling did not settle when the ship was under way, the device consisted of perforated tubes fixed lengthways on the ship's bottom, through which air containing an atomized solution of organotin poison in kerosene was blown. This promised effectiveness and economy because it was only used when needed. It failed mainly because the tubes were always being ripped off, but results from undamaged ships suggested that the hypothesis that poison need only be dispensed in port was faulty.

Another device which has been mentioned repeatedly in the literature is the electrolysis of seawater to produce chlorine, which is highly toxic to marine fouling. Lovegrove and Robinson (47) described experiments on this subject. Anodes carrying up to 200 mA would protect an area which fell from 1.6 to 0.65 m^2 over 28 weeks. The cost of putting the necessary density of anodes on a hull would be prohibitive, but the system has been suggested for keeping seawater intakes clean. Chlorine is commonly added to the seawater intakes of power stations and refineries to prevent mussel fouling (48).

Nearer to antifouling paint is the idea of incorporating organotin poison into a sheet of rubber (49). The inventors have claimed lives, estimated by extrapolation, from 1.5 years (at 4 parts tributyl tin oxide per hundred of rubber in 1 mm thickness) to 7.8 years (8 parts per hundred at 3 mm). The idea certainly works, but the life has yet to be verified, and there is as yet no basis for assessing its practicality on a large scale.

A quite different approach from that of poisoning is to deter the organisms by ultrasonic vibration (50). Despite evidence that this worked, it was not commercially successful, no doubt because insufficient power could be used in practice. A similar reason was responsible for the failure of radioactivity to be useful as an antifouling agent (51).

We return, therefore, to antifouling paint, and in the next section consider the mechanism of its action.

8-6. ANTIFOULING PAINTS: HOW THEY ARE MADE AND HOW THEY WORK

Since most research into antifouling paints has been done on cuprous oxide, and it is still the most important antifouling poison, most of this section is devoted to it. Indications of how other poisons differ from cuprous oxide are given later.

The first question is whether antifouling paints function by deterring settlement or by poisoning the newly settled organisms. In the case of copper, the latter is undoubtedly the case, and since other poisons act in paint roughly according to their killing powers relative to copper, it most probably applies to them, too. The evidence for copper was provided by Crisp and Austin (52), who found that, if given the choice, barnacle larvae prefer to settle in pits rather than on a smooth surface, irrespective of whether the pits are coated with antifouling paint, though significantly fewer survived in the painted pits. Wiseley (53) found that given the alternative of painted or unpainted surfaces, the larvae of the bryozoan Bugula settled in greater numbers on the painted surface, but were later found to be detached and dead, whereas on the unpainted surface they were alive and adherent. He concluded that antifouling paints exercise a toxic, but not a repellent, effect.

The next question is whether the toxicity is due to contact with the surface or to poison dissolved in the water. That poison dissolved in the water is effective is shown by the "adjacency effect," namely that nontoxic areas adjacent to antifouling paint are relatively free from fouling (54). This can be seen in the case of the diffusion cells (Fig. 8-2). To prove that contact with the surface is of no significance is, however, a different matter, and has not been done.

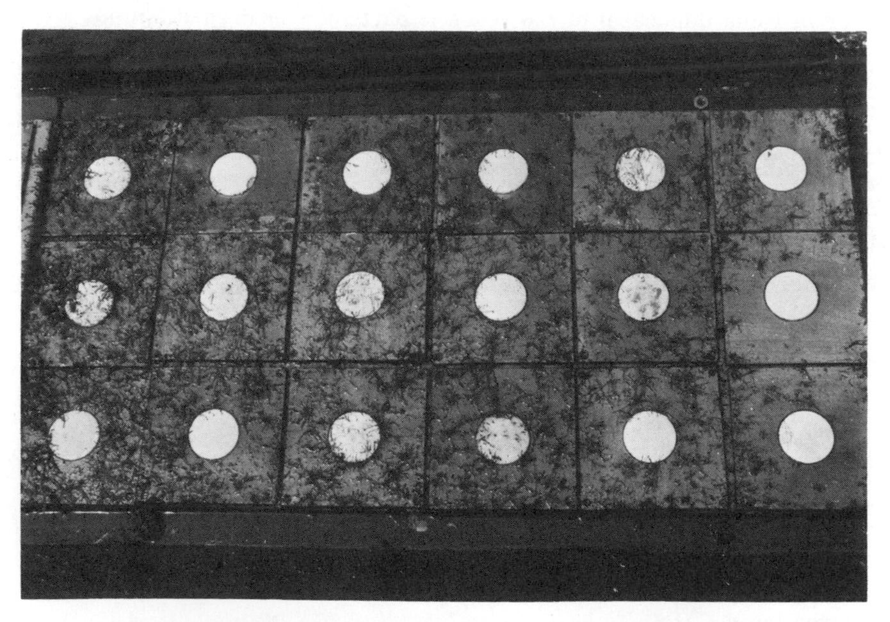

FIG. 8-2. A bank of poison diffusion cells after exposure on the "flats" facet of a Turtle raft.

It is clear that the poisoning effect must be confined to a very thin layer on the surface. Suppose that a concentration of poison of 1 $\mu g/cm^3$ is required, and a paint is leaching at 10 $\mu g/cm^2/day$ (which are plausible figures), then the rate of renewal of the water over a surface of 1 cm^2 can only be 10 cm^3/day, which at a layer thickness of 1 mm would correspond to a linear speed of 100 cm/day. The calculation is in fact much more complicated than this, but the conclusion remains, that a paint dissolving at a useful rate can effectively poison only a very thin layer of water. Crisp and Meadows (55) calculated by analogy with heat diffusion that the diffusion layer, i.e., the layer containing a measurable concentration of poison, could only be 0.07 mm thick at a surface with water flowing past at 10 cm/sec.

The layer of poisoned water is therefore of the same order of thickness as that of the spores and larvae. These must succumb before they can grow, or they will become big enough to emerge from the poisoned layer, and neither the animals nor the plants draw sustenance from their bases.

Poisons are assessed according to their solubility and toxicity. A solubility in seawater of the order of 1 $\mu g/cm^3$ is required. If too soluble, the poison dissolves out too quickly, if not soluble enough, it cannot produce an adequate concentration.

The estimation of toxicity involves some test to determine the concentration needed to kill a certain proportion of organisms, say 50% in a certain time (LD_{50}). Usually the poison is tested alongside a standard, for example, mercury. Such a test is somewhat arbitrary because in practice a complete kill is required, and a poison is not necessarily the worse for being slow to act. Nevertheless, it is a useful guide to promising compounds.

Whereas poisons are assessed according to their concentrations, paints are assessed according to their leaching rates. This is because the concentration of poison in the diffusion layer is very difficult to measure. It was established independently by the Woods Hole workers (42) and by Harris (54) that the "critical" leaching rate for copper, that is the rate required to keep off all fouling, is 10 $\mu g/cm^2/day$. The surface concentration of poison corresponding to this has not, however, been established, and there is therefore an unbridged gap between the toxicity and leaching rate tests.

It may be thought that the surface concentration of poison must be at most the saturated concentration, but this can be misleading. To say that aerated seawater at normal temperatures and pH 8.15 is saturated by 1 $\mu g/cm^3$ of copper is correct at equilibrium, but equilibrium conditions do not necessarily apply when the poison is diffusing from the film. Moreover, the pH in slime films at the surface may be low, which would increase the equilibrium solubility of copper within the slime film, and possibly make it more available for antifouling (62). Whatever the reason, it is a fact that if the leaching rate of copper is high enough, all algal fouling can be kept off, whereas in vitro a concentration of copper as high as 2.5 $\mu g/cm^3$ will prevent normal growth but still allow a stunted growth of Ectocarpus (63).

A. CUPROUS OXIDE PAINTS: FACTORS AFFECTING THE LEACHING RATE

The dissolution of cuprous oxide in seawater has been attributed to a simultaneous attack by hydrogen and chloride ions (42). The solubility is a linear function of $[H^+]$, and at pH 8.15 the equilibrium [Cu] in aerated seawater is 1.0 $\mu g/cm^3$. In paints the pH affects the dissolution of acidic resin as well as Cu_2O, and in the opposite direction. But the effect on the copper is predominant, and for practical purposes one may say that in the region of pH 8 the log of the leaching rate is linearly proportional to pH, so that a rise of 0.1 pH unit lowers the leaching rate by 25%.

The rate of solution of cuprous oxide, in the region of seawater salinity, is linearly proportional to $[Cl^-]^2$. Copper paints therefore leach much more slowly in fresh water than in the sea.

Temperature has an Arrhenius law effect, the lot of the leaching rate being proportional to T^{-1}. At constant pH, the leaching rate rises approximately 5%/deg.

The speed of movement of the water has an effect because of the reduction in the thickness of the diffusion layer. Between 0 and 0.5 m/sec the effect is large; thereafter it is smaller. Wilkie and Edwards (56) did experiments with painted cylinders in tubes with seawater passing at 0.5 and 8 m/sec in turbulent flow, and compared the periods needed to extract the same amount of copper. They found varying degrees of acceleration, depending on the alternatives of fast and slow circuits and raft storage, the figures lying between 1.5 and 3.2 times for the 16 times increase in speed; the acceleration factor tended to fall as the amount of copper lost rose. A rough guide would be about 2 times acceleration from raft to ship.

The question of whether cathodic protection can affect the leaching rates of copper antifoulings has often been discussed. Van Londen (60) observed that no effect has ever been seen on rafts or on ships, but in the laboratory he found acceleration factors of 2 to 8 times, using both continuous contact and soluble matrix types of antifouling containing rosin, but no acceleration with continuous contact type containing no soluble binder. Sussex (35) stated that cathodic protection had no effect on the raft performance of antifoulings, and criticized van Londen's experiments because he thought they were conducted at too high a current density. Anderton (61) has reported cases of cathodic reduction of cuprous oxide to copper at pores in the anticorrosive film.

Sulfide pollution of waters can precipitate insoluble black copper sulfide on the surface, and in these conditions copper antifoulings are unsuitable.

The rate of solution of cuprous oxide itself has been estimated (42) to be 250 $\mu g/cm^2/day$. When bound in a paint film, the leaching rate of cuprous oxide is reduced below this uneconomic figure, in accordance with the concentration of cuprous oxide in the film, the nature of the paint medium, and the length of time for which leaching has progressed. This raises the whole question of formulation, which is considered in the next section, but the influence of the medium is considered generally here.

With the exception of a very small class of paints using a very high concentration of cuprous oxide, all antifoulings contain rosin in some form or another. This resin has the essential properties of being a film former and being acidic (one carboxyl group per molecule, acid value about 160 mg KOH/g). At the high pH of seawater, it can dissolve, and the solution of cuprous oxide and rosin occur simultaneously. In practice, a simple rosin film would be far too soft and would dissolve far too quickly, so that it is usually handled as a metallic soap, or is blended with insoluble resins. Barnes (57) measured the rate of solution of rosin in the sea from mixtures with various inert resins (ester gum, linseed/rosin-modified phenolic varnish, rosin-modified phenolic resin, hydrogenated ester gum) and found that it was proportional to the rosin content of the film down to about 65% below which it declined more than proportionately, and below 33% hardly any was lost. The rate of solution of rosin itself was 171 $\mu g/cm^2/day$.

The function of the rosin raises the question of the mechanism of solution of copper from antifouling paints, and here there is no consensus of opinion. The Woods Hole workers (42) distinguished between "contact leaching" paints on the one hand, in which the cuprous oxide was only just bound together by an insoluble medium and dissolved by direct contact with seawater, making channels from particle to particle, and "soluble matrix" paints on the other hand, in which cuprous oxide and rosin dissolved simultaneously, and there was a steady state of leaching controlled by the rate of solution of the rosin. Neither picture is real, however. Almost all "contact leaching" paints contain rosin, and the leaching rate of antifoulings is hardly ever steady. Rather, it declines continuously with time.

The building up of a diffusion barrier of precipitated copper salts, and the skeleton of undissolved resin and insoluble pigment left after the surface layers of cuprous oxide have dissolved, would be expected in all cases to produce a leaching rate that declined with time. Barnes (58) showed that mechanical removal of the surface residue increased the leaching rate two or three times. Marson (59) constructed a theoretical model of a contact-leaching paint with an insoluble binder, and derived the rate of decline of leaching rate from the rate of building up the surface diffusion barrier of residual resin. His model, however, carried the implication that the leaching rate was, up to the point of exhaustion, independent of film thickness. This is not true, as Fig. 8-3 shows.

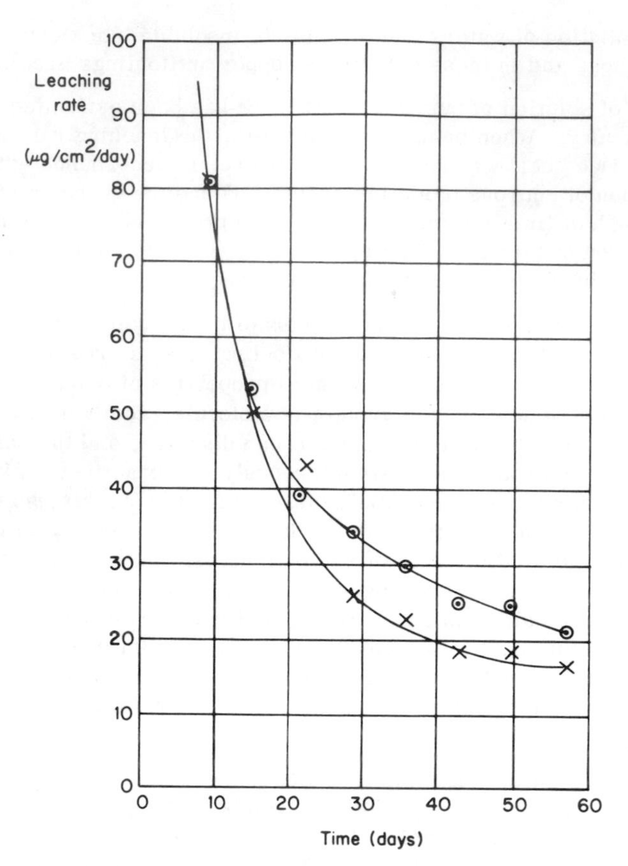

FIG. 8-3. Leaching of copper paints in seawater at pH 8. Cu_2O = 49% of dry film by volume in an insoluble binder. Dry film thicknesses (given in micrometers): 23—lower curve; 32—upper curve.

Van Londen (60) produced photographic evidence of the leached layer, which corresponded in thickness to the loss of copper from the film. He introduced the idea of "total soluble matter" (cuprous oxide and rosin) as a parameter, and said that the structure of "continuous contact" of cuprous oxide particles was made up of both the particles themselves and the rosin in the film. If the fraction by volume of total soluble matter in the film was too low, the structure of continuous contact was incomplete, and release of copper could occur by diffusion through the film.

To sum up, several different mechanisms of poison release have been put forward and claimed to apply to different classes of paint. The existence of classes is beyond doubt. Cuprous oxide will leach at a useful rate from a totally insoluble binder, provided that its concentration in the film is sufficient; at lower concentrations the film must contain some fraction of soluble matter in the binder. This classification is, however, of little relevance to real paints, nearly all of which contain some rosin, irrespective of the concentration of cuprous oxide. The rosin has two functions, (a) to dissolve, making pathways for the seawater to reach the cuprous oxide particles and (b) to provide carboxyl groups, which can ionize and promote the diffusion of cations, such as copper, which can jump from group to group. The copper dissolves mainly from the upper layers downward and leaves behind a skeleton which acts as a diffusion barrier whose thickness increases with time; it also diffuses from the lower layers at the same time, because leaching rates, even in the early stages of leaching, are greater with greater film thicknesses.

B. THE FORMULATION OF COPPER ANTIFOULINGS

Traditionally, copper antifoulings are divided into four groups of which the first three carry commonly used names. Their compositions have never been defined, but the following is a guide:

a. North Atlantic

Containing 100-200 g/liter Cu_2O. Suitable for low intensities of fouling only, in which case it may well last 12 months or more.

b. Tropical

Containing 150-250 g/liter Cu_2O. Will resist moderate intensities of fouling and has a life under those conditions of 6 to 9 months.

c. Supertropical

Containing 300-400 g/liter Cu_2O, and usually a "boosting" poison (mercury, organotin, etc.). Suitable for quite high intensities of fouling, with a life of 6 to 12 months, depending on the intensity.

d. High-Strength Antifoulings

These are used when lives of 12 months or more are required and fouling may be intense. They may contain up to 1500 g/liter of Cu_2O and boosting poisons as well.

The first three groups are clearly "soluble matrix" types, the last is usually "contact leaching."

Rather than a list of formulas, the necessary properties are considered:

1. Leaching Rate

Apart from the general proposition that with low-copper types a very large proportion of the medium must be soluble in seawater, whereas with contact-leaching types only one-third or more need be soluble, there is no advice possible except to test the paint, preferably by raft performance as well as measured rate of solution.

2. Color

The staining power of cuprous oxide is low, and once the upper layers have dissolved the red color of the paint disappears and is replaced by a pale green due to precipitated copper salts. Insoluble colored pigments may be included (red iron oxide is the commonest) which will therefore show in their true colors after a short time of immersion. It is not unknown, especially with carbon black, for them to affect the leaching rate.

3. Stability

Cuprous oxide is a basic pigment. It readily forms the divalent Cu(II) ion on storage in paint, and may cause gelation of reactive resins.

4. Hardness

Soluble-matrix types are necessarily soft after immersion. The formation of rosin soaps, e.g., of Zn or Ca, reduces the softness, and other resins may be added. The hardening effect of these is great only when they are present in a fair proportion, and therefore when the concentration of cuprous oxide is high.

5. Compatibility with the Undercoat

This is not only a matter of adhesion and absence of wrinkling, but also of the possible effect of constituents in the undercoat on the leaching rate, e.g., bituminous materials may diffuse into the antifouling and slow its action.

C. POISONS OTHER THAN COPPER

Traditionally, mercury has been used to boost the action of copper, but its price and the advent of other poisons have driven it out, and apart from a small amount of phenylmercury compounds (which are more effective than inorganic mercury at the same cost), it is hardly used.

Zinc has too low a toxicity to be useful on its own, being about one-quarter as toxic as copper. It is, however, relatively cheap, and zinc oxide has often been used together with cuprous oxide. An interesting use of zinc is in zinc silicate paints, that is, zinc dust in very high concentrations in an alkali silicate binder. Such paints have been shown to have an antifouling effect (70) but they are weak against algae. By incorporating other poisons, it is possible to make very effective and long-lived antifoulings, and therefore to produce completely zinc-silicate schemes for ship bottoms.

The main reason for using poisons other than copper is that copper is somewhat deficient in its action against algal fouling, which has become particularly important with the introduction of big tankers and bulk carriers. Most of these other poisons are toxic to fouling of all sorts.

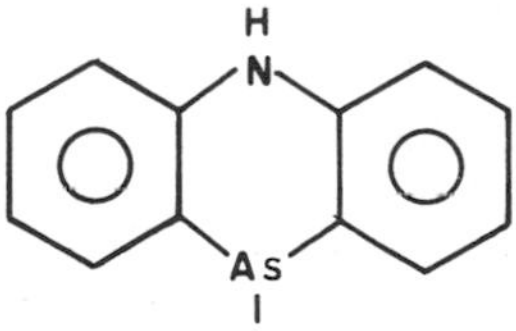

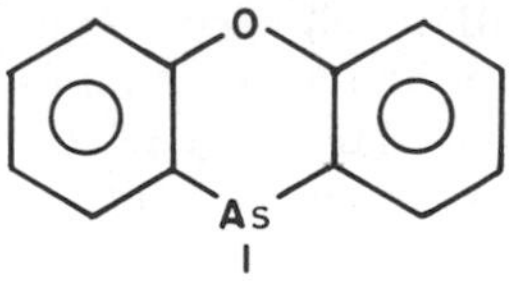

H N As I

O As I

Inorganic arsenic was widely used in the early days, but is ineffective. Organic arsenical compounds, on the other hand, are very effective indeed. Compounds of the phenarsazine (Structure 8-1) and phenoxarsine (Structure 8-2) radicals have the right combination of toxicity and solubility. Unfortunately, they are highly irritant to the eyes and nasal membranes, so that although they constitute little danger of systemic poisoning, they are a problem to apply.

Tri-n-butyltin $(C_4H_9)_3Sn$- and triphenyl tin $(C_6H_5)_3Sn$- radicals are also very effective against all types of fouling, and have been widely used (64). They are toxic to man, but not to a degree which makes them a hazard, provided elementary precautions are taken.

Tri-n-butyl lead $(C_4H_9)_3Pb$- and triphenyl lead $(C_6H_5)_3Pb$- compounds may be classified as of similar toxicity to fouling as the organotins, though perhaps to a lesser degree. They have been extensively evaluated (65) but not used much. Unlike tin, lead is a cumulative poison; therefore the toxic hazards on application are greater.

All the poisons mentioned so far have been metallic or organometallic. The search for purely organic poisons has gone on for decades, but so far only two have been used to any considerable extent: DDT, which is specifically toxic to barnacles and nothing else; and TMT (tetramethyl-thiuram-disulfide), a rubber accelerator which has a general toxicity of the same order as copper and is fairly cheap. Neither, however, has been employed to anything like the extent of copper, mercury, organic arsenical, or organic tin compounds.

Accounts of the investigation of poisons have been given by Harris (54), who tested 96 poisons and found only four to be of promise (TMT was not reckoned promising), and Miller (66), who investigated hundreds and found only four (including tributyl tin and phenylmercury compounds) to be effective against all forms of fouling, and another 11 to be effective against many, but not all, forms of fouling.

The evaluation of antifouling poisons is quite a complicated business. The first thing is to test the toxicity, irrespective of other properties. For this purpose it is desirable to eliminate any influence of the paint medium, because the poison may well require a different medium from that suitable for other poisons. The author and his colleagues found, for example, that tributyl tin compounds are particularly well suited for chemical combination with a polymer to form a binder which is heavily loaded with copper or zinc oxides (67), a mechanism which had not been used with other poisons. Miller (66) adsorbed the poisons on charcoal or silica blocks and immersed them in the sea. Christie and Crisp enclosed them in porous pots (68). A similar method, enclosing the poisons in cells closed by microporous membranes, has been used by the author and his colleagues and is illustrated in Fig. 8-2.

Such a test is valuable for a start, but if suffers from two disadvantages: it is not quantitative, because the rate of solution of poison is not controlled, and it takes up a great deal of time and space. The measured solubility helps to overcome the first uncertainty, but the second remains.

Toxicity tests in the laboratory may be done with precise concentrations, but the tests are, as previously discussed, rather arbitrary, and the range of organisms that can be used is limited. Nevertheless, the information is valuable. Harris (54) published the results of tests against crustaceans and algae, covering a wide range of inorganic, organometallic, and organic compounds; Vind and Hochman (69) tested many organotin compounds against marine borers, with results of general significance for fouling prevention.

Once a promising compound has been found, namely, one with a good toxicity to fouling, acceptably low toxicity to people, and the right order of solubility and cost, it must be tested in a variety of paint media, both alone and in conjunction with other poisons, for effectiveness on raft panels and then on ships.

D. THE REQUIREMENTS AND CURRENT LIMITATIONS OF ANTIFOULING PAINTS

Thanks to cathodic protection and improved anticorrosive treatment, ships can stay out of dock for two years instead of the nine months that was the average time about 1960. The main obstacle to universal acceptance, and to the adoption of even longer times is the growth of fouling.

Suppose that a copper antifouling containing 800 g/liter of Cu_2O is applied at the rate of $4.4 m^2$/liter, that is 250 μ wet film, the amount of cuprous oxide per square centimeter is 18,200 μg. If we assume, which is probably near the truth, that a leaching rate of 20 $\mu g/cm^2$/day is required to keep off all fouling including algae, then if the paint leaches steadily at this rate, it would last 910 days, or two and a half years. In actuality, it barely lasts a year. The reason is that the leaching rate is not steady. In order to keep an adequate rate in the later stages, when the diffusion barrier has built up on the surface, the initial rate must be much higher, that is, there is a wasteful loss of poison. When the leaching rate has fallen well below 20, there is still a good deal of poison left, so there is waste at the tail end also.

This calculation takes no account of the fact that much of the poison loss when the vessel is steaming is unnecessary. If the amount of poison needed were compared with that used, it would be no more than about 20%.

Thus, the efficiency of antifouling paints is very low. On the other hand, the cost of fouling is so high that it is well worthwhile to waste 90% of the poison if the remainder serves to extend the docking period significantly.

So there is no need to bewail the economics. The problem is that even using stronger poisons, thicker films, and greater poison concentrations, it is very difficult to attain an antifouling life of even 18 months, reliably, on worldwide steaming.

What possibilities are now visible to produce longer-lived, high-strength antifoulings?

The use of stronger poisons, requiring lower leaching rates, is one answer. In the early days of organotin compounds it was hoped that they provided this answer, but the advantage turned out to be much less than was hoped for. As compounds become even more toxic to fouling, they are likely to be more toxic to people, unless very specific poisons can be found which, for example, selectively attack the attachment mechanisms of fouling larvae and spores.

Elimination of the leached layer, that is, total erosion of the film, is difficult to control, particularly as it is apt to be very dependent on the salinity of the water. It would also be disastrous if a ship lost all its antifouling on entering a river. However, an erodable coating on top of a conventional hard type, to act as a second line of defense, is a possible answer.

Slow hydrolysis of the pigment to form the active poison is a theoretically possible answer. At present, dissolution of the medium is part of the control mechanism. If the pigment controlled its own rate of dissolution, it could be incorporated in a very high concentration without the initial wasteful loss. This is one approach to the ideal of a steady leaching rate.

Another line of approach is to alter the mechanism of poison diffusion through the film in order, perhaps, to allow the leached layer to grow without presenting a diffusion barrier. This requires a better understanding of the diffusion mechanisms which operate in present-day antifoulings. A better understanding of the physical chemistry of antifouling paint films is needed.

8-7. DECORATIVE AND PROTECTIVE COATINGS ABOVE WATER

In the area of decorative and protective coatings above water, the problems, and consequently the products, are largely similar to those encountered in painting structural steelwork on land, which are discussed elsewhere in this Treatise. We therefore mention here only those problems peculiar to ships.

A. THE BOOTTOP

The boottop is the area between the light and deep load lines that is exposed both to the atmosphere and to the water. Many bottom paints, for example bituminous types, are unsuitable for this area because of lack of decorative color and atmospheric durability. Conventional decorative paints lack the resistance to water and, in particular, to alkali generated by cathodic protection. Due to the ready availahility of oxygen, cathodic action occurs readily in this region.

Another problem is fouling, mainly by green weed which flourishes in highly aerated conditions. This may be dismissed by saying that a satisfactory antifouling boottop is unobtainable except at a cost which is outside the reach of merchant shipping.

Until recent years the problem of painting the boottop was quite simply abandoned as too difficult. Most boottops were appallingly rusty. No paint could endure long on such a surface under such conditions, so the cheapest were chosen, which did not help.

The advent of shotblasting and shop priming changed this, because it became possible to apply paints which perform well on well-prepared steel, but badly on rusty steel. "Better boottops" is the cry, and it has been answered as follows.

1. Chlorinated rubber and amine-cured epoxies give good corrosion protection, resist cathodic protection, and enable bright colors to be used. These have proved the most popular of high-performance boottop paints. Coal-tar epoxies have all these virtues except the bright color. They may

be made at best a dull red, and weather to a brownish appearance, but they do stop the rusting, which is the most unsightly thing. Attempts to overcoat them with brightly colored finishes are hazardous due to uncertain adhesion.

2. Zinc silicates are good in every respect, except for their gray color; they resist mechanical damage better than organic paints. If overcoated with suitable schemes, they provide a good rust-resisting foundation, but the schemes must indeed be suitable. Many cases of detachment from zinc silicate have been encountered (71).

B. LOSS OF ADHESION UNDER MOIST CONDITIONS

It is a common observation that many paint schemes which adhere perfectly when dry may lose adhesion between coats when wet. Usually the adhesion is regained on drying out, and the cycle may be repeated frequently. Only if the system is exposed to some stress during the wet cycle does the loss of adhesion become translated into a failure, such as blistering or peeling.

Bullett (72) found that alkyd paints, when dried, contain a water-soluble, gummy material that tends to collect at the interface between coats, and when exposed to moisture, causes loss of adhesion. Oleoresinous films suffered less from this defect than alkyds.

The author (73) directed attention to the underlying coat rather than the topcoat because of the fact that loss of adhesion is often associated with long drying of the first coat before overcoating. This is commonly attributed to its becoming "too hard," and strong solvents are added to the overcoating paint in order to soften the first coat and permit the topcoat to "bite." This does not work. The real reason for long drying being detrimental (assuming the surface to have been cleaned free of airborne contaminants) is that many paints develop more hydrophilic surfaces on long drying. When the system is soaked in water, some of it permeates the topcoat and competes with the molecules of the topcoat for adhesion to the underlying coat. If the underlying coat has a very hydrophilic surface, the water may displace the topcoat, causing loss of adhesion.

In order to prevent this, paints should be chosen which do not develop very hydrophilic surfaces on drying. Simple drying oils are in this category, and may be safely overcoated after many weeks' drying (unfortunately they need many weeks' drying to harden, and therefore are little used in ship paints).

The problem may also be tackled from the point of view of the topcoat. Neither swelling by water nor permeability to water are very important in this connection, but the mechanical properties of the film are important. In general, a degree of brittleness is allowed, greater than is common in decorative paints. A brittle film will not have the ultimate in atmospheric durability, but fortunately this is not required, since repainting is usually done within two years. Whereas a brittle paint might not last as long as a flexible one on a test fence, it is much less likely to peel off in massive sheets, to be rolled into tubes like wallpaper, and sent back to the unfortunate paint chemist.

C. HIGH-PERFORMANCE COATINGS ABOVE WATER

Traditionally the crew has occupied the voyage time by painting the superstructure, hull, and decks. This is still done, but the advent of large tankers and bulk carriers run by small crews has brought a demand for paints which will last without this repair work for the period between dry-dockings, about two years.

There have been several proposed answers to this problem, and it is by no means settled. In the United States Navy, it has been shown that a zinc rich primer and a silicone-alkyd finish reduced the man-hours for maintenance over 275 days by nearly two-thirds (71).

Whether it is worth paying the very high price for the extra durability conferred by a silicone-alkyd is doubtful. Loss of gloss and chalking are not the major reasons for repainting ships' weatherwork. Indeed, loss of gloss is barely noticeable because all surfaces when dry are covered with a thin film of salt from the dried sea water. Undoubtedly the main reason is rust staining, which may often cover a large area even though the amount of actual metal corroded is very small. The use of zinc-rich primers is certainly the best answer to this. The money is best spent on surface preparation and application of a zinc-rich primer followed by a good thickness of tough primer to present the best possible resistance to mechanical damage. The decorative topcoat may then quite satisfactorily be a relatively cheap type such as alkyd, which has quite a satisfactory outdoor durability and presents little difficulty in repainting.

8-8. THE CORROSION AND PROTECTION OF SHIPS' CARGO TANKS

In 20 years the corrosion and protection of ships' cargo tanks has been transformed from an insoluble problem to a large business for the paint industry. The agencies of the transformation were zinc silicates and epoxy resins.

A. CORROSION IN OIL TANKERS

The nature of the problem of corrosion in oil tankers was described by Robinson and Fleming (74). Water is the immediate corrosive agent.

It may be present as ballast, as residues from ballasting during carriage of cargo, and as residues from tank cleaning. In all cases it will be seawater. The cargo itself carries dissolved oxygen, and in gasoline, for example, oxygen is seven times as soluble as in seawater. White oils are associated with general corrosion; black oils leave behind a waxy film which will protect the steel, but the inevitable detachment in places during tank cleaning creates the situation of small anodes and large cathodes, which is classical for pitting corrosion. Robinson and Fleming showed the results of surveys of three white-oil tankers, which after eight years' service required 600 to 940 tons of steel renewing at costs of £150 to 200,000 each. Surveys of four black-oil tankers showed they required £26 to 48,000 worth of repairs each after 12 years.

The conventional remedies for seawater corrosion were inapplicable before about 1950. Anticorrosive paints would not resist the solvent action of the cargoes, and cathodic protection was inapplicable except in ballast tanks because the electrolytic path was not continuous.

1. The Evolution of Oil Tank Coatings

Briggs and Francis (75) have described the experience of the U.S. Navy on this subject up to 1963. From 1948 to 1952 they successfully used Saran, a copolymer of polyvinylidene chloride and acrylonitrile. However, this suffered from a defect which made it practically inapplicable to commercial use, namely a flash point of -1°C. (Being a thermoplastic coating, it clearly depended upon its solubility in strong solvents only for its applicability and resistance to hydrocarbons.) Saran therefore proved to be a blind alley, but it did serve to show the value of coating ships' tanks. During the same period, Esso began the use of zinc silicate, and shortly afterward Shell started with epoxies. In 1957 the U.S. Navy began trials with various coatings, including zinc silicates, epoxies, coal-tar epoxy, polyurethanes, polysulfide, and Saran, and decided that zinc silicate and epoxy were the best. This opinion has been endorsed by commercial practice.

Zinc silicate has been the almost exclusive province of two American companies, Amercoat and Humble, the latter an associate of Esso. Colberg has described the Esso experience up to 1956 (76), using vinyl, epoxy, and inorganic zinc silicate, as a result of which they chose the latter. Two types have been marketed, the first requiring a wash of curing solution, the latter being "self-curing." The chemistry of these coatings has been described by Pass and Meason (77). They have the advantage of using no inflammable solvent, except in the curing agent, which initially was low-flash but was soon reformulated to remove this drawback. The surface preparation of the steel must be very good, a blast to white metal; the danger of flash rusting has led to the frequent use of dehumidifying equipment in the tank ventilation.

While the introduction of zinc silicates was largely due to Esso, which controlled a tanker fleet and one of the paint suppliers, epoxies must be mainly credited to Shell. They were not the first to introduce the resins, but they again controlled a tanker fleet and a supplier, in this case a supplier of resins, not paints. As a result, the supply of epoxy tank coatings has been much more widespread through the marine paint industry. These resins have the ability to cure at ambient temperatures (15°C or above), producing films which resist both the seawater and the oils. The Shell experience up to 1967 has been described by McMahan, Edwards, and Dunbar (78). They concluded that an epoxy coating had a greater initial cost than a system of cathodic protection in ballast tanks and an inhibitive wash in non-ballast tanks, but eventually produced a cost saving. For clean oil service they recommended either a zinc epoxy primer at 75 μ and one coat of high-build epoxy at 150 μ, or two coats of high-build epoxy at 250 μ, with a preference for the former. They also preferred a solventless or high-solids coating, based on liquid epoxy resins, to a lower-solids type on solid resins.

The introduction of coal-tar epoxies made it possible to use them for black oil cargoes at a considerable cost saving. They may also be used for white oils if overcoated with straight epoxy paints.

Another source of economy has been the selective painting of tanks. Corrosion has been found to be worst in the upper spaces of cargo tanks (where oxygen is freely available) and on the upper surfaces of the bottom plating and stiffeners. In many cases, therefore, the coating has been confined to the deckhead and bottom of the tank, and for about 2 m down and up, respectively (79).

The original epoxy paints were mostly formulated on solid resins, and required three or four coats to build up the necessary thickness. High-build paints have been made in two ways, the first being to use liquid resins, the second to use a high concentration of inert extender pigment in a paint based on solid resins. In either case, the required thickness can be built up in two coats or, ideally, in one. The highly pigmented paint has the advantage of being cheaper, but it contains much more solvent than the liquid epoxy type. However, this is a difference of degree, not of kind, and where complete absence of solvent is required, for reasons of health or safety, neither is acceptable. A truly solventless epoxy paint can be made, but its application is a problem.

Like zinc silicates, epoxide resin paints must be applied on surfaces prepared by gritblasting, though they are not quite so demanding of the ultimate in freedom from rust. Unlike zinc silicates, they may be applied over the common types of organic binder shop primers.

The excellent solvent resistance of epoxies is one of their important features in oil tank coating, but since seawater is the main corrosive agent, it is most important to achieve resistance to it, and this requires careful formulation. The author (80) showed that the use of an excess of amine or polyamide curing agent, or of a slow-evaporating, water-miscible solvent, would drastically reduce the water resistance of an epoxy paint. These conclusions were also reached by Wray and Tator (81).

2. The Economics of Painting Oil Tanks

In the painting of oil tanks, the paints themselves, the blasting of the steel, and the application of the paints under the difficult conditions in tanks, are all costly and must be justified by ultimate savings due to the reduction in corrosion.

A very detailed analysis of the matter was published in 1964 by Talma (82), covering trials with eighteen 21,000-ton tankers. Cathodic protection was found to be too limited in its effect; treatment with inhibitor after tank cleaning was effective but had to be done after every voyage. Painting with epoxies was therefore chosen as the best method of protection. Talma took into account the cost of steel renewals, including the delays of the ship to carry them out and the cost of cathodic protection in the center tanks for uncoated ships. He compared this with the cost of painting, cathodic protection in wing tanks, and the much lower cost of steel renewals for coated ships. Then he calculated that with money at 5% interest, the expected saving per ship would be 600,000 guilders over a 20-year life, and 100,000 guilders over a 16-year life. His company concluded that for a 20-year life the cost of painting was certainly justified, and for a 16-year life it depended on the nature of the ship.

Purlee, Leyland, and McPherson (83) made a calculation for a hypothetical 30,000 tonner in continuous clean oil service, comparing conventional methods of corrosion control (cathodic protection of ballast tanks) with renewal of steel and complete zinc silicate coating of all tanks after five years. The economic factors were the costs of steel renewal, renewal of cathodic protection, docking fees, loss of cargo in the corrosion scale, costs of painting, and the ultimate value of the ship. By discounted cash flow, assuming money earning 8% and corporation tax at 54%, they calculated that the savings would pay for the coating 6.6 years after coating (11.6 years of the ship's life).

Clearly, much depends on how well the coatings actually perform, and facts on the life before renewal is necessary are not easily found. A good indication is given, however, by the fact that the classification societies allow reductions in scantlings (steel thicknesses) for ships with coated tanks. Narter (84) described how the American Petroleum Institute started research on the subject in 1955, and by 1960 had sufficient experience to approach the classification societies for reduction in scantlings when coatings were applied; in 1963 this was granted by the American Bureau of Shipping, and other classification societies have similar rules. This means that in addition to the other economies, tank coating allows the use of less steel in construction, which is a direct initial saving and a continual saving in the weight of ship, which has to be propelled.

B. THE COATING OF TANKS IN CHEMICAL PRODUCT CARRIERS

While most tank coatings have been applied to the carriage of hydrocarbon oils, there is a rapidly growing market in the carriage of chemicals, which include animal and vegetable oils and fatty acids. For many years these cargoes have been carried in the "deep tanks" of ships, but now whole ships are being devoted to them, and new problems arise.

Zinc silicates are necessarily excluded from many of these tanks because of their lack of resistance to acids and alkalis. They are, however, very resistant to the most powerful neutral organic solvents, and some ships have one or several tanks coated with zinc silicate for the carriage of these cargoes. The rest have organic coatings or are lined with stainless steel.

Mineral acids are included in many cargo lists, but it is very doubtful whether any paint coating on mild steel would be safe to carry them; any break in the coating would lead to rapid dissolution of the exposed steel. Stainless steel, or possibly massive linings of rubber or plastics, are almost certainly required.

There remains a very wide range of cargoes which could be carried in painted tanks, but the existing epoxy and polyurethane paints will not resist the most aggressive of them. Another limitation is the shop primer. When the carriage of benzene became important in the mid 1960s, it was found that some of the most popular shop primers, such as the wash primer types, failed disastrously; the benzene swelled and permeated the epoxy tank coating, and although it did not do so enough to cause failure over bare steel, it caused severe blistering over the shop primer. It is therefore necessary to make shop primers which will not detract from the resistance properties of the tank coatings; to do so, while maintaining the good protection against weathering which is the reason for using a shop primer, is not easy. The shop primers and tank coatings for chemical product carriers are the main subjects of research in the field of ships' tank coatings at the beginning of the 1970s.

8-9. SURFACE PREPARATION AND THE APPLICATION OF MARINE PAINTS

This section deals with the underlying reason for the great improvement in the protective painting of ships during the 1950s and 1960s.

A. THE REMOVAL OF SCALE AND RUST FROM STEEL

Hot-rolled steel, used for ships' plating, carries a thin film of millscale, which is made up of layers of iron oxide, the iron increasing in oxidation number from the inner to the outer layers. The scale is a good electronic conductor and is cathodic to the steel, so that it will promote corrosion; moreover, it is brittle and when stressed or exposed to a corrosive environment it tends to crack off, leaving small anodic areas to carry the current from the large cathode.

A century ago steel ships took a long time to build, the steel weathered out of doors until the millscale had gone, and after wirebrushing, the rusty surface was coated with red lead in oil and again dried for a long time. This was about the only prescription for painting rusty steel, and it worked quite well.

The system broke down when the speed of building ships was increased. No longer was the steel weathered to complete removal of all scale, nor was there time for the drying of the red lead in oil. Quick drying primers over incompletely descaled steel gave bad results. This came to a head during World War II when steel was short and could not be held in stock for long. At that time the Bakelite Corp. produced a remarkably far-sighted report for the U.S. Navy (85). They discussed the merits of chemical, flame, electrolytic, abrasive, and impact methods of cleaning steel, and came down in favor of wet sandblasting. In the event, dry blasting was the main method used on a worldwide scale, when the conclusions of this report were implemented 20 years later.

There is little point in discussing other methods for removing scale and rust from steel. Chemical and electrolytic treatments, except for pickling in tanks, have proved ineffective or impracticable. Flame-cleaning, which depends on the differential expansion of scale and steel, is ineffective against rust, which is nearly always present. Abrasive cleaning (e.g., emery disking) is the only practical alternative to blasting, but is much less effective. The remarks on the effects of inefficient blasting that follow apply a fortiori to abrasive cleaning.

The corrosive action of millscale on steel immersed in seawater is easily appreciated; that of rust is somewhat more subtle.

The rusting of steel is closely associated with the pollution of the atmosphere as well as its relative humidity. These facts were first established by Vernon and elaborated by Fancutt and Hudson (86); the matter is also discussed extensively by Evans (87). In industrial areas, sulfur dioxide and particulate carbon are the main agents. They are responsible for rates of corrosion that may be three times those in rural areas. By the sea, sodium chloride borne on the wind is responsible for corrosion. Fancutt and Hudson recorded rates of rusting rising from 0.2 mils/year inland to 37.7 mils/year 50 yards from the sea in Nigeria. Shipyards have a combination of industrial and marine atmospheres that makes for the highest rates of rusting and the most destructive types of rust.

Rust is destructive to paint because it has a fairly porous structure that occludes the contaminants which have given rise to it, and in the presence of water, rust produces an electrolyte which allows corrosion to continue. All paint films can allow water and oxygen to permeate at rates sufficient to allow this process, even though they may reduce its rate by comparison with free corrosion. The formation of further corrosion product produces an increase in volume which eventually disrupts the paint film (rust-nibbing) and allows even more rapid corrosion. This process was first analyzed by Mayne (88), who found cyclic temporal variations in the performance of paints over rusty steel and correlated the times of bad performance with high concentrations of iron (II) sulfate in the rust. Thus, panels 7.5 × 5 cm exposed in January to March carried 26 mg of $FeSO_4$ in the rust, whereas when exposed in May to July they acquired only 6 mg. This was attributed to the smoke from domestic fires during the winter months, the rust catalyzing the oxidation of SO_2 to SO_3. The $FeSO_4$ was mainly present in the tightly adherent rust; when the panels were bent to flake off the rust, only one-quarter of the $FeSO_4$ came with it. This observation was very important because it explained the known ineffectiveness of wirebrushing; any method which fails to remove the very lowest layers of rust fails to produce a good surface for painting.

Chandler (89) followed this by spreading "synthetic rusts" containing different amounts of iron (II) chloride and sulfate on bright steel and allowing them to stand in atmospheres of different relative humidities. He found that the chloride and sulfate ions stimulated the corrosion, and in addition, the soluble salts, being hygroscopic, reduced the critical relative humidity below which corrosion did not occur.

The difficulty of removing contaminants from rusty steel was described by Singleton and Wilson (41), who showed that appreciable amounts of soluble iron salts remained on the surface of ship's plate even after gritblasting. The effect of this on exposure to air may be seen in Fig. 8-4. They recommended the use of fine grit with a wide range of particle sizes.

In the blasting and coating of old ships, removal of corrosion product is the important thing. Fresh corrosion of the blasted surface may occur and

FIG. 8-4. Shotblasted-steel after standing 24 hr in the laboratory. Left-hand panel cut from new steel of ships' plate quality. Right-hand panel cut from an old plate taken from a ship's hull.

is generally blasted off before painting. Bullett and Dasgupta (90) showed the deterioration in paint performance due to exposure of the blasted steel up to 32 hr before painting, and recommended a 4-hr maximum exposure outdoors, less in severely contaminated atmospheres. Rowlands (91) found a considerable deterioration in the performance of paint on gritblasted surfaces weathered 8 hr before painting by comparison with surfaces painted immediately.

The conclusion is that the best paint performance is given by surfaces blasted to remove all surface contaminants and painted immediately.

In the building of new ships a different requirement is met. The steel is quickly and economically blasted before fabrication, using centrifugal shotblasting machines; blasting the fabricated structure is much slower and more expensive. The steel must therefore go through the fabrication process, which may last weeks, months, or even years after blasting. It is therefore necessary to paint the steel immediately after blasting, and since the final painting scheme would be damaged and would interfere with the processes of fabrication, the paint must be a thin film of primer only. This is known as a prefabrication or shop primer.

B. SHOP PRIMING SHOTBLASTED STEEL

The impetus for shipyards' adopting shotblasting for the preparation of their steel came from the demands of the big tanker companies, and became a practical proposition with the invention of centrifugal blasting machines. Wilson and Zonsveld of Shell presented an important paper in 1962 (92) in which they analyzed the types of abrasives and the nature of the surfaces they produced. They recommended the application of a zinc dust cured-epoxy shop primer. This was based upon Dutch experience, that country being in the lead in this area at that time.

The requirements for a shop primer are very exacting. It must dry in about two minutes so that when the plate emerges from the automatic painting machine and goes onto the rollers, it may be quickly picked up and stacked. The shop primer must be applied at about 15 to 25 μ dry film thickness in order not to produce excessive fumes when burned during plate cutting and welding. At that thickness, it must suffer minimal breakdown during atmospheric exposure for periods up to about one year. It must resist the deformation due to plate forming and must not reduce the quality of welds. It must accept the final paint schemes without causing blistering or detachment in service, and it may have to meet heavy stresses, such as cathodic protection of underwater areas or aggressive cargoes in tanks.

It is not surprising there is no material that wholly meets all these requirements. A fascinating picture of the development of thought on this subject is provided by two issues of The Propeller, published by the author's company (93). In 1960 the disadvantages of having to coat an exposed shop primer seemed to outweigh the advantages, and it was recommended that shotblasted plates be uncoated during fabrication and flame-cleaned before painting. This was consistent with a recommendation by Hudson, Stanners, and Miller (94). Unfortunately their tests, which led them to recommend a two-month weathering period after blasting, were carried out by exposure in an unpolluted marine atmosphere. In shipyards this procedure led to the formation of a strongly adherent rust that could not be removed by any means other than blasting, so the advantage of the original blasting was lost. In 1962 (93) this was recognized, and the possible shop primers were reviewed, namely, wash primers, oleoresinous primers, zinc-dust types, and aluminum flake in binders drying by solvent evaporation. All these have been used, but the two most successful have been the phenolic-modified wash primers and the zinc dust cured-epoxy types. The choice between these has been made on various grounds, but the prevailing attitude toward zinc dust has always been a factor —its excellent protection against corrosion is offset by its greater tendency to produce toxic fumes when burned and its liability to shed topcoats, particularly coal-tar epoxies.

If zinc silicates are to be used in the final protective scheme, they cannot be applied over shop primers based on organic binders. Therefore, zinc silicate shop primers are used. These may present even worse overcoating problems than organically bound zinc-dust paints. They are, however, being increasingly used for their excellent corrosion protection.

Neither of the two main types of shop primers is suitable for use in tanks carrying certain aggressive chemical products. This new requirement has produced new materials such as isocyanate-cured epoxies. It is more difficult to obtain the rest of the desirable properties with new materials, however this is a field of current research.

C. THE EFFECT OF ABRASIVE BLASTING ON PAINT CONSUMPTION

A steel surface blasted with shot or grit is rough, and a paint film will not cover it evenly. If the minimum specified thickness is to be maintained over the peaks, the paint consumption per unit apparent area will be greater than on a plane surface.

Rowlands (91) tested a bare plate primer (A/C No. 655) over gritblasted plates of varying profile (peak-to-valley height) and found that an extra two coats (30 to 45 μ per coat) above the normal three was required to protect profiles up to 150 μ.

Truelove (95) tested thin (shop primer) and thick film schemes, measuring the thickness over peaks. He found that with thin films the ability of the paint to "follow" the profile determines the thickness, and this ability is encouraged by thixotropy. With thick films, whose upper surfaces are ideally plane, the reduction in thickness over peaks is directly proportional to the maximum amplitude of the profile, thus a 150-μ profile gave a 60-μ reduction, the same order as that found by Rowlands.

Daniel (96) introduced the idea of "surface volume," that is, the volume included by unit area of the rough surface, bounded at the top by a plane touching the highest peaks. In the case of thick films this volume measured the extra paint needed per unit area. The volume could be estimated by a magnetic thickness gauge. In the case of thin films the estimation of the thickness over peaks was not possible by this means. A feature of his work was the extremely low thickness (<2.5 to 5 μ) found over the peaks of shop-primed steel, much lower than the figures quoted by some other authors.

Bullett and Dasgupta (90) introduced an instrument for measuring the surface roughness by means of the leakage of air past a gasket placed on the surface.

In spite of all this work, the estimates of extra paint consumption and thickness over peaks on practical jobs is usually unsatisfactory, the magnetic gauge being almost universally used, and the interpretation of the results very dubious. Perhaps the most useful outcome of the work has been to bring out the fact that a rough blasted surface is very expensive to protect properly and that there are advantages to using fine grits. A maximum profile of 100 μ is now a commonly accepted specification (97, 98).

D. MARINE PAINT APPLICATION

The greatest volume of ship painting is done either in drydock or at sea in the open air, with all the disadvantages of rain, fog, ice, and salt water spray. The wonder is not that the paint sometimes peels off, but that it mostly stays on.

The progression in marine paint application has been from brushing (which has now been largely superseded by rolling) to cold airless spraying. A typical roller job is shown in Fig. 8-5. Airless spray is much quicker and has made the application of high-build paints an actuality, but the equipment is fairly sophisticated. The greatest volume of paint is therefore applied by roller.

Automatic spraying, air-assisted and airless, is almost universal in shop priming practice.

The assembly of prefabricated sections weighing tens or hundreds of tons into ships is now coming into practice. This enables application of the full painting scheme to be made under cover in controlled atmospheres. This method requires quick-drying, high-build paints, which in the case of two-component paints may mean a short life after mixing; twin-feed, airless spray machines have been made for this purpose (99).

Van Londen (100) tested three anticorrosive systems, applied by various combinations of brush and spray over shotblasted steel, dry and moist. He recommended brushing the first coat and spraying the rest. (Others have done this before and since, but to no avail. Invariably, all coats are either rolled or sprayed.) He found that moisture on the surface had little effect on subsequent rusting, but did encourage blistering, and that airless spraying, unlike brushing, did not displace the moisture.

FIG. 8-5. Application of paint by roller to a ship in drydock.

E. UNDERWATER SURFACE PREPARATION AND PAINT APPLICATION

The cost of drydocking a large vessel is so high that there is great interest in the possibility of avoiding it when drydocking is for painting only and not mechanical repairs.

Since the late 1950s a company in Marseilles has offered an underwater cleaning service by frogmen (101). Others have done this, and a later development has been to use a brush fixed on a small boat; this does the job more quickly (102).

Underwater cleaning is adequate if the purpose is to remove fouling, but so far there has been no indication that antifouling paints are revived for long by this treatment, so its value is limited. If a fresh coat could be applied underwater it would be a great advance. This has been done suceessfully by hand application, but very slowly, and a machine is necessary. At the time of writing this is being developed (102).

Provided the surface is well cleaned, antifouling paint applied underwater will adhere to a previous coat. If bare steel is exposed, an anticorrosive applied underwater will not provide protection for any useful time unless the surface is cleaned to bright metal. This has been done experimentally, but so far it is too slow and expensive for practice.

8-10. THE PAINTING OF SMALL BOATS

The problems of corrosion and fouling are similar whatever the size of the vessel, but there are certain differences of practice which make the subject of small boats worth a separate treatment from that of big ships.

First, the materials of construction of small boats are much more various, including Fibreglas-polyester, wood, steel, zinc-sprayed steel, aluminum, cement, and rubber. Second, the standard of perfection required is much higher. Third, the price which the owner will pay for painting, in terms of both labor and money, is also much higher.

There is not much published work on the subject. Reference is made here to one book (103) and one commercial brochure (104) for general information.

A. MATERIALS OF CONSTRUCTION

Fibreglas-polyester has replaced wood for most hulls. Being resistant to decay and marine borers, it requires decoration rather than protection, and the only painting problem is that of adhesion. This arises from two causes, the first being the residues of mold-release agent which lie on the surface. These must be completely removed, but simple washing with solvent is not adequate. It is best to use an aqueous emulsion cleaner. The removal of the mold-release agent is shown by the water-wettability of the surface. The second cause of adhesion problems is in the nature of the clean polyester surface itself. Many paints will not adhere to this kind of surface, probably because of the presence of weak surface layers, according to the general theory of Bikerman. A special primer which strengthens these layers will ensure the adhesion of any type of paint.

Wood is still quite extensively used in small boats, and in addition to the oleoresinous and alkyd primers, enamels, and varnishes, which are traditional, polyurethanes are popular because of their hard films. There is a danger of poor adhesion of subsequent coats when the weathered finishes are repainted, especially with the moisture-cured types, but this is overcome when the weathered surfaces are well rubbed down, and fortunately this is the usual practice on boats. Wooden hulls are often reinforced by coating with Fibreglas cloth, which is impregnated with several coats of epoxide resin paint, which acts as an adhesive as well as a protective film.

Steel yachts present the same painting problems as ships, and are treated in much the same way, except for the higher standards of finish. Sometimes, however, the steel is coated with zinc by metal spraying or hot-dip galvanizing, and the normal treatment is first with wash primer, followed by the normal underwater primers, and above water by zinc chromate primer.

Aluminum must be degreased and preferably etched by means of a phosphoric acid preparation. It is then treated as for zinc. A particular point to watch is that it may suffer severe corrosion from antifoulings rich in copper, particularly if the vessel lies in shallow stagnant water for appreciable periods; the copper dissolves in the seawater, plates out on any bare aluminum surfaces, and forms a couple which produces pitting of the aluminum. For boats with aluminum hulls or fittings (such as outboard motors), the antifouling should be low in copper or preferably free from it altogether; the same applies to mercury in antifoulings. The problems of using aluminum in ships have been discussed by Lees (105).

There is a movement to construct boats from steel-reinforced cement, which is a difficult surface to paint. Zinc silicate is a suitable coating, and a zinc silicate anticorrosive and antifouling scheme is very successful. An alkali-resisting paint such as a chlorinated rubber is suitable when a decorative effect is required or when conventional antifoulings are to be applied.

Rubber is a very difficult material to paint, and both antifoulings and decorative paints with elastomeric binders promise to be most successful for this purpose. The high polymer binders give low-solids content, but since the substrate is incorrodible, thick films are not necessary.

B. STOPPERS AND FILLERS

While stoppers and fillers are used in all types of painting work, they are particularly important on yachts because of the high standard of finish required and the emphasis on a smooth hull. The materials and techniques are similar to those in other classes of work, ranging from stoppers which are knifed into bolt holes and seams to brushing cements to fill in hair cracks and smooth out large surfaces. If they are to be followed by paint systems containing strong solvents such as polyurethanes, epoxies, or vinyls, the stoppers and fillers must also be solvent resistant. The problem is to obtain the correct consistency with the smallest quantity of volatile material, in order that shrinkage may be minimized. The most popular binders are epoxies and polyesters.

Because of the movement of planks, there is a requirement for an elastomeric type of stopper, which may be based on polyurethane or other types of oligomer, that can be applied as a two-component mix with the minimum of solvent and form the elastomer in situ. This type of material has also proved popular for bedding down metal keels.

The formulation of stoppers and fillers is quite a difficult branch of paint technology. Much testing is required to make sure that they are easy to apply, will not shrink, crack or fall out, and will take the overcoating paint system and retain it under service conditions.

C. ANTIFOULINGS

The purpose of a yacht antifouling, like that on big ships, is to maintain a smooth hull. Since the aim in racing is to obtain the maximum possible speed, the standard of smoothness is very high, and antifoulings are often smoothed down with abrasive paper before launching. This requires a hard film, and vinyl types are popular for this purpose.

The high standard of antifouling performance required is made somewhat easier to obtain by the fact that long life is seldom needed, periods of 6 to 12 months being usual.

One respect in which yacht antifoulings may receive different treatment from those on ships is that they may be exposed to the air for long periods. The conventional rosin-based, soft-film antifouling is quite unsuitable for this treatment. In the sun the resin is degraded and the film becomes powdery or may crack. Vinyl/rosin and other hard film types are generally resistant to exposure, but one fact must be noted, that when an antifouling has been immersed and dried out, it seriously loses efficiency unless it is wetted and scrubbed before re-immersion. This is probably because the film of slime on the surface dries and produees a barrier to leaching of the poisons.

Another difference in yacht requirements is in the popularity of bright-colored antifoulings. When a big ship is sailing, the bottom proper is seldom seen and the color is of little importance. A yacht moves up and down much more, and since it has very little boottop (the difference between laden and unladen weight being small), the bottom is often on view, and is part of the total decorative effect.

Copper oxide, with its dull red color, is an obstacle to the production of bright colors. Fortunately, it rapidly loses its color on immersion in seawater, due to dissolution of the upper layers, and acquires the pale green color of basic copper salts. Brightly colored pigments may therefore be mixed with it, and the paint, though dull when dry, becomes quite bright when immersed. If the concentration of copper oxide is high, however, the color is never very bright.

Colorless poisons such as tributyl tin and phenyl mercury compounds will enable bright-colored antifoulings to be made, but not without difficulty. Mercury compounds on their own are only effective at concentrations which are extremely costly. Tributyl tin compounds, in simple addition to antifouling paints, tend to have short active lives, particularly if fresh water is encountered, in which they are much more soluble than in salt water. Zinc oxide in a tributyl tin-containing polymer gives a very effective white antifouling (67) but colored pigments may interfere with its action.

8-11. THE EVALUATION OF MARINE PAINTS

The evaluation of marine paints is the essence of our subject. Antifoulings apart, there is no marine paint which is not similar to some paint designed for industrial or decorative use. Its value as a marine paint is determined entirely by the extent to which it has been tested and found suitable for marine conditions.

A. ANTICORROSIVE SHIP BOTTOM PAINTS

There are two types of approach to the problem of anticorrosive ship bottom paints:

1. Direct testing of the anticorrosive value of the paint system by ship trials, raft trials, and laboratory trials, which may use normal or accelerated conditions of corrosion.

2. Analytical testing of corrosion resistance by consideration of individual properties such as permeabilities.

De Wolf and van Londen (39) considered the merits of ship and raft testing and concluded that they are complementary. Ship trials are done under service conditions (which has the advantage that all the service factors are present, some of which might be overlooked and omitted from an artificial test) and the motion of the ship induces quicker breakdown. But with ship trials the conditions are uncontrolled, or poorly controlled, and the results are seen only when the ship docks. Raft trials are done under known conditions, and the results may be seen at frequent intervals, but the breakdown is slower. De Wolf and van Londen found that the results of both kinds of test were in agreement.

It might be added that if paint is to be sold, there is no substitute for a trial on a customer's ship.

The disadvantage common to both types is the time taken to obtain results. If a proper evaluation, including repainting, is to be made, a bottom scheme cannot be considered satisfactory with less than five years' testing. For this reason much work has been done to devise laboratory tests in order to obtain results more quickly.

Brown (106) argued that laboratory evaluation was better than service testing because the conditions during a service trial are unknown and only one set of conditions applies to any one trial, whereas in the laboratory, the conditions may be chosen and varied. He advocated improving the sensitivity of measuring rate of breakdown rather than accelerating the rate, and for this purpose measured the electrical resistance of the coated panel.

Most workers in the field have taken Brown's view of accelerated tests. Barton (107) showed that the optimum concentration of SO_2 in an atmospheric corrosion test, to give the correct relative rates of corrosion of Fe, Zn, Cu, and Al, was 0.01%. If the concentration of SO_2 were too high, the corrosion of steel would be reduced, due to suppression of the hydrolysis of $FeSO_4$ to form H_2SO_4, whereas the corrosion rate of copper would be increased, due to the formation of $CuSO_4$ instead of basic copper sulfates. Similarly with salt solutions, the use of very concentrated solutions of sodium chloride accelerated the corrosion of metals that formed basic chlorides as corrosion products, for example, zinc. But the use of concentrated solutions of sodium chloride did not accelerate the corrosion of iron, which regenerates NaCl in forming its corrosion product. In order to obtain the relative rates of Zn and Fe found in natural marine corrosion, the degree of acceleration Barton could obtain was very small, though it could be increased by raising the temperature to 35°C.

Faced with this uncertainty about the validity of accelerated tests, most workers have preferred to use seawater as the corrosive medium and concentrate on early detection of failure. The acceleration of breakdown by rapid movement through the water has, however, been investigated by Wormwell, Nurse, and Ison (108), who constructed a rotor which would move panels at 20 knots, and in 50 days obtained quite a good correlation with raft tests that had lasted 200 days. They concluded, however, that many paint systems would require even more intensive conditions to obtain breakdown, and as an alternative, suggested the development of a more sensitive test for paint failure.

Bacon, Smith, and Rugg (109) measured the change with time in the electrical resistance of coated steel billets immersed in seawater. They found that most coatings started off with a resistance of about 10^9 Ω/cm^2 which decreased on immersion, at rates which could be divided into three categories:

Good: resistance fluctuates but remains above 10^8 Ω/cm^2.

Fair: resistance declines to 10^7-10^8 Ω/cm^2 in 30 days and eventually to below 10^5.

Poor: resistance falls rapidly and is below 10^5 Ω/cm^2 in 30 days.

They estimated that the fall in resistance would predict a failure in less than one-fifth of the time needed for it to appear visibly. This is a very attractive idea, but in the author's experience it has not been confirmed. By the time a definitely declining trend in the resistance has been established, corrosion is visible, so as a means of prediction it has been valueless. Brown (106), however, as mentioned before, has found it to be of value.

Wormwell and Brasher (110) measured the electrode potentials of bare and painted steel in artificial seawater. They found painted steel to be more noble than unpainted, and they found the potential to move in time toward that of the corroding metal. They specifically warned, however, against drawing conclusions from the potentials in the first few days of the test, and they found that the beginning of the final decline of potential in the base direction coincided with the appearance of rust spots. Since they considered this to be the significant point, its predictive value is negligible.

The rather depressing conclusion from all this is that the only reliable way to test a ship bottom scheme for corrosion resistance in the laboratory is to apply it to steel in the prescribed way, suspend the specimen in stirred, aerated seawater, and wait for results. When doing this test, the antifouling should be omitted, because the poison will rapidly saturate the water, and in the case of copper, will produce an artificially stimulated corrosion.

When it is necessary to test a scheme for resistance to cathodic protection, the specimen must be well coated, free from gaps, and immersed in stirred, aerated seawater with an arrangement for keeping the anodic products separate from the main tank. The required potential is applied and the current measured at intervals. If after several months there is no serious breakdown, a small area should be scraped down to bare metal and the test continued, with particular attention to adhesion adjacent to this area. Testing procedures have been described by Boynton (111) and Routley (112).

The foregoing has concerned the direct testing of paints for their protective properties. A great deal of work has been done on the analysis of these properties, measuring water absorption, permeability to water and ions, swelling, and other individual features of the interaction of paints and seawater. The author has provided a short survey and list of references in a previous publication (113). To this should be added the following. Gentles (114) adopted the method of Brasher (115) for measuring water absorption by the change in capacitance of a condenser whose dielectric was the paint film. The estimate of water absorption by this method may be in agreement with the gravimetric estimate, in which case the water is assumed to be distributed randomly; if it is higher, the water is thought to be distributed in capillaries, and such a distribution is associated with poor corrosion resistance. Mayne and his colleagues (25) have studied the permeation of ions through varnish films, which is fundamental to the resistance inhibition of corrosion, and have discovered that the processes are extremely complex. In addition to resistance inhibition, paints with anticorrosive pigments exercise a chemical inhibition of the electrode processes. This is surveyed by Mayne (25), and reference should also be made to the work of Bharucha (116), who devised a method of distinguishing between inhibition by interference with the anodic reaction or the cathodic reaction, or by electrolytic resistance; the medium, however, was not seawater but a 3% solution of water in triacetin, chosen for its high electrical resistance.

Eventually this type of work may provide reliable alternatives to the direct testing of corrosion resistance, but it has not done so yet.

B. ANTIFOULING PAINTS

As with anticorrosives, the testing of antifoulings may be divided into the direct testing of paints against fouling settlement and the laboratory testing of individual properties such as toxicities of poisons and their rates of solution.

Van Londen (60) compared raft and ship trials, to the effect that raft trials have the advantage that the conditions are known, but the water flow is slow, and the settlement is seasonal. The latter can be overcome by immersing panels at intervals, the former by aging on a rotor. He described one which would carry the panels at 25 or 40 knots, and claimed that the type of leaching in the apparatus resembled that on a ship rather than that on a raft.

Because of their accessibility and controllability, raft trials have been adopted by most workers to choose paints which are ultimately tested on boats and ships. Details of techniques of raft testing have been given by Hunt and Sparrow (117). Antifouling paints should be applied evenly and at controlled thicknesses on the anticorrosive paints over which they will be used. The panels should not be too small, because of the statistical nature of the fouling settlement, 20 cm square being about the useful minimum. The water should be subject to some sort of flow and it should be unpolluted. The raft should be designed to allow adequate light to reach the panels in order to encourage algal growth. (In the "Turtle" pattern, some panels are deliberately turned toward the light and others away from it in order to encourage weed and animal fouling, respectively.) The settlement of fouling should be monitored by means of nontoxic panels which are replaced at intervals. Because of the change of fouling pattern from week to week, panels should be inspected frequently.

Ship trials are subject to the vagaries of the ship's voyage, and the only control that can be exercised is in application. The quantities of paint applied per unit area should be measured and the application should be even. Because the fouling conditions encountered on all the voyages between dry-dockings are often unknown, all trials should be comparative, a control paint of known performance being applied alongside the trial. Since different fouling conditions may alter the relative performances of antifoulings, decisions should never be made on the basis of one ship trial.

The laboratory testing of antifoulings is mainly concerned with the "leaching rate test," which has been described by the Woods Hole workers (42), Harris (54), Barnes (118), and Partington and Dunn (119). Specimens coated with the paint are exposed on a raft or in some storage vessel containing seawater, and are periodically removed, shaken with a small amount of seawater which is analyzed for copper or other poisons, and returned to storage. This provides instantaneous rates, from which a curve of leaching rate against time may be constructed.

This procedure has been criticized by Partington and Dunn on the grounds that the leaching rate measured is not necessarily the same as the prevailing during storage, and indeed the conditions of storage greatly affected the measured leaching rates, so that in some cases a low rate of loss during storage would correspond to a high measured leaching rate. For this reason, workers at International Paints Ltd. (120) have measured leaching rates throughout the life of the paint, the leaching being always under the same regime — small specimens of paint are rotated in bottles of seawater, which are changed every three days and analyzed. This method is sound but takes up much time and space.

The leaching rate test is the only means of scientifically determining the effect of formulation variables on antifouling paint performance. Raft testing will tell whether a variation is beneficial, but it will not tell why.

The leaching rate test provides information, but it does not save time by comparison with a raft test. If the solubility of the poison is increased, the leaching rate will be accelerated, and Marson (121) has claimed the successful use of an accelerated test using sodium glycinate/sodium chloride solution as the leaching medium. However, this is open to the same objection as other accelerated tests, that one feature is accelerated and the balance — in this case, the relative leaching of poison and medium— is not necessarily maintained when different media are tested.

The leaching rate test depends upon analysis of concentrations of poison in seawater that are well short of saturation, which means 0.1 $\mu g/cm^3$ or less. This can be done satisfactorily for copper, zinc, and mercury (though mercury presents a problem in that it rapidly disappears from the solution due to bacterial action). Tributyl tin compounds are much more difficult to determine, and the method described by Rivett (122) is useful, though not precise. It depends upon the fact that the green alga Chlamydomonas will grow in seawater containing poisons up to a certain concentration, at which point a small increase causes a very rapid fall in the rate of growth. This "critical concentration" is about 5×10^{-9} g/cm^3. The solution of unknown concentration is diluted successively and used as a growth medium for the alga. Provided the dilutions span the critical concentration, this can be estimated.

It is important to understand that this is a means of assessing rates of solution of poisons, not the relative toxicities of poisons to marine fouling. The susceptibility of Chlamydomonas to different poisons is not necessarily a guide to the relative efficacy of those poisons against other marine algae or animal fouling species.

C. WEATHERWORK PAINTS

The types of weatherwork paint used are essentially similar to those used on structural steel work on land, except that the slow-drying red lead in oil is very seldom used because of the speed of ship construction. The properties required for weatherwork paints on ships differ in some points of detail.

A great deal of emphasis is laid on the gloss retention of finishing paints for use on land; this is much less important in marine use because all surfaces are either wet (and therefore glossy) or become rapidly covered with a layer of salt (and therefore matte). The paint chemist may therefore be disappointed to learn that a finish which on his tests has shown excellent gloss retention is pronounced no better than the standard in service. Heavy chalking or change of color due to loss of gloss are unacceptable defects, but loss of gloss itself is not very important on exterior marine work.

Anticorrosive primers should be tested by exposure to a marine atmosphere, or to an inland atmosphere with frequent applications of seawater. They should be applied not only over clean steel, clean shop primer, and rusty steel, but also over the same surfaces contaminated by sea salt. Painting over salt is unsound in principle and will cause breakdown by rusting and blistering, but primers do differ in their resistance to this form of misuse. There is no general guide to the type of primer which is more resistant to painting over salt, so an empirical test is necessary.

Peeling due to loss of adhesion between coats when wet is a common form of failure. It arises from two causes: excessively long drying of a paint which is not tolerant to such treatment, or painting over salt. Every paint should therefore be tested by overcoating with itself and likely following coats after different intervals of drying. The seams are soaked in water and may be tested by simple probing with a knife. The British Admiralty produced a testing machine which used a jet of water to effect separation between coats, and since the force of the jet can be controlled, it gives a roughly quantitative test for peeling tendency (123).

The tendency to peel depends not only on loss of adhesion, but also on the mechanical properties of the paint, in the sense that a soft, flexible film is much more likely to peel than a hard, brittle one. Flexibility tests such as are common for industrial paints are therefore inappropriate for marine paints. Short of excessive brittleness, which could cause failure by cracking, chipping, or powdering off, a marine paint should be brittle to a degree which will discourage peeling. By this means the bad effects of painting over salt may be minimized. This is the best that can be done because the osmotic action of salt between coats is unavoidable and cannot be overcome by paint formulation.

ACKNOWLEDGMENTS

Where possible, references have been made to published works. Where no reference is given, the information has been provided from the author's own knowledge and that of his colleagues in the International Paint Co. Ltd., to whom he gratefully acknowledges his debt. The opinions are his own.

REFERENCES

1. H. U. Sverdrup, M. W. Johnson, and R. H. Fleming, The Oceans, Prentice-Hall, 1942.
2. F. S. Russell and C. M. Yonge, The Seas, Frederick Warne, 1928.
3. A. L. H. Gameson, New Scientist, April 30, 1964, p. 295.
4. F. M. Fowkes, Ind. Eng. Chem., 56, 40 (1964).
5. H. Lackenby, Proc. Inst. Mech. Engrs., London, 176, 981 (1962).

6. H. Sasajima and E. Yoshida, Intern. Shipbuilding Progr., 2, 441 (1955).

7. B.S.R.A. experiments on the "Lucy Ashton," Trans. Inst. Naval Arch., London, Part I, 40 (1951); Part II, 350 (1953); Part IV, 525 (1955).

8. G. Aertssen, "Sea Trials on the 'Lubumbashi,'" Trans. Royal Inst. Naval Arch., London, 502 (1957); 174 (1960).

9. H. J. S. Canham and W. M. Lynn, Trans. Royal Inst. Naval Arch., London, 104, 1 (1962).

10. H. J. Lageveen-van Kuyk, Shipping World and Shipbuilder, 1347 (August, 1967).

11. A. Logan, Trans. N.E.C.I., Newcastle-upon-Tyne, 76, S. 61 (1960).

12. A. Peterkin, Nature, 227, 598 (1970).

13. M. O. Kramer, J. Am. Soc. Naval Engrs., 25 (February, 1960).

14. R. B. Couch, U.S. Navy Dept., David W. Taylor Model Basin, Washington, D.C., Report 789, August, 1951.

15. P. Ffield, Trans. Soc. Naval Arch. and Marine Engrs., 62, 688 (1954).

16. E. Uusitalo, Extended Abstracts of Papers Presented at the Second International Congress on Metallic Corrosion, N.A.C.E., New York, 1963, p. 15.

17. W. P. Iverson, Corrosion Prevent. Control, 15, (February, 1969).

18. W. S. Patterson, Trans. N.E.C.I., Newcastle-upon-Tyne, 68, 93 (1952).

19. G. H. Booth, A. W. Cooper, and A. K. Tiller, J. Appl. Chem., 13, 211 (1963); 15, 250 (1965); Brit. Corr. J., 2, 22 (1967); 2, 222 (1967).

20. E. F. Corcoran and J. S. Kittredge, Conference on Marine Corrosion and Fouling Problems, Scripps Institute of Oceanography, University of California, April, 1956.

21. D. M. Brasher, Brit. Corr. J., 2, 95 (1967); 4, 122 (1967).

22. K. N. Barnard, Can. J. Res., F26, 374 (1948).

23. J. H. Morgan, Cathodic Protection, Leonard Hill, London, 1959, p. 194.

24. L. T. Carter and J. T. Crennell, Trans. Inst. Naval Arch., London, 97, 413 (1955).

25. J.E.O. Mayne, Brit. Corr. J., 5, 106 (1970).

26. S. Guruviah, J. Oil Colour Chemists' Assoc., 53, 669 (1970).

27. P. J. Gay, J. Oil Colour Chemists' Assoc., 32, 488 (1949).

28. C. F. T. Young, The Fouling and Corrosion of Iron Ships, The London Drawing Association, 1867.

29. A. S. Cushman and H. A. Gardner, The Corrosion and Preservation of Iron and Steel, McGraw-Hill, New York, 1910, p. 215.

30. M. Ragg, Die Schiffsboden-und-Rostschutzfarben, Union Deutsche Verlagsgesellschaft, Berlin, 1925.

31. R. Robinson [Former (Present) Chief Chemist of International Paints, Ltd.], private communication, 1970.

32. F. Fancutt and J. C. Hudson, J. Oil Colour Chemists' Assoc., 30, 135 (1947).

33. J. H. Greenblatt, K. N. Barnard, C. W. Reyno, and D. G. Smith, Naval Research Establishment, Dartmouth, Nova Scotia, August, 1964.

34. J. H. Deterding, M. W. Singleton, and R. W. Wilson, paper to the First International Congress on Marine Corrosion and Fouling, Cannes, 1964.

35. A. G. Sussex, "Third Inter-Naval Corrosion Conference," London, 1969.

36. The Formulation of Anti-corrosive Compositions for Ships' Bottoms and Underwater Service on Steel, British Iron and Steel Research Association, London, 1950.

37. T. F. Birkenhead, D. F. Bowerman, and B. S. Karten, J. Paint Technol., 42, 525 (1970).

38. J. R. Brown, Materials Report 61-E, Pacific Naval Laboratory, Esquimalt, B.C. Defence Research Board of Canada, 1961.

39. P. de Wolf and A. M. van Londen, "Raft Trials and Ship Trials with some Underwater Paint Systems," Report No. 43C, Netherlands Research Centre T.N.O. for Shipbuilding and Navigation, Delft, 1962.

40. R. P. Hanson and E. G. Tucker, Trans. Inst. Eng. Shipbuilders Scot., 109, 82 (1966).

41. D. W. Singleton and R. W. Wilson, Brit. Corr. J., Supplementary Issue, 12 (1968).

42. "Marine Fouling and Its Prevention," Contribution No. 580, Woods Hole Oceanographic Institution U.S. Naval Institute, Annapolis, Maryland, 1952.

43. "Hydrological and Biological Co-operative Research," Directorate for Scientific Affairs, Organisation for Economic Co-operation and Development, Paris, 1966.

44. A. C. A. Holzapfel, Trans. Inst. Naval Arch., London, 46, 252 (1904).

45. Tin and Its Uses, No. 53, 12 (1961), Tin Research Institute, Greenford, Middlesex.

46. Shipbuilding and Shipping Record, March 10, 1960, p. 315.

47. T. Lovegrove and T. W. Robinson, Proceedings of the First International Biodeterioration Symposium, Southampton, 1968, p. 617. Elsevier, New York, 1960.

48. D. H. Tullis, L. C. Neill, and A. T. Henderson, J. Inst. Petrol., 45, 155 (1959).

49. N. F. Cardarelli, J. W. Born, and C. H. Stockman, Marine Fouling Prevention by Sheet Rubber, B. F. Goodrich & Co., January 30, 1968.

50. A. Little, J. Commerce, Shipbuilding and Engineering Edition, London (October 13, 1955).

51. C. O. Morley, H. J. Clarke, H. J. M. Bowen, and M. H. M. Arnold, J. Oil Colour Chemists' Assoc., 41, 445 (1958).

52. D. J. Crisp and A. P. Austin, Ann. Appl. Biol., 48, 787 (1960).

53. B. Wiseley, Australian J. Marine Freshwater Res., 14, 44 (1962).

54. J. E. Harris, J. Iron Steel Inst., London, 297P (1946).

55. D. J. Crisp and P. S. Meadows, Proc. Roy. Soc., London, B15, 500 (1962).

56. E. T. Wilkie and A. C. Edwards, J. Appl. Chem., 14, 155 (1964).

57. H. Barnes, J. Iron Steel Inst., London, 157, 193 (1947).

58. H. Barnes, J. Iron Steel Inst., London, 169, 360 (1951).

59. F. Marson, J. Appl. Chem., 19, 93 (1969).

60. A. M. van Londen, Report No. 54C, Netherlands Research Centre T.N.O. for Shipbuilding and Navigation, Delft, 1963; also J. Oil Colour Chemists' Assoc., 52, 141 (1969).

61. W. A. Anderton, J. Oil Colour Chemists' Assoc., 52, 711 (1969).

62. N. I. Hendey, Nature, 160, 97 (1947); and J. Royal Microscop. Soc., 71, 1 (1951).

63. G. T. Boalch, unpublished work as International Paints Ltd. Research Fellow at the Marine Biological Laboratory, Plymouth.

64. C. J. Evans, Tin and Its Uses, No. 85, 1970, Tin Research Institute, Greenford, Middlesex.

65. International Lead Zinc Research Organisation Inc., Final Report, Project No. LC-81, March 31, 1970, Battelle Memorial Institute, Columbus, Ohio.

66. S. Miller, Antifouling Potentials of Pesticidal Materials, Marine Laboratory, University of Miami, 1961.

67. A. S. Deeks, D. W. Hudson, D. M. James, and B. W. Sparrow, paper to the Second International Congress on Marine Corrosion and Fouling, Athens, 1968.

68. A. O. Christie and D. J. Crisp, Ann. Appl. Biol., 51, 361 (1963).

69. H. P. Vind and H. Hochman, "An Evaluation of Organotin Compounds as Preservatives for Marine Timbers," Technical Report R-188, U.S. Naval Civil Engineering Laboratory, Port Hueneme, California, 1962.

70. A. Pass and M. J. F. Meason, J. Oil Colour Chemists' Assoc., 44, 417 (1961).

71. L. S. Birnbaum and W. J. Francis, paper to the Third Inter-Naval Corrosion Conference, London, 1969.

72. T. R. Bullett, J. Oil Colour Chemists' Assoc., 46, 441 (1963).

73. D. M. James, J. Oil Colour Chemists' Assoc., 39, 39 (1956).

74. J. G. Robinson and K. Fleming, in Corrosion Problems of the Petroleum Industry, Society of Chemical Industry, London, 1960.

75. W. H. Briggs and W. H. Francis, Combatting Corrosion in Ships' Tanks, Soc. Naval Arch. and Marine Engrs., New York, 1963.

76. G. W. Colberg, "Tank Coatings — Clean Oil Tankers," paper to the A.P.I. Division of Transportation, Absecon, N.J., June 14, 1956.

77. A. Pass and M. J. F. Meason, J. Oil Colour Chemists' Assoc., 43, 897 (1965).

78. R. C. McMahan, G. A. Edwards, and R. E. Dunbar, "The Evaluation and Implementation of a Cargo Tank Coating Program," paper to N.P.V.L.A. Marine Coatings Conference, New York, 1967.

79. A. Logan, Trans. Inst. Marine Engrs., London, 70, 153 (1958).

80. D. M. James, J. Oil Colour Chemists' Assoc., 43, 391 and 653 (1960).

81. R. I. Wray and K. Tator, Mater. Protection, 2, 30 (1963).

82. P. Talma, "Earning Powers and the Protection of Tank Spaces" (Shell Tankers, Rotterdam), paper to Congress Europort, 1964.

83. E. L. Purlee, W. A. Leyland, and W. E. McPherson, Marine Technol., 380 (October, 1965).

84. A. N. Narter, paper to N.P.V.L.A. Coatings Conference, Atlantic City, 1965.

85. "Final Report on Methods for Cleaning Ship Bottoms," Research and Development Department, Bakelite Corp. Report OSRD No. 1310, Division 11, National Defense Research Committee of the Office of Scientific Research and Development, Washington, D.C., April 2, 1943.

86. F. Fancutt and J. C. Hudson, "The Protection of Structural Steel Against Atmospheric Corrosion," Publ. of the Intern. Assoc. for Bridge and Structural Engineering, Zurich, 16, 185 (1956).

87. U. R. Evans, The Corrosion and Oxidation of Metals, Arnold, London, 1960.

88. J. E. O. Mayne, J. Appl. Chem., 9, 673 (1959).

89. K. A. Chandler, Brit. Corr. J., 1, 264 (1966).

90. T. R. Bullett and D. Dasgupta, J. Oil Colour Chemists' Assoc., 52, 1099 (1969).

91. J. C. Rowlands, J. Iron Steel Inst., 199, 329 (1961).

92. R. W. Wilson and J. J. Zonsveld, Trans. N.E.C.I., Newcastle-upon-Tyne, 78, 277 (1962).

93. "The Propeller," No. 6 (1960), No. 14 (1962), International Paints Ltd., London.

94. J. C. Hudson, J. F. Stanners, and A. G. B. Miller, Trans. Inst. Marine Engrs., London, 71, 269 (1959).

95. R. K. Truelove, J. Oil Colour Chemists' Assoc., 49, 57 (1966).

96. H. G. Daniel, Trans. Royal Inst. Naval Arch., London, 3 (2), 241 (1969).

97. Protection of Iron and Steel Structures from Corrosion, British Standards Institute, CP 2008 (1966).

98. Recommended Practice for Protection of Ships' Underwater and Boot-topping Plating from Corrosion and Fouling, British Ship Research Association, Wallsend, Northumberland, 1966.

99. "Spraying Equipment for Epikote Resin High Solids Systems," Resin Rev. (September, 1969), Shell Chemicals Ltd., London.

100. A. M. van Londen, "An Investigation into the Influence of the Method of Application on the Behaviour of Anticorrosive Paint Systems in Seawater," Delft. Report No. 46C, Netherlands Research Centre, T.N.O. for Shipbuilding and Navigation, 1962.

101. "Frogmen at the Service of the Merchant Marine," Cie. Phocéenne Sous-marine, Marseille.

102. "The Propeller," No. 34 (1968), No. 37 (1970), The International Paint Co. Ltd., London.

103. G. Bryant and C. Jasper, How to Paint Boats and Yachts, Crosby Lockwood, London, 1963.

104. International Yacht Painters' Manual, International Yacht Paints, Canute Road, Southampton.

105. D. C. G. Lees, Chem. Ind., 78, 949 (1954).

106. J. R. Brown, "Paints for Ship Bottoms: the Case for Laboratory Evaluation," paper to the Vancouver Meeting of N.A.C.E., February 10, 1960. Defence Research Board of Canada, Pacific Naval Laboratory, Esquimalt, B.C.

107. K. Bartoň, "Acceleration of Corrosion Tests on the Basis of Kinetic Studies of the Rate Controlling Combinations of Factors," First International Congress on Metallic Corrosion, Butterworths, London, 1963.

108. F. Wormwell, T. J. Nurse, and H. C. K. Ison, J. Iron Steel Inst., 247 (1948).

109. R. C. Bacon, J. J. Smith, and F. M. Rugg, Ind. Eng. Chem., 40, 161 (1948).

110. F. Wormwell and D. M. Brasher, J. Iron Steel Inst., 129 (June, 1949).

111. F. R. Boynton, "Verträglichkeit des Kathodischen Schutzes mit Anstrichen," Jahrb. Schiffbautechnischen Gesellschaft, 61, 414 (1967); "The Propeller," No. 5 (1960), International Paints Ltd., London.

112. A. F. Routley, Paint Technol., 31, 28 (1967).

113. "Convertible Coatings," Oil Colour Chemists' Assoc. Paint Technology Manuals, Part 3. Chapman and Hall, London, 1962.

114. J. K. Gentles, J. Oil Colour Chemists' Assoc., 46, 850 (1963).

115. D. M. Brasher, J. Appl. Chem., 4, 62 (1954).

116. N. R. Bharucha, J. Oil Colour Chemists' Assoc., 44, 515 (1961).

117. "The Propeller," No. 3 (1959), No. 20 (1964), International Paints Ltd., London; B. W. Sparrow, "Essais, sur radeaux, de peintures de carènes de navires," Peintures, Pigments, Vernis, 42, 4 (1966).

118. H. Barnes, J. Oil Colour Chemists' Assoc., 31, 455 (1948).

119. A. Partington and P. F. Dunn, J. Oil Colour Chemists' Assoc., 44, 869 (1961).

120. "The Propeller," No. 2 (1959), International Paints Ltd., London.

121. F. Marson, J. Oil Colour Chemists' Assoc., 47, 323 (1964).

122. P. Rivett, J. Appl. Chem., 15, 469 (1965).

123. J. C. Kingcome, J. Oil Colour Chemists' Assoc., 44, 237 (1961).

Chapter 9

COATINGS FOR METAL CONTAINERS

G. W. Gerhardt and G. W. Seagren

Mobil Chemical Company Pittsburgh,
Pennsylvania

9-1. INTRODUCTION

The subject of coatings for metal containers can be quite varied, since food containers holding as little as two ounces of product might possibly be considered in this category as well as tank cars or stationary storage tanks holding several million gallons of product. While the coatings used on large tanks are often similar to those used on small containers, it is generally better to consider in the latter group only those metal containers up to and including 55-gallon steel drums, and in the former group the very large mobile units and stationary storage containers. This chapter deals only with the smaller containers.

In addition to containers that are made completely of metal, many others made of glass, plastic, or cellulose require that certain components be made of metal. In this category are metal closures (1), which are used primarily on glass or plastic bottles, the metallic components of metal/paper composite cans, and units made from foils, such as various pouch and bag constructions. For the purpose of this work, the container and its coatings are considered if the organic coating is applied to the metallic components of the container.

Coatings are used on metallic containers for a number of reasons (2-5). The primary purpose is one of protecting the contents of the container (6), but the identification or advertising features which can be displayed on the exterior of the container sometimes become the controlling factors in deciding what types of coatings can be used. In considering the protective qualities of a coating, sometimes the product must be protected by the organic coating from possible contamination from the container itself, while other times it is important to protect the container so that it does not corrode excessively. In the case of decorative coatings, the over-all marketing image of the container must be considered, as well as the handling qualities of the unit. Thus, in high-speed filling operations, the containers must move in contact with each other or in contact with machine guide rails without sticking or marring of the exterior coating. One other significant function of a coating is to supply a seal. This is particularly important for those containers which are heat-sealed, and, in the case of closures, where the coating may constitute the gasket or must be the adhesive coat for the gasket.

The general term "coating" in the container industry thus encompasses all functional organic film formers with the exception of inks. The formulation, application, and properties of inks are generally significantly different from those of coatings and, therefore, the subject of inks must be a separate discussion (5).

9-2. TYPES OF CONTAINERS

Metal containers are of such wide variety and style that it is best that they be described briefly.

A. FOOD AND BEVERAGE (SANITARY) CONTAINERS

The most common sanitary can is the three-piece can, consisting of a body and two ends. The seam of the body may be a simple dry-lock seam, it may be soldered, it may be organically cemented, or it may be welded. The method used to make a seam is important because different degrees of heat are required for each type of seam. The ends are usually applied by an operation known as double seaming. In this operation, the circumference of the end and the end of the body are curled together. This is usually a high-speed operation in which high pressures are exerted on both sides of the seam. A few cans, such as evaporated milk cans, have the ends soldered to the body; however, these cans seldom have organic coatings on them.

The other major category of sanitary cans is the two-piece can in which the body and one end are one continuous piece of metal. The other end is double seamed to the can body after filling.

B. PAINT CANS

Most paint cans consist of a bottom, a body, a ring at the top of the can, and a cover that fits with friction into the ring.

C. AEROSOL (PRESSURE) CANS (7, 8)

The three-piece aerosol can consists of a bottom, a body, and a dome. Construction is generally as stated in the section on sanitary cans, except that provision must be made for higher strength of the container. In addition to the three components mentioned, two other metal components generally require coating, namely, the valve cup and the over cap. The over cap protects the valve during shipping and often carries some of the advertising.

Two-piece aerosol cans are produced by drawing the can body in such a way that either the bottom is an integral part of the side wall, or the dome is an integral part of the side wall. These cans, of course, like the three-piece can, also need a valve cup and an over cap.

D. PAILS AND DRUMS

Pails and drums are usually of three-piece construction and are manufactured from cold-rolled steel or hot-rolled steel. The seams of the bodies are usually welded.

E. COMPOSITE CONTAINERS

Composite containers are made from laminates of metal foil, paper, and organic films in a variety of combinations. These laminates are wound spirally or convolutely over a mandrel to produce tubes, which are later cut into the desired length of the final container. There are many variations possible in the construction of composite cans. When foil is used as the inner ply it usually has a slip coating applied to permit ready removal from the mandrel.

Also, in some cases, metal foil is used as the outside ply, and thus, is printed and varnished. In most cases, metal ends are used for composite containers. These ends are double seamed onto the laminate body.

F. CLOSURES

A wide variety of closures are produced (1), but they can generally be classified in four groups:

1. Screw caps are closures with a thread formed into the closure.

2. Roll-on closures are formed from a coated metal cup with a smooth side wall. At the time the closure is put on the bottle, the threads are rolled into the closure.

3. Press-on closures constitute the group of the pry-off closures, bottle crowns, and over caps used on aerosol containers. Each of these types of press-on closures will have its own individual requirements, but the fabrication usually is not as severe as on the other type closures.

4. Lug caps are made by forming a cup from precoated sheets and turning under the edge of the cup to produce lugs, which serve as threads to close onto the container threads. These lugs are usually made on a very tight radius, and thus, represent a fairly severe fabrication requirement.

G. COLLAPSIBLE TUBES

Collapsible tubes are generally made by an extrusion process and any coatings are applied to the formed container. Very little fabrication occurs during the filling and capping of the tube, but usually severe deformation occurs when the consumer is using the tube. This must be taken into account when formulating coatings for this end use.

H. PANS AND TRAYS

This category includes unit containers (containers which hold a single serving), serving trays, such as heat-and-serve units, and others. Most of these containers are made from foil by a simple forming process. Cover stock may be foil or plastic film.

I. POUCHES AND BAGS

Many pouch constructions require essentially complete barriers to oxygen, water vapor, and possibly other gases. In these cases, aluminum foil is usually used as a component of the pouch, and laminating coatings, heat-seal coatings, or decorative coatings may be applied to the metal foil.

J. SQUARE CANS

Cans with square or rectangular shapes generally have no organic lining in them. Some have coatings on the outside, but such applications do not represent any problem unique to this type of container.

K. MISCELLANEOUS CONTAINERS

In addition to the containers mentioned above, there are a large variety of containers classified as general line cans. Many are used for specialty food products and for nonfood products, such as household items, automotive products, tobacco, medical supplies, etc. These, however, do not involve any unique manufacturing methods. Organic linings, if required, and exterior coatings or decoration must withstand the environment to which the coated container is subjected.

9-3. FACTORS INFLUENCING COATING SELECTION

At first glance, it is difficult to imagine why so many different coatings are used in the metal container industry. However, it must be realized that there are at least three basic factors which influence the choice of the container coating:

1. Requirements dictated by the product packed.

2. Requirements dictated by the container construction.

3. Requirements dictated by the manufacturing methods used in making or filling the container.

A. REQUIREMENTS DICTATED BY THE PRODUCT PACKED

1. Foods

Probably the major factor in considering coatings for the packaging of food is whether the coating will impart a flavor or odor to the product. A corollary of this is whether the coating might absorb a flavor or odor from the product. This consideration includes the factor of solvent removal as well as the cure or proper bake of the coating. In some cases, it has also been observed that a coating will absorb an odor from some extraneous product which it might contact during manufacture of the container. Subsequently, this extraneous odor is given up to the packaged product and makes the product unacceptable commercially. This, too, must be taken into consideration.

The status given the coating by the Food and Drug Administration and the U.S. Department of Agriculture must be established prior to use of coatings which contact foods and beverages. The USDA establishes the guidelines for the packaging of meat and poultry food products. The FDA establishes the guidelines for coatings for all other food products. A series of FDA guidelines has been published in Section F, Title 21, Code of Federal Regulations. Coatings not covered by existing regulations can be used only after a successful petition for that composition.

With one or two exceptions, food products are acid. Mildly acid products are not very difficult to package, but strongly acidic products, such as sauerkraut, rhubarb, and others, present some serious challenges. Consideration must also be given, when coating a tinplate can, to those products which contain strong detinning components, for example, spinach.

Products which contain sulfur-bearing proteins sometimes yield low-molecular-weight, sulfur-bearing compounds during processing. In contact with tin or iron, these sulfur compounds produce a black stain on the metal, which is probably a tin or iron sulfide. In cases of extreme buildup, the black deposit may even get mixed into the food product. It is essential that the coating act as a barrier to the sulfur compounds in packing products such as meats, fish, corn, beans, etc.

Most food products require a sterilization, either immediately prior to packing, or after the container has been packed and sealed. For those products which are presterilized, the container must also be presterilized. Depending on the product packed, the coating may have to withstand temperatures from 63°C for 45 min for products such as beer, to as high as 121° C for 90 min for some meat products. During the sterilization cycle, the coating should not lose adhesion to the metal or become degraded in any way.

Color stability of the packaged product is important. Ideally, there should be no absorption of color from the product into the coating, nor should there be any degradation in color of the product because of the presence of a coating. Typical products which may contribute color into the coating are tomatoes and spinach. Other products may show a color change because of a "reducing atmosphere" in the container. For example, if a slightly acidic product is packaged in an unlined or imperfectly lined tin container, a "reducing atmosphere" will prevail and the color of the product may be bleached. Unwanted bleaching may occur with dark colored products such as purple plums, etc. In the case of these dark colored products, it is necessary that the container be completely lined to prevent bleaching. On the other hand, products, such as pears or applesauce, which may show the browning effect when packed in a fully lined container, generally have part of the container unlined to produce the "reducing atmosphere," and thus, maintain the color as packaged.

Fatty products may present some problems because the fat might be a good solvent for some of the coating materials. It is essential that the coating not soften in contact with the fatty product to prevent gouging it out when removing the product from the container.

Many meat and fish products when processed in a container will adhere to the walls of the container unless there is sufficient fatty material to prevent that adhesion. In those cases where the product does not contain sufficient fat, it is necessary to add a release agent to the coating to permit easy removal of the product from the can. This is a particular requirement for many ham cans, dog food cans, and others.

Products which contain essential oils, such as spice oils, peppermint oils, etc., are sometimes difficult to hold because the essential oil appears to act as a solvent and plasticizer for the coating, and, once the oil has been absorbed into the coating, the integrity of the lining may be destroyed. One particular place in which this shows up in the exterior coating on the container is in the case of spice cans, where contact with the fumes and dusts from the spices may lead to yellow discoloration of the white coating used on the exterior.

2. Nonfoods

Nonfoods comprise the group of paints, petroleum products, chemicals, aerosol (pressure can) products, etc. No specific guidelines can be set up because these products vary so widely. For aqueous products, the pH range may go from very acid to very alkaline. In the organic products, there is a wide variety of solvents. Probably the most troublesome product is one which contains both organic and aqueous components, such as some cleaners, since the organic component may solvate the coating and the aqueous component may lead to corrosion of the container.

B. REQUIREMENTS DICTATED BY CONTAINER CONSTRUCTION

1. Metal Substrates

Metallic containers are generally fabricated from steel, chrome-coated steel, tinplate, aluminum, or alloys which are primarily aluminum (9). Actually, some cans contain as many as three metals in the various component parts.

The categories of steel substrates include the light-gauge stocks, commonly called blackplate, as well as the heavier-gauge stocks, commonly called cold-rolled steel. In this category must also be included various types of chemically pretreated steels which have on their surface a chemical compound, usually a chromium or phosphorous compound of the steel itself, or some other metallic element. Surfaces which have been chemically pretreated generally are most resistant to corrosion and generally give better organic coating adhesion. The so-called "tin-free steels," which consist mainly of a chemically treated steel, are grouped in this category.

Chrome-coated steel (10) consists of a surface coating of chromium and chromium oxides on a steel base. This type of substrate is often called "tin-free steel," but should be considered in a category all by itself, since the surface is primarily a plating of chromium, rather than a chemical compound as noted above.

Tinplate (11) consists of a coating of tin on a steel base. In order to prevent rapid oxidation of the tin, the tinplate producers deposit a controlled oxide film on the surface. The oxides and the methods of producing them have been varied over the years; the most common one today is referred to as a cathodic dichromate treatment.

Aluminum substrates (12, 13) may vary from essentially pure aluminum, such as is found in many foils, to the aluminum alloys which are used where high strength is required. Aluminum and its alloys may be used with or without chemical pretreatment, depending upon the alloy and the end use. Generally speaking, the greater the amount of the alloying element, usually magnesium, or the more critical the end use, the more necessary it is to have a chemical pretreatment on the surface.

Practically all of the metal substrates as furnished to the coater have a layer of oil or lubricant on their surface. Substrates intended for direct application of the organic coating without prior washing or pretreatment generally have a very thin layer of a lubricant, usually dioctyl sebacate, to aid in handling and prevent scratching. Surfaces which are to be pretreated immediately prior to coating may carry much more lubricant on the surface and the lubricant must be removed prior to chemical pretreatment in order to permit proper reaction of the pretreating chemicals. Also, in those cases where a unit is drawn prior to coating, it is necessary to remove the drawing lubricant prior to chemical pretreatment or the application of the organic coating.

2. Fabrication

Many containers or container components are fabricated after the metal has been coated. Typical mild fabrication constitutes the forming of container ends, side-seamed container bodies, trays, pans, etc. Intermediate fabrication is typified by closures such as crowns, lug caps, and screw caps. In this category of intermediate fabrication, the circumferential beads used in reinforcing can body side walls can be included. The ring and the plug, which constitute the top of a typical paint can, are in the category of severe fabrication. Valve cups used in pressure containers represent another severe fabrication problem. The drawn cans, such as the sardine can, the snack pack can, etc., can also be classified in this category. An equally difficult fabrication, but of an entirely different nature, is the scoring and rivet deformation that takes place during the production of most easy-open can ends. All of the above fabrication operations are possible using precoated stock.

Operations such as the drawing and ironing of metal containers wherein the side wall of the can may end up only one-half the thickness of the starting metal, are almost impossible to make from precoated flat sheet. Such containers are usually fabricated in the presence of large amounts of lubricants. The lubricants are subsequently removed and the fabricated container is then coated.

3. Gaskets

In the construction of a can, a gasketing material is usually used on the can end to help seal the joint between the can end and the can body. The gaskets commonly used do not present any particular problem to the supplier of the organic coating.

Gaskets for closures may be preformed from sheets of cork, rubber, polyethylene, or other plastic polymers. These preformed gaskets are inserted into a closure with an adhesive or are simply crimped in. Egg albumin has been the classic adhesive for cork gaskets for many years, but the plastic-type gaskets require other adhesives. The only requirement of the organic linings when preformed gaskets are used is that the adhesive adhere to it without deteriorating the lining.

Gaskets for closures may also be formed in place by extrusion of a ring or strip of gasket compound onto the area requiring a gasket. This type of gasket compound is generally a plastisol (dispersion of a resin in a nonvolatile solvent) or a rubber latex. The gasket is fused by baking or drying. The selection of the organic coating is very critical because the post-baking operation may cause attack and deterioration of the basecoat by the nonvolatile dispersant in the liquid gasket or the heat of post-baking. In most cases, the formed-in-place gasket must adhere strongly to the coating in order to prevent seepage of the product out of the container.

C. REQUIREMENTS DICTATED BY MANUFACTURING METHODS USED IN MAKING OR FILLING THE CONTAINER

1. Coating Application Techniques

Sheet stock is generally coated with a direct roll coater and the coated sheets are passed through an oven on wickets to dry or cure the coating. The wicket construction is such that the backside of the sheet rests against a steel support, which consists of metal tubing or rod bent into the shape of a "W." This method of conveying metal requires that any coating in contact with the wicket be resistant to softening under heat when travelling through the oven.

Coil stock is generally coated with a reverse roll coater or with a gravure roll. Low-strength foils must be supported on idlers through the oven, and consequently, only one side of the stock can be coated prior to entry into the curing unit. Higher-strength webs can be coated on both sides with the web hanging, unsupported, in the shape of a catenary in the curing oven.

Prefabricated containers for which coating prior to fabrication is not practical have the linings spray applied to the interior. Cylindrical containers are held in a rotating chuck while spraying to ensure complete coverage. Coating the exterior of a prefabricated container is usually done by direct roller coating of the container held on a spindle. Exterior roll coating is most general for small containers, but spray coating may be employed for large containers or for small containers where full coats are required.

2. Drying and Curing

Since application of coatings to containers or container stock is a high-speed operation, the coating must be dried or cured in a relatively short time, usually less than 10 minutes. Baking conditions will vary with the substrate and the coating. Those coatings which dry by simple solvent evaporation need relatively low temperatures, while those which chemically cure on the metal often require metal surface temperatures as high as 260° F.

The types of heating are hot air, induction heat, and radiant heat. The former is by far the most generally used because of its low cost. Induction heating has limited use because only steel substrates can be heated economically by this method, and most coating and curing installations must be capable of handling both steel and aluminum. Radiant heat with infrared sources offer some possibilities where radiation adsorption is uniform; color affects cure rate. Ultraviolet and electron radiation cure are currently under extensive investigation. This latter type of cure has great possibilities because only the coating itself is directly affected, rather than having heat or other energy transferred to the metal substrate. Some of the current disadvantages are the cost of the equipment and the limitation of coating composition to those that are radiation curable.

Since most coatings contain an appreciable amount of volatile ingredients, the problem of effluent control from the coater or the drying unit is extremely important. Not only is there solvent evaporated from most coatings, but many of them also show a small weight loss, especially those which cure by condensation with evolution of low-molecular-weight condensation products. Depending upon the composition of the effluent, it may be an explosion hazard, a health hazard, or simply a nuisance because of odors. Combustion techniques are the only ones fully accepted to handle the volatiles from the wide variety of coatings and solvents used in the container industry. The fact that a combustion unit requires additional fuel above that supplied by the effluent itself, makes this method of effluent control unattractive, and much effort is being expended on making organic coatings with reduced emissions.

3. Mobility and Tack Resistance

In the application of any coating, the time between application and the time the coated pieces can be stacked, coiled, or moved in a conveyor system is of prime importance. In those cases where flat stock is coated and stacked or coiled, the coating must not remain soft or tacky at the stacking temperature, which may run as high as 80°C. The same kind of requirements dictate that individual units be tack free so that there is no pick-off of coating on a conveyor unit.

The actual handling of the container during the filling and processing operations may require special mobility characteristics. In the case of closures, many times the insertion of the gasket requires a heating operation and the coating on the exterior must not pick off during this heat cycle. Also, many closing systems use a steam flush, which warms the conveyor system and the closures, and requires that the warm closures move freely during this operation.

The production of spiral-wound composite containers results in the winding mandrel becoming hot because of friction of the wraps on the steel mandrel. The slip coating used on the innermost layer of the composite container must have good lubrication properties at slightly elevated temperatures.

Stamping or forming operations usually require a highly lubricated sheet. In most cases, an extra or external lubricant is applied to the coated stock immediately prior to stamping, but even then, incorporation of a small amount of lubricant into the coating aids in fabrication properties. This is called an internal lubricant and may be selected from a large variety of materials such as waxes, oils, or other organic compounds which provide the requirement for control of friction.

The important consideration in the use of lubricant is that it serves the intended purpose without affecting the integrity of the coating.

9-4. TYPES OF ORGANIC COATINGS

A. OLEORESINS

Oleoresins are the oldest type of coating used in packaging. Their use was a natural extension of the varnishes made for many other protective coating applications.

Oleoresins are, as the name implies, combinations of an oil, or an oil derivative, and a resin. Depending upon the drying conditions, the list of oils ranges all the way from drying to nondrying oils, but normally the drying oils, such as tung, linseed, and dehydrated castor, are most widely used. The composition of the resin component has also varied widely over the years and may consist of naturally occurring resins, polymerization products of unsaturated hydrocarbon materials, phenolic resins, or combinations of any or all of these. As would be suspected, the more oil that is used, the more flexible the coating, but the slower the drying period. Compositions of approximately 50% drying oil and 50% resin have the required drying speed, adhesion, flexibility, and product resistance. Metallic driers can be used to speed up cure.

The chief use of oleoresins is for the manufacture of fruit and vegetable cans. These coatings constitute most of the typically gold colored coatings seen on such cans. When used without pigmentation, they are commonly referred to as "R" (regular) or "F" (fruit) enamels. When used with zinc oxide at levels of 10 to 20% zinc oxide and 90 to 80% oleoresin, the coating is called a "C" (corn) enamel. The zinc oxide is used to react with low-molecular-weight sulfur compounds produced by foods which contain sulfur proteins. The reaction product of zinc oxide and the low-molecular-weight sulfur compound produces a light colored zinc compound, probably zinc sulfide, which is essentially the same color as the zinc oxide, and thus, does not change the appearance of the coating during the processing or storage life of the container. If no zinc oxide is used in the coating, the low-molecular-weight sulfur compounds migrate through the film and develop an undesirable black tin or iron sulfide on the metal surface.

One other major use of oleoresins is as a basecoat for cans which require a two-coat system. Many beverage and fruit cans require such a two-coat system and oleoresinous coatings provide a satisfactory basecoat for the vinyl topcoats.

Oleoresins are also the major type of coating used on the interior of crowns which use a cork gasket.

Oleoresins also constitute the coating most used on many miscellaneous type containers where mild corrosion resistance is required, e.g., petroleum products, dry products, etc. For these types of mildly corrosive or noncorrosive products, steel or backplate is often used. Normal atmospheric storage prior to or after filling the container would result in unsightly corrosion of an uncoated steel surface. Corrosion is minimized by the application of a thin oleoresinous coating as a rust-preventative to both sides of the metal.

The major advantages of oleoresins are their low cost and general applicability for food containers. They are low cost because of the availability of low-cost oils and resins and their ability to be used at relatively high solids (40 to 50%) in low-cost solvents, such as aliphatic or aromatic hydrocarbons.

The major disadvantages of oleoresins are their limited flexibility and their failure to withstand many products. Their limited flexibility becomes a factor during the fabrication of the container parts. This becomes especially critical depending upon the exact composition, cure, and age of the coated metal. The ability to make a coating of this kind, which cures to a certain point and then stops, is difficult, since most of them are sensitive to aging in the presence of atmospheric oxygen. The failure of oleoresins to hold some products is a function of the fact that they are made from oils, and thus, generally contain a significant number of ester linkages. Products which might tend to hydrolyze ester linkages will lead to disintegration of the film. The fact that most fruits and vegetables are only slightly acid permits the use of oleoresins in containers to hold these products. Products which are very sensitive to oily flavors cannot be packaged in contact with oleoresins.

One other limit to oleoresins is their relatively slow cure. Applications which permit a cure of 6 min or longer present no problem. However, high-speed cures of less than 30 sec are difficult to attain because the cure of the coating involves some oxidation which, for these coatings, is relatively slow. Incorporation of activators or oxidants to permit high-speed curing has not reached practicality.

B. HYDROCARBONS

Coatings in which hydrocarbon resins constitute the major portion of the film are limited in number. The most important group of such coatings constitutes the polybutadiene resins. By controlled polymerization techniques, the proper molecular configuration can be obtained. Subsequent oxidation or modifications made during the polymerization can be used to introduce oxygenated polar groups. Only by the development of polar groups during the resin preparation, or in the baking of the coating on the metal, can one obtain adequate adhesion to metallic substrates.

The main advantage of polybutadiene resin coatings is that they are generally relatively low cost and can compete with oleoresinous coatings on a cost/performance basis. The advantage over the oleoresinous coatings is that they are produced from basic organic chemicals and are less subject to variations in price and quality, such as may be the case with oleoresinous coatings made from agricultural products.

The chief disadvantage of polybutadiene resin coatings is their relatively poor flexibility. Modifications to improve flexibility can be made, but such modifications generally result in the loss of the chemical resistance, which is due to the nature of the hydrocarbon chain, or in the loss of the attractive economics.

The main use of polybutadiene resins in the metal container field is as a base coat for the interior of beer and beverage containers. They have satisfactory flavor properties for this application, and since they are essentially always topcoated, the poor fabrication noted above is not a serious deterrent to their use.

Hydrocarbon resins other than polybutadienes are used chiefly as modifiers of other resins; their most common use is in oleoresinous coatings.

C. ALKYDS AND POLYESTERS

In the container field as well as in many other end use applications, phthalic alkyds constitute a large portion of the coatings used. Typical phthalic alkyds have been used for years as coatings for the exterior of a wide variety of containers. They can be made to yield high-gloss varnishes which dry rapidly. Their use as food container linings has been limited because a trace of phthalic acid, or one of its low-molecular-weight derivatives, will impart an odor or flavor to many products.

Because the common alkyd made from orthophthalic acid, a polyol, and an oil or oil derivative has limitations due to the phthalic acid and the oil used, many efforts have been made to develop more chemically resistant polyesters. The so-called oil-free alkyds and the use of acids other than orthophthalic has prompted the industry to use the terminology "polyester" to cover the entire group of coatings.

By proper selection of raw materials and by varying the reaction conditions, a wide variety of polyesters can be produced. One of the main uses for these newer products is in those instances where better color or color retention is needed in a coating than is obtainable with a composition containing a drying oil. One limitation of these compositions is that either the molecular weight of the basic resin must be relatively high, or a curing additive must be used to produce a satisfactory coating. If the molecular weight of the resin is made high initially, then poor solvency, even in strong, expensive solvents, results. If a reactive ingredient is used with a polyester, then solution stability or actual extent of cure on the metal surface are major concerns.

The principal uses for alkyds are for color coats and varnishes. In this use, the alkyds are often modified with 10 to 20% of a nitrogen-containing resin, such as a urea/formaldehyde, to permit lower bake temperatures or to yield a less yellowing film. For example, some alkyds use semidrying or nondrying oils or fatty acids derived from them to ensure better color of the base resin. Such resins are usually slow in curing, and by blending with a urea/formaldehyde resin, curing can be accelerated.

In the preparation of varnishes, particular attention must be paid to the inking qualities, lack of color, speed of dry, and fabrication. The inking properties are of particular interest because many times it is desired to varnish an ink without having previously dried the ink. The ability to "wet ink" is not as much a function of the resinous ingredients as it is the composition of the ink and the solvents used in the varnish. Occasionally, it is necessary to add agents such as low-molecular-weight silicones to prevent bleeding and striation of the ink.

Many other modifications of alkyds are possible. Reaction of the alkyd with polymerizable monomers is a common method of producing a harder or a faster curing resin. Styrene, acrylic, and other similar monomers may be used. Styrenated alkyds produce fast-drying, high-gloss, hard finishes particularly useful on prefabricated articles, such as drums.

Another common modifier for alkyds for metal containers is the series of vinyl copolymer resins containing free hydroxyl groups. These are used at 10 to 20% levels to impart better flexibility to an alkyd than can be obtained by selection of the acid or alcohol. Flexibility is obtained at the sacrifice of some economy, since use of a vinyl resin usually lowers the solvency of the coating.

An increased use of oil-free polyesters is appearing in those applications where high-heat resistance is required. Particular problems on fabrication and on applied cost are being encountered, but progress is being made.

D. PHENOLICS

The original use of phenolic resins, the reaction product of a "phenol" and formaldehyde, was as modifiers in oleoresinous coatings, and as such they had to be made to be compatible with drying oils. In order to make phenolic resins compatible with oils, it was essential to have solubilizing groups in the resin and to have a minimum number of free hydroxy methyl groups remaining from the reaction between the phenolic body and formaldehyde. A number of such resins are still made and widely used and are the ones referred to in the section on oleoresins.

The phenolics that are oil compatible do not generally self-condense, and consequently, are not considered as film formers by themselves. However, if a phenolic resin produced from a phenol and a formaldehyde has a significant number of hydroxy methyl groups on the polymer, then these will co-react and form a film. Such types of resins are generally made by alkaline catalysis.

Phenolic resins that are film formers can be used by themselves or as modifiers for vinyl resins, epoxy resins, and others. Their use as modifiers is discussed under the other resinous subheadings. When used by themselves, they sometimes give wetting problems on metallic surfaces, and it is common practice to add 5 to 10% of a high-molecular-weight compatible resin to give more viscosity to the solution and permit good plate wetting. One common resin used with phenolics is polyvinyl butyral; in addition to improving wetting, this resin will increase the flexibility slightly.

The main use for phenolic coatings is in those applications requiring excellent corrosion resistance or resistance to sulfur-staining products. This makes these coatings ideal for both the interior and exterior of containers such as fish cans. They may be used unpigmented or with aluminum pigment, which helps to prevent or mask sulfide staining.

The probable reason for the effectiveness of phenolic coatings in providing stain and corrosion resistance is the capability of the phenolic resin to react with the metal substrate and form with it an organometallo compound with excellent adhesion.

One other property of a phenolic coating is that when completely cured, a film resistant to a wide variety of organic chemicals is produced. This makes phenolic coatings of particular use in the pail and drum industry, where the coating must resist any solvent action of the product being packed and must prevent the product from being contaminated by any rust from the container. In these applications, the coating can be used unpigmented, but it is more often pigmented with an iron oxide or with a titanium dioxide to give the typical red or buff colored coatings.

Phenolic coatings are also used on other containers that require high product resistance, such as collapsible tubes, some aerosol cans, latex paint cans, and others.

The one disadvantage of phenolic coatings is their extreme brittleness. When cured to their maximum extent, at which point they have their best solvent resistance, they are considered to be completely crosslinked, and thus, have very little flexibility. If used in very thin films (0.5 to 2.0 mg/in.^{2}), they can be coated on tinplate or steel substrates and have passable fabrication on fish cans, container ends, and others. When used in heavier coats, they will generally crack off, even during mild fabrication. One reason why they can be used in many pail and drum applications at relatively heavier film weights (20 to 40 mg/in.^{2}) is that pails and drums are not subject to as much deformation during their production or use, as in the case with a can or a closure. However, there is a definite failure in most phenolic coatings if a coated drum is dented severely.

One other characteristic of phenolic resins is that they are generally gold in color; thus, production of a colorless coating is impossible. Efforts to make a colorless phenolic and to improve the fabrication of phenolics have had only limited success. To make a phenolic resin which does not discolor during crosslinking, the phenolic group must be blocked. If one blocks the phenolic group, the reactivity of the resin and the corrosion resistance of the resin is generally reduced. Blocked phenolics are on the market. These, however, are usually used as modifiers for other resins, typically epoxies. One advantage they have over a nonblocked phenolic is that in addition to better color, they generally have better resistance to alkalis.

No major breakthrough in making a flexible phenolic with all of the other phenolic properties has been made. Generally, by the time one introduces other groups into the molecule to impart flexibility, the typical phenolic resin properties of solvent resistance and corrosion resistance are lost. It is generally easier to impart some flexibility to a phenolic by formulation with an epoxy or with a compatible vinyl polymer.

E. NITROGEN RESINS

Although nitrogen resins never constitute the predominant proportion of a coating for metal containers, they are modifiers for so many other resins used in this field that they require special mention.

The nitrogen resins comprising the group prepared from urea, melamine, and benzoguanamine reacted with formaldehyde constitute the major nitrogen resins used in coatings. To make them compatible with other resins, it is customary to run the amine-formaldehyde reaction in the presence of an alcohol, such as butanol, to introduce sufficient hydrocarbon groups to make them soluble in common organic solvents. During reaction of the nitrogen resin with the other components of the coating, the alcohol is removed along with the other solvents.

The properties which make urea, melamine, and benzoguanamine resins attractive are that they have a wide range of compatibility, are colorless, and impart increased drying speed to most coatings. In this respect, they are most useful in a slightly acidic system, such as produced by an alkyd, since the crosslinking reaction between the nitrogen resin and the other resins proceeds most rapidly under acidic conditions. One other property which is of significance is that these resins impart hardness and toughness to a coating, thus making alkyds modified with these resins quite useful as exterior coatings for a wide variety of metal containers. The fact that they are colorless precludes little, if any, effect on the dye or pigment used in preparing colored exterior coatings. The only possible disadvantage of these coatings is their relatively high cost and their tendency to embrittle the coating if too high a percentage is used as a modifier.

Polyamides, as a group, are generally too insoluble to be of any significance as container coatings. By the time the structure or the molecular weight of the resin is established to have good film properties, their solubility is quite low.

Compositions in which acrylamide is copolymerized with other monomers are generally considered in the class of acrylic resins (q.v.).

Resins containing free amine groups are generally used only as reactive constituents to cure epoxy resins and similar materials.

F. ACRYLICS

Acrylic resins encompass the polymers made from acrylate or methacrylate acids or esters, acrylamide, and most often a wide variety of other copolymerizable monomers. By varying the actual composition and the molecular weight of the resin, a wide variety of products can be obtained. By including monomers with reactive pendent groups, such as carboxyl, amide, hydroxyl, and glycidyl, one can obtain polymers which can be formulated to crosslink during application and bake.

The outstanding property of acrylic resins is the fact that most of them are colorless and will remain colorless after aging. Their lack of color makes them most useful in the formulation of coatings that must withstand high temperatures because of subsequent bakes or processes to which the container is subjected. Examples of such containers are meat cans and aseptically processed cans.

One other property which makes acrylic resins of interest is their ability to adhere vinyl plastisols and to not show any deterioration of that bond with age. The use of acrylics, usually blended with other resins such as vinyls, epoxies, and others, as basecoats for linings for closures with plastisol gaskets is of interest.

The major advantages of acrylic resins have been emphasized above. From the good points these resins have and the fact that they can be polymerized under fairly well controlled conditions, one may wonder why they are not universally used. There are several reasons why the use of acrylic resins is limited, namely, odor, fabrication, and corrosion resistance.

Most acrylic resins contain traces of monomers, or the polymer may break down on subsequent heating to regenerate monomer. The odor threshhold level of most acrylic monomers is very, very low, and the odor is objectionable to most people. Consequently, workers applying the coatings prefer not to use them. Furthermore, if there is any breakdown of the polymer in the coating, monomeric constituents may get into the product pack and have a deleterious effect.

The property of fabrication must be considered along with the property of toughness or hardness. Most acrylics, when they are made flexible enough to withstand a fabricating operation, are too soft to be handled on commercial equipment. Furthermore, when coated metal emerges from the oven, it must have sufficient tack resistance so that metal does not stick to the adjacent sheet.

Acrylics are not particularly noted for their corrosion resistance. This may be due to the fact that most of them are made from esters, and consequently, they are probably a little more porous to aqueous and similar corrosion media. The corrosion resistance of the film can be increased by adding other resins, but usually a gold colored film results and one loses the advantage of the colorless acrylics.

G. EPOXIES

Epoxies are used in two different ways in container coatings. Epoxy esters tend to have properties similar to those of oleoresins and alkyds, and consequently, they can be considered in that manner. Epoxies formulated with nitrogen resins, such as urea/formaldehyde, phenolic resins, and amine resins, have considerably different properties than the epoxy esters.

1. Epoxy Esters

In making an epoxy ester, an epoxy resin is used as the polyol and is reacted with oils or acids in a manner similar to that by which an alkyd is made. Compositions generally fall in the range of 60 to 40% epoxy and 40 to 60% acid. An epoxy ester varnish can be made by reacting: 60 parts epoxy resin, 1000 epoxide equivalent, and 40 parts dehydrated castor oil fatty acids. Depending on the exact composition, the drying rate can be accelerated with metallic driers or with nitrogen resins such as urea/formaldehydes. Solvency of the esters is such that the major portion of the solvent can be an aromatic hydrocarbon.

Two properties of epoxy esters that give them an advantage over oleoresins or alkyds are: (a) a high order of toughness and mar resistance: and (b) a higher order of resistance to hydrolysis by mild acids or alkalis. The toughness and mar-resistance qualities probably result from the condensed aromatic structures in the epoxy resins. The increased resistance to hydrolysis is probably the result of the fact that the ester groups are spaced along the chain in such a way that it makes them less available to attack.

The disadvantages of epoxy esters are mainly associated with their failure to remain completely colorless. Color is imparted not only by the fatty acids when they are put into the composition, but also by the inherent nature of the epoxy resin, which has a tendency to discolor slightly on high bakes or upon aging.

Epoxy esters have found use as varnishes for crowns, other types of closures, and can ends. They are generally used where an alkyd or an oleoresin does not perform quite well enough. The advantage of using them as a crown varnish is that most beer bottles using crowns must pass through a pasteurizer after being filled. Most pasteurizing solutions contain a quaternary salt, which acts both as a bacteriacide and a mild detergent. Epoxy esters are resistant to the action of the pasteurizing bath, while most of the common alkyd varnishes will disintegrate or lose adhesion.

2. Thermosetting Epoxies

Epoxy resins, because of their epoxide and hydroxyl groups, will react with nitrogen resins, such as urea/formaldehydes and melamine formaldehydes, with polyamines and with phenolic resins made by the condensation of a phenolic body with formaldehyde. The reaction of the epoxy resin with an amine will occur at room temperature or slightly elevated temperatures, dependent on the exact composition of the amine. When the epoxy is reacted with a resin containing hydroxy methyl groups, such as the urea/formaldehydes and phenolics, then the reaction is usually considered to proceed through the hydroxy methyl group, and temperatures in the area of 150°C and higher are needed to carry the condensation far enough to give a usable film.

When formulating with nitrogen resins containing hydroxy methyl groups, the amount of nitrogen resin may vary up to 40% of the composition. Properties will vary depending upon the reactivity of the nitrogen-containing resin and the resin ratio; higher contents of nitrogen resin produce more brittle films. Most of these compositions are essentially colorless and can be formulated into a wide variety of colored coatings with good color control. The coatings are hard and tough and have very good resistance to alkaline materials.

Epoxy resins formulated with phenolic resins can vary in composition through the entire range if the phenolic is also a film former by itself. If the phenolic is not a film former, then compositions usually run in the same range as with the nitrogen resins (10 to 40%). When using typical phenolic resins, one obtains gold colored films; if blocked phenolics are used, the films are generally much less colored and are often almost colorless. Unblocked phenolics generally impart superior corrosion resistance to the composition. Blocked phenolics do not have quite the corrosion resistance, but have less color and better resistance to mild alkalis.

Epoxy resins usually react so fast with amines that it is necessary to mix the two components immediately prior to use. This is particularly true of the low-molecular-weight amines such as tetraethylene pentamine. Some of the higher-molecular-weight amines, which may be polyamides with free amine groups left on the molecule, impart shelf stability. The advantage of epoxy amine compositions is their low temperature cure. The disadvantage of this type of composition is that many amines are difficult to handle from a safety standpoint.

Epoxy/nitrogen resin compositions find widespread use where a tough, adherent, colorless coating of good product resistance is needed. They can be used as size coats at light coating weights, as color coats, and as varnishes. They can be used on paint cans and aerosol cans where alkaline products may be encountered. They can be used on both tinplate and aluminum with good success as basecoats or single coats in contact with products which are very flavor sensitive, such as beer.

Epoxy phenolics differ from the epoxy/nitrogen resin compositions discussed above in only three characteristics: (a) they are generally not as colorless; (b) they are generally not as resistant to alkalis; and (c) they have superior corrosion resistance. Taking these three differences into consideration, they may have many uses such as mentioned for the epoxy/nitrogen resin compositions. The epoxy phenolics do find more widespread use where corrosion resistance is a factor. Thus, they make excellent linings for pails and drums in that they have good product resistance, but are more flexible than the unmodified phenolics. Also, in the field of sanitary cans, they are used extensively for the highly acidic products which cannot be contained in a can lined with an oleoresinous composition.

Epoxy amine compositions have very limited use in the container field. Mention is made of them because various possibilities exist depending upon the actual amine or amine resin blended with the epoxy.

H. VINYLS

1. Introduction

In the field of organic coatings, vinyl resins generally comprise the group of polymers and copolymers made from vinyl chloride, vinyl acetate, and vinylidene chloride. Many other unsaturated compounds containing the CH_2=CHX group are technically vinyl compounds, but traditionally the term of vinyl has come to be defined as noted above.

Vinyl chloride or vinylidene chloride homopolymers are generally materials which have very low solubility in organic solvents. Copolymers of vinyl chloride, vinylidene chloride, vinyl acetate, or others, generally have much higher solubility and can be used in solution. The increased solubility is probably due to a reduced tendency of the resin to exist in a crystalline form. One of the most common vinyl copolymers is that of vinyl chloride and vinyl acetate, although many other monomers have been copolymerized into this type of resin.

2. Thermoplastic Vinyls

Polyvinyl acetate and its derivatives constitute one class of thermoplastic vinyl resins.

Polyvinyl acetate resins in a wide variety of molecular weights are available. As coatings for metals, they have limited use; their derivatives, such as polyvinyl alcohol and polyvinyl acetal resins, are much more useful. Most polyvinyl alcohols are actually water soluble, and thus, have applications limited only to water-based systems.

Partial etherification of polyvinyl alcohols with aldehydes leads to the acetal resins known as the formal, acetal, and butyral polymers. The acetal resins are much less water sensitive than either the polyvinyl acetates or alcohols.

Most resins based on acetal derivatives of polyvinyl acetate are colorless and will remain colorless even under relatively high drying temperatures. They represent an excellent base resin for corrosion protection. Probably the most widespread use of the acetal series of coatings for metallic containers depends on the compatibility of the resins with phenolic resins. In such compositions they impart better plate wetting and better flexibility to the phenolic coating.

Polymers of vinyl chloride, vinyl acetate, and/or vinylidene chloride constitute the major group of thermoplastic vinyl resins. Since resins made from the three main constituents have relatively poor adhesion to substrates, it is necessary to use other copolymerizing monomers. Thus, small amounts of maleic acid, or the half-ester of maleic acid, acrylic acids, and others, introduce sufficient polarity to give some degree of adhesion to metallic substrates. Adhesion to oleoresinous coatings, alkyds, and others, is developed by the introduction of hydroxyl groups. The most common method for introducing hydroxyl groups is to make a polymer that includes vinyl acetate and to hydrolyze part of the acetate linkages.

The properties of the thermoplastic series of vinyl polymers that make them attractive are their high degree of flexibility and the fact that most of them are flavor and odor free. Many of the other properties that are needed in coatings can be imparted by proper selection of the modifying resin, and many compositions exist in which the vinyl resin is modified with phenolics, ureas, alkyds, epoxies, or acrylics. In formulating with other resins, care must be taken that the flavor properties and the flexibility properties are not sacrificed by inclusion of too much modifier.

The disadvantages of this class of resin is that they require active solvents for solution and yield workable solutions at relatively low per cent solids. In many solutions, the cost of the solvent is equivalent to the cost of the resin in the mixture. One other property which limits the use of vinyl chloride and vinylidene chloride copolymers is that when they are subjected to high temperatures, or lower temperatures for prolonged periods, they will decompose to form traces of corrosive chlorine compounds, probably HCl, and the resin film will turn yellow or brown. This decomposition is catalyzed by metals such as iron and, consequently, care must be exercised when baking these coatings on ferrous-based substrates. One other property which limits the use of the resin is that by definition they are thermoplastic and, as such, cannot be used at temperatures at which the resin would flow or in situations in which the coating is subjected to water at temperatures above approximately 100°C.

Thermoplastic vinyl coatings are used where the flavor properties and the flexibility properties are needed for the particular container. One of the largest uses is as the topcoat or beverage contact surface in beverage containers. Closures which require a flexible, product-resistant lining can be successfully formulated from this type of coating. The good flexibility of the coatings makes them ideal for fabricated containers or those which require resistance to deformation during use, such as closures, and collapsible tubes. The fact that these coatings are thermoplastic makes them particularly useful as heat-seal coatings, particularly for foil packages.

3. Thermosetting Vinyls

Some vinyls in commercial production contain carboxyl, hydroxyl, or epoxy groups along the polymer chain. These groups are reactive with selective modifiers during baking of the coating. The most common modifiers are epoxy resins and urea/formaldehyde resins, since they react at normal baking temperatures. A typical basic formula is:

70% vinyl copolymer containing hydroxyl groups;

20% low-molecular-weight epoxy resin; and

10% urea/formaldehyde resin.

In none of these instances is there infinite baking latitude. The reactions normally require a minimum of 120°C. The maximum is generally limited because at temperatures much above 205°C, degradation of the vinyl polymer occurs.

The flavor properties and the flexibility characteristics of the thermoplastic vinyls are carried over to the thermosetting vinyls, but with certain reservations. The flavor and odor properties will depend upon the modifier and the level of modifier needed to produce the crosslinking. Flexibility, per se, is usually reduced because the polymer is partially crosslinked, but this can usually be balanced by addition of a plasticizer. Many esters and polyesters are available as plasticizers.

The greatest advantage that the thermosetting vinyls have over the thermoplastic vinyls is that they are more resistant to softening under heat. This permits their use in applications where an applied film must contact a hot surface without softening or sticking. Such applications are encountered many times when using sheet ovens in which the coating on the backside of the sheet contacts heated wicket supports as it travels through the oven. The fact that thermosetting vinyls do not soften under heat makes them particularly useful for containers requiring a high-temperature (120°C) sterilization procedure. When subjected to temperatures of this order of magnitude for periods of as high as 90 minutes, they do not soften, mar, or discolor due to water absorption.

Typical end uses for thermosetting vinyls are in the closure field, particularly as the coating for the exterior. Production of many colors is possible with the thermosetting vinyls, and the exterior coatings can be applied prior to the interior linings, which often require a high bake in order to attain maximum product resistance.

4. Dispersion Vinyls

Dispersion vinyls include those vinyls normally used to formulate organosols and plastisols. They are generally homopolymers of vinyl chloride and are essentially insoluble in common organic solvents at room temperature. Their particular advantage is that in the presence of high-boiling organic compounds they can fuse with the added compounds to produce useful coatings. The fusing agents most commonly used are monomeric or polymeric ester plasticizers, but high boiling solvents and some resins also act as fluxing agents. Variations in fusing temperatures can be obtained depending upon the composition of the vinyl resin and the fluxing agents, but most compositions require a minimum of 177°C for satisfactory fusion.

Tough films can be produced by a composition such as:

60% vinyl dispersion resin;

20% soluble modifying resin; and

20% plasticizer.

The advantages of dispersion vinyl coatings are that they can be applied in heavy films, when required, because they usually require less volatile material to obtain a workable fluid than does a solution vinyl coating. They can be formulated to have good flexibility, good adhesion to metal, and good product resistance, including resistance to high-temperature sterilization. Probably their most outstanding property is their extreme toughness. This property permits their use where rough handling of the finished product is a requirement. The coating may mark slightly when rubbed against metal, but it does not usually tear, nor is it harmed in any other way, even though the rub mark may show.

The disadvantages of vinyl dispersions are mostly associated with the fact that they are dispersions containing discrete solid particles. Care must be exercised in formulating with dispersion resins so that solvation of the resin does not occur during manufacture or storage of the liquid formulation, but will give complete fusion during the baking operation. A dispersion of solid particles in a liquid can also lead to viscosity problems not normally encountered with solution vinyls. In the production of the liquid formulation, it is necessary to partially solvate the resin particle so that it remains in suspension between manufacturing and application.

One other characteristic of dispersion vinyl resins is that, in their production, wetting agents and suspending agents are often needed. Frequently these agents remain on the surface of the resin particle. Most of these suspending agents are water sensitive, and if the total formulation is too water sensitive, the coating will not resist high-temperature water sterilization process. Another phenomenon often associated with the dispersion vinyls is that complete fusion during baking of the coating does not occur. This leads to a grainy looking film. Often the film has a slight haze. The combination of a large amount of wetting or suspending agent and incomplete fusion makes it difficult in most cases to produce completely transparent films.

Dispersion vinyls find their main use where excellent product resistance coupled with excellent flexibility is required. This makes them particularly useful for closures, drawn cans, and can ends. These coatings are widely used as basecoats for closures that contain fused-in-place plastisol gaskets. It is generally necessary to have good adhesion of the gasket to the basecoat, and dispersion vinyls generally provide good adhesion. Some aerosol can linings are made from vinyl dispersions. Another example of a system requiring excellent product resistance is the dry cell battery shell. The interior coating on many shells is made from a vinyl dispersion.

I. CELLULOSE DERIVATIVES

Cellulose derivatives constitute the group of cellulose esters and ethers. The chief resins used are nitrocellulose, cellulose acetate, proprionate, butyrate, or combinations of these, and the ethyl celluloses.

The property of the cellulose derivatives that makes them the most useful is the property of imparting good gloss to a coating. Nitrocellulose is the only resin of this group that is used to any extent as the major resinous component of a coating. The other esters and ethers are used as additives in coatings such as alkyds, etc.

To improve solubility, impart specific film properties, or lower the cost of the formulation, blends of nitrocellulose with other resins are normally used. The most common modifiers are alkyd resins; often, up to 50% of the resin composition will consist of the modifiers.

The nitrocellulose coatings find their major use as decorative coatings for aluminum foil. A wide variety of colors can be produced. They have good gloss and can be applied on high-speed equipment with relatively short drying times and have good resistance, depending upon the colorant, to light exposure.

J. MISCELLANEOUS RESINS

1. Natural Resins

Naturally occurring resins, such as the gums, asphalts, rosin, and others, are seldom, if ever, used as the main constituent of a coating for metallic substrates. They are used as components of oleoresinous coatings to impart hardness and increase drying speed. Some of the dark colored natural resins are used to impart a gold color to coatings.

2. Silicones

Silicones do not constitute the main resin in metal container coatings. However, some low-molecular-weight silicones find use as additives to impart selective wetting, slip, or lubrication. Care must be exercised when using silicones, since they generally migrate to the surface of the coating and can transfer readily if coated metal is stacked or coiled.

9-5. NONRESINOUS CONSTITUENTS OF COATINGS

A. SOLVENTS

Practically all volatile organic compounds, with the exception of chlorinated compounds, can be used as coating solvents. The most common solvents are the hydrocarbons, aliphatic, naphthenic and aromatic, alcohols, esters, and ketones. Most of these solvents are produced in high purity, which gives the coating formulator the ability to produce a constant solution viscosity independent of the batch of solvent used. The common solvents cover a wide evaporation rate, which permits selection depending upon the method of application, be it spray coat, roll coat, or other means. The particular reason for not using chlorinated compounds is that most container coatings are baked, and in most cases, the solvent can contact the heating element. Under these conditions, there is always a possibility that the volatile chlorine compound will decompose to form a volatile toxic material. They are therefore not used in this type of application.

Since a large quantity of solvent is evaporated into the atmosphere from facilities applying coatings for containers, there is concern about the pollution aspects of these solvents. Hazard, per se, and odor are not generally factors, because the solvents are so diluted with air by the time they are exhausted they do not present such problems. However, it has been established that certain organic compounds are photochemically reactive and can produce eye irritants in the atmosphere. The most common photochemically reactive solvents are the unsaturated, or branch chain, ketones and the higher aromatic hydrocarbon homologs. Unsaturated hydrocarbons are also susceptible to photoreaction. Because of this solvent reactivity, several geographic regions have established limitations on the use of certain solvents. These limitations will probably become standard nationwide in the United States. It is doubtful that the metal container industry can ever become completely free of control regulations concerning solvents, so considerable effort is being spent in solvent-removal studies, such as incineration, and in the formulation of coatings which contain only minor amounts of solvents or other volatile ingredients so that the exhaust cleanup procedures are minimized.

The obvious answer to the solvent problem is to utilize water as the carrier for the coating. Progress is being made, but the physical characteristics of water make coatings dispersed in it difficult to handle. Resins that are soluble in water are often difficult to insolubilize after they are applied to the metal. Since most metal container coatings require some water resistance, this makes formulation of the coating ingredients difficult. If the resinous ingredients are simply dispersed or suspended in water rather than dissolved, this leads to many problems involved in application and fusion of the resinous ingredients. The use of water also limits one in adjusting the evaporation rate of the solvent, much as would be the case if only one organic solvent were available. Also, considerably more heat is required to evaporate a unit of water compared with an equivalent unit of organic solvent because of the difference in the heats of vaporization.

B. PIGMENTS AND DYES

Many metal containers use colored coatings on the exterior. The exact colorant that is used depends upon the coating vehicle and the end use for the container. No specific guidelines can be given, except to note that it is important to remember that most food container components get relatively high bakes during curing of the coating, and many containers are subject to water immersion during sterilization of the packed product. It is also important that there be no change in color because of the actual application or handling of the container. In addition, since most containers require good fabrication properties and good chemical resistance, it is important that the pigment not detract from the desirable qualities built into the resinous vehicle.

For interior coatings, there is generally no serious restriction for nonfood containers, but for food containers it is important that the pigment as well as the organic vehicle be acceptable for contact with the product. For food contact, the following types of pigment can be used:

1. Aluminum

Aluminum pigmentation is used to brighten the interior surface or to hide any imperfections or stain produced on the metal substrate because of a reaction of the substrate with the packed product during packing or storage, An FDA-acceptable grade of aluminum pigment must be used if food contact is expected.

2. Titanium Dioxide

Titanium dioxide is the chief white pigment used where a porcelain enamel type of appearance is desired. Many green vegetables packaged in a white lined container will produce a green stain, which is a function of the organic part of the coating, not a function of the titanium dioxide. The white coating will, however, make the stain more visible than when using no colorants or other colorants.

3. Zinc Oxide

Zinc oxide is used as a sulfur acceptor in "C" enamels to prevent sulfur staining of the container substrate. Zinc oxide can be used in the United States in contact with foods, provided certain extraction limits are not exceeded. In a few countries, zinc oxide is not permitted as a major constituent of a coating contacting food.

4. Iron Oxides

Iron oxides, particularly the red iron oxides, are used to develop build and produce hide in a few coatings, particularly phenolic-type coatings used on pails and drums.

5. Lakes of FD&C Pigments

FD&C pigments are permitted ingredients of food container linings, but they are all water soluble and thus have very limited use. Lakes of FD&C pigments are produced commercially, and these are more water resistant than the unlaked pigments. They must be dispersed like a typical pigment by a grinding technique. Their use is limited because they do not have perfect water resistance. They produce a hazy, rather than a truly transparent, film. They are also relatively expensive per unit of color intensity.

C. LUBRICANTS AND RELEASE AGENTS

There are few, if any, metal containers which are made without a lubricant on the exterior or interior surface. During coating of the metal, scratching of the coil or sheet must be reduced to a minimum. During fabrication, sufficient lubricant for handling on high-speed drawing or forming equipment must be on the surface. During filling and shipping of the metal container, abrasion must be minimized so that the containers will move without abrading off the exterior decoration or coating, or, in extreme cases, even wear through the metal itself.

In some cases it is also desirable to have a lining in the container which will permit easy removal or release of the product from the container. This is particularly true in the case of meat products which have a tendency to adhere to the container lining when the meat product is processed or sterilized with heat in the container.

Lubricants or release agents are often applied as a separate coating operation by the company that applies the coating, but sometimes the lubricant is applied by the fabricator immediately prior to forming the container. While such post lubrication generally leads to the best efficiency with regard to lubricant use, it represents a separate operation, and often it is desirable to incorporate the lubricant or release agent into the final coating on one or both sides of the metal. Incorporation of a lubricant or release agent into the organic coating has the advantage of supplying extra lubrication to high-speed forming operations. However, use of such internal lubrication is seldom sufficient by itself and must be supplemented with a lubricant applied prior to fabrication. Some other problems encountered with the use of the lubricant or release agent directly in the organic coatings are that one seldom attains 100% efficiency of the lubricant because some is undoubtedly dispersed throughout the coating. Problems are also often encountered in producing a stable liquid coating because the lubricants have a tendency to separate from the coating during storage. It is very difficult, and sometimes impossible, to obtain a lubricant that will come to the surface of the coating during application and bake (that is, be incompatible in the dried film), but be completely miscible with the resins in the solution or dispersion coating.

Many lubricants are in use commercially because the lubricant must be selected so that it does not deteriorate any property of the coating or interfere with the packaged product. Thus, a lubricant that might be entirely satisfactory with an oleoresinous coating, may give no lubrication or slip property to a vinyl resin coating. Typical lubricants range from the hydrocarbon waxes to polyethylenes, esters, high-molecular-weight alcohols, and naturally occurring waxes of mixed chemical composition. One of the most effective meat release agents is a high-molecular-weight amide (14). This compound probably functions so effectively because it is not readily absorbed into most coatings, nor is it readily miscible with meat products, even at elevated temperature processing. It thus remains on the container surface and supplies a parting film.

REFERENCES

1. The Packaging Institute, Inc., New York, N.Y., Source Book for Closures, 1967.
2. R. D. McKirahan and R. J. Ludwigsen, Mater. Protection, 7, 29 (1968).
3. Federal Specification PPP-C-29c, Packaging and Packing of Canned Subsistence Items, Feb. 20, 1967.
4. D. E. Brody, Paint Varnish Prod., 54, 28-33 (1964).
5. E. A. Apps, Paint Manuf., 32, 304-7, 347-54, (1962).
6. S. R. Cecil and J. G. Woodruff, Food Technol., 17, 131-138 (1963).
7. M. A. Johnsen, Aerosol Age, 7, No. 6, 20, No. 7, 29, No. 8, 39, No. 9, 39 (1962).
8. M. A. Johnsen, Aerosol Age, 15, No. 8, 20, No. 9, 26, No. 10, 30, No. 11, 40 (1970).
9. C. E. Scruggs, J. L. Krickl, and N. J. Linde, MBAA Tech. Quart., 8, 63-71 (1971).
10. E. J. Smith, Chromium-Coated Steel for Container Applications, presented at meeting of American Iron and Steel Institute, May 25, 1967. New York, N.Y.

11. Steel Products Manual, Tin Mill Products, American Iron and Steel Institute, 1963.

12. Aluminum Data Books, Reynolds Metals Co., Aluminum Co. of America, and Kaiser Aluminum and Chemical Co.

13. W. A. Anderson, Mod. Packaging, 40, 131-136 (1967).

14. Albert C. Edgar and Hiram T. Spannuth (to Wilson & Co., Inc.), U. S. Pat. 2,735,354 (Feb. 21, 1956).

Chapter 10

COATINGS FOR NUCLEAR POWER PLANTS AND FACILITIES

M. J. Masciale

Mobil Chemical Company
Edison, New Jersey

10-1. INTRODUCTION

Energy requirements that are projected for the future indicate that we must employ atomic energy to produce electric power or else face depletion of our fossil-fuel resources, coal, oil, and gas (10). Consequently, both conservation and economic considerations will require us to use nuclear energy to generate the electricity that supports our civilization. Ecological considerations can also play a significant role in dictating the use of nuclear power to help us reduce many forms of environmental pollution.

A nuclear power plant is similar to a conventional power plant; each type uses steam to drive a turbine generator that produces electricity. The heat energy of steam is converted into mechanical energy in the turbine, and a generator converts the mechanical energy into electric energy. The turbine functions equally well, no matter where the steam comes from. Of course, conventional plants burn coal, oil, or gas, and the heat from the combustion of these fossil fuels boils water to make steam. In nuclear plants, no burning or combustion takes place; nuclear fission is used instead. Heat from the fission reaction is transferred to the water by a heat exchanger, producing steam. Therefore, the fission reaction in a nuclear plant serves the same purpose as the burning of fossil fuels in a conventional power plant.

The selection of coatings for use in nuclear power plants is based upon requirements for normal operations as well as accidental conditions. Coating manufacturers evaluate materials in accordance with current standards to determine radiation resistance, decontamination properties, and serviceability under design basis accident conditions. Criteria for the testing, selection, and application of coatings in nuclear power plants are described in this chapter.

In the discussion of coatings for nuclear power plants that follows, the coating exposures described are restricted to nuclear plants utilizing one of two types of reactors, i.e., a boiling water reactor or a pressurized water reactor. Both are light-water-moderated nuclear reactors, which comprise the majority of nuclear plants in operation today.

Protective coatings are being used extensively in nuclear power plants for corrosion protection, for maintaining appearance, and to aid in the removal of radionuclide soils (contamination). Normally, the corrosive atmosphere at any nuclear installation is considered mild, since most nuclear facilities are located in rural or suburban areas. However, when facilities are located adjacent to oceans, bays, or general salt-water corrosive atmospheres, atmospheric corrosion is a significant factor in the selection of coating systems. Coatings are used more significantly for protection of steel, concrete surfaces, and nonferrous surfaces from: (a) contamination by radioactive nuclides and subsequent decontamination processes; (b) ionizing radiation, chemical, and water sprays; (c) high-temperature and high-pressure steam; and (d) abrasion.

10-2. COATING REQUIREMENTS AND TEST CRITERIA

A coating used in a nuclear complex may be exposed to widely varying levels of chemical attack and radiation. Severe exposure areas are characterized by relatively large quantities of radioactive materials, high radiation fields, high probability of contamination by radioactive nuclides, frequent exposure to high humidity, elevated temperature, acids, alkalis, steam, and possibly constant immersion in deionized water. Typical areas in this category include cubicles for head-end operations (initial steps) in spent-nuclear-fuel reprocessing facilities; radioactive waste disposal systems; underwater storage and fuel dismantling facilities and hot-cell facilities for the separation and isolation of radioisotopes, certain solid state studies, and metallurgical and ceramic work with irradiated materials.

Reactor containment facilities in a nuclear power plant present a special problem of severe exposure. In the improbable event of a reactor incident, in which coolant and spent fuels would be vaporized and expelled from the reactor core into the containment shell, coatings would be subjected to gross contamination from radioiodine and other fission products, such as radioxenon and radiokrypton, steam, high temperatures, and vaporized fuels. Some typical conditions of nuclear, chemical, and physical exposure are described by Watson and West (2) as follows.

A. RADIATION

Throughout a nuclear complex, coatings are exposed to varying total dosages of ionizing radiation. In general, such levels vary from less than 1 rad to 1×10^9 rads, and possibly higher. Nuclear radiation is defined as that emitted from atomic nuclei in various nuclear reactions, including alpha, beta, and gamma radiation and neutrons. A rad is defined as the basic unit of absorbed dose of ionizing radiation. A dose of 1 rad means the absorption of 100 ergs of radiation energy per gram of absorbing material.

For severe or moderate exposures to radiation, a coating is usually selected on the basis of its radiation resistance properties. Obviously, film integrity must be maintained, otherwise the coating cannot function properly as an anticontamination and anticorrosion medium. To evaluate these and other properties, accelerated radiation exposure tests are necessary. In radiation testing, a source commonly used is a cobalt 60 irradiator of an intensity of 6×10^5 rads/gamma radiation or greater. It takes a duration of approximately 70 days to arrive at a level of 10^9 rads using this source of radiation. When studies are conducted in deionized water, the temperature is generally held constant at approximately 60°C throughout the test. Parkinson (3) summarizes the radiation damage phenomenon as follows: "Organic materials, including coatings, dissipate the energy of gamma radiation and energetic particles through ionization and electronic excitation. Both modes of energy dissipation lead to broken valance bonds in the form of an electrically charged site or an unpaired electron. Both species are quite reactive and chemical reaction products can be expected in variety of types and yields."

When exposed to radiation, saturated hydrocarbons evolve considerable quantities of hydrogen, whereas aromatic compounds dispose of the absorbed energy in nondestructive electronic transitions and give off very little gas. In solid or polymeric hydrocarbons, the bonds ruptured in the formation of hydrogen frequently link between molecules, forming crosslinks and increasing the hardness of the material. Halogenated hydrocarbons evolve the halogen or the corresponding halogen acid in the presence of moisture, the result of breaking the carbon-halogen bond.

Under irradiation, a coating may undergo cracking, blistering, softening, chalking, discoloration, embrittlement, or a combination of these phenomena. Organic plasticizers, if present in a coating formulation, migrate to the surface of a paint in minute globules and are probably partially decomposed. Following withdrawal from an exposure to ionizing radiation, coatings slowly harden, crack, and deteriorate. Gaseous products continue to evolve from the exposed materials for as long as six months after withdrawal from a radiation field.

Watson and Parkinson (3) further report that inert inorganic fillers in organic materials reduce radiation-induced effects. On the other hand, irradiation in air or water frequently causes greater decomposition or damage than irradiation carried out in an inert atmosphere. The radiolytic decomposition of water produces several oxidizing species, including peroxide radicals, and results in the oxidation of many organic materials.

Additives such as pigments, plasticizers, and other coating ingredients, as well as the type of surface on which the coating is applied, influence the radiation stability of the coating. Coatings that are highly pigmented are generally more resistant to radiation than similar gloss coatings containing lesser amounts of pigments.

The color of a paint cannot be related by present-day empirical tests to its resistance to radiation. All colors studied are darkened or discolored by exposure to radiation. Initially, the surface discolors and, as the exposure increases, the depth of discoloration also increases. White pigmented epoxies and modified epoxies turn yellow after exposure of about

5×10^8 rads and progressively darken to brown after exposure of about 2×10^9 rads (2).

Studies indicate that for many coating materials the effect of radiation is essentially a curing process characterized by an increase in hardness, a decrease in solubility, and sometimes, initially, by an increase in strength. This is especially noticeable in an epoxy coating's superior resistance to deionized water while being irradiated. A moderate amount of radiation may be beneficial to certain materials, but ultimately, in an intense radiation field, they lose tensile, shear and impact strengths, and elasticity and finally become brittle. The eventual stability of a polymer depends on its chemical structure. Aromatic structures with their resonance stabilization lend stability to polymers; hence, epoxies and modifed epoxies show a great deal of stability in a radiation field. Epoxy resins, when cured, are generally hard, extremely tough, and chemically inert. These resins are above average in radiation resistance, having withstood exposures 10^{10} rads, and yet appear to be in a serviceable condition. This is very likely because of their rigidity and aromatic content (4).

The resistance of organic coatings to radiation can be predicted to some degree from the data available from polymers and plastics. However, because of the complicating factors introduced by pigments, plasticizers, and other coating ingredients, a coating's resistance to radiation should be obtained only by empirical testing.

Radiation Exposure Guide, Table 10-1, defines the general level of exposure (radiation) that is anticipated in various areas of a nuclear complex.

Oak Ridge National Laboratories has evaluated vinyl, epoxy, modified epoxy, polyester, polyurethane, and inorganic zinc coatings by exposing them to gamma radiation in air and in deionized water. Generally, the epoxy and the modified epoxy coatings have proved to be more resistant to gamma radiation than other generic organic types tested. Epoxy or modified epoxy coatings appear to remain serviceable after an exposure of 10^{10} rads in air. This is equivalent to many years of service, even in a severe radiation environment. In deionized water, this same group of coatings fails in the range of 2.5×10^8 rads to about 5×10^9 rads.

Vinyl paints are surprisingly resistant to gamma radiation, with tolerances ranging from 2.2 to 6.6×10^9 rads in air and from 1 to 8.7×10^8 rads in deionized water. Vinyls, therefore, can be used for moderate radiation exposures and, in many instances, for severe exposures for short periods of time.

Both epoxy and vinyl coatings are exceeded in radiation resistance by the inorganic zinc coatings. The ranking of zinc coatings in resistance to radiation is sometimes negated in practice because of poor resistance to

TABLE 10-1. Radiation Exposure Guide (5)

Facility of area	General level of exposure (rad)		
	Floor	Wall	Ceiling
1. Spent-nuclear-fuel processing plant			
a. Processing cells	$5 \times 10^{8} - 5 \times 10^{9}$	5×10^{8}	5×10^{8}
b. Research and analytical labs	$1 \times 10^{3} - 1 \times 10^{4}$	1×10^{3}	1×10^{3}
c. Decontamination and waste disposal facilities	1×10^{4}	1×10^{3}	1×10^{3}
d. Canals, fuel receiving, and storage	$5 \times 10^{9} - 1 \times 10^{10}$	1×10^{10}	
e. Spent-nuclear-fuel carriers	$5 \times 10^{8} - 1 \times 10^{9}$	$5 \times 10^{8} - 1 \times 10^{9}$	$5 \times 10^{8} - 1 \times 10^{9}$
f. High-level laboratories	$1 \times 10^{9} - 5 \times 10^{9}$	1×10^{9}	$5 \times 10^{8} - 1 \times 10^{9}$
g. Off-gas ductwork and filtration system	$5 \times 10^{8} - 1 \times 10^{9}$	5×10^{8}	5×10^{8}
2. Nuclear reactor sites			
a. Containment structures	1×10^{2}	1×10^{2}	1×10^{2}
b. Canals for fuel storage and handling	$7 \times 10^{9} - 1 \times 10^{10}$	$7 \times 10^{9} - 1 \times 10^{10}$	
c. Fuel disassembly and examination facilities	$>2 \times 10^{9}$	$>2 \times 10^{9}$	$>2 \times 10^{9}$
d. Laboratories	1×10^{3}	1×10^{2}	1×10^{2}
e. Off-gas ductwork and filtration system	$5 \times 10^{8} - 1 \times 10^{9}$	5×10^{8}	5×10^{8}

chemicals and poor decontamination properties, but an inorganic zinc primer is often used in conjunction with epoxy or other organic topcoats in high radiation areas.

Gamma Radiation Resistance of Several Protective Coatings, Table 10-2 shows typical levels of radiation that a variety of general coatings have withstood up to their failure dosage.

B. DECONTAMINATION

Contamination by radioactive substances is believed to occur through chemisorption, by ion exchange with free surface ions, by physical adhesion, or by migration of the radioactive nuclides into cracks and crevices. Radioactive ions become attached to the surfaces of metals and other structural materials in a few minutes. Some ions are fixed by the visible and invisible oxide films on metals, but after extended times (months), some contaminants may diffuse through the oxide films into the metal (2). Generally, decontamination can be defined as a highly effective cleaning process. It is the practice or art of removing radioactive materials (soils, chemical compounds in solutions, reaction products, etc.) from surfaces. The purpose of decontamination is to render the affected areas harmless to unprotected workers and the biological environment, and to salvage costly equipment in work areas.

In nuclear installations, ferrous and nonferrous surfaces in various areas must be protected from radiation deposits. As a case in point, if a bare concrete surface is contaminated, there is no practical way to decontaminate the surface except to remove the contaminated concrete. Suitable coatings applied over ferrous and nonferrous surfaces provide resistance to radiation as well as abrasion resistance. It is, therefore, imperative that all areas where radioactive contamination is possible must be protected with suitable coatings.

Decontamination testing provides methods for the quantitative evaluation of the ease and degree of contamination and decontamination of protective coatings. Essentially, decontamination testing measures the ratio of the original beta-gamma activity versus the activity after decontamination. This is identified as the decontamination factor, or D. F. The larger the D. F. number, the better the decontamination characteristics are.

In a decontamination test prescribed by Oak Ridge National Laboratories (5), the radioactive contaminant used is normally a solution of mixed fission products obtained by the dissolution of spent uranium fuel previously irradiated to about 20,000 MW days per metric ton and cooled for at least 90 days and no more than 3 years. The solution of mixed fission products is required to have a beta-gamma activity of 5×10^6 disintegrations per

TABLE 10-2. Gamma Radiation Resistance of Several Protective Coatings[a]

Type	Manufacturer[b]	Substrate	Exposure[c] demineralized water		Exposure in air[c]	
			Dose (rads)	Effect[d]	Dose (rads)	Effect[d]
Modified phenolic	1	Steel	$>5.0 \times 10^9$		$>5.0 \times 10^9$	
Phenolic	2	Steel	3.4×10^9	C	$>3.6 \times 10^9$	
Phenolic		Concrete	3.4×10^9	C, D	2.4×10^9	C, D
Epoxy	2	Steel	$>4.3 \times 10^9$		$>3.6 \times 10^9$	
Epoxy		Concrete	3.4×10^9	C	$>2.9 \times 10^9$	
Urethane	2	Steel	$>4.3 \times 10^9$		$>3.6 \times 10^9$	
Urethane		Concrete	$>4.3 \times 10^9$		$>1.9 \times 10^9$	
Alkyd	2	Steel	1×10^8	B	$>3.6 \times 10^9$	
Alkyd		Concrete	1×10^8	B	$>2.9 \times 10^9$	
Epoxy	3	Concrete			$>3.3 \times 10^9$	
Vinyl	3	Steel	2.5×10^8	B	2.0×10^9	B, C
Vinyl	3	Steel	2.5×10^8	B	2.0×10^9	B, C
Epoxy[e]	3	Steel	8.0×10^8 [f]	B	$>9.5 \times 10^9$	
Modified phenolic	4	Steel	$>4.5 \times 10^9$		4.59×10^9	C

Modified phenolic		Concrete	3.3×10^{9}	B, C		
Epoxy[e]	4	Steel	$>4.5 \times 10^{9}$		4.59×10^{9}	C
Epoxy		Concrete	$>4.0 \times 10^{9}$		4.1×10^{9}	A
Epoxy[e]	4	Steel	9.0×10^{8}	B	$>4.5 \times 10^{9}$	
Vinyl	5	Steel	7.0×10^{8}	B	$>1.5 \times 10^{9}$	
Epoxy	5	Steel	3.8×10^{8}	B	$>1.5 \times 10^{9}$	
Phenolic	6	Concrete	2.3×10^{9}[f]	A	9.0×10^{9}	A
Phenolic		Steel	4.0×10^{8}[f]	B	1.0×10^{10}	D
Epoxy[e]	7	Steel	1.6×10^{9}[f]	A, B	$>9.0 \times 10^{9}$	
Vinyl	7	Steel	3.8×10^{8}[f]	B	5.0×10^{9}	C
Chlorinated rubber	7	Steel	3.8×10^{8}[f]	B	$<6.3 \times 10^{9}$[g]	C, D
Epoxy[e]	7	1-Steel	1.6×10^{9}[f]	A, C	$>9.0 \times 10^{9}$	
Epoxy[e]		2-Steel	3.8×10^{9}	A, B		
Vinyl	7	Steel	1.1×10^{8}[f]	B	5.0×10^{9}	C
Epoxy	7	1-Steel	1.6×10^{9}[f]	A, C	$>9.0 \times 10^{9}$	
Epoxy		2-Steel	1.2×10^{9}	B, C		
Polyurethane	7	Steel	7.8×10^{9}	C	$>9.0 \times 10^{9}$	
Epoxy[b]	7	1-Concrete	3.8×10^{8}[f]	B	$>9.0 \times 10^{9}$	

TABLE 10-2. — Continued

Type	Manufacturer[b]	Substrate	Exposure[c] demineralized water		Exposure in air[c]	
			Dose (rads)	Effect[d]	Dose (rads)	Effect[d]
Epoxy		2-Concrete	2.0×10^9	A, B		
Epoxy[e]	7	1-Concrete	1.1×10^{8}[f]	B	1.9×10^{9}[d]	C, D
Epoxy[e]		2-Concrete	1.7×10^9	B		
Chlorinated rubber	7	1-Concrete	3.8×10^{8}[f]	B	7.8×10^9	C, D
Chlorinated rubber		2-Concrete	9.5×10^8	B		
Vinyl	7	1-Concrete	3.8×10^{8}[f]	B	2.0×10^9	B, C
Vinyl		2-Concrete	2.0×10^9	B, C		
Epoxy	7	1-Concrete	5.9×10^{8}[f]	B	$>9.0 \times 10^9$	
Epoxy		2-Concrete	1.2×10^9	B		
Polyurethane	7	1-Concrete	3.1×10^{8}[f]	B	$>9.0 \times 10^9$	
Polyurethane		2-Concrete	$>8 \times 10^8$			
Epoxy	7	Concrete	3.4×10^8	B	$>4.5 \times 10^9$	
Modified phenolic	7	Steel	1.2×10^9	B, C		
Modified phenolic		Concrete	$>8 \times 10^8$			
Inorganic	8	Steel[h]	1.0×10^{9}[f]	E	$>7.5 \times 10^9$	
Epoxy	8	Steel[h]	1.0×10^{9}[f]	B	$>7.5 \times 10^9$	

Inorganic	8	Steel[h]	$>7.5 \times 10^9$		$>7.5 \times 10^9$	
Inorganic	8	Steel[i]	$>7.5 \times 10^9$		$>7.5 \times 10^9$	
Polyurethane	9	Steel	1.0×10^8	B	2.8×10^9	A
Polyurethane		Concrete	$>8.0 \times 10^8$			
Polyurethane	9	Steel	1.0×10^8	A, B	2.8×10^9	A

[a]Radiation source: ^{60}Co with an intensity of 6×10^5 rad/hr. Temperature: 40 to 50°C (9).

[b]Manufacturer's identity can be obtained from Oak Ridge National Labs (ORNL), Report 4572, pp. 258-261.

[c]The coatings were inspected for radiation damage at various exposure levels (1×10^8, 3×10^8, 5×10^8, and each additional 5×10^8 rads exposure thereafter). The values listed represent the cumulative dose that had been received at the time adverse effects were observed. If no effects are entered, the exposure test is continuing.

[d]Radiation effects: A, chalked; B, blistered; C, embrittled; D, loss of adhesion; E, "sweating."

[e]Polyamide-cured epoxy.

[f]Previously reported in ORNL-4422. Continuing test results will be shown in later ORNL reports.

[g]Earlier failure not detected. Appearance indicates that resistance is less than that given.

[h]Sandblasted steel.

[i]Pickled steel.

minute per 0.1 ml. The detector used is normally a gamma-sensitive device, such as sodium iodide crystal, with photomultiplier and amplifier coupled to a pulse-height analyzer. Disintegration rates of the various radioactive nuclides are calculated from the recordings posted by the instrument.

All specimens are scanned and counted at the same geometry, with both the original count and the count following each decontamination step being recorded. Specimens are decontaminated by placing each one in a stirred solution of the decontamination reagent. Each specimen is subjected in succession to three decontamination steps as follows:

a. Water Wash

Stir in tap water for 10 min and air dry. The ratio of the original radioactivity (counts per minute) to the radioactivity detected (counts per minute) after the water wash is recorded as the water D. F.

b. Decontamination with Room Temperature (25°C) Acid

Stir for 10 min in an aqueous mixture of oxalic acid and determine the acid D. F., which is defined as the ratio of the radioactivity (counts per minute) after the tap water wash to the radioactivity (counts per minute) after the acid wash.

c. Decontamination with Heated (80°C) Acid

Continuing to use the same specimens, perform the third decontamination with hot oxalic acid (80°C). The decontamination factor achieved with the heated acid is a ratio of the radioactivity (counts per minute) after the room temperature acid wash to the radioactivity (counts per minute) after the heated acid wash.

The over-all D. F. for the three decontaminations is the ratio of the radioactivity (counts per minute) before the tap water wash to the radioactivity (counts per minute) after the heated acid wash. The over-all D. F. is the value commonly used to measure the effectiveness of a coating with regard to decontamination properties.

Oak Ridge National Laboratories has conducted decontamination studies and their findings of four generic types of coatings are as follows:

1. Epoxies. The D. F. s that have been determined on a wide variety of epoxies range from 7 to 590, more generally in the lower range.

2. Vinyls. The results on a few vinyl materials range from 14 to 156.

3. Polyurethane. Infinitely high decontamination values have been determined on two polyurethane coatings that have been evaluated.

4. Chlorinated Rubber. Although Oak Ridge has not checked on chlorinated rubber products in their tests, they have commented that independent

laboratories have shown one particular chlorinated rubber formulation to have a value of 278, which certainly compares well with other generic types of coatings.

To achieve a high degree of decontamination, both chemical and physical methods are usually used. Surface contaminants are removed chemically by strong solvents (acids, or alkalis) that simultaneously prevent their redeposition. Physical methods include abrasion, erosion, jet impingement, and ultrasonic cleaning. In the contamination-decontamination cycle, coatings serve not only as a barrier, but also as a sacrificial surface, thus often eliminating some, or all, of the chemical or physical treatments.

A typical procedure for cleaning or decontaminating painted surfaces is reported (2), using the following successive steps, each of which is more drastic than its predecessor:

1. Remove loose or lightly held contaminants by pressure spraying with a steam lance or water jet, using either water or steam with any good detergent. If the surface is dusty, vacuum clean prior to the aqueous step.

Alternatively, it may be advantageous to use surface cleaners or emulsion or foam cleaners rather than a steam lance, not only because they may be more efficient, but also because they are used at low pressure and there is less chance of dislodging radioactive particles into the atmosphere.

2. Scrub with water and the cleaner described in (1), above, using manipulator-held fiber brushes, sponges, or an inexpensive rotating household floor polisher equipped with fiber, steel wool, or cloth pads. Adhesive strips may also be applied and pulled from the surface to scavenge insoluble particles.

3. Repeat step (2), above, at room temperature and at 80°C, using a peroxide-inhibited decontamination solution (0.4 M $H_2C_2O_4$, 0.05 M NaF or NH_4F, and 0.3 M H_2O_2). This solution is effective on painted, plastic, and metallic surfaces and does not produce a corrosive effect, as does the stronger oxidizing acids.

4. Repeat step (2) by using household scouring powders to grind away or abrade the surface to a depth of from 1 to 3 mils. A solvent that will dissolve or soften the coating may also be used in this step. The extent to which solvents are used may be restricted, however, depending on the toxicity and flammability of their vapors (2).

For some applications, it is important to have a coating or coating system that has good resistance to aromatic solvents in addition to all-round chemical resistance. Chlorinated rubber coatings and vinyl coatings are obviously unsuitable for the purpose, but solvent-resistant, epoxy-resin-based coatings would certainly be suitable.

Walker (6) reports that the most important single variable in governing the ease of decontamination of coatings is the polymeric base used in the paint, although type and amount of plasticizer are also very important. Walker evaluated coatings both in the glossy condition and after abrasion, when the surface gloss had been removed. In general, the retention values of the abraded specimens were higher than those of the unabraded, ones but the differences were not great. Walker has found that chlorinated rubber coatings, when suitably plasticized, provide one of the best bases for a paint with good decontamination properties. He further reports that coatings containing amino crosslinking agents invariably absorb large amounts of active material and retain a considerable proportion of this on subsequent decontamination. Of the crosslinking agents commonly used in the cure of epoxide resin coatings, Walker reports amine adducts lead to coatings showing both the lowest initial contamination and final retention value after decontamination. These results are somewhat in conflict with results obtained in the United States (7) where other amino crosslinking agents such as polyamides were tested. The survey of results from an independent laboratory shows that polyamide-cured epoxies provided decontamination results in the same range as amine adduct epoxy coatings.

In the past, vinyl, chlorinated rubber, and, in some cases, alkyd coatings were used in nuclear facilities because of their acceptable decontamination properties. More recently, epoxy coatings are replacing the vinyl and chlorinated rubber coatings because properly formulated epoxy coatings will provide satisfactory decontamination properties as well as superior abrasion resistance, high-temperature service, solvent resistance, and radiation resistance in deionized water. Chloride bearing materials are often prevented from coming in contact with stainless-steel components, and consequently, vinyl and chlorinated rubber coating applications are restricted in many areas of nuclear plants.

C. DESIGN BASIS ACCIDENT

In the design and operation of the light-water-moderated nuclear power system, such as a pressurized water reactor (PWR) or a boiling water reactor (BWR), consideration must be given to a Design Basis Accident (DBA) and the subsequent events that might lead to a fractional release or expulsion of a fraction of the fission product inventory of the core to the reactor containment facility (1). Engineered safety features, principally a reactor containment facility, are provided to prevent a release of fission products to the biological environment during and after this improbable event. Large areas of the reactor containment facility are painted with a protective coating for the purpose of corrosion protection as well as ease of decontamination. If severe peeling, flaking, or chalking caused significant portions of the coating to be discharged into a common water reservoir, the performance of the safety systems could be seriously compromised by the plugging of strainers, flow lines, pumps, spray nozzles, and core coolant channels. If coating failure occurred during the DBA, the performance of the safety systems could be seriously compromised. Therefore, it is important that coatings withstand the harsh DBA exposure conditions, as well as meet the other stated criteria that are unique to nuclear plant operation.

The applied coatings are expected to be resistant to radiation exposures of 1×10^9 to 3×10^9 rads and temperatures up to 160°C and higher in the presence of borated suppression spray solutions or demineralized water. Watson, Griess, Row, and West (1) report that the coatings are also exepcted to remain in place during (and following) a DBA, even if it occurs as late as the last day of 40 years of service. In addition, it is desirable for the coatings to be easily decontaminated.

Once a DBA has occurred, another safety feature, a suppression spray system, is actuated inside the containment structure to prevent a recurrence, to lower pressure and temperature, and to possibly "getter" certain fission products, particularly iodine (1). Since large areas of the reactor containment facility are painted with a protective coating, it is important that coatings be compared and evaluated for use in reactor containment structures by stringent standard tests that simulate exposure conditions as closely as possible. These conditions include exposures of temperature, pressure, chemical solutions, and radiation. Examples of temperature-pressure versus time curves for the containment atmosphere following a DBA are presented in Fig. 10-1 (PWR) and Fig. 10-2 (BWR). These figures are given only as illustrations of variations of temperature and pressure following a DBA in the PWR and BWR types. However, each containment design will have its own criteria, and coatings should be evaluated for the actual DBA criteria of any specific design. The general thermal stability of a protective coating can be evaluated by using either Fig. 10-1 or Fig. 10-2, as appropriate, or the particular curves from the latest safety analysis report for the containment facility in question.

During a DBA, a protective coating may be exposed to water spray and to chemical spray solutions. Tests of the coatings for a specific application should be conducted with the solution to be used in that application. Generally, BWRs utilize water sprays whereas PWRs utilize spray solutions that may contain one or more chemical additives (see Table 10-3).

Testing is normally conducted in a dynamic autoclave, wherein steel- and concrete-coated specimens are exposed by the test apparatus to a simulated DBA condition specified by the temperature-pressure curve of the particular PWR or BWR. The apparatus normally provides steam during the first few minutes of operation and it can produce the maximum DBA temperature within 20 minutes. It should not be vented during heat-up to

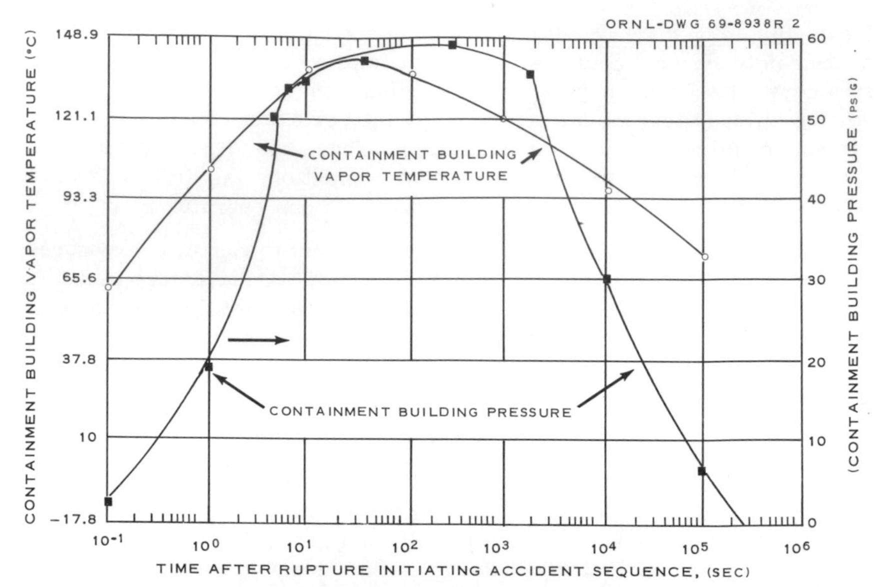

FIG. 10-1. Typical DBA curve for TWR containment facilities.

remove noncondensables. The chamber (autoclave) should be constructed so as to allow the test specimens to be exposed to total immersion, liquid/vapor interface, and to spray, and should be constructed of materials that are corrosion resistant to test solutions used. Typical examples of equipment components and arrangements used are presented in Fig. 10-3 as a guide by which DBA conditions can be simulated. It should be noted that high-temperature steam is involved and that adequate safety measures are mandatory to protect the personnel operating such equipment. The composition of the test solution, and the temperature, pressure, and time requirements should be in accordance with the design established for the intended service.

It may be concluded from such tests that the epoxy or modified epoxy systems are the best of the air-dried systems tested. Blistering and peeling caused by heat and chemical attack is the principal cause of failure of vinyls, chlorinated rubbers, and alkyds.

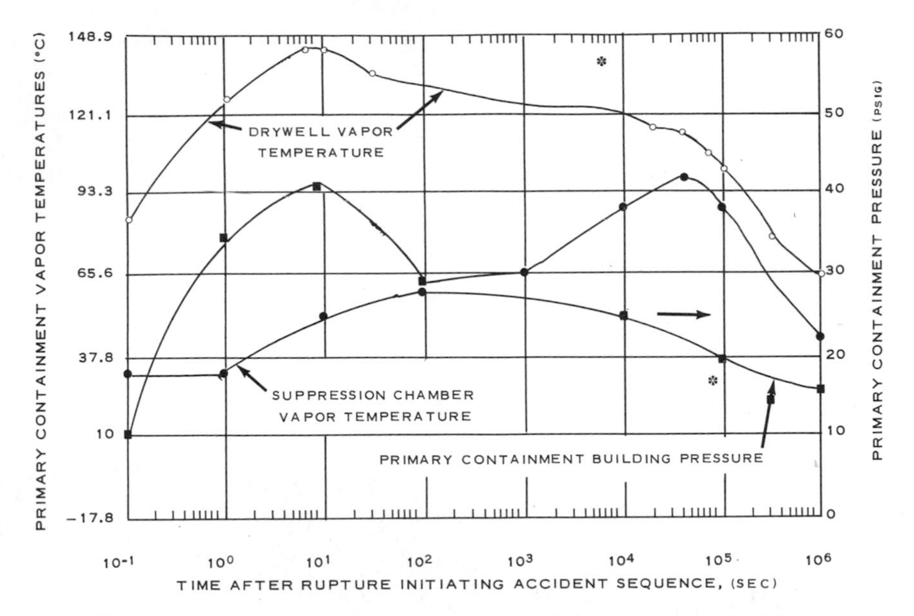

FIG. 10-2. Typical DBA curve for BWR containment facilities.

TABLE 10-3. Typical Spray Solution Additives (PWR)

Chemical compound	Concentration	Purpose
Sodium borate	2000 to 4000 ppm borom	Reactivity control
Sodium thiosulfate[a]	$\leq 2\%$ by weight	Reactant for iodine
Sodium hydroxide[b]	≤ 0.2 normal	Alkaline reactant for iodine
Boric acid	2000 to 4000 ppm boron	Reactivity control

[a]In conjunction with boric acid, with or without sodium hydroxide.

[b]In conjunction with boric acid, with or without sodium thiosulfate.

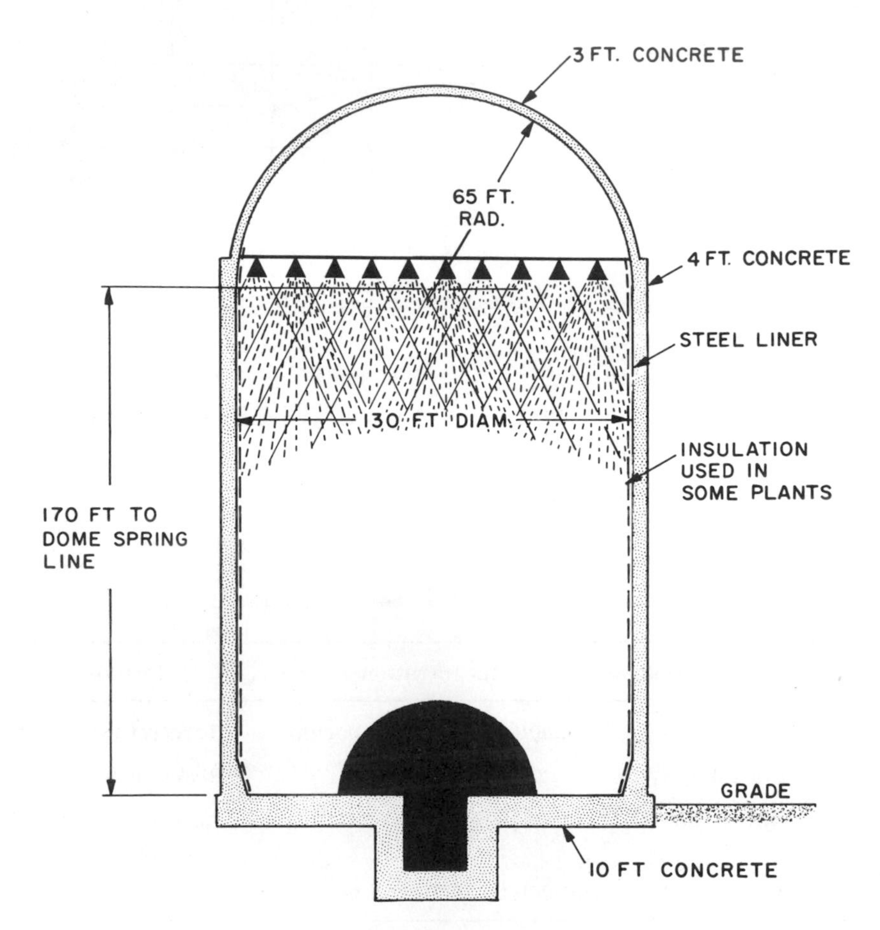

FIG. 10-3. Typical TWR containment structure (ORNL drawing 68-9457B).

It is reported (1) that the use of an inorganic zinc prime coat appears to degrade a given epoxy or modified epoxy system. The subject of the use of inorganic zinc coating as a prime coat on interior containment surfaces is a subject of much debate in the industry today. For example, it has been contended that so long as the coating system does not peel or flake following a simulated DBA test, the coating should be regarded as satisfactory. In this case, some epoxy or modified epoxy coatings perform satisfactorily when applied over an inorganic zinc coating. On the other hand, others consider that blistering is to be regarded as a failure and, therefore, the use of an inorganic zinc primer that allows for minute topcoat blistering is unacceptable.

Since there are conflicting reports about the use of inorganic zinc primers in containment areas, a suitable test program is essential to develop valid empirical data to reinforce the choice of coating system used. *

---

*It has been reported (7) that an ethyl silicate inorganic zinc with a polyamide epoxy topcoat, as well as an epoxy-polyamide coating system (epoxy primer and topcoat), satisfactorily withstood simulated design basis accident testing at an independent laboratory with no signs of blistering or other deleterious effects. Following this DBA testing, the same specimens were forwarded to Oak Ridge National Laboratories for further exposure to radiation testing in deionized water up to a level 1×10^9 rad. (This testing was begun on September 2, 1970, and was completed in November, 1970.) The resulting report from Oak Ridge National Laboratories showed that specimens were satisfactory after their testing in deionized water.

It should also be noted that under DBA conditions, hydrogen evolution from inorganic zinc-rich primers applied to the containment walls has been reported from test work conducted at Oak Ridge National Laboratories. Thus far, many regard the quantity of hydrogen evolved from inorganic zinc primers to be negligible when considering explosive limits. Further studies at elevated temperatures are under way.

The results of DBA testing have demonstrated two points. First, the technology currently exists to produce an acceptable coating; second, some coatings previously considered for use in nuclear plants fail quite drastically. The evaluation of coatings under simulated DBA conditions, therefore, is critical, and explains the great amount of testing and consultation in the industry.

D. OTHER TESTS

Other test requirements for nuclear plant applications are defined in ANSI-N-5.9-1967 (5) and ANSI-N-101.2-1972 (8). The principal features are coating performance, test procedures, and evaluation methods including: (a) screening of coating systems prior to testing; (b) fire evaluation test; (c) thermal conductivity determination; and (d) repairability and maintenance tests. It does not serve to dwell on these test procedures and evaluations any further, since they are readily defined in referenced standards. The testing procedures and results described in this chapter are receiving the major emphasis in the field of nuclear coatings today and give reason for their discussion. Further efforts by ASTM and ANSI are under way to update and clearly define methods of test and quality assurance requirements relating to coatings for nuclear facilities.

10-3. CONSIDERATIONS FOR COATINGS SELECTION

The major requirements for coatings in nuclear power plants have been described, and, although there are many specific requirements for coatings, the more general requirements can be listed as follows:

1. Radiation resistance in air (above 10^9 rads).
2. Radiation resistance in deionized water (above 10^9 rads).
3. Decontamination.
4. Design basis accident.
5. Resistance to continuous immersion in deionized water (to 82° C).
6. Abrasion resistance.

There are many generic types of coatings being used in nuclear power plants today, the most common of which are:

1. Inorganic zinc-rich primers.
2. Epoxy and modified epoxy coatings.

3. Epoxy and phenolic surfacing compounds.
4. Inorganic silicate topcoats.
5. Alkyd primers and topcoats.
6. Latex block fillers, latex primers, and latex/finish coats.
7. Vinyl coatings.
8. Chlorinated rubber coatings.

A brief description of each of these generic types follows.

A. INORGANIC ZINC-RICH PRIMERS

A zinc-rich coating is characterized by a very high degree of pigmentation, normally 85% of metallic zinc in the dried paint film. One essential property is that the paint film is electrically conductive and that it is in electrical contact with the steel substrate. These properties are obtained when a high portion of metallic zinc particles are in electrical contact with each other in the paint film and with the steel. An inorganic zinc-rich primer has properties similar to that of a galvanized coating, and when it is applied on an electrically conductive media, the zinc-rich coating will give electrochemical protection.

The inorganic binder is essentially a solution of silicates. The inorganic zinc-rich primer is unaffected by most organic solvents, is nonflammable, provides good sacrificial release of zinc to provide cathodic protection, has relatively high heat resistance (in excess of 400°C), and is resistant to thermal shock cycles. The inorganic zinc coatings are normally applied by spray techniques. They adhere well to blasted steel and give excellent protection against corrosion and abrasion throughout the preconstruction phase of a nuclear plant.

The most common type of silicate coating used is a self-curing coating based on alkyl silicates. In the liquid form these paints might be grouped among the organic systems, but in a dried film form they definitely are of the inorganic type, exhibiting all of the properties characterizing inorganic coatings. The binders are essentially modified alkyl silicates and consist of partially hydrolyzed members of the series methyl through hexyl, or glycol ether silicates.

In a nuclear facility, inorganic zinc-rich coatings are top-coated. Organic topcoats are usually applied, since they are more easily decontaminated. It is generally agreed that the use of an inorganic zinc during the construction phase allows for excellent corrosion protection of the steel. In addition, such systems are more tolerant of abusive handling procedures, since the abrasion resistance is superior to that of nonzinc-containing organic primers.

B. EPOXY AND MODIFIED EPOXY COATINGS

The catalyzed epoxy coatings are composed of an epoxy resin and suitable catalyst (activator, hardener, or curing agent). Epoxy resins most commonly used in room-temperature-cured coatings are the reaction products of various proportions of bisphenol A and epichlorohydrin. Many resins are formed by this chemical reaction and they vary in degree of polymerization, solubility in various solvents, and physical form, but all have terminal epoxide groups. These epoxide groups are capable of reacting at room temperature with various amines to further polymerize the epoxy resin and yield a coating which is hard and tough and adheres to a wide variety of surfaces with resulting excellent chemical and solvent resistance. Amine adduct and polyamide curing agents are most commonly used. The amine adduct curing agents are formed by partially reacting polyalkyl amines with epoxy resins. The amine adduct curing agents produce coatings of superior chemical and solvent resistance to those made using polyamides. Polyamide curing agents are formed from the condensation of di or poly amines, such as ethylene diamine or diethylene triamine with a dimer acid. A dimer acid is the mixture of dibasic and polybasic acids obtained by the polymerization of di-unsaturated fatty acids, such as linoleic acid. Polyamide-cured epoxy coatings are not quite as resistant to solvents or chemicals as the amine-adduct-cured systems, but they do have better application properties, better wetting properties, and are more flexible in the cured film. The polyamide curing agents are practically harmless in respect to contact dermatitis (skin irritations) and are more tolerant to applications in high humidity.

The modified epoxy (epoxy-phenolic) is another variety of an epoxy coating in which phenolic groups are introduced into the backbone of the resulting polymer. The amine adduct epoxy and the modified epoxy (epoxy-phenolic) perform similarly when tested under nuclear environments. Each is capable of being formulated to meet specific requirements. Generally, the unmodified epoxy coatings have application advantages over the modified (epoxy-phenolic) coatings in that they are more tolerant to adverse field painting conditions of high humidity, varying temperatures, and prolonged drying periods between coats.

The cured epoxy or modified epoxy resin is a hard thermoset material in which the epoxy resin molecules are joined in a three-dimensional network of linkages and crosslinkages. It is no wonder that these materials are primarily used in nuclear plants for their excellent resistance to radiation, decontamination properties, and abrasion resistance. It is important to note that epoxy or modified epoxy coatings can vary significantly due to formulation differences. Therefore, certain epoxy coatings may be quite suitable for deionized water service, while other epoxy coatings are not. The architect/ engineer is normally aware that each coating formulation must be tested to determine whether it meets the criteria for the particular service intended.

C. EPOXY AND PHENOLIC SURFACING COMPOUNDS

In many critical areas of nuclear power plants, concrete surfaces must be sealed to prevent contamination from radioactive nuclides. It has already been stated that such contamination of concrete is most difficult to remove. Surfacing compounds have been formulated to fill hollow areas, air pits, holes, and other surface imperfections. The required film thickness of surfacing compounds will range from several mils up to $\frac{1}{8}$ in. as required. Normally, surfacing compounds are thick mastic materials that can be applied in a number of ways, depending upon the trade and skill of the applicator. Generally, the surfacing compounds are epoxy or phenolic materials filled with suitable inert material(s). The physical properties of the surfacing compounds normally exceed those properties specified for concrete. Consequently, such factors as compressive strength, tensile strength, modulus of elasticity, flexural strength, linear shrinkage, shear bond, coefficient of linear thermal expansile , decontamination properties, radiation studies, design basis accident studies, bond strength, impact resistance, and abrasion inded have been investigated and measured. In independent study of all of these properties (7) on a proprietary epoxy surfacing compound is reported showing excellent results.

Epoxy or phenolic surfacing compounds are normally used only in critical areas, such as floors, wainscots, and walls where contamination with radioactive materials is anticipated. Otherwise, conventional products, such as latex block fillers and cementicious fillers, are commonly used for nonnuclear areas.

D. INORGANIC SILICATE TOPCOATS

Inorganic topcoats are composed of essentially the same type vehicle as an inorganic zinc primer with substitution of selected inert fillers in place of metallic zinc. The use of inorganic silicate topcoats is extremely specialized and they are used only where organic topcoats have certain shortcomings. An inorganic topcoat is generally recommended for use only over an inorganic zinc-rich primer. Architects and engineers will selectively require limited use of such a topcoat depending upon the requirements for high-temperature resistance and thermal conductivity.

E. ALKYD PRIMERS AND TOPCOATS

Alkyd resins are polymeric esters prepared by the reaction of polyhydric alcohols and polybasic acids or their anhydrides. The trade often refers to short-oil or long-oil alkyds. Long-oil alkyds contain a relatively high percentage of oil, while short-oil alkyds contain a relatively small percentage of oil. The long-oil alkyds are softer, more flexible, and slower drying, while the short-oil alkyds are harder, fast drying, and more brittle. Coating manufacturers often use a blend of different types of alkyds to achieve desired properties.

The alkyds as a group offer a durable film with very good weather resistance. A high-quality alkyd will offer moderate resistance to radiation in air and will seal a substrate from radioactive particles at ambient conditions. Alkyds do not have good resistance to strong chemicals and solvents, and consequently, are not normally used where frequent decontamination procedures are anticipated.

Alkyds are a preferred coating for most noncritical areas within a nuclear facility due to their ease of application, over-all good durability, and aesthetic properties.

F. LATEX PAINTS

In general, when latex emulsion paints are compared with solvent-based resinous paints of the same type, they have many advantages: faster drying, easier application, improved cleanup of application tools with tap water, no fire hazard, improved resistance to mild alkali, resistance to fats and oils, and less color change. Latex paints do have disadvantages in that they are subject to freezing and may be damaged by the freeze-thaw process. They should not be applied below 10° C. At these temperatures poor film formation can be expected due to poor coalescing action.

Many variations of latex coatings can be formulated. They are generally of three chemical compositions: (1) styrene-butadiene; (2) polyvinyl acetate; and (3) acrylic. Many variations of these basic types can be formulated, and they can be plasticized with different plasticizers to give a variety of properties, such as flexibility, stability, etc.

Latex coatings will normally be used in nonnuclear areas. Latex concrete block fillers are used commonly on much of the concrete surfaces to fill small voids prior to topcoating with latex or alkyd finish coats.

G. VINYL COATINGS

The binders used in most vinyl coatings are copolymers of vinyl chloride and vinyl acetate resins. Polyvinyl acetate is a relatively soft, readily soluble resin, while polyvinyl chloride is a hard, tough resin. Polyvinyl chloride and polyvinyl acetate can be reacted to produce a copolymer vinyl chloride-vinyl acetate resin. A great variety of copolymers can be formulated with different solubilities and chemical resistance.

Vinyls are hard, brittle resins that must be plasticized to provide satisfactory coatings. Vinyl coatings dry solely by evaporation of solvent. Since no reaction takes place, the vinyl coating does not become hard and brittle on aging, but remains tough and elastic. Recoating is readily accomplished because the solvents in subsequent coats of vinyl soften the old vinyl, thus providing good bond between coats. Vinyls are one of the better chemical-resistant and durable coatings. However, they tend to soften above 65° C, and consequently, are not specified for use within a nuclear reactor containment area due to possible blistering and peeling at elevated temperatures.

H. CHLORINATED RUBBER COATINGS

Coatings based upon chlorinated rubber resins are obtained by reacting and saturating natural rubber with chlorine. Chlorinated rubber coatings can be typified as lacquer-type resins. Films are obtained by dissolving the resins in suitable solvents and letting the applied film dry hard solely by solvent evaporation. The chlorinated rubber resin exhibits good resistance to a wide variety of chemicals and dries and hardens quite rapidly. A mixture of resin with various modifiers or plasticizers is necessary to obtain wetting, flexibility, and other qualities. Chlorinated rubber coatings (unmodified, not the chlorinated alkyd variety) are similar to vinyl coatings in that they exhibit good resistance to radiation and are readily decontaminated. They do, however, soften above 65° C, and consequently, are not normally specified for use within containment areas.

The description of the above generic types of coatings are based on current technology. Naturally, with improved formulations, the properties of any generic type of coating can be altered.

It has been the author's observations that architects and engineers consider the coating systems shown in Table 10-4, Nuclear Coatings Guide, most frequently when preparing coating specifications for a nuclear facility.

TABLE 10-4. Nuclear Coatings Guide

Coating system	Design basis accident (DBA)	Decontamination	Continuous immersion in distilled water (up to 82°C)	Radiation resistance in air	Radiation resistance in water
Inorganic zinc primer	x		x	x	x
Epoxy or modified epoxy coating over inorganic zinc primer	x	x		x	x
Epoxy or modified epoxy primer and topcoats	x	x	x	x	x
Epoxy or phenolic surfacing compounds	x	x		x	
Inorganic silicate topcoat over inorganic zinc primer	x			x	x
Vinyl primer and topcoat		x		x	
Chlorinated rubber primer and topcoat		x		x	

REFERENCES

1. C. D. Watson, J. C. Griess, T. H. Row, and G. A. West, Protective Coatings for PWR and BWR Reactor Containment Facilities, ANS National Meeting, Invited Paper, Los Angeles, Calif., June 28-July 2, 1970.

2. C. D. Watson and G. A. West, "Protective Coatings (Paints)," in Decontamination of Nuclear Reactors and Equipment (J. A. Ayres, ed.), Ronald Press, New York, 1970.

3. C. D. Watson and W. W. Parkinson, "Irradiation of Bituminous Materials," in Bituminous Materials: Asphalts, Tars, and Pitches (A. J. Hoiberg, ed.), Vol. I, Wiley (Interscience), New York, 1964.

4. R. W. King, N. J. Broadway, R. A. Mayer, and S. Palinchak, "Polymers," in Effects of Radiation on Materials and Components (J. F. Kircher and R. E. Bowman, eds.), Reinhold, New York, 1064, pp. 84-166.

5. US Standard, Protective Coatings (Paints) for the Nuclear Industry, USASI Standard N5.9-1967, Am. Inst. Chem. Engr., 1968.

6. P. Walker, "The Formulation and Testing of Paints for Use in Radioactive Areas," J. Oil Colour Chemists' Assoc., 49, 2 (Feb., 1966).

7. Mobil Chemical Co., Internal Report, Nuclear Coatings, Maintenance and Marine Coatings Dept., Edison, N.J.

8. Proposed American National Standard, Protective Coatings (Paints) for Light Water Nuclear Reactor Containment Facilities, Proposed ANSI N101.5-1970 Standard, April, 1970.

9. Oak Ridge National Laboratory Report No. 4572, Chemical Technology Division, Annual Progress Report for period ending May 31, 1970, pp. 258-261.

10. United States Atomic Energy Commission, Division of Technical Information, Nuclear Power Plants, 1969 (rev.).

Chapter 11

"UNIVERSAL" COLORANTS

Lyman P. Hunter

Bennett's
Salt Lake City, Utah

11-1. INTRODUCTION

It is recommended that before studying this chapter, the reader review Chap. 10, "Surface Active Agents" by Thomas N. Ginsberg, which appears in Vol. 1, Part III of this series.

This chapter discusses "universal" colorants as produced for use in dispensing machines for the retail trade. Quotation marks are used around the word "universal" since even the best of these colorants may require some modification of the paint base in order to obtain satisfactory tinting. Even if a completely "universal" colorant were available today, there is the possibility that a new paint finish might be developed tomorrow that would require changes in either the base paint or the colorant.

Colorants have been developing slowly, and are still imperfect. In one way or another they all represent compromises. Nevertheless, what has been accomplished today would have seemed impossible a few short years ago.

11-2. HISTORY

In the early third of this century, color systems had practically no impact on the coatings industry. Paints for the retail trade were factory tinted, and any variations from the standard lines were left to professional painters. Color pigments were ground in raw linseed oil and packaged in 1-lb, $12\frac{1}{2}$-lb, and 25-lb tins for the professional trade. They were usually ground on the old-time flat stone mills, which, unlike the mills of the Gods, did not always grind fine, but were certainly exceedingly slow.

Rather than promoting color, the efforts of the paint industry were directed toward limiting its use. The U. S. Department of Commerce, with the wholehearted cooperation of the industry, sponsored voluntary agreements reducing the number of tints that any manufacturer would produce. In this way, the industry was actually moving away from, rather than toward, the use of color for trade sales paints.

Then, in the early 1930s, a few farsighted men began to sense a change in the public attitude on the subject. Color printing processes were improved and color photography was developing, both for stills and movies. It appeared that there was a potential wave of demand that the industry could ride if means could be developed for easy marketing.

At least three efforts had been made prior to this time to produce tinted paints at the point of sale through the use of colorants in small packages. The old Devoe-Raynolds Company had produced colorants in tubes for many years, and the Bradley & Vrooman Company of Chicago had patented a base and tinter system as early as July 16, 1912. Also, Rice's "Barreled Sunlight" was marketed as a white base with small tubes of colorant to be added to the individual cans of paint. However, these were only partially successful. They were undoubtedly ahead of their time, and they did not attempt to produce wide color coverage.

One point that was emphasized by Bradley & Vrooman has been verified many times over. That was the fact that factory-tinted paints will change color as the product ages in the can. This is particularly true in the deeper colors, and is probably due to slow flocculation of pigments. The net result is that paints that are custom tinted at the time of sale will usually match better than factory-tinted paints.

11-3. FLUID-TINTING COLORS

One decisive step of progress was made when "fluid-tinting colors" were introduced on the market. These were made with a bodied oil vehicle, instead of raw linseed oil. Whereas the color pigments in raw oil were short in consistency and frequently difficult to disperse, the "fluid-tinting colors" had a limited flowing property that made them easy to stir into paints without prior thinning with turpentine or rubbing up with a portion of the paint.

11-4. COLOR SYSTEMS

The first expanded color systems for the retail trade, introduced in the late 1930s, were based on these "fluid-tinting colors." Packages were originally glass jars, and later, one-shot metal tubes, followed still later by plastic tubes, and finally, plastic fin-type packages. The colorant was carefully standardized and packaged in measured quantities according to the size of paint can in which it was to be used.

Technical problems were not too severe, since retail paint lines were relatively simple, and the grinding liquids and suspending agents to keep pigments from settling were already known and on the market..

Once the paint industry had developed practical methods for producing an unlimited range of colors at the point of retail sale, the previous conservative attitude toward simplification was reversed and color systems covering the complete range of the spectrum rapidly appeared. Many were roughly patterned after Munsell (1) or Ostwald (2), with adjustments made to weight them in the lighter or pastel areas and in the best selling, warmer color tints.

11-5. DISPENSING MACHINES AND EMULSION FINISHES

In the 1950s, two factors that greatly altered the requirements and consequent formulation of colorants became prominent.

One was the development of dispensing machines for retail stores that could accurately measure bulk colorant into cans of base paint. These machines required a colorant that was essentially nondrying in its bulk form in order to avoid skinning in the containers and clogging of the orifices. Originally these dispensers were high-priced automatic machines that were too expensive to be practical. However, they were quickly followed by low-cost manual machines that have gained steadily in favor.

The second factor was the rapid rise in sales of water-thinned latex emulsion paints. With the appearance of these finishes on the market, questions immediately arose concerning methods of tinting them. Some technical men felt that the logical approach was to tint emulsion-type finishes with water paste colorants, but it was rapidly discovered that the same water-dispersed colorants would not tint all emulsion paints equally well.

The emulsion paint systems were so complicated, with so many variables that were often in such delicate balance, there seemed to be no way of establishing specific rules or fixed guidelines for the formulation of colorants to be used with them.

11-6. "UNIVERSAL" COLORANTS

The final answer came from the merchandising group in the industry. They insisted that a "universal" colorant must be developed that would tint both oil-type, solvent-thinned paints, and also emulsion-type, or water-thinned, finishes.

This required a new approach and a much more detailed study of problems, not only of initial dispersion of color pigments, but also of agglomeration, flocculation, and flood and float. The selective wetting of pigments by certain portions of the vehicle, and the effects that various pigments might have on each other were further complicating factors.

After several years of study, this technology is still in its infancy, and even now it is only possible to speak in general terms.

11-7. PIGMENT SURFACE STUDIES

From time to time, vehicles for "universal" colorants have been offered on the market, but these are at best only a partial solution to the problem. The pigment itself may be as much as, or even more of, a controlling factor than the vehicle.

An early microscopic study made in our author's laboratory gave interesting indications of this. Several different pigments were ground in a very long oil alkyd, to which sufficient surfactants had been added to provide miscibility with water.

One paste, using ultramarine blue, stirred very readily into a commercial polyvinyl acetate emulsion paint. In fact, it was impossible to obtain a sharp line of demarcation between colorant and base paint by deliberately attempting to streak the colorant in.

Red and yellow iron oxides also stirred in quite readily, while black iron oxide tended to flocculate.

Phthalo blue behaved in an entirely different fashion. The most vigorous agitation could not produce a dispersion of the blue into the water-thinned paint. The pigment paste would collect into spheres approximately 5 to 7 μ in diameter.

Since the red oxide dispersed well in water-thinned finishes, as a matter of interest it was combined intimately in advance with the phthalo blue, with the thought that this could physically separate the blue particles to the extent that they might tint. When this combined colorant was added to an emulsion paint, the red oxide dispersed readily into the base paint, while the phthalo blue again collected into spheres. These were somewhat smaller than in the original attempt, but nevertheless, still tightly held together so that the pigment did not tint.

The blue particles seemed to cling tenaciously to a portion of the vehicle, creating a situation whereby the surface tension of the pigment-vehicle combination was sufficient to form spheres in the water-surfactant carrier. Jettmar (3), in his studies with the electron microscope, has confirmed this tendency of phthalo blue to selectively adsorb portions of the vehicle system.

Another study of two quinacridone red colorants, using a different vehicle system, showed that in emulsion paints, one developed the same type of spheres that had been observed with phthalo blue pigments. In this case, the spheres were 1 to 2 μ in diameter and these in turn had a tendency to flocculate into groups somewhat like small clusters of grapes.

The second quinacridone red, from a different manufacturer, was free from this effect. It is possible that a sufficient surfactant search might have corrected this problem with the first pigment.

11-8. THE HLB FACTOR

In 1962, papers were published by Atlas Chemical Industries (4) providing tables of hydrophilic-lipophilic balance (HLB) for both surfactants and pigments.

Another paper on the same subject by Pascal and Reig was printed in 1964 (5).

While the HLB tables for surfactants and pigments undoubtedly have significance, only partial correlation with practical experience can be found.

In general, the inorganic pigments have a higher HLB than the organics, and these same inorganic pigments cause fewer problems in the production of "universal" colorants. Beyond this generalization, however, radical exceptions must be made. Chrome yellow, for example, with a listed HLB of 18-20, is more troublesome than red iron oxide, with an HLB of 13-15. Lampblack, although inorganic, has an HLB of 10-12, which is lower than most of the organics, and in practical experience it is distinctly a problem pigment.

Certain of the organic pigments with a low HLB may require high HLB surfactants for maximum dispersion in water emulsion paints. HLB is of necessity a figure that averages all of the factors involved, while tinting properties may depend on specifics.

Pigments vary in their average polar-nonpolar surface characteristics, but many, and possibly most pigments, may have such polarity variations on the surface of individual particles. This may be one factor leading to the lack of other than broad correlation between HLB tables and practical experience.

Beyond this, pigment particle size, degree of agglomeration, degree of milling, and HLB of the total ultimate vehicle of the tinted paints, as well as the character and composition of individual components of this vehicle undoubtedly have a strong influence.

11-9. PIGMENTS FOR "UNIVERSAL" COLORANTS

As has been said, the inorganic pigments offer fewer obstacles in the production of "universal" colorants than the organic.

Chromium oxide green is among the least troublesome of the practical commercial pigments on the market. Its particle size and surface characteristics are such that it is easily ground, it does not have a tendency to flood vehicles, it will disperse readily in either water or solvent-thinned paints. Of course, precautions must be taken to avoid pigment settling. Nevertheless, if all other pigments would behave as consistently well as chromium oxide, the production of "universal" colorants would be relatively easy.

As mentioned before, the red and yellow iron oxides are not too difficult.

Black oxide tends to flocculate, yet certain specific grades will be equal in performance to the red and yellow oxides.

Chrome yellows and molybdate oranges probably rank next, but here there are differences even in various lots of the same pigment from the same manufacturer. The reasons for this are not clear, but the colorant producer should recognize that these chrome pigments are more marginal in their behavior than the oxides previously mentioned and require greater care in the vehicle balance (see Sec. 11-10 regarding restrictive legislation).

Lampblack, as its HLB figure might indicate, will behave more like the organic pigments than the inorganic.

The tinting properties of the organic color pigments are more difficult to control than those of the inorganic. Part of this effect is undoubtedly due to their small particle size.

They are frequently agglomerated into tight clumps that require careful grinding. In the case of phthalo blue, this has been studied and reported by Carr (6). The grinding process is of greatest importance in the preparation of "universal" colorants, since without the proper reduction of particle (agglomerate) size, the effects variously described as flooding, floating, flocculation, or, as Dr. Carr suggests, "optical effect," are distinctly pronounced. Also, if the grinding is incomplete, reagglomeration or flocculation may occur after aging four to eight weeks.

Another problem with organic pigments, which is particularly noticeable with phthalo blue, is the readjustment, previously mentioned, of the total vehicle in a tinted paint, with some portions of the vehicle being selectively attracted to the surface of the pigment particle. This effect was described and beautifully illustrated by Jettmar's optical and electron microscopic studies (3).

In some cases, the use of a lake rather than a pure toner can change surface and particle size characteristics enough to reduce tinting and dispersion problems.

11-10. GUIDELINES FOR PIGMENT SELECTION

A sound general rule is to select and use, as far as possible, pigments that inherently are not sensitive to minor variations in formulation.

A second such rule is to avoid, as far as possible, the use of pigment combinations where surface characteristics and particle size vary widely. To do so will invite differential flocculation with subsequent color problems during application of the finish. An example of such a combination would be phthalo blue with red or yellow oxide.

A third generalization is to hold the number of pigment combinations to a minimum.

A fourth is to avoid, as much as possible, the use of pigments that are far apart in hue. If this is not done, any slight variation in standardization or subsequent flocculation of any one of the pigments in the tinted paint is visually emphasized and obvious.

All pigments used for "universal" colorants must have at least satisfactory, and preferably, maximum exterior durability, both in mass tone and tint tone. It is necessary to check this durability in blends with other pigments as well as single tints with white.

For example, both phthalo blue and red iron oxide have excellent out-of-door permanence alone, but when a tint of a combination of the two in the purple range is exposed, the blue tends to fade out, leaving a pink that bears little resemblance to the original color. A very slight modification of a blue tint with red oxide, or a slight modification of the red with blue, might be satisfactory, since changes with weathering would be less noticeable to the eye.

Similarly, an olive green tint made by combining phthalo blue and yellow iron oxide is not suitable. It is better to use chromium oxide and modify it for the desired tone.

A green tint made from a combination of chrome yellow and black is hazardous, since under certain atmospheric conditions, particularly high humidity and acid pollution of the air, the yellow may fade out rapidly, leaving a simple gray tint. Recently, new silica-coated chrome yellows and molybdate oranges have been placed on the market. These are substantially resistant to this type of failure. However, recent federal, as well as some local, legislation would seem to make the use of any chrome yellow or molybdate orange pigment illegal because of debatable toxicity. The siliceous coating would certainly reduce whatever hazard may have existed, but current legislation makes no provision for taking this into account.

Pigments for "universal" colorants should not be soluble in any portion of the vehicle. Even a slight solubility can result in a buildup of crystals during storage through variations in temperature. These crystals of concentrated soluble material can, in turn, create spots of serious color bleeding when the paint finish is applied.

Also, the pigments for "universal" colorants should be free from soluble salts, which could inactivate anionic dispersants.

11-11. VEHICLES FOR "UNIVERSAL" COLORANTS

There are four general classes of vehicles for "universal" colorants.

The surfactants are a major controlling factor in all four. The preferred surfactants are nonionic because of their reduced reactivity and consequent greater stability. The higher-molecular-weight materials are more water soluble, while those of lower molecular weight are more oil soluble. In some cases, a combination of the two may be more effective than a single one of medium molecular weight.

Attempts have been made to produce "universal" colorants by using surfactant plus volatile thinner for the vehicle with suspending agents to keep the pigments from settling. Among other difficulties with this type of colorant was loss of thinner and drying out when it was used in the retail store dispensing machines. Pigment settling was another problem that was never fully conquered with this type of colorant.

A second type of colorant relies on glycols in combination with surfactants for the vehicle. Many glycols are available, such as ethylene, diethylene, propylene, and polypropylene glycol. Again, the colorant producer must decide on his compromise in selecting any one, or a blend, of the glycols available. This type of colorant, when properly made, will show good dispersion in most water-thinned paints, but a modification of some solvent-thinned paint bases may be necessary in order to get full colorant development upon tinting. Also, luster of solvent-thinned semigloss and gloss finishes may be affected.

Manufacturers of this type of colorant will usually recommend that no more than six to eight ounces of colorant should be added to each gallon of paint. This may be due to the fact that the glycols are hygroscopic, and in a humid climate may pick up enough moisture from the air to interfere with the drying of the paint.

A third type of colorant vehicle relies on oil, or a very long-oil alkyd, in combination with dispersants. This has been troublesome because of skinning and drying in dispensing machines, and was not successful with some organic pigments or lampblack because of difficulty in getting dispersion in emulsion paints.

Still another type of colorant has been made from a water-soluble, or water-dispersible, oil or resin in combination with surfactants. There are many possible ways of producing such oils or resins, and the manufacturer must carefully decide on how much water sensitivity he wants and the degree of drying he can build into his vehicle without developing skinning and drying in dispensing machines.

Organic pigments may require a more water-dispersible vehicle than is used for the inorganics.

Properly formulated, this type of colorant can be suitable for dispensing machines if the volatile content is low. Since it is film forming, larger ratios of colorants to base paint can be used.

11-12. SCREENING "UNIVERSAL" COLORANT FORMULATIONS

At present, the only way known to us to determine suitability of a "universal" colorant is to put it through a careful and time-consuming screening process.

The technique of testing is most important. The laboratory test must be substantially more severe than actual application conditions in order to screen out colorant formulations that might be marginal, i.e., good enough to pass under favorable conditions, but still weak enough to give trouble in the field.

As a preliminary test, we like to make side-by-side drawdowns of two samples made from standard paint tinted to a medium level. This test should be made with both solvent-thinned and water-thinned standard paints. A pint of the paint plus colorant is first hand-stirred for 3 min. Then $\frac{1}{2}$ oz of paint is taken from the can and the remainder is put on a Red Devil Shaker for 4 min.

Both the hand-stirred and the shaken samples are then drawn down side-by-side on a sheet that has both sealed and unsealed areas. The drawdown blade has a 0.006 in. gap which gives a 0.003 in. wet film thickness.

The finger rub-up is a very important part of this test. It is made after most of the thinner has evaporated, but before the film over the sealed portion is fully set. In the case of a flat paint, this can be determined when the drawdown begins to lose its gloss. The rub-up is made with ten circular rubs with the ball of the finger.

If tinting according to this test is good, no trouble in the field should be expected. A marginal result indicates that a further large-scale panel test, as described below, should be made, and possibly further development work must be done.

In some cases where minor difficulties are encountered with a commerical paint, a small amount of surfactant added to the base paint formulation will bring results sufficiently good that no difficulties should be encountered in the field.

Preliminary screening should be followed by large panel tests over both sealed and unsealed surfaces with both brush and roller. These panels can be divided for 5-, 10-, and 15-min lap tests. A final heavy streak across the panel is made with the brush or roller.

Another important final part of the screening process for "universal" colorants is to check them against commercial paints taken from the dealer's shelves.

11-13. PRODUCTION GUIDELINES

Going back to the industry's early studies on tinting problems, which began long before emulsion paints complicated our situation, certain general rules were hammered out that are more than ever applicable today:

1. Pigments, both in the base paint and in the colorants, should be ground to the maximum practical degree of dispersion. To do less is to invite possible flocculation or "sweating in" upon aging, and either one can cause color problems.

Most colorant pigments can be ground by using a combination of high-speed dispersers and sand mills. Paste consistencies and batch sizes should be so controlled that the maximum benefit can be obtained from the high-speed disperser. The degree of grind desired will probably call for closed sand mills and a slow rate of feed into the mill, which may require more than the normal amount of cooling water to avoid overheating. This, in turn, may call for recirculating cooling water through a cooling tower. All of this effort at the present time has not been sufficient for certain difficult pigments, such as phthalo blue, which require ball milling or its equivalent (6).

While the Hegman gauge may be a satisfactory test for grind for most inorganic pigments, many of the organics will require a supplementary check against a reserve standard paste of the colorant using the method described in the first part of Sec. 11-13.

2. Agents used for viscosity or suspension control in either the colorant or the base paint must be examined with great care. Many of these are actually flocculating agents.

3. Where possible, vehicles used in the base paint should be of similar composition. This is not always possible because sometimes we deliberately select products of near incompatibility for special effects.

4. For in-plant tinting, good agitation during the tinting process is most desirable. The old-time post agitator, with its slow-turning propeller blade, was rather poor. Many people prefer to do all factory tinting with a modern high-speed disperser.

However, in such factory tinting, temperature must be watched with great care. Many dispersants in water-thinned paints begin to lose their effectiveness as temperatures pass 38°C and are quite ineffective at 50°C. The thickener used in the emulsion paint may also become ineffective at these temperatures. If the tinting process takes very long, it may be necessary to use the high-speed disperser in on-and-off spurts to avoid raising the temperature too high.

5. Finally, the standardization of "universal" colorants requires standard comparison base paints of both solvent- and water-thinned types, and a good supply of retained "universal" colorant that has been checked sufficiently to be considered standard.

11-14. SUMMARY

The manufacture of colorants, and particularly "universal" colorants, for the coating industry is still an art, relying heavily on experience. There are so many variables involved that there is little hope at the present time of outlining precise rules or charts for their formulation. The best that can be done is to provide guidelines gained from many years of experience, and to point out specific pitfalls that can be sources of trouble in the open market.

REFERENCES

1. Munsell Color Co., Munsell Book of Color, Baltimore.
2. Center for Advanced Research in Design, Color Harmony Manual, 645 N. Michigan Ave., Chicago, Ill. 60611.

3. W. Jettmar, J. Paint Technol., 41 (537), 559 (October, 1969).

4. Atlas Chem. Ind., Inc., The Atlas HLB System, 1962.

5. R. H. Pascal and F. L. Reig, Official Digest Fed. Soc. Paint Technol., 36 (375), 838 (August, 1964).

6. W. Carr, J. Paint Technol., 42 (551), 695 (December, 1970).

Chapter 12

LEAD-FREE DRIER SYSTEMS

Charles A. Burger

Interstab Chemicals, Inc.
New Brunswick, New Jersey

Much has been written on the subject of driers and the theoretical concepts surrounding the complex chemical and physical process of film formation involving unsaturated vegetable oils and the wide variety of synthetic polymers. What have been neglected are the practical aspects of drier technology and usage. It is for this reason that coatings chemists are often frustrated by what appear to be insurmountable problems when they must convert lead-containing paints to essentially nonlead systems in order to comply with various legislated controls.

The coatings industry is confronted not only with federal regulations, but also municipal ordinances, which make it an offense to sell, supply, or offer for sale any coating for use in or around the household if it contains lead metal in excess of 0.5% or, in some areas, as low as 0.06%, based on the total weight of the contained solids or the dried film. Such paints and similar surface coating materials will be banned as a hazardous substance if shipped in interstate commerce.

These restrictions mean that paint manufacturers must avoid incorporating lead compounds, such as driers and lead-based feeder products, traditionally used to prevent loss of dry during shelf storage. However, paint manufacturers cannot afford the risk of knowingly or purposely adding lead to their products when it is possible that the threshold level may have been reached by contamination and impurities introduced by other raw materials in the formulation.

The prospect of reviewing and revising drier systems is not one welcomed by the average paint manufacturer. However, reformulating enables the coatings chemist to replace, with confidence, a familiar chemical, such as lead drier, by "Zirco," an equally familiar and accepted catalyst, which has been used successfully for 20 years for this very purpose.

Zirco drier catalyst 6% is an oil-soluble, polymeric zirconyl complex of a synthetic organic acid. When administered orally to white rats it has an LD_{50} of 6800 to 10,000 mg/kg body weight. The intraperitoneal LD_{50} is 1700 to 3400 mg/kg body weight of albino rabbits.

According to FDA 21, CFR 121.2514, Zirco is permitted as a constituent of resinous and polymeric coatings where these coatings are used as the food contact surface of articles intended for use, among other things, for holding of food. It is neither GRAS nor prior sanction material, but separately investigated and considered a food additive. It may very well be the least toxic of any of the coatings raw materials in use today.

In order to understand the application possibilities of this metal complex, it is important to consider the role of zirconium in its relationship to other driers. It is normally used with other driers in coating systems which rely upon the catalytic effect of certain metallic salts to convert what can be best described as a pourable mixture of pigments, vehicle, and solvents to a durable protective and decorative film.

To begin with, it should be stressed that zirconium is not a drier per se, but a drier catalyst, and its function is to extend the functional capacity of standard drier metals. In so doing, it exerts a pronounced effect upon the chemical reaction which occurs during the film formation process. This activity affects both air-dried and heat-cured coating formulations.

Conventional drier systems utilizing cobalt, manganese, calcium, iron, cerium, and lanthanum metal combinations perform not as a unit, but with each metal contributing its own individual characteristic to film properties. When combined with cobalt, zirconium causes this oxidation catalyst, which normally is considered a surface drier, to extend its catalytic activity throughout the thickness of the film. In this way, drying is achieved without the assistance of any other metal such as lead. Zirconium has an equally beneficial and bifunctional effect upon manganese. It increases the desirable through-drying properties of manganese and simultaneously increases the surface-drying potential of this metal. Similar synergistic effects are evident with other metals such as calcium, iron, and rare earths.

Calcium salts have been employed successfully for many years with manganese and cobalt as a replacement for lead. The function of calcium is best described as an auxiliary drier that will produce some improvements in the physical properties of the dried paint film such as gloss, hardness, and, in some instances, through-dry. Calcium is also useful as a wetting agent and dispersion aid if incorporated with the pigments during grinding.

However, calcium is not universally dependable as a replacement for lead, and its function and applicability is somewhat limited. Zirconium, on the other hand, because of its synergistic influence on all drier metals, including calcium, has proved to be a consistently dependable lead drier replacement.

The importance of a proper balance of drier metals in any given coating formulation cannot be overemphasized. This fact is especially critical when zirconium systems are devised. Optimization of zirconium should be sought in order to obtain maximum benefits; usually, this is best achieved by laboratory evaluation. Such an approach can be time consuming unless some guidelines are provided. Actually, there is no precise mathematical equation that permits computation of the exact amount of zirconium metal required for the replacement of lead. Fortunately, most paint formulators have been exposed to the Zirco concept of lead replacement and, therefore, it is not a totally new or unfamiliar experience. An orderly laboratory evaluation, using gradient levels of zirconium metal consisting of 0.03%, 0.06%, 0.08%, 0.10%, 0.15%, and 0.20% based on vehicle solids is generally recommended. As a rule, the major part of the primary drier system remains intact, except for the exclusion of lead.

There are cases where a drier system is over complicated and can be simplified by eliminating unnecessary metals that may not be contributing anything to drying performance. Usually a two- or three-metal component system will suffice. For example, cobalt/zirconium or manganese/zirconium modification of either of these systems with calcium in the range 0.05 to 0.20% may provide additional benefits in the curing schedule. Frequently a combination of cobalt/manganese/zirconium will provide a more desirable balance of activity. Again, much depends upon the vehicle system and the performance requirements of the coating. Also, the replacement of lead by zirconium may result in some benefits that would be difficult to achieve by any other method except, perhaps, a total reformulation of the coating.

Some typical examples of lead drier replacement and comparison of drying performance are presented in a cross section of trade sales paints (see Tables 12-1 to 12-4).

While experienced formulators will generally agree that lead drier has served a necessary and useful purpose in promoting through drying of the paint film, they also recognize that this very characteristic, as a strong sustaining polymerization catalyst, is a prime contributing factor to many film failures and deficiencies. Among some of the film failures are embrittlement, poor impact resistance, and a subsequent reduction in long-term flexibility and adhesion. Moreover, lead has the inherent disadvantage of being the least compatible of all the metallic soaps and, due to its high stoichiometric basicity, is likely to react with oils and resins to form insoluble lead soaps. This will adversely affect clarity in clear varnishes and, in pigmented systems, will produce hazing and loss of gloss. Yellowing and sulfide staining in whites and pastels is a typical phenomenon stemming from high percentages of lead drier.

TABLE 12-1. Linseed Oil House Paint

Raw and bodied oils pigmented with TiO_2, ZnO, and magnesium silicate		
Concentration of drier metal based on vehicle solids		
	Lead system	Zirconium system
	Manganese, 0.02%	Manganese, 0.02%
	Lead, 0.50%	Zirconium, 0.10%
Drying time in hours (3 mils wet film)		
Dust free[a]	5.0	5.0
Thru hard[a]	>24.0	17.0

[a]Improved Gardner Circular Drying Time Recorder.

TABLE 12-2. Interior Alkyd Enamel

Based on a medium-oil soya alkyd resin.		
Concentration of drier metal based on vehicle solids		
	Lead system	Zirconium system
	Cobalt, 0.06%	Cobalt, 0.06%
	Manganese, 0.02%	Zirconium, 0.10%
	Lead, 0.30%	
	Calcium, 0.07%	
Drying time in hours (3 mils wet film)		
Dust free[a]	2.0	2.0
Tack free (touch)	7.0	5.0
Thru hard[a]	20.0	13.0

[a]Improved Gardner Circular Drying Time Recorder.

Just the opposite is true of zirconium, which has traditionally been used to overcome these deficiencies without any sacrifice in the drying performance of the coating. Long-term use of zirconium in all types of architectural and industrial finishes has repeatedly demonstrated that not only will it maintain or improve the curing schedule, but very often zirconium will also enhance gloss and gloss retention and eliminate such defects as wrinkling, hazing, yellowing, loss of adhesion, and embrittlement of the film after application.

One of the critical questions arising from substitution of lead drier is whether drier loss will occur during shelf storage of certain pigmented systems. Tests indicate that zirconium does not cause loss of dry in those paints where drier loss has never been a problem. If formulators separate only those coatings in which feeder driers were used as insurance from those where a real problem existed and where only lead minimized the difficulty, they find that the over-all situation may not be as critical as first feared.

Probably one of the most common causes of loss in drying activity is through absorption of driers by pigments. Since zirconium is a poor pigment-wetting compound, it is relatively free from absorption by pigment and, therefore, is always available to the mechanism of the drying process. Even if a small portion of the primary drier is lost through pigment absorption or as a result of some other chemical or physical phenomenon, there is normally sufficient drier remaining for zirconium to catalyze and thus maintain, to a large degree, most of the original drying schedule.

In many cases it is possible to compensate for some of this loss of primary drier activity by increasing the cobalt metal concentration as much as 50% above the level generally required for adequate drying; in combination with zirconium, the risk of wrinkling or a sacrifice in the long-term protective performance of the film due to oxidative destruction or deterioration is eliminated. Another technique is the use of calcium at a level of 0.1 or 0.2% metal based on vehicle solids incorporated in the grind. Calcium at these levels will serve a dual function by aiding pigment dispersion and by acting as a sacrificial metal, being preferentially absorbed by the pigments and thereby reducing the tendency to absorb cobalt or manganese. Iron naphthenate is also useful in this respect, but only in formulations where the color it contributes can be tolerated. In addition, there is a large selection of pigment dispersants and wetting agents which will help minimize drier absorption by increasing the capacity of the pigment to absorb vehicle instead of the drier.

Overgrinding should be avoided and not continued beyond the time when adequate dispersion of the pigment is achieved. Temperature of the batch should be kept as low as possible during processing and all driers, except where calcium or iron are incorporated as a grinding aid, should be added last, after the batch has been let down with all remaining ingredients and cooled to as close to room temperature as is feasible. Whenever possible, lower-oil-absorption-demand pigments and extenders should also be used.

All of these techniques, either singly or in combination, have been employed in commercial production to eliminate drier loss and are valid approaches to maintaining drying activity during prolonged storage.

Even under the best conditions it is not unusual for a coating to show some lengthening of the drying time by possibly several hours. Unless a dry time specification is involved where maximum limits are imposed, a paint which reaches a tack-free state within a 24-hour period is acceptable. Loss of dry becomes a problem only when surface tackiness persists for several days.

Experience indicates that the most critical drier loss problems will occur within two months after the coating has been packaged. In some extreme cases, the problem may become evident within one week. If no loss in drying activity takes place within the first four to six months, the probability that a critical situation will develop suddenly is remote.

There is no reliable accelerated laboratory method for predicting drier loss. The best procedure, however time consuming, is to store the paint at prevailing temperatures in the range of 21 to 30°C and to determine periodically over a 12-month period whether there is any change in the drying schedule. Some attempt should be made to control test conditions in terms of temperature and humidity so that environmental variables can be kept to a minimum.

Attempts to accelerate aging by storing paint for one to four weeks at an elevated temperature of perhaps 49°C may produce erratic data. First of all, there is no relationship with which to make a prediction that so many days or weeks at 49°C is equivalent to X number of months or years under actual storage conditions in a paint store or warehouse. Too many side reactions may take place during sustained elevated temperatures that would never occur under average storage conditions. Such reactions result in erroneous conclusions that loss in drier activity can be anticipated. The best that can be said for such a test is that if no loss of dry occurs following storage at elevated temperatures then perhaps drier stability will not be a problem.

The difficulty involving drier absorption does not lend itself to a simple solution because of variations in types of vehicle used. This is especially true with pigments that are prime offenders, notably, carbon blacks, iron oxides, iron blue, and toluidine red. It should be pointed out that the extent of drier absorption can be affected by deviations in the composition of each raw material. In some instances, the moisture content of a raw material would be sufficient to alter absorption characteristics.

Aside from all the technical aspects of drier reformulation, another concern of the paint manufacturers is that in order to comply with lead restriction legislation, the cost of producing a gallon of paint may be higher.

Considering the drier situation only (excluding lead-containing pigments), where zirconium is substituted for lead, the cost of a gallon of paint will not necessarily increase. The cost can and will vary in either direction and, in many instances, there will only be minimal changes. In fact, taking a fresh look into drier systems may prove very enlightening. To illustrate, over the years, many drier combinations have become unnecessarily expensive and complicated. The usual procedure has been that when a problem arose where a production batch didn't dry properly or other film defects appeared requiring some fast remedial action, the additive, whether it was in the form of more drier or some other control chemical, automatically became a permanent part of the batch ticket or formula, thus another added cost.

When these drier systems are reviewed and reformulated with lead-free combinations designed for optimum performance, the paint manufacturer may find to his surprise that careful selection of drier metals and concentrations has actually resulted in a lowering of the drier cost per gallon of paint. There will inevitably be a number of cases where drier costs will remain unchanged or will increase, but increases will not be such that they produce an unbearable burdern on manufacturing costs.

Tables 12-5 to 12-14 represent typical examples of actual drier cost comparisons per gallon of paint when conversions from lead to lead-free systems were adopted in commercial production coatings.

No discussion of zirconium would be complete without raising the question of whether zirconium-based products will someday face the same scrutiny by environmentalists as lead-based compounds. It is highly improbable that zirconium compounds will ever be considered hazardous in coating formulations, for zirconium and its salts generally have low systemic toxicity. Zirconium carbonate, zirconium oxide, and zirconyl chloride are useful in dermatitic ointments and antiperspirant preparations.

According to Dangerous Properties of Industrial Materials (3rd ed., N. Irving Sax, ed., D. Van Nostrand, p. 1249), "Zirconium is not an important poison and as far as is known, the inherent toxicity of zirconium compounds is low."

In summary, any coating normally dependent upon the catalytic activity of driers to promote the oxidation-polymerization mechanism of film formation will respond favorably to zirconium. Careful otpimization of the standard drying metals in combination with zirconium compounds will produce a coating film whose protective and decorative function is equivalent, if not superior, to those combinations employing lead soaps. This can be accomplished without a drastic escalation in cost and, in many cases, the drier cost per gallon of paint is reduced.

TABLE 12-3. Architectural Enamel

Based on medium-long-oil soya alkyd.		
Concentration of drier metal based on vehicle solids	Lead System	Zirconium system
	Cobalt, 0.06%	Cobalt, 0.06%
	Lead, 0.60%	Zirconium, 0.08%
Drying time in hours (3 mils wet film)		
Dust free[a]	3.5	3.5
Tack free (touch)	5.5	5.0
Thru hard[a]	26.0	18.0

[a]Improved Gardner Circular Drying Time Recorder.

TABLE 12-4. Clear Spar Varnish

Concentration of drier metal based on vehicle solids	Lead system	Zirconium system
	Cobalt, 0.03%	Cobalt, 0.04%
	Manganese, 0.02%	Zirconium, 0.08%
	Lead, 0.55%	
Drying time in hours (4 mils wet film)		
Dust free[a]	3.5	2.5
Tack free (touch)	5.0	4.0
Thru hard[a]	16.0	12.0

[a]Improved Gardner Circular Drying Time Recorder.

TABLE 12-5. Clear Alkyd Brushing Varnish

Metal on vehicle solids (%)			
Lead system		Zirconium system	
Co	0.03	Co	0.03
Mn	0.03	Mn	0.03
Ca	0.03	Zr	0.08
Pb	0.55		
Drier cost/gal			
$0.039		$0.033	

TABLE 12-6. Interior Semigloss Enamel

Metal on vehicle solids (%)			
Lead system		Zirconium system	
Co	0.05	Co	0.02
Pb	0.60	Zr	0.08
Drier cost/gal			
$0.023		$0.0165	

TABLE 12-7. Flat Alkyd Paint

Metal on vehicle solids (%)			
Lead system		Zirconium system	
Co	0.04	Co	0.04
Pb	0.60	Zr	0.06
Drier cost/gal			
$0.0126		$0.010	

TABLE 12-8. Gloss Enamel

Metal on vehicle solids (%)			
Lead system		Zirconium system	
Co	0.10	Co	0.10
Ca	0.09	Ca	0.05
Pb	0.90	Zr	0.10
Drier cost/gal			
$0.079		$0.064	

TABLE 12-9. Alkyd Sash and Trim Enamel

Metal on vehicle solids (%)			
Lead system		Zirconium system	
Co	0.06	Co	0.06
Pb	0.60	Zr	0.08
Drier cost/gal			
$0.0398		$0.0397	

TABLE 12-10. Spar Varnish

Metal on vehicle solids (%)			
Lead system		Zirconium system	
Co	0.03	Co	0.04
Mn	0.02	Zr	0.08
Pb	0.55		
Drier cost/gal			
$0.030		$0.0297	

TABLE 12-11. Medium Oil Alkyd Enamel

Metal on vehicle solids (%)			
Lead system		Zirconium system	
Co	0.15	Co	0.10
Mn	0.03	Mn	0.03
Pb	1.02	Ca	0.10
		Zr	0.10
Drier cost/gal			
$0.076		$0.076	

TABLE 12-12. Epoxy Ester Varnish

Metal on vehicle solids (%)			
Lead system		Zirconium system	
Co	0.03	Co	0.03
Pb	0.20	Zr	0.04
Drier cost/gal			
$0.019		$0.022	

TABLE 12-13. Linseed Oil House Paint

Metal on vehicle solids (%)			
Lead system		Zirconium system	
Co	0.01	Co	0.01
Mn	0.01	Mn	0.01
Pb	0.40	Zr	0.10
Drier cost/gal			
$0.020		$0.033	

TABLE 12-14. Long-Oil Alkyd Enamel

Metal on vehicle solids (%)			
Lead system		Zirconium system	
Co	0.03	Co	0.02
Pb	0.60	Ca	0.05
		Zr	0.06
Drier cost/gal			
$0.028		$0.032	

There is a detailed information available regarding the Zirco approach to lead-free drier systems based on 20 years of extensive experience with a broad range of commercially produced coating formulations (see general references).

GENERAL REFERENCES

1. P. G. Stecker, ed., The Merck Index, 8th ed., Merck & Co., 1968.
2. N. I. Sax, ed., Dangerous Properties of Industrial Materials, 3rd ed., Van Nostrand, Princeton, N.J., 1968, p. 1249.
3. C. A. Klebsattel, Protective and Decorative Coatings (J. J. Mattiello, ed.), Vol. I, Wiley, New York, 1941.
4. H. F. Payne, Organic Coating Technology, Vol. I, Wiley, New York, 1954.
5. C. R. Bragdon, Film Formation, Film Properties and Film Deterioration, Wiley (Interscience), New York, 1958.
6. W. von Fischer, Paint and Varnish Technology, Reinhold, New York, 1948.
7. H. W. Chatfield, The Science of Surface Coatings, Van Nostrand, Princeton, N.J., 1962.
8. C. R. Martens, Technology of Paints, Varnishes and Lacquers, Reinhold, New York, 1968.

AUTHOR INDEX

Numbers in parentheses are reference numbers and indicate that an author's work is referred to although his name is not cited in text. Underlined numbers give the page on which the complete reference is cited.

K

L

M

SUBJECT INDEX

N